Technology
Entrepreneurship

Technology Entrepreneurship
Creating, Capturing, and Protecting Value

Thomas N. Duening, Ph.D
El Pomar Chair of Business and Entrepreneurship
Director, Center for Entrepreneurship
University of Colorado at Colorado Springs

Robert D. Hisrich, Ph.D
Garvin Professor of Global Entrepreneurship
Director, Walker Center for Global Entrepreneurship
Thunderbird School of Global Management

Michael A. Lechter
Adjunct Professor, Ira A. Fulton School of Engineering
Arizona State University
President and CEO, Michael A. Lechter, PC
President and CEO, TechPress, Inc.
General Counsel, Pay Your Family First, LLC

AMSTERDAM • BOSTON • HEIDELBERG • LONDON • NEW YORK • OXFORD
PARIS • SAN DIEGO • SAN FRANCISCO • SINGAPORE • SYDNEY • TOKYO
Academic Press is an imprint of Elsevier

Academic Press is an imprint of Elsevier

30 Corporate Drive, Suite 400, Burlington, MA 01803, USA

525 B Street, Suite 1900, San Diego, California 92101-4495, USA

84 Theobald's Road, London WC1X 8RR, UK

Notices

Knowledge and best practice in this field are constantly changing. As new research and experience broaden
our understanding, changes in research methods, professional practices, or medical treatment may become
necessary.

Practitioners and researchers must always rely on their own experience and knowledge in evaluating and
using any information, methods, compounds, or experiments described herein. In using such information or
methods they should be mindful of their own safety and the safety of others, including parties for whom
they have a professional responsibility.

To the fullest extent of the law, neither the Publisher nor the authors, contributors, or editors, assume any
liability for any injury and/or damage to persons or property as a matter of products liability, negligence or
otherwise, or from any use or operation of any methods, products, instructions, or ideas contained in the
material herein.

Library of Congress Cataloging-in-Publication Data

Duening, Thomas N.
   Technology entrepreneurship : creating, capturing, and protecting value
  / Thomas Duening, Robert Hisrich, Michael Lechter.
      p. cm.
  Includes bibliographical references and index.
   ISBN 978-0-12-374502-6 (alk. paper)
  1. Technological innovations–Economic aspects. 2. High-technology industries–Management.
3. Entrepreneurship. 4. Intellectual property. 5. New business enterprises–Management. I. Hisrich,
Robert D. II. Lechter, Michael A. III. Title.

  HC79.T4D84 2010
  658.5'14-dc22

                         2009026251

British Library Cataloguing-in-Publication Data
A catalogue record for this book is available from the British Library.

For information on all Academic Press publications
visit our Web site at www.elsevierdirect.com

Printed in The United States of America
09 10 11 9 8 7 6 5 4 3 2 1

Contents

Preface. xvii
Acknowledgments. xxiii
About the Authors. xxv

PART I   FUNDAMENTALS OF BUSINESS AND ECONOMICS

CHAPTER 1   Fundamentals of Business 3

  1.1 What is a Business?. 6
      1.1.1 Business as an Organized Activity 7
      1.1.2 Business as a Purposeful Activity 7
      1.1.3 Business as a "Going Concern". 8
      1.1.4 Value. 8
  1.2 Value and the Market 9
      1.2.1 Value Creation 9
      1.2.2 Value Protection. 11
      1.2.3 Value Capture. 11
      1.2.4 Value Proposition 11
  1.3 Business and Society 12
      1.3.1 Business and Social Responsibility 14
      1.3.2 The Mechanics of a "Going Concern" 15
  1.4 Business Dynamics 17
      1.4.1 Change and Competition 17
      1.4.2 Change and Globalization. 18
      1.4.3 Engineering Creative Solutions. 18
  1.5 The Future of Business 21

CHAPTER 2   Fundamentals of Economics in a Global Context 29

  2.1 Economic Resources. 30
      2.1.1 Goods and Services. 31
      2.1.2 Resource Allocation 31
      2.1.3 Product Distribution. 33
  2.2 Supply, Demand, and Pricing 33
      2.2.1 Price and Market Equilibrium. 34
      2.2.2 Price and Competition 36

2.2.3 Price and Gross Margin 37
2.2.4 Inflation 38
2.2.5 Price in a Compromised Market 39
2.2.6 Interest and the Price of Money 40
2.3 Economic Systems 41
2.3.1 Socialism 41
2.3.2 Communism 41
2.3.3 Mixed Market Economies 42
2.3.4 The U.S. Economic System 44
2.4 International Trade 46
2.4.1 Exporting and Importing 47
2.4.2 Balance of Trade 47
2.4.3 Balance of Payments 47
2.5 International Monetary System 47
2.5.1 International Monetary Fund 48
2.5.2 The World Bank 49
2.6 Elements of Global Trade 49
2.6.1 The WTO 50
2.6.2 Trade Blocs and Free Trade Areas 51
2.6.3 Currency Exchange Rates 52

CHAPTER 3 **Technology Ventures in a Global Context** **59**
3.1 Globalization and Technology Ventures 61
3.1.1 Technology Venture Drivers 62
3.1.2 Capital Intensity 63
3.1.3 Knowledge Intensity 66
3.1.4 Accelerated Pace of Change 67
3.1.5 The Network Effect 68
3.2 Value Creation in the Global Economy 72
3.2.1 Creating Market Value 72
3.2.2 Creating Enterprise Value 73
3.2.3 Value Protection in Technology Ventures 75
3.3 Value Capture in the Global Economy 78
3.3.1 Capturing Market Value 78
3.3.2 Capturing Enterprise Value 80
3.4 Global Market Entry Strategies 81
3.4.1 Exporting 82
3.4.2 Licensing 82
3.4.3 Joint Ventures 83
3.4.4 Strategic Alliances 83
3.4.5 Trading Companies 84
3.4.6 Countertrading 84
3.4.7 Direct Ownership 84

PART II LEGAL STRUCTURE AND CAPITAL

CHAPTER 4 Legal Structure and Equity Distribution 95
4.1 Ownership and Liability Issues 97
    4.1.1 Limited versus Unlimited Liability 98
    4.1.2 The Extent of Limited Liability 99
4.2 Choice of Legal Structure 101
    4.2.1 Sole Proprietorship 102
    4.2.2 General Partnership 103
    4.2.3 Limited Partnership 105
    4.2.4 Corporation 108
    4.2.5 Limited Liability Company 116
4.3 Limited Liability Entities—A Comparison 118
    4.3.1 Expense 119
    4.3.2 Shareholder Options 119
    4.3.3 Taxation 119
    4.3.4 Profits and Losses 119
    4.3.5 Partnerships 120
4.4 Equity and Equity Types 120
    4.4.1 Equity and Stocks 120
    4.4.2 Common Stock 121
    4.4.3 Preferred Stock 122
    4.4.4 Preferred Stock Distributions 122
    4.4.5 Convertibility 123
    4.4.6 Participating Preferred Stoch 123
    4.4.7 Voting Rights 124
    4.4.8 Founder's Stock 124
4.5 Equity Distribution in the Nascent Venture 125
    4.5.1 Employee Stock Options 127

CHAPTER 5 Capital and Deal Structuring 139
5.1 Role of Capital in Technology Ventures 141
5.2 Capital Sources 142
    5.2.1 Angel Investors 143
    5.2.2 Venture Capital 144
5.3 Equity and Debt Financing 146
    5.3.1 Equity Financing 146
    5.3.2 Costs of Equity Financing 150
    5.3.3 Debt Financing 151
    5.3.4 Institutional Lender Requirements 152
5.4 Loan Rates, Payment Methods, and Lender Types 152
    5.4.1 Small Business Administration Loans 155
5.5 Fund-Raising Tools and Techniques 156

        5.5.1 Private Placement Memorandum 158
        5.5.2 Subscription Agreement 158
        5.5.3 "Elevator Pitch" 159
    5.6 Alternatives to Debt and Equity Financing 159
        5.6.1 SBIR Program 159
        5.6.2 Small Business Technology Transfer Program 160
        5.6.3 Bootstrap Financing 161

CHAPTER 6   Exit Strategies for Technology Ventures 171
    6.1 Acquisition 173
        6.1.1 Due Diligence 174
        6.1.2 The Acquisition Deal 177
    6.2 Mergers 178
    6.3 Venture Valuation 180
    6.4 Going Public 183
        6.4.1 Advantages of Going Public 183
        6.4.2 Disadvantages of Going Public 184
        6.4.3 Timing 186
        6.4.4 Underwriter Selection 186
        6.4.5 Registration Statement and Timetable 188
        6.4.6 The Prospectus 188
        6.4.7 The Red Herring 189
        6.4.8 Reporting Requirements 191

PART III INTELLECTUAL PROPERTY AND CONTRACTS

CHAPTER 7   Intellectual Property Management and Protection 203
    7.1 Intellectual Property and Technology Ventures 205
    7.2 Intellectual Property Protection 205
        7.2.1 Recognizing Intellectual Property 207
        7.2.2 Record Keeping 207
        7.2.3 Record-Keeping Procedures 210
    7.3 Trade Secrets 211
        7.3.1 Trade Secret Protection 212
        7.3.2 Maintaining Trade Secrets 212
    7.4 Patents 215
        7.4.1 Patentability 216
        7.4.2 Requirements for Novelty and Nonobviousness 216
        7.4.3 Exclusive Right 217
        7.4.4 The Patent Application Process 218
        7.4.5 The Patent Examination Process 221
        7.4.6 Patent Pending 222
    7.5 Patent Ownership 222

7.6 International Patent Protection. 222
7.7 Copyrights. 224
   7.7.1 Considerations with Respect to Software 225
   7.7.2 Notice. 226
   7.7.3 The Term 226
   7.7.4 Copyright Registration 226
   7.7.5 Copyright Ownership. 227
7.8 Mask Works. 228
7.9 Trademarks 228
   7.9.1 Acquiring Trademark Rights. 229
   7.9.2 Registering a Trademark. 229
   7.9.3 Principal Register 231
   7.9.4 Supplemental Register 231
   7.9.5 Application Based on "Intent to Use". 232
   7.9.6 The Strength of a Mark 232
   7.9.7 Choosing a Mark. 234
   7.9.8 Term of the Registration. 235
   7.9.9 Maintaining Trademark Rights 235

CHAPTER 8 Contracts. 247
8.1 Sources of Contract Law 249
8.2 Contract Formation. 250
   8.2.1 The Offer 250
   8.2.2 The Counteroffer 251
   8.2.3 Acceptance. 252
   8.2.4 Revocation of Offer or Acceptance 253
   8.2.5 Consideration 253
8.3 Defenses against Contract Enforcement 253
8.4 Performance and Breach. 254
   8.4.1 Damages 256
   8.4.2 Rescission and Restitution 257
   8.4.3 Specific Performance 257
   8.4.4 Quasi-contract. 258
   8.4.5 Reformation 258
8.5 Anatomy of a Contract 258
   8.5.1 Preamble. 259
   8.5.2 Recitals 259
   8.5.3 Definitions 259
   8.5.4 Performance 259
   8.5.5 Ownership and Use of Intellectual Property 261
   8.5.6 Consideration 261
   8.5.7 Representations and Warranties. 262
   8.5.8 Indemnity 263
   8.5.9 Term and Termination 264

8.5.10 Miscellaneous Provisions 264
8.6 Operating Agreements 265
8.6.1 Employment Agreements 266
8.6.2 Noncompete Agreements 266
8.6.3 Confidentiality Agreements 266
8.6.4 Consulting and Development Agreements 268
8.6.5 Maintenance and Support Agreements 269
8.6.6 Manufacturing Agreements 269
8.6.7 Assignment Agreements 270
8.7 License Agreements 270
8.7.1 Patent Licenses 271
8.7.2 Know-How Licenses 271
8.7.3 Trademark Licenses 271
8.7.4 Franchise Agreements 272
8.7.5 Technical Services Agreements 272
8.7.6 Distribution Agreements 272
8.7.7 VAR and OEM Agreements 273
8.7.8 Purchase Agreements 273

CHAPTER 9   Negotiating Fundamentals 285
9.1 Negotiation Fundamentals 287
9.1.1 Process, Behavior, Substance 288
9.1.2 Preparation 288
9.1.3 Mindset 288
9.1.4 Emotions 290
9.1.5 Position versus Interest 290
9.1.6 Establishing Your BATNA 291
9.2 Negotiation Approaches 293
9.3 Integrative versus Distributive Bargaining 294
9.3.1 Integrative Bargaining 294
9.3.2 Distributive Bargaining 295
9.4 Negotiation Outcome Types 297
9.4.1 Lose-Win Negotiating 297
9.4.2 Lose-Lose Negotiating 297
9.4.3 Win-Win Negotiating 298
9.5 Negotiating Gambits 299
9.5.1 Opening Gambits 299
9.5.2 Bargaining Gambits 300
9.5.3 Closing Gambits 302
9.5.4 Vendor Tactics 302
9.5.5 Negotiations after the Fact 303
9.6 Negotiating Contracts 303
9.6.1 Sales Contracts 303
9.6.2 Complex Project Contracts in Technology 305

PART IV TECHNOLOGY VENTURE STRATEGY AND OPERATIONS

CHAPTER 10 Launching the Technology Venture 317
    10.1 The Business Plan 319
        10.1.1 Writing a Business Plan 320
        10.1.2 Company Information 322
        10.1.3 Product/Service Description 323
        10.1.4 Competitive Analysis 324
        10.1.5 Market Analysis 324
        10.1.6 Industry Analysis 326
        10.1.7 The Management Team 328
        10.1.8 The Marketing Plan 329
        10.1.9 Financial Projections 330
    10.2 Networking 331
        10.2.1 Serendipity 333
        10.2.2 Using the Internet to Network 333
        10.2.3 The Primary Objective of Networking 334
    10.3 Resource Aggregation 334
        10.3.1 Capital Resources 335
        10.3.2 Human Resources 335
        10.3.3 Organizational Resources 336
        10.3.4 Technology Resources 337
    10.4 New-Venture Operations 337
        10.4.1 Performance Standards 338
        10.4.2 Information and Measurement 338
        10.4.3 Quality Standards 339
        10.4.4 Managing for Quality 341

CHAPTER 11 Going to Market and the Marketing Plan 351
    11.1 Marketing 353
        11.1.1 The Marketing Concept 355
        11.1.2 Market Segmentation 355
    11.2 Product 358
        11.2.1 Product Planning and Development 358
        11.2.2 Product Life Cycle 360
    11.3 Price 361
        11.3.1 Economic Dimension of Pricing 361
        11.3.2 Pricing Objectives 361
    11.4 Distribution 363
        11.4.1 Functions of Physical Distribution 364
        11.4.2 Modes of Transportation 364
        11.4.3 Types of Transport Services 366

11.4.4 Managerial Practices in Physical
Distribution. 366
11.5 Promotion. 367
11.5.1 The Marketing Mix. 368
11.6 The Marketing Plan. 373
11.6.1 The Marketing Budget 374
11.6.2 Determining a Marketing Budget. 375
11.6.3 The Marketing Message 376

CHAPTER 12 Financial Management and Control **387**
12.1 Accounting—Definition and Practices 389
12.1.1 The Accounting Cycle 390
12.1.2 The Accounting Equation 392
12.1.3 Working with Accountants 392
12.2 Finance—Definition and Practices 393
12.2.1 The Financial Manager. 393
12.2.2 The Financial Plan 394
12.3 Financial Statements 395
12.3.1 Sales Forecast. 396
12.3.2 Income Statement 399
12.3.3 Cash Flow Statement 402
12.3.4 Balance Sheet 402
12.4 Financial Statement Analysis. 405
12.4.1 Breakeven Analysis. 405
12.4.2 Ratio Analysis. 406
12.4.3 Types of Ratio Analyses 407
12.5 Financial Management 409
12.5.1 Start-up Costs. 409
12.5.2 Working Capital. 409
12.5.3 Accounts Receivable 410
12.5.4 Credit and Collections 411
12.5.5 Inventory Management 414
12.5.6 Purchase of Capital Assets 414
12.5.7 Payment of Debt 414
12.5.8 Payment of Dividends 415
12.6 Capital Budgeting 415

CHAPTER 13 Venture Management and Leadership **425**
13.1 Entrepreneur Managers. 427
13.2 Basic Management Skills 428
13.2.1 Analytical Skills 428
13.2.2 Decision-Making Skills 430
13.2.3 Communication Skills 431

13.2.4 Conceptual Skills 432
13.2.5 Team Building Skills. 433
13.3 Entrepreneurial Leadership. 434
13.3.1 Influence 435
13.3.2 Leadership versus Management. 436
13.4 Effectuation and Entrepreneurial Expertise 437
13.5 Entrepreneurial Ethics 439
13.6 Entrepreneurial Strategy. 442
13.6.1 Real Options Approach 443
13.6.2 Resource-Based Theory 444
13.7 Competitive Strategy Model 445
13.7.1 Differentiation 445
13.7.2 Cost Leadership 446
13.7.3 Niche Strategy 446

PART V MANAGING RISK AND CAREER DEVELOPMENT

CHAPTER 14 Venture Risk Management 459
14.1 Venture Risk Management 461
14.2 Hazard Risks 462
14.2.1 Product Liability. 463
14.2.2 Liability Insurance 463
14.3 Operational Risks 464
14.3.1 Venture Governance 464
14.3.2 Investor Relations 468
14.3.3 Human Resources Management. 469
14.3.4 Legal Risk Management 471
14.4 Laws Affecting Start-up Ventures 474
14.4.1 Law of Torts. 474
14.4.2 Law of Sales 474
14.4.3 Law of Bankruptcy. 475
14.4.4 Law of Negotiable Instruments 475
14.5 Obtaining and Working with Legal Counsel. 476
14.6 Managing Failure Risk 478
14.6.1 Developing Resilience 478
14.6.2 Overcoming Cognitive Biases 479

CHAPTER 15 Your Entrepreneurial Career. 491
15.1 Your Entrepreneurial Career. 493
15.1.1 The Individual Entrepreneur 493
15.1.2 Conceptual Model of Entrepreneurial Careers 494
15.1.3 Finding a Mentor 496

15.2 The Entrepreneurial Personality 496
    15.2.1 Characteristics of Growth-Oriented Entrepreneurs. 497
    15.2.2 Emotional Intelligence 498
15.3 Intrapreneurship 499
15.4 Five Entrepreneurial Careers 501
    15.4.1 Paul Baran 501
    15.4.2 Marc Andreessen 502
    15.4.3 Pierre Omidyar. 502
    15.4.4 Craig Newmark 503
    15.4.5 Robert Metcalfe 504
15.5 The Franchising Option 505
    15.5.1 Franchising: A Brief History. 506
    15.5.2 Uniform Franchise Offering Circular 506
    15.5.3 Ten-Day Rule 507
    15.5.4 Franchising Agreement 507
    15.5.5 Advantages of Owning a Franchise 509
    15.5.6 Disadvantages of Owning a Franchise 510
    15.5.7 Franchisor Disclosure 511
15.6 Launching Your Entrepreneurial Career. 512

Appendix. 521

Index 545

Preface

This textbook has been organized around the central concept of "value." Although, on the surface, there does not appear to be anything mysterious or ambiguous about the concept, the reality is a bit different. In fact, a special edition of the *Academy of Management Review* was dedicated entirely to the topic of value creation. In the lead article to that edition, the editors noted: "There is little consensus on what value creation is or on how it can be achieved."[1] Imagine that. While most of us believe we know what "value creation" means, there is "little consensus" on its meaning. We wrote this textbook to help faculty and students wrestle with and come to terms with the concept of value and value creation and the process of building a successful technology venture.

ORGANIZATION AND STRUCTURE

In this text, we've attempted to create a flow and structure that encapsulates creating, capturing, and protecting value. One of the key innovations in this text is the sequence in which the material is presented. Most entrepreneurship textbooks begin with value creation and move on to value protection and value capture. We have inverted that logic in this text based on our many years of experience in teaching entrepreneurship to technology students as well as forming new technology ventures. Our experience indicates that technical students are the least interested—at the beginning—in the material associated with value creation. The concepts of selling, marketing, customers, and distribution are not deemed important at first to most technical students. Therefore, it is not desirable to begin an introductory course in technology entrepreneurship with a focus on value creation.

Instead, our experience has shown that technical students are very interested in the techniques and strategies involved with capturing and protecting the value of what they create. It is only after they become aware of how much value can be created by observing tried and tested techniques for deal structuring, equity distribution, and intellectual property protection that they become interested in learning how to market and sell their products and services. This logical inversion of the standard sequence of topics has been tested in our classrooms at the undergraduate and graduate levels with great success. We are sure that you will find the presentation sequence to be counterintuitive at first, but far more effective in the long run. The figure below provides an overview of how this textbook has been structured.

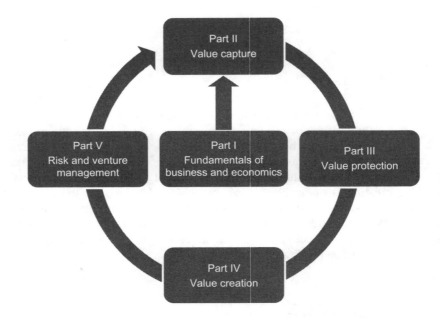

Part I of this text is a unique feature not found in other textbooks in technology entrepreneurship. It focuses on key foundational concepts in business and global economics. The reason for this section is to ensure that technology-oriented students are made aware of the logic and rationale for business ventures and entrepreneurial ventures in particular. The global economy is now an essential part of nearly any technology venture. As such, students should develop a fundamental understanding of the playing field and the many opportunities and threats that exist as a result of global competition.

Part II begins the exploration of value by focusing on value capture. The concept of value capture applies principally to what is referred to as "enterprise value." As a result of serving markets effectively, technology entrepreneurs build the value of the enterprise over time. Chapter 4 ranges over the topics of legal structure and equity distribution among the founders. Chapter 5 addresses the issues associated with raising capital and deal structuring. Chapter 6 focuses on exit strategies for technology ventures. Together, these chapters provide an extensive and detailed introduction to distributing equity on founding, using equity to raise necessary capital, and gaining a return on invested capital via particular exit strategies.

Part III addresses value protection. Value protection focuses in part on the protection of the intellectual property that resides within a technology venture. Chapter 7 examines in detail the techniques associated with intellectual property identification, protection, and defense. Since contracts are an important element of value protection for a technology venture, Chapter 8 covers contracts, including terms and conditions, penalties and remedies, and contract structure and duration. Finally, contracts must be negotiated, especially complex contracts that are often associated with technology ventures and their counterparties. Chapter 9 examines the techniques and nuances associated with contract negotiations and managing complex contracts.

Part IV focuses on value creation. Value creation consists primarily of those functions of a technology venture that have to do with founding the venture and with marketing, selling, and distributing its products and services. We follow Sarasvathy and others in our understanding and application of the concept of value creation. That is, we take it as a given that entrepreneurs don't simply serve preexisting markets. Rather, they are actively engaged in creating the markets they serve. Sarasvathy's notion of "effectuation" is central to this view. Effectuation, if it can be summed up in a phrase, presents the entrepreneur as someone who transforms amorphous resources and nondescript markets simultaneously. The action of the entrepreneur to create value includes both the mobilization of disparate resources and the transformation of a market.

Chapter 10 examines the issues surrounding founding a venture, focusing on the aggregation of initial resources required to operate. Chapter 11 focuses on go-to-market strategies and the challenging topics—to technical students—of sales and distribution. As we are now discussing an operating venture, Chapters 12 and 13 focus on the issues inherent to an operating and growing concern. Chapter 12 focuses on financial management and control of the venture, and Chapter 13 addresses venture management and leadership.

Finally, in Part V, the topics of risk management and entrepreneurial career development are covered. Chapter 14 covers venture risk management, including the topics of financial risk, legal risk, and personal risk. Many people erroneously assume that entrepreneurs are high-risk takers. While it's true that most entrepreneurs take risks, they are not more or less risk averse than the general population. Instead, entrepreneurs have learned to become adept "risk minimizers." They have learned to take enough risk out of a deal to make it tolerable for them to move forward. We spend substantial time on the topic of personal risk management. That is, we cover strategies associated with the stresses and emotional strains of being an entrepreneur and operating a technology venture. Developing coping skills, including personal resilience, is essential to the success of the entrepreneur and of the venture itself.

Chapter 15 surveys the joys and challenges of an entrepreneurial career. It is difficult for most technology entrepreneurs to realize that they could be making more money and have greater security working in a large firm. In fact, most technology entrepreneurs launch their ventures primarily because they simply do not want to work for someone else, not because they want to get rich. Still, the promise of fortune lies in the back of every entrepreneur's mind, and likely is a major source of motivation through the tough times. Finding a mentor, developing a network of supporters, and realizing that failure is often a significant part of ultimate success are important elements in the career of the technology entrepreneur. Learning how to work with a mentor and establish networks of key and trusted advisors is part of what is covered in this final chapter.

ADDITIONAL FEATURES

In addition to the unique content flow and structure of this text, we've embedded a number of teaching and learning objects to aid comprehension and classroom management.

- Each chapter contains several "Tech Tips." These are encapsulations of key lessons discussed in the chapter. Students and faculty should take time to discuss the Tech Tips boxes, making sure to understand each of them.

- We have also included "Tech Micro-cases" in each chapter. These are short case studies featuring real-world examples of topics and concepts developed in the chapter. At the end of each chapter, we have provided a set of study questions to help in chapter review and learning. Also included are two exercises per chapter to get students out of their seats and actively engaging the chapter material. Some exercises are meant to be conducted in the classroom and others outside the classroom.

- Each chapter includes a set of Web resources that students and teachers can explore for more information on topics covered in the chapter.

- Key words are bolded in the chapters and defined in the "Key Terms" section at the end of each chapter.

- The middle chapters each contain an in-depth case study pertaining to the chapter material.

- A model business plan (for USuggest.com) is provided as an appendix.

- Finally, for those who are interested in exploring more deeply the research and other supporting material used in the chapters, the endnotes are provided.

TEACHING AND LEARNING SUPPORT

For faculty adopting this text for classroom use, a solutions manual, set of PowerPoint lecture slides, and additional questions/exercises are available by registering at www.textbooks.elsevier.com. Additional learning resources for students will be posted from time to time at the book web site, www.elsevierdirect.com/9780123745026.

During the development of this book, many things have occurred to reinforce our belief that it will make a significant contribution to technology entrepreneurship teaching and learning. Much has been discovered over the past decade about how entrepreneurs think and how they approach opportunities. We have tried to incorporate much of this research in the curricular flow of the material presented in the text, and in the teaching tools associated with it.

We feel certain that this textbook will be useful to the growing ranks of faculty members who teach technology entrepreneurship. We also feel that the students exposed to technology entrepreneurship for the first time through this text will ignite lifelong interest in the entrepreneurial career path. On one side, we see great challenges ahead for the nation, the global economy, and for individuals and their families. On the other side, we hope that the prepared minds of aspiring technology entrepreneurs will realize that great challenges are often signs of great venture opportunities. We need them now, more than ever. And we would regard our labors to be well spent if several of the students who study this text join the ranks of the successful technology entrepreneurs who have preceded them.

Thomas N. Duening
Robert D. Hisrich
Michael A. Lechter

May 22, 2009

ENDNOTE

[1] Lepak DP, Smith KG, Taylor MS. Value creation and value capture: a multilevel perspective. *Acad Manage Rev* 2007;**32**(1):180–194.

Acknowledgments

We are grateful for the support, encouragement, and feedback we received during the production of this text. Several people deserve special mention for their unwavering and tireless support of this project. Susan Garrison has been steadfast in managing the flow of content among the three authors and in keeping everyone on schedule. Sunanda Vittal was invaluable as an editor and guide to organizing each of the chapters. Carol Pacelli assisted in the preparation of many of the chapters. Falyne Chave, Anetta Hunek, and Sarah Liggett provided substantial research support and assistance in developing the end-of-chapter exercises. Tiffany Duening provided early versions of the cover design for this book, and Kathleen Harrison assisted in cover design and created a poster for a conference. Several of the case studies were originally drafted by students at Arizona State University, including Brandon Woodward, Kimberly Keith, Patricia Johnson, and Kellie Parisek. We also thank Joe Hayton from Elsevier Publishing for seeing the potential in this project and for supporting its development. Others from Elsevier kept the project moving and added their expertise to its final look, including Maria Alonso, Sarah Binns, and Eric DeCicco. Of course, any remaining errors, either of omission or commission, remain the sole responsibility of the authors.

In addition, a number of teachers and scholars participated in reviewing this text during its development. They are listed below in alphabetical order:

> - Alan L. Tharp, North Carolina State University
> - Aron Spencer, New Jersey Institute of Technology
> - Arun Nevader, University of California, Berkeley
> - Audrey MacLean, Stanford University
> - Bob White, Western Michigan University
> - Danny Warshay, Brown University
> - Denis K. Koltsov, Lancaster University
> - Edward F. Gehringer, North Carolina State University
> - Jeffrey A. Martin, University of Texas, Austin
> - John Callister, Cornell University
> - Jonathan Burgstone, University of California, Berkeley
> - Joseph Stanislao, Montana State University
> - Patrick Crowley, Washington University
> - Robert Crockett, California Polytechnic State University, San Luis Obispo
> - Tom Mason, Rose-Hulman Institute of Technology

About the Authors

Dr. Thomas Duening is the El Pomar Chair of Business and Entrepreneurship at the University of Colorado in Colorado Springs (UCCS). He is also an associate professor of management in the UCCS College of Business and Administration. Duening is a 1991 graduate of the University of Minnesota with a PhD in higher education administration and an MA in philosophy of science. He began his academic career as the assistant dean for the University of Houston's College of Business Administration. There, he was also a faculty in its Center for Entrepreneurship & Innovation.

Duening launched his first technology venture while a graduate student. His international consulting firm served the electric utility industry with information products centered on the issue of health effects associated with electric and magnetic fields (EMF) from high voltage power lines. He left in 1991 to assume the assistant dean position in Houston.

After his 9-year stint as assistant dean, Duening founded several more companies. With a partner, he founded U.S. Learning Systems in 1998. The firm provided e-learning content to providers around the country. U.S. Learning Systems was acquired in 1999 by Edgia.com. Duening left Edgia in 2002 to launch the Applied Management Sciences Institute. As part of this firm, he co-wrote three business textbooks *Managing Organizations, Business: Principles, Practices, and Guidelines,* and *Management Skills*. He also wrote two popular trade books, *Managing Einsteins* and *Always Think Big*.

Duening next founded INSYTE Business Services Group to study best practices in business process outsourcing. The result of this effort was two trade books: *Business Process Outsourcing: The Competitive Advantage*, and *The Essentials of Business Process Outsourcing*. Both published by John Wiley & Sons in 2004 and 2005, respectively. As he was conducting the research for these books, Dr. Duening co-founded InfoLabs India, Pvt. Ltd., a business process outsourcing firm based in Bangalore, India. The firm provides outsourcing services to a wide range of publishing companies. It was acquired by ANSRSource, Inc., in 2008.

In 2004, Duening joined Arizona State University's Ira A. Fulton School of Engineering as Director of its Entrepreneurial Programs Office. In this role, he taught courses in Technology Entrepreneurship to engineers at the graduate and undergraduate levels. Dr. Duening's ongoing research is in the areas of enterprise innovation, and entrepreneurship, including entrepreneurship education. He is often a featured speaker at campus and community events, and he has consulted broadly with entrepreneurial ventures as well as multinational corporations. Duening is the author of numerous journal articles and 12 books on business and management.

Robert D. Hisrich is the Garvin Professor of Global Entrepreneurship and Director of the Walker Center for Global Entrepreneurship at Thunderbird School of Global Management. He is also president of H&B Associates, a marketing and management consulting firm he founded and has been involved in the founding of several successful technology ventures.

Professor Hisrich received his BA from DePauw University, his MBA and Ph.D. degrees from the University of Cincinnati, and honorary doctorate degrees from Chuvash State University (Russia) and the University of Miskolc (Hungary). Prior to joining Thunderbird, Dr. Hisrich held the A. Malachi Mixon, III Chaired Professor of Entrepreneurial Studies at the Weatherhead School of Management, Case Western Reserve University. Dr. Hisrich was a Fulbright Professor at the International Management Center in Budapest, Hungary in 1989. In 1990-91 he was again named a Fulbright Professor in Budapest at the Foundation for Small Enterprise Economic Development, where he also held the Alexander Hamilton Chair in Entrepreneurship. Dr. Hisrich has held visiting professorships at the University of Ljubljana (Slovenia), the Technical University of Vienna (Austria), the University of Limerick (Ireland), Donau University (Austria), Queensland University of Technology (Australia), the University of Puerto Rico, and the Massachusetts Institute of Technology.

He has authored or co-authored twenty-five books including: *International Entrepreneurship: Starting, Developing and Managing a Global Venture*, *Entrepreneurship: Starting, Developing, and Managing a New Enterprise* (translated into 13 languages and soon to be in its eighth edition), *The 13 Biggest Mistakes that Derail Small Businesses and How to Avoid Them*, and *The Woman Entrepreneur*. Dr. Hisrich has written over 350 articles on entrepreneurship, international business management, and venture capital, which have appeared in such journals as *The Academy of Management Review*, *California Management Review*, *Columbia Journal of World Business*, *Journal of Business Venturing*, *Sloan Management Review*, and *Small Business Economics*. He has served on the editorial boards of *The Journal of Business Venturing*, *Entrepreneurship Theory and Practice*, *Journal of Small Business Management*, and *Journal of International Business and Entrepreneurship*. Besides designing and delivering management and entrepreneurship programs to U.S. and foreign businesses and governments, particularly in transition economies, Dr. Hisrich has instituted academic and training programs such as an MBA program in Hungary, a high school teachers entrepreneurship program in Russia, an Institute of International Entrepreneurship and Management and the Zelengrad Business College in Russia, and an Entrepreneurship Center in China.

Michael A. Lechter, attorney, certified licensing professional (CLP) and entrepreneur, CEO of TechPress Inc., a publishing and literary agency company, CEO of Michael Lechter PC, and Adjunct Professor in the Entrepreneurial Program in the Ira A Fulton School of Engineering at Arizona State University, is the bestselling author of *OPM: Other People's Money: How To Attract Other People's Money For Your Investments— The Ultimate Leverage* (2005), and *Protecting Your #1 Asset: Creating Fortunes from Your Ideas* (2001).

An internationally known expert in the field of intellectual property, his clients have included everything from breweries to fast food companies, casinos, professional sports teams, major software companies, semiconductor and medical device manufacturers, and venture capitalists to start-ups. When asked what he does for a living, he typically replies, "I build forts and fight pirates."

Michael has been the architect of strategies for building businesses—using both conventional and unconventional forms and sources of "Other People's Money and Resources." His experience in representing both venture capitalists and start up and emerging businesses, and experience as an angel investor himself, provides a unique perspective to the subject of building a business.

Michael is also the author of *The Intellectual Property Handbook,* TechPress (1994), coordinating editor of *Successful Patents and Patenting for Engineers and Scientists,* IEEE Press (1995), and contributing author to the *Encyclopedia Of Electrical And Electronics Engineering,* Wiley (1999), and *Licensing Best Practices: The LESI Guide To Strategic Issues and Contemporary Realities,* John Wiley & Sons, Inc. (2002). Over the years he has also written monthly columns for INC.com, IEEE-USA Today's Engineer, and Washington Technology.

Michael has been an active member of the Licensing Executives Society (LES) USA/Canada, serving as a trustee (1996–2000), and as Computer and Electronics Industry Sector Chair (1992–1996). He has been a LES USA/Canada delegate to LES International since 2001, and has served as chair or vice chair of a number of LESI committees.

He has lectured extensively throughout the world on intellectual property law and entrepreneurship. Upon request of the House Judiciary Committee he has submitted testimony to the Congress of the United States, and has participated in various United Nations and foreign government proceedings on intellectual property law and technology transfer.

Michael is also the owner of Cherry Creek Lodge LLC, a resort/dude ranch in the Tonto National Forest of Northern Arizona (www.CherryCreekLodge.com), a study in rustic elegance where modern comfort meets the old West. The Cherry Creek Lodge specializes in corporate and family retreats.

Fundamentals of Business and Economics

Fundamentals of Business

After studying this chapter, students should be able to:

- Understand the logic and mechanics of a business enterprise
- Appreciate the role of value in business and economics
- Identify roles and responsibilities of business in society
- Explain business as a human activity
- Evaluate elements of a going concern versus a failed venture
- Identify factors essential to a dynamic venture

TECH VENTURE INSIGHT

It's Hard to Be a CEO

Mark Zuckerberg is the 23-year-old chief executive of Facebook, the popular social-networking web site used by millions of people around the world. Zuckerberg founded Facebook as a 19-year-old college student in 2004. The company enjoyed almost instant success as millions of users flocked to the site, and Silicon Valley venture capital firms competed with each other to provide funding. In 2007, Microsoft invested $240 million in the venture at a valuation of $15 billion.

Zuckerberg moved Facebook's headquarters to Palo Alto, California, shortly after he launched the company. At first, Zuckerberg was able to maintain a "college student" life-style, living and working out of a rented house with the other Facebook engineers. When he finally moved into a Silicon Valley office, Zuckerberg would arrive mid-morning in his flip-flops and work late into the night. It was a laid-back style that was not suited to the growing business.

To adjust to the challenges of running a fast-growing technology company, Zuckerberg turned to other industry giants for consultation. He became friends with Marc Andreessen who founded the Internet browser company Netscape when he was just 22 years old. The two of them met often to discuss business and strategy issues. Zuckerberg also had frequent conversations with Silicon Valley investor Roger McNamee. McNamee had become a mentor to his young protégé. Zuckerberg lamented that being CEO of his own company "is hard—I do sometimes whine to Roger about it."

In March 2008, Zuckerberg finally reached into the Internet company talent pool and hired 38-year-old Sheryl Sandberg from Google to become Facebook's CEO. Sandberg was one of the first 300 employees at Google, and helped turn the company into a global advertising power. Together, the two confronted the business challenges that the youthful Zuckerberg may not have had the knowledge or experience to tackle on his own.

Source: Adapted from Vara V, Facebook CEO seeks help as site grows up, The Wall Street Journal, March 5, 2008, A1; Kafka P, Facebook's new grownup: Google Exec Sheryl Sandberg, Silicon Valley Insider, March 4, 2008.

INTRODUCTION

Mark Zuckerberg is the youthful CEO of the popular social-networking web site, Facebook. As his company grew rapidly, Zuckerberg discovered that his background and skills were increasingly inadequate to the business challenges he faced. Confounded not only by the technical issues that inevitably arise in a rapidly growing venture, he also encountered increasingly complex business and strategic issues. As the opening Technology Venture Insight highlights, after four years of operation, Zuckerberg finally hired a more seasoned and business-savvy executive to help him achieve his business goals.

This is a common path for many technology entrepreneurs. While they are experts in a narrow technology field, they often find their business skills less than adequate to cope with a growing number of employees and the pressures of global competition. This textbook starts with the premise that many technology-oriented students simply are not exposed to the fundamental principles of business operations. To address this problem, we have devoted the first three chapters of this textbook to business and economics topics to familiarize technology students with basic business concepts, as well as the challenges faced by technology ventures in the modern global economy.

Business activity pervades societies and cultures around the globe. The nomadic reindeer herder on the steppes of Mongolia delivers hides to a market in the capital city of Ulan Bator to supplement a family income. The fashion designer in Hong Kong sends out the latest designs to an eager audience in New York. And the dairy farmer in Wisconsin has lunch while listening to up-to-the-minute broadcasts on the price of feed grains and other commodities. Business is all-encompassing and globally connected, yet usually only modestly understood by most people.

In modern times, the pervasive influence of business plays out in entire television networks, such as CNBC, which broadcasts every tick of the major stock markets as they make their daily up and down movements. Newspapers, such as *The Wall Street Journal*, *Financial Times*, and *Investor's Business Daily*, are dedicated entirely to business and economic news. Blogs, such as TechCrunch and Killer Startups, report continuously about new technology ventures being launched, acquired, going public, or closing their doors for the last time.

Despite this increased awareness of business activity, most technology-focused education and training programs do not include lessons on fundamental business principles and practice. Technology students, however, are exposed to compulsory instruction in literature, history, composition, and other of liberal arts subjects, but few are required to enroll in basic business courses. This is a pity since most technology-oriented students will typically end up working *in* a business. In fact, a recent review of the Standard & Poor's 500 largest companies in the United States revealed that fully 23% are led by individuals with an undergraduate engineering degree. The next closest undergraduate major is business, with a 15% representation.[1]

Of course, it is possible to learn some basic principles of business and management by working within a company and observing how it operates and is managed by others.[2] Unfortunately, this approach has two critical drawbacks. First, it takes a long time to both *perform* in your technical job and at the same time *observe* enough business situations, decisions, and consequences so that you may extract meaningful lessons. Second, those who manage and lead the business may not be doing a very good job—and the lessons learned may not be helpful in your career. If your workplace is managed by individuals who have minimal formal management or leadership training, they may be less effective than they could be. Thus, limiting your learning of business and management to observation alone may, in the end, only serve to suboptimize *your* effectiveness.

The focus of this book is on technology ventures—how they start, operate, and sometimes exit profitably. To understand technology ventures (or any other type of venture), it is critical that you understand fundamental business concepts. Fortunately, you don't necessarily need to enroll in a formal business degree program to learn important basic principles. In fact, a limited number of fundamental business concepts can be learned and absorbed relatively quickly, giving you a leg up on the next step toward higher levels of mastery. It is our hope that you will continue your learning beyond the basics described here, by reading business publications, watching business programming on television, and thinking about the business issues that are described in those media.

TECH MICRO-CASE 1.1

Mastering the Fundamentals

Consider the example of a professional golfer like Tiger Woods. Mr. Woods would not be the extraordinary golfer he is today without having previously struggled with and mastered the fundamentals of golf. In his case, that means he had to learn the mechanics of the golf swing, the placement of the hands on the golf club shaft, the focusing of the mind before each shot, and myriad other details. As these *fundamentals* became *habits* through *practice*, Mr. Woods was able to take them for granted and shift his attention to other factors of the game. Having mastered the fundamentals, he is, for example, now able to experiment with hand placement on golf shots that present unusual challenges, such as those out of rough grass, behind a hazard, or those that require a certain type of spin to be placed on the ball. Experiments such as these can only take place *after* the fundamentals have been mastered. Tiger Woods is a multi-millionaire, with many opportunities in his life. Without his world-leading mastery of the fundamentals of golf, none of those other things would be options for Mr. Woods.[3]

Section 1.1 is intended to guide you, the technology-oriented student, through what we have determined to be the fundamentals of business. If you have already taken basic business courses, undoubtedly you will find some familiar material here. In that case, use the chapter as a reminder of the fundamentals, in the same way star athletes need to be reminded from time to time to "go back to the fundamentals." If you are being exposed to these concepts for the first time, your goal will be to learn and then practice the fundamentals until they become habits.

The first chapter aims to initiate a personal interest in the business side of the technical field that you have been studying. Whether you are an engineer, scientist, or studying other technical disciplines, there is a business aspect that generates the capital to pay for lab equipment, computers, salaries, etc. Getting to know this side of your technical field of study will enable you to become more valuable to your employers, and it will prepare you someday to make the entrepreneurial leap into your own technology venture.[4] In this chapter, we will examine business from various perspectives: logical, ethical, and societal. These perspectives overlap in a number of ways, but together form a fundamental understanding that can be applied to nearly any type of technology venture that you may dream up. Let's begin by examining the basic question: What is a business?

1.1 WHAT IS A BUSINESS?

The term "business" generates different reactions among different people. Some of the leading practitioners of business and venture creation regard business as "fun," "exciting," and even "playful."[5] They are successful in part because business is a passion for them, and they don't make a distinction between their "work" and their "play." To others, business is an activity that they would prefer to avoid—and some, such as the founders of socialist economics, even regard business and businesspeople as a temporary stage in the development of economies. Although not as prevalent as they were a few decades ago, communist and socialist countries and economies tended to view business leaders and businesspeople as a passing stage in the history of cultural and social development (for more on this perspective, see, for example, Ref.[6]).

Business, in fact, has become a global phenomenon over the past decade, and modern national and international policies promote the flow of goods, services, and capital with relative freedom around the world.[7] Today, even the formerly communist countries, such as Russia and China, have adopted decidedly pro-business policies. In other words, business is global in scope and is here to stay. As such, it is useful to begin with a definition. Webster's dictionary defines business as "a usually commercial or mercantile activity engaged in as a means of livelihood." For our purposes, a business is defined as:

> *An organized and purposeful human activity designed to create value for others and to exchange that value for something else of equal or greater value (usually, money), and that is intended to continue to provide such value over time as a going concern.*

1.1.1 Business as an Organized Activity

Notice that we define business as an **organized activity**, where an individual or a group combine and deploy resources, such as land, labor, and capital, to use toward a productive activity. It is often easy to overlook the importance of the ability to organize and deploy resources for economic or social gains. After all, we are all born into a world of large organizations—schools, automobile manufacturers, banks, grocery stores, and many others. These large organizations may appear as if they have always existed to serve the purposes that they do. Remarkably, nearly every single one of the large organizations that provide basic personal belongings and needs did not exist a mere 150 years ago. They all were founded by an individual entrepreneur or a group of partners who gathered the necessary resources and launched a venture. Ford Motor Company was launched by Henry Ford in 1901.[8] The H.J. Heinz Company was launched in 1869 by Mr. Henry Heinz when he was 25 years old.[9] General Electric was established by Thomas Edison and others in 1892.[10] And the list goes on. Remember, every large organization that today provides you with the staples and fashionable items that you need and want was, at one point in time, a brand-new venture with only limited prospects for success.

1.1.2 Business as a Purposeful Activity

We also define business as a **purposeful** activity, which implies that businesses are founded on the belief that an innovation can be converted to value that is wanted or needed by a market or markets. The innovations that underlie a new business idea can be widely varied. Some innovations are centered on new products, or extensions of existing products. Take, for instance, the original Apple computer introduced by Steve Jobs and Steve Wozniak to the U.S. market in March 1977. That product was truly revolutionary, and was the first to make the power of the computer available to the average American.[11] Its graphical user interface (GUI) was just one of many features that differentiated the Apple computer from all other efforts, putting computing power into the hands of millions of average users. Since then, the company and its computers have undergone myriad transformations, each of which was an extension of the original founding act and product. Jobs and Wozniak launched Apple in 1976. Today, Apple, Inc. is a global enterprise that continues to build value according to the purposes on which the company was founded more than 30 years ago.[11]

1.1.3 Business as a "Going Concern"

A third aspect of our definition of business is that it is a **going concern**, distinct from a project or hobby. A going concern means that the business will continue into the indefinite future, with no clear end date or precise definition of "success." A business is usually said to be successful if it continues to make profits over time—usually increasing profits relative to industry averages. The manner in which a business makes profit is deemed its **business model**. Business models vary across and within industries. This concept will appear again and again throughout the textbook. For now, it is sufficient to know that we define the term "business model" as the way the business makes money.

In contrast, consider short-term projects undertaken by Boy Scout or Girl Scout troops to raise funds. The cookie sale, chili cook-off, or hot dog stand organized by these groups have some of the characteristics of a business, where an aggregation and deployment of resources are required, along with adding value to those resources in a manner that other people are willing to pay *in excess* of the costs. However, several elements are missing here that sets these undertakings apart from our definition of business. Namely, these activities are not intended to continue indefinitely, and they have a limited duration based on a date or fundraising goal. In other words, there is no intent on the part of the organizers to establish a going concern. A business, on the other hand, intends to make profits from the application of its resources in the short term (quarterly), and over time (annually, and year over year). Intending to establish a going concern means that business founders are interested in establishing a **system**, whereby the venture's resources are organized to continue to create valuable output that produces profits into the indefinite future.

1.1.4 Value

The final term to recognize in our definition of business is **value**. Our definition has only a single, very specific meaning. In the context of business, value is defined by the market.[12] That's it. If the market says something is valuable, there is no need for further examination. This straightforward, market-focused definition of the term "value" is occasionally disconcerting to technically oriented people. They have been trained to associate value with highly precise experimental designs, or with exquisitely crafted algorithms, or with unique solutions to complex problems. In the domain of engineering and the sciences, these concepts represent valuable contributions to research and practice. In the domain of business, however, the "best" product or service does not always have the most value to a market. A good way to understand "value" from the perspective of a market is through this simple equation:

$$\text{Value} = \text{Price} \times \text{Quality}$$

This equation expresses the concept that most markets recognize a trade-off between price and quality, i.e., higher-priced goods usually have higher quality, while lower-priced goods have lower quality. Some markets will trade higher quality for a lower price simply because there is scant demand for higher quality or there is no interest in paying more for higher quality. In contrast, other markets are interested in higher quality and do not mind paying a higher price to obtain it. Take automobile purchases, for instance, that are available in a wide price range and across different quality levels. The individual seeking to purchase a Hyundai Elantra will not likely be tempted by the Cadillac dealer across the street, and vice versa.

Now that we have explored the definition of *what* a business is, let us turn our attention to *how* a business converts disorganized resources into value for a market.

1.2 VALUE AND THE MARKET

Markets are comprised of people, who have the freedom to make choices about how they spend their time, money, and energy. Collectively, these individual free choices comprise a market. Building a business means developing a mechanism for consistently and continuously providing value to a market. Of course, that means the business must tune into the choices the market is making, and will make in the future. Many businesses find early success with products or services that appeal to a certain market, only to find that over time the market changes and the formerly popular offerings are no longer in demand.[13] Hence, it is very important for a business to understand that markets shift their definition of value over time, and in the context of competitor products.

1.2.1 Value Creation

There are probably as many ways to create value as there are people on this planet. Consider the Turkish fisherman who sets out each morning into the Black Sea to catch bass. He brings the fish he catches back to the open air market near the boat docks where housewives, professionals, and others eagerly scan the day's catch for a tasty dinner meal. The shoppers on the dock constitute the "market" for Black Sea bass. The individuals constituting that market find value in the fisherman's ability to catch the bass to the extent that they do not want to go out and catch their own. The market will also be interested in whether the bass provided by the fisherman are of adequate quality. Prices for the bass will reflect the quality of the catch and the extent to which the market has other Black Sea bass or substitute product choices.

For his part, the fisherman has a going concern business venture that consists of setting out daily to bring fish back to his market. His resources are his boat and fishing gear, fishing know-how, and retail market space. On their own, none of these resources would provide the market with the bass that it needs. It is the entrepreneurial activity of the fisherman to organize and deploy these resources that leads to the creation of value for the bass-eating market.

Let's take the example of value creation to another domain. Consider the case of Chad Hurley, Steve Chen, and Jawed Karim, three entrepreneurs who set out to create a new type of Internet company in Menlo Park, California, which is located in the heart of Silicon Valley and is the birthplace of some of the most rapidly growing technology companies in history. These three individuals were veterans of technology companies, having been principals in the online payment service known as PayPal. From their garage in Menlo Park in December 2005, they went on to create one of the fastest growing companies of all time. They did this by aggregating the resources centered on the Internet and by employing creativity, leading to the creation of YouTube.

By July 2006, YouTube reported more than 100 million videos being viewed every day from its site, with as many as 50,000 videos added each day. The brilliant aspect of this concept is that most of the resources that form the "product" of YouTube are provided by people who have no ownership or other interest in the venture. YouTube is part of the popular Web 2.0 revolution that began with companies such as FaceBook and MySpace. Each of these web sites, as well as many others, gather resources from users themselves in order to build a community of users. Once a community is built, the entrepreneurs then determine how to develop a profitable business model. Some Web 2.0 sites sell advertising to those companies that want to reach the community of users. In the case of YouTube, advertising is the main source of revenue. In October 2006, a mere 18 months after its establishment, YouTube was acquired by Google for $1.65 billion.[14]

As these stories illustrate, the resources required to launch a venture can vary greatly, and they can be obtained from a wide range of sources. The Turkish fisherman required tools, physical labor, and a retail space to operate his venture. The YouTube founders required capital, computing power, and brain power to operate their venture. The fundamental thread that ties the two vastly different companies together is their focus on serving a market need. In the case of the fisherman, the need is obvious: people need food to eat. In the case of YouTube, the need is less obvious. Do people really need to watch thousands of videos each day? Probably not, but the value created by YouTube ($1.65 billion in 18 months) far exceeds what the individual fisherman can ever hope to achieve.

1.2.2 **Value Protection**

Technologists create value for markets in multiple ways. Computer scientists invent new software programs that enable greater communication and efficiency. Geneticists invent new forms of living matter that ensure a growing and hungry world population has the food it needs. Engineers solve problems, build roads and bridges, and cure our ailing environment. Each of these types of ventures develops intellectual property. Understanding how to protect the intellectual property that one creates, or that is created by the venture, is an important part of success in technology entrepreneurship. Intellectual property can be nearly anything that an individual or group of individuals in a venture produce or enact. That is, it includes both the products and processes of the venture.

1.2.3 **Value Capture**

Value capture refers to the process of exiting a venture via a sale, merger, or public stock offering. There are a number of events that precede a successful exit from a venture. Along the way, the technology entrepreneur must distribute equity, raise capital, and incentivize employees and others to contribute their talents and energies to ensure that the venture grows. Technology entrepreneurs at the center of this all need to structure the growth of their enterprise to help it prosper, and at the same time they must maintain enough personal ownership in order to stay motivated. If sufficient equity is retained through the fundraising and distribution to others, the technology entrepreneur, upon exit, will capture the value that has been created in the venture.

1.2.4 **Value Proposition**

"Value" has myriad definitions, and entrepreneurs can develop successful ventures with widely different **value propositions**. A value proposition is the story that a venture tells its market about what it intends to provide. For example, YouTube's value proposition is: "Broadcast yourself"—a simple statement, while not necessarily appealing to everyone, is the foundation of the online video-sharing rage. Similarly, the value proposition for the fisherman might be something like: "The freshest Black Sea bass."

Value propositions are important for a venture because they not only communicate what the venture intends to provide, but also help guide the decision-making process. For example, the value proposition for well-known consumer products company Procter & Gamble (P&G) is "Touching lives, improving life."[15] This value proposition functions as a goalpost for P&G scientists and product developers on how to structure their research and development resources and investment. P&G introduces hundreds of new products to markets around the world each year. The firm's value proposition guides decision making so that consumers do not get confused about the firm's intent and offerings.[16]

New technology ventures also benefit from having a well-articulated value proposition to help steer them through the various stages of venture development. Dell Computer, for example, had a potent value proposition when it was founded in 1984 by Michael Dell, a college student at the University of Texas. Mr. Dell established his company on the belief that, by selling personal computer systems directly to customers, he could better understand customers' needs and provide the most effective computing solutions to meet those needs. Dell Computer built on this vision over the years, and is now among the 500 largest companies in the United States, employing more than 70,000 people worldwide.[17]

1.3 BUSINESS AND SOCIETY

The role of business in society has been the subject of debate and discussion for centuries. As far back as the Roman Empire, politicians, business entrepreneurs, and common folk debated the role of business in creating wealth and prosperity.[18] More recent history has seen this debate crystallize around two dominant modes of thought: capitalism and socialism.

Capitalism forms the economic bedrock of the Western world, including the United States, the European Union, and Australia, as well as much of Southeast Asia. The principles of capitalism uphold freedom and individual choice as the drivers of prosperity. Business is seen as a noble undertaking where humans freely enter into exchange relationships to maximize their individual and, as a result, their collective prosperity.[19]

Socialism, on the other hand, is the primary alternative to capitalism, and predominates in countries such as China, Russia, and much of Latin America. Of course, there are a number of variants of socialism. The most extreme form is communism. China and North Korea, for example, are two of the few remaining communist economies, although China has been moving further and further from the traditional communist approach. Communism was a major force in the twentieth century and was based on the maxim coined by Karl Marx: "From each according to his ability, to each according to his need." In short, a socialist economy presupposes that centralized planning of business activity can lead to the most equitable distribution of goods and services. Under communism, the individual subordinates himself to the needs of the State. In its extreme form, that meant individuals would be told where to work, how long to work, and how much to produce.[20]

Under more moderate forms of socialism, however, individuals have greater freedoms to choose their professions and even to start companies. The State is involved in coordinating business activity and the distribution of wealth. Socialist economies, as opposed to communist ones, use indirect measures to influence business activity, including taxation, special incentives, and welfare programs. Socialist countries like Sweden, for example, have comparatively high corporate tax rates. Businesses pay high taxes to the government, which then uses this money to provide free college

1. Hong Kong	34. Norway	67. Colombia	101. Brazil	133. Ukraine
2. Singapore	35. Slovak Republic	68. Romania	102. Algeria	134. Russia
3. Ireland	36. Botswana	69. Fiji	103. Burkina Faso	135. Vietnam
4. Australia	37. Czech Republic	70. Kyrgyz Republic	104. Mali	136. Guyana
5. United States	38. Latvia	71. Macedonia	105. Nigeria	137. Laos
6. New Zealand	39. Kuwait	72. Namibia	106. Ecuador	138. Haiti
7. Canada	40. Uruguay	73. Lebanon	107. Azerbaijan	139. Sierra Leone
8. Chile	41. South Korea	74. Turkey	108. Argentina	140. Togo
9. Switzerland	42. Oman	75. Slovenia	109. Mauritania	141. Central African
10. United Kingdom	43. Hungary	76. Kazakhstan	110. Benin	Republic
11. Denmark	44. Mexico	77. Paraguay	111. Ivory Coast	142. Chad
12. Estonia	45. Jamaica	78. Guatemala	112. Nepal	143. Angola
13. The Netherlands	46. Israel	79. Honduras	113. Croatia	144. Syria
14. Iceland	47. Malta	80. Greece	114. Tajikistan	145. Burundi
15. Luxembourg	48. France	81. Nicaragua	115. India	146. Rep. of Congo
16. Finland	49. Costa Rica	82. Kenya	116. Rwanda	147. Guinea Bissau
17. Japan	50. Panama	83. Poland	117. Cameroon	148. Venezuela
18. Mauritius	51. Malaysia	84. Tunisia	118. Suriname	149. Bangladesh
19. Bahrain	52. Uganda	85. Egypt	119. Indonesia	150. Belarus
20. Belgium	53. Portugal	86. Swaziland	120. Malawi	151. Iran
21. Barbados	54. Thailand	87. Dominican Rep.	121. Bosnia and	152. Turkmenistan
22. Cyprus	55. Peru	88. Cape Verde	Herzegovina	153. Burma
23. Germany	56. Albania	89. Moldova	122. Gabon	154. Libya
24. The Bahamas	57. South Africa	90. Sri Lanka	123. Bolivia	155. Zimbabwe
25. Taiwan	58. Jordan	91. Senegal	124. Ethiopia	156. Cuba
26. Lithuania	59. Bulgaria	92. The Philippines	125. Yemen	157. North Korea
27. Sweden	60. Saudi Arabia	93. Pakistan	126. China	
28. Armenia	61. Belize	94. Ghana	127. Guinea	**NOT RATED**
29. Trinidad and	62. Mongolia	95. The Gambia	128. Niger	Congo Democratic
Tobago	63. United Arab	96. Mozambique	129. Equatorial	Republic
30. Austria	Emirates	97. Tanzania	Guinea	Iraq
31. Spain	64. Italy	98. Morocco	130. Uzbekistan	Montenegro
32. Georgia	65. Madagascar	99. Zambia	131. Djibouti	Serbia
33. El Salvador	66. Qatar	100. Cambodia	132. Lesotho	Sudan

EXHIBIT 1.1

2008 Index of Economic Freedom.

Source: The Wall Street Journal, January 15, 2008.

education for everyone. While that may seem like a great benefit to you as a college student trying to pay spiraling education costs, there are consequences to government intervention in business activity.[21]

Most economies, whether purporting to be socialist or capitalist, are actually a mix of both. The index of economic freedom (Exhibit 1.1) is a good gauge of how much individual freedom exists within a given economic system.[22]

It may surprise you to learn from this figure that the United States ranks only in fifth place on this index. Contrariwise, it probably comes as no surprise that a repressive regime such as that in North Korea places that nation last. Other major economic powers, such as China (126) and Russia (134), have fared well economically over the past few decades despite their low rankings on the freedom index. It will be interesting to watch how these nations perform in the years to come and whether they can sustain growing economies with relatively low rankings on economic freedom.

The "freedoms" considered in making this list include:

- Business freedom
- Trade freedom
- Fiscal freedom
- Government size
- Monetary freedom
- Investment freedom
- Financial freedom
- Property rights
- Freedom from corruption
- Labor freedom

1.3.1 Business and Social Responsibility

The role of business has been a constant source of debate in society, and given its pervasive influence on individual and collective well-being, it will probably always remain a lively topic of conversation and discourse. The role of business has been a constant source of debate in society, and given its pervasive influence on individual and collective well-being, it will probably always remain a lively topic of conversation and discourse.

For example, in recent times, there has been increasing pressure on business to accept more responsibility for the environment and to develop the so-called "green" or "sustainable" business practices.[23] In part, this is a response to the popular belief that the earth's climate is warming due, in large measure, to the carbon output produced by economic activity.[24] Notice that the actions proposed by groups that support the "green" cause are not necessarily those that businesses would undertake purely on economic factors alone. Thus, these groups want business to respond to political as well as market forces.

In the capitalist Western world, the social responsibility of business has been debated in terms of:

- *Social obligation*: This school of thought adheres to the principle of the "business of business is business." Based on the work of some famous economists, most notably Milton Friedman, Gary Becker, and Ronald Coase from the Chicago School of Business, it is sometimes referred to as the "Chicago School" of thought.[25] The most prominent light among these luminaries is Milton Friedman, who was a staunch advocate of human freedom and free markets. He argued persuasively throughout his lifetime that market forces should dictate how business behaves. Any attempt by business leaders to act other than for the sake of maximizing shareholder value would amount to a form of "taxation without representation."[26] According to this perspective, the social obligation of business is to bring goods and services to market as efficiently as possible. Competition assures that consumers will have choices, and will be able to purchase those goods and services that are within their ability to pay.

- *Social reaction*: This perspective maintains that business exists within a society and cannot be immune to prevailing social mores and cultural pressures. Those who adopt this view point out that the policies and legal conditions that make business activity possible are socially constructed. As such, businesses should recognize that they exist because of the policy apparatus, and should be reactive to the changes in mores over time. The social reaction view would deem that companies are obliged to change their practices and that more environmentally friendly business practices are necessary.

- *Social responsiveness*: This view holds that businesses are uniquely positioned within society to not just react to social changes, but to lead them. Proponents believe that the power of large corporations is, in some cases, very similar to the power wielded by national governments. Therefore, General Motors Corporation, for example, were it a country, would be among the richest nations on earth. Clearly, this line of thinking promotes the idea that the level of economic power that large firms possess carries with it the ability to influence the behavior of people within the company and beyond. With such power, companies should lead social causes and be on the forefront of positive change.

No matter which of these perspectives you find to be closest to your own personal beliefs, it will no doubt continue to be debated and refined. But there is no escaping the fact that business is a human activity, and it does have massive implications for the welfare and sense of well-being for people all around the globe.

1.3.2 The Mechanics of a "Going Concern"

We have defined business using the term "going concern" to differentiate it from a short-term project or a hobby. Establishing a going concern usually requires establishing systems that are **repeatable** and **scalable**. Repeatable simply means that the processes designed to produce valuable output for a market can be repeated indefinitely. Scalability means that the repeatable processes can handle increasing market demand over time.

Repeatable processes

The principle of a going concern is most evident in repeatable processes that comprise production in a factory. It is useful for anyone interested in technology ventures to take a tour of a local production facility—no matter what it produces. Most factories today have integrated automation, robotics, information technology, and human labor to create highly efficient and repeatable business processes. Modern approaches to manufacturing emphasize "lean" production processes—where costs are continuously examined and worked out of the system to maximize profits.

Although factory floor operations offer a vivid example of the mechanics of repeatable business processes, this principle can be applied across any type of business venture. Consider a software development business, such as Google or Microsoft. Here, repeatable processes underlying production are not as evident as they may be on a factory floor. Rather, they are embodied primarily in the policies, both explicit and implicit, that govern the day-to-day work activity of these companies. Thus, it is important to note that the forces that comprise the mechanics of a going concern are both **physical**, i.e., the machinery and tools of the factory floor, and **tacit**, i.e., the policies and rules that workers follow.

Physical forces are normally explicitly causal. The assembly line is a good example. Here, workers must conform and adapt their activities directly to the causal forces of the line. The speed of the line, its physical configuration, and the tools available to complete assigned tasks all require the worker to adapt to physical features, and there is little room for *interpretation*.

On the other hand, the tacit forces that shape production—the policies and rules that govern the workplace—are subject to interpretation. As such, many creators of technology ventures find this part of developing repeatable processes the most difficult to implement. Technically oriented individuals and companies often have not been trained formally in management or communication sciences. Yet these skills are extremely important for developing and instilling repeatable processes. Additionally, included within this challenge is the need to align the organization toward common goals and objectives, provide individualized and substantial incentives that motivate people to perform at high levels, and promote openness and collaboration for future growth.[27]

Scalability

Ventures must be able to adjust their output to meet market demand. A technology venture that was not scalable, for example, would not be able to meet demand if customers favored its products and services. This would be a critical error for the founders and investors of the company.

Scalability relies on a variety of factors, including qualified labor to operate machinery or deliver services, an access to capital to purchase new equipment, and the presence of standard operating procedures that limit variability of output across different production lines or service providers. Managing growth and scaling at an appropriate pace requires a delicate balance that many young technology ventures face. Growing too fast may lead to quality problems that turn customers away. Growing too slowly may lead to opportunities for competitors to race into the market and steal potential new customers.

Creating the mechanisms that underlie a going concern is a challenge faced by every technology entrepreneur. While establishing the physical features of systematic and repeatable processes may be relatively straightforward, the management and leadership elements can be challenging. Chapter 13 in this textbook explores in greater detail the management and leadership skills that are required to operate a technology venture.

1.4 BUSINESS DYNAMICS

During the late 1990s, entrepreneurship programs in business and engineering schools were being overrun with applicants. Everyone wanted to learn the techniques that would help them gain access to the riches that were being created by dot-com ventures. The decade of the 1990s was punctuated repeatedly by stories of soaring stock prices of dot-com companies, instantaneous riches for a legion of Silicon Valley technology mavens, and highly visible public offerings of youthful companies whose values sometimes rose to rival those of the largest industrial companies in the world. These heady times created a generation of optimistic technology entrepreneurs. In 1999, there were 457 IPOs in the United States, most of which were Internet or technology ventures. Of the 457 offerings, 117 of them doubled their price per share on their first day of trading. By way of contrast, there were only 71 IPOs in 2001 and not a single one experienced a first-day doubling in share price.[28]

All too soon after it began, however, the "new economy" that many believed was the reason for the wealth they were creating came crashing down around them. The dot-com crash that began in March 2000 was precipitous and unnerving. The confidence that had been building throughout the previous decade was eroded in mere months. The stock market index most closely associated with technology ventures—the National Association of Securities Dealers Automated Quotation system, or NASDAQ—sank nearly 80% from its high of 5,046.86 recorded on March 11, 2000. NASDAQ crumbled from that high to a low of 1,114.11 on October 9, 2002. What took a decade or more to build took a little more than two years to wipe out.

Business dynamics, thus, change, and the rate of change in technology-oriented industries and businesses can be dizzying. The dot-com crash is still a vivid memory for many current technology venture creators. Nearly $5 trillion in wealth was wiped out during the crash, and many formerly high-flying entrepreneurs left for more secure jobs in established high-tech companies.

1.4.1 Change and Competition

The forces that impinge on technology ventures are many, and in the past few years they have become global in scope. The primary force acting on technology-related businesses is the rapid pace of technological change. Companies that enter a market with a new technology are likely to encounter staunch competition if the market proves lucrative. Most technology companies attempt to defend their market position, at least in the short term, by developing intellectual property protection, such as patents, trade secrets, and trademarks.

Today, it is increasingly difficult to build a sustainable competitive advantage based solely on a unique bit of intellectual property.[29] Patents and other intellectual property vehicles are difficult and costly to protect, especially for technology firms that pursue global markets.[30] It is also costly to develop unique intellectual property in many industries, making it difficult to recover those costs during the protected lifetime offered by a patent. Drug companies, for example, must invest tens of millions of dollars to develop a new pharmaceutical, and then additional tens of millions to win approval to distribute the product to the public. A patent protection may last 17 years. The drug company cannot afford to waste time if it wants to earn a positive return.[31]

1.4.2 Change and Globalization

In addition to the rapid pace of change and the difficulties inherent in establishing and protecting intellectual property, there are a number of other factors that produce the dynamism in today's technology industries. One of these is globalization. The global economy that we have become used to is, in fact, a very recent phenomenon.[32] The primary creator of the global economy was the Internet, and the connectivity to the global workforce that it has enabled. While global trade is as ancient as the Old Silk Road carved out by Marco Polo, the connectivity of global trade and the fluid policy environment in which it now takes place is unprecedented in human history.[33]

Technology companies have certainly played a major role in creating the infrastructure for global trade, and they will continue to do so for many years. According to Internet World Stats, a little over 1 billion people now are connected to the Internet (Exhibit 1.2). That figure represents just 16% of the total world population. In other words, more than 80% of the earth's people do not have access to one of the most important technologies of the modern age. Clearly, there is plenty of room for intrepid technology entrepreneurs to continue to serve these vast markets.

1.4.3 Engineering Creative Solutions

Not only are there immense opportunities for technology entrepreneurs to serve vast global markets, there are also a large number of complex challenges crying out for solutions. No doubt the major problems facing human beings in the coming decades will present opportunities for commercially minded engineers and scientists to develop venture-based solutions. The challenges facing engineers, for example, have been discussed and debated by the U.S. National Academy of Engineering (NAE). The NAE asked a distinguished international panel of scientists to identify

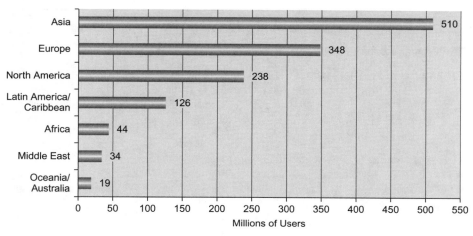

EXHIBIT 1.2

Internet Users in the World (December 2007).

Note: Total World Internet Users Estimate Is 1,319,872,109 for year-end 2007. Copyright © 2008, Miniwatts Marketing Group—www.internetworldstats.com

the "Grand Challenges for Engineering."[34] The panel identified the following 14 challenges:

1. Make solar energy economical
2. Provide energy from fusion
3. Develop carbon sequestration methods
4. Manage the nitrogen cycle
5. Provide clean access to water
6. Restore and improve urban infrastructure
7. Advance health informatics
8. Engineer better medicines
9. Reverse-engineer the brain
10. Prevent nuclear terror
11. Secure cyberspace
12. Enhance virtual reality
13. Advanced personal learning
14. Engineer the tools of scientific discovery

Each of these Grand Challenges will undoubtedly require radical thinking, including creative sources of financing, use of materials, and approaches to markets—especially those that are traditionally underserved—and equally creative use of labor and other production factors.

TECH MICRO-CASE 1.2

The Grameen Bank Project

Creative sources of financing are beginning to appear in remote regions of the world. The Grameen Bank, for example, was founded in 1976 in the village of Jobra, Bangladesh.[1] It specializes in providing "micro-loans" to the poor people of the region to help them set up and operate their own small businesses. The Grameen Bank Project (Grameen means "rural" or "village" in Bangla) came into operation with the following objectives:

- Extend banking facilities to poor men and women
- Eliminate the exploitation of the poor by money lenders
- Create opportunities for self-employment for the vast multitude of unemployed people in rural Bangladesh
- Bring the disadvantaged, mostly the women from the poorest households, within the fold of an organizational format that they can understand and manage by themselves
- Reverse the age-old vicious circle of "low income, low saving and low investment," into a virtuous circle of "low income, injection of credit, investment, more income, more savings, more investment, and more income."

Today, the bank is over 90% owned by the people it serves. The bank's founder, Dr. Muhammad Yunus, was awarded the Nobel Peace Prize in 2006 for his efforts in providing financing to the underserved masses of Central Asia.

Creative material uses are occurring more frequently as companies strive to manage product life cycles in response to calls for sustainable economics. Construction companies are using sustainable products to build homes, offices, and urban infrastructure.[35] Energy companies are aggressively seeking sustainable alternatives that will release them from dependence on fossil fuels.[36] Airlines, furniture manufacturers, soft drink makers, and many, many other types of companies are looking to engineers and scientists to invent and commercialize the sustainable materials of the future.

Companies are also adopting unique approaches to markets to establish their respective brands. Some of the largest markets in the world are emerging among the vast populations of India, China, and Africa, where new wealth that is creating middle-class lifestyles for millions presents incomparable opportunities for companies that are creative and adaptive enough to expand businesses there. For example, the Chinese government has put out a call to multinational corporations to serve its growing middle class.[37] Many companies find themselves shut out of the Chinese market because they are unable to get their costs low enough to offer affordable products there. In response, consumer product companies such as Colgate-Palmolive, Anheuser-Busch, and L'Oréal have shifted their approach to consumers in the Chinese market, with each developing products or investing in products that meet the unique needs of the Chinese people.

Another way to enter these emerging markets is exemplified by IBM. IBM has made its Indian operations one of its most important locations for delivery of services to clients worldwide. Nearly one-sixth of IBM's global workforce is now based in India.[38] Utilizing the labor forces that have demonstrated world-class talent is a trend that is likely irreversible. Companies today are using outsourcing as a strategic ploy to take advantage of the talented labor that exists in all corners of the world. Intel, for example, is building a $300 million assembly and test facility in Viet Nam. The site is expected to employ more than 1,000 people when it is operational. Intel also has a sales and marketing office in Ho Chi Minh City, providing sales and support at the original equipment manufacturer (OEM), developer, and end-user levels. Set up in 1997, this office drives Intel's initiatives, technologies, products, and services into the marketplace, creating demand and promoting Intel's role in the Internet.[39]

1.5 THE FUTURE OF BUSINESS

This chapter introduces you, the technical student, to some fundamental concepts of business. The global economy is evolving so rapidly today that it may be overly naive to think that the fundamentals conveyed in this chapter will hold true over the course of the usable life of this textbook. However, there are some business fundamentals that simply do not change, regardless of the changing world. For example, value creation and distribution to a given market will continue to be definitive of business activity. Profit motivation as a driver of innovative value creation will continue to be an important factor as well.

What will change in the coming years is the manner in which value is created, the markets available for any given unit of value, and the business models that will be used to generate profits. Think about the vast changes that have occurred in industrial economies over the past century. The early part of the twentieth century was characterized by vast migrations from agricultural and rural lifestyles to cities and factory or mass production lifestyles. This industrial age lifestyle provided steady work and stable income for millions of people in the industrialized world. During the latter half of the twentieth century, the stability of that lifestyle began to erode as emerging nations built factories that were able to thrive based on cheaper labor and fewer regulatory obligations. Factory work shifting to low-labor-cost regions around the world was only the beginning of the globalization of economics.

The explosion of information technologies in the latter half of the twentieth century led to the information revolution. Chief among the enabling technologies of the information age, of course, is the Internet. Now nearly global in reach, the Internet has enabled unparalleled value creation opportunities, and has opened new markets as well. The practice of business process outsourcing (BPO) was largely enabled by the spread of broadband technologies.[40] As large companies shifted information-related work to knowledge-worker hotspots like Bangalore and Manila, increasing numbers of people in those regions joined the burgeoning global middle class. With their new found wealth came demands for the goods and services common to middle-class people everywhere: automobiles, appliances, electronics, furniture, and a wide variety of foods. These emerging regions became new market targets for companies that had previously seen them as too poor or too remote to serve. Now, companies from General Motors to P&G are scrambling to understand the unique product, service, and distribution needs of these global markets.

SUMMARY

This chapter is designed for engineers and scientists who have had little or no prior exposure to business, management, or economics. Although the chapter title "Fundamentals of Business" suggests a comprehensive overview, in reality that would be impossible in a single chapter. Business is all-pervasive in modern society and exceedingly complex.

We defined business as "an organized and purposeful human activity designed to create value for others and to exchange that value for something else of equal or of greater value (usually, money) and that is intended to continue to provide such value over time as a going concern." The main themes of this definition were identified as "organized activity," "purposeful," "going concern," and "value."

The focus of this textbook is on value creation, value protection, and value capture. These themes will be expanded in greater detail in respective sections of this textbook. In brief, the job of the technology entrepreneur is to create value for a market, protect that value via intellectual property rights and other means, and capture that value through the growth of the enterprise and retention of ownership. The venture's value to customers is summarized in what we refer to as the "value proposition."

The role of business in society has long been debated. This issue came to the fore again in 2008 when many business failures around the world led people to question how best to govern and control business activity. As the question affects how one thinks of business and entrepreneurship, we consider some alternative answers that have been proposed.

The three proposed answers to the question of the role of business in society were expressed as social obligation, social reaction, and social responsiveness. These three perspectives are very different and, depending on which one you subscribe to, will influence the type of technology venture you pursue.

As a going concern, entrepreneurial ventures must strive for respectable processes that are also scalable. That means the business will be able to serve growing demand and pursue new markets.

Finally, this chapter discussed several of the dynamic forces that are affecting business in the twenty-first century. These include accelerated change and competition, globalization of a wide range of industries, and the Grand Challenges identified by the NAE.

STUDY QUESTIONS

1. What are the two primary elements of a going concern?

2. What does it mean to say that a business is a purposeful activity?

3. How does a business determine whether or not its products and services are valuable? Define the relationship between price and quality, and their effects on value.

4. Explain the role of a value proposition for a business. What is the value proposition for YouTube?

5. Describe how a capitalist economy differs from a socialist economy.

6. This chapter defined three different perspectives regarding the social responsibilities of business. Name each and briefly explain each perspective.

7. This chapter identified two primary dynamics driving technology ventures in the modern age. Define each and briefly identify how technology ventures can or should deal with each of these elements.

8. How should a person master the fundamental concepts of business? (HINT: Use the Tiger Woods example as part of your explanation.)

EXERCISES

1. Name at least five of the so-called "Grand Challenges" identified by the National Academy of Engineering.

2. Pick one of the Grand Challenges and identify one or two business ideas that address the challenge.

KEY TERMS

- **Business as a going concern:** Businesses are organized to pursue profits in the short term and over time. They differ from projects or hobbies in that they do not have a predetermined end date or singular definition of success.

- **Business as an organized activity:** Business requires that resources be organized to deliver value to a market. Resources include, but are not limited to, land, labor, and capital.

- **Business as a purposeful activity:** Business requires that the resources that have been organized be put to purposeful use.

- **Business model:** The way a business makes money.

- **Physical factors:** Technology venture managers must adjust physical assets to ensure that its production processes are repeatable and scalable.

- **Repeatable:** One of the mechanical elements of a business is to make the production or service providing processes repeatable.

- **Scalable:** One of the mechanical elements of a business is to make the production or service providing processes scalable. That is, as the business grows it should be able to expand its repeatable processes to meet greater customer demand.

- **Social responsibility as social obligation:** This is a perspective on the social responsibility of business that is associated with the Chicago School of thought. It holds that business should focus on making profits for owners as its primary concern.

- **Social responsibility as social reaction:** This perspective holds that business is possible via the rules of society. As such, a business should respond to changing social mores and adapt its practices accordingly.

- **Social responsibility as social responsiveness:** This perspective holds that business is a social leader, and that companies should be early exemplars of new and improved social practices.

- **Tacit factors:** Technology venture managers must adjust the policies and operating procedures to ensure that its production processes are repeatable and scalable.

- **System:** Business systems are put in place to create the going concern element of a business.

- **Value:** In business, value is determined by a market or markets, and is often expressed via the equation: Value = Price $\times$ Quality.

- **Value proposition:** The value that a business brings to a market is expressed in its value proposition.

WEB RESOURCES

The web sites below are intended as destinations for your further exploration of the concepts and topics discussed in this chapter:

1. http://en.wikipedia.org/wiki/Business: This is the Wikipedia section on business. It has a very thorough list of related topics that you can also link to and explore in case any particular area of business is of more interest to you than others.

2. http://www.cnbc.com: This is the web site for the leading business and finance channel on television. CNBC is known for its continuous stream of stock prices and wide range of programming on contemporary business and industry topics.

3. http://www.wsj.com: *The Wall Street Journal* may be the most well-known business publication in the world. If you want to keep track of developments in business and economics on a daily basis, a subscription to this newspaper will help you do that.

4. http://www.ft.com: *The Financial Times* is a London-based publication that has more international news coverage than *The Wall Street Journal*.

5. http://www.businessweek.com: *Business Week* is a print publication that has a very useful and informative web site. Updated daily rather than weekly like its print brethren, the *Business Week* web site is a good source of analysis on current business events.

6. http://www.fortune.com: *Fortune* magazine is differentiated from *Business Week* in its focus on longer stories about business and the leaders who make business work.

7. http://www.forbes.com: *Forbes* magazine was founded by the Forbes family and is currently headed by former U.S. presidential candidate, Steve Forbes. Its articles focus on economic issues more than those featured in *Business Week* or *Fortune*.

8. http://www.economist.com: *The Economist* is a European publication that provides perspective on global business and economic issues. It is a useful counterbalance to U.S.-centered publications like the ones listed above.

9. http://money.cnn.com/magazines/business2/: This web site is a remnant of the dot-com era, called Business 2.0. The "2.0" label was intended to indicate that a new form of business was emerging in what some called the "new economy." That conceit is no longer held by most people, but this web site still covers a lot of really interesting businesses.

10. http://www.redherring.com: *Red Herring* magazine is a leading technology venture publication.

ENDNOTES

[1] Orsak GC. Engineers: the new leadership class. *Electron Bus* 2006; **February**:6.

[2] Van der Klink MR, Streumer JN. Effectiveness of on the job training. *J Europ Ind Train* 2002;**26**(2–4):196–9.

[3] Sirak R. The Golf Digest 50. *Golf Digest* 2008; **March**:19.

[4] The hard work behind soft skills: closing the gap between technical and business expertise. ESI International, Inc., white paper, p. 4.

[5] Branson R. *Losing my virginity: how I've survived, had fun, and made a fortune doing business my way*. New York: Three Rivers Press; 1999.

[6] Marx K, Mandel E, Fernbach D. *Capital: a critique of political economy*. New York: Penguin Classics; 1993.

[7] Fernald JG, Greenfield V. The fall and rise of the global economy. *Chicago Fed Letter* 2001; **April**:1–4.

[8] Watts S. *The people's tycoon: Henry Ford and the American century*. New York: Knopf Publishing; 2005.

[9] Lentz S. *It was never about the ketchup: the life and legacy of H.J. Heinz*. Garden City, NY: Morgan James Publishing; 2007.

[10] Stross RE. *The wizard of Menlo Park: how Thomas Alva Edison invented the modern world*. New York: Three Rivers Press; 2008.

[11] Linzmayer O. *Apple Confidential 2.0: the definitive history of the worlds' most colorful company*. San Francisco, CA: No Starch Press; 2004.

[12] Anderson JC, Kumar N, Narus JA. *Value merchants: demonstrating and documenting superior value in business markets*. Cambridge, MA: Harvard Business School Press; 2007.

[13] Moore G. *Crossing the chasm*. New York: HarperCollins; 2002.

[14] Farzad R. A deal that paid for itself. *Business Week* 2006; **October 30**:38.

[15] http://www.pg.com.

[16] Stringer S. Connecting business needs with basic science. *Research Technology Management* 2008; **January/February**:9–14.

[17] Holzner S. *How Dell does it*. New York: McGraw-Hill Publishing; 2005.

[18] Whittaker CR. *Frontiers of the Roman Empire: a social and economic study*. Baltimore, MD: The Johns Hopkins University Press; 1997.

[19] There are several classic texts that articulate this point of view, including: Smith A. *The Wealth of Nations*. New York: Bantam Books; 2003; Friedman M, Friedman R. *Free to Choose*. Fort Washington, PA: Harvest Books; 1990; and Hayey F. *The Road to Serfdom*. New York: Routledge Publishing; 2007.

[20] Service R. *Comrades! A history of world communism*. Cambridge, MA: Harvard University Press; 2007.

[21] Wyman P. *Sweden and the "Third Way": A macroeconomic evaluation*. Farnham, Surrey, UK: Ashgate Publishing; 2003.

[22] The Wall Street Journal, January 15, 2008.

[23] Esty DC, Winston AS. *Green to gold: how smart companies use environmental strategy to innovate, create value, and build competitive advantage*. New Haven, CT: Yale University Press; 2006.

[24] Gore A. *An inconvenient truth*. New York: Viking; 2007.

[25] van Overtveldt J. *The Chicago School: how the University of Chicago assembled the thinkers who revolutionized business and economics*. Chicago, IL: Agate B2 Publishing; 2007.

[26] Friedman M. *Capitalism and freedom*. 40th Anniversary ed.: University of Chicago Press; 2002.

[27] Ivancevich JM, Duening TN. *Managing Einsteins: leading high-tech workers in the digital age*. New York: McGraw-Hill Publishing; 2001.

28 http://www.investopedia.com/features/crashes/crashes8.asp.

29 Tierny TJ, Lorsch J. Creating competitive advantage in the knowledge economy. *Leader to Leader* 2002; **October**:41-7.

30 Prowse P. IP protection at what cost?. *Engineering Management* 2007; **August**:30-1.

31 Mullin R. Drug development costs about $1.7 billion. 2003; **December 15**:8.

32 Friedman TL. *The world is flat: a brief history of the 21st century*. Picador Publishing; 2007.

33 Stiglitz JJ. *Globalization and its discontents*. W.W. Norton & Company; 2003.

34 See "Introduction to the Grand Challenges for Engineering" at http://www.engineeringchallenges.org/cms/8996/9221.aspx.

35 Kibbert CJ. *Sustainable construction: green building design and delivery*. John Wiley & Sons; 2007.

36 Kruger P. *Alternative energy resources: the quest for sustainable energy*. John Wiley & Sons; 2006.

37 Gaidesh O, Vestring T. Capturing China's middle market. *The Wall Street Journal* 2007; **November 17**:A10.

38 Gupta AK, Wang H. How to get China and India right. *The Wall Street Journal* 2007; **April 28**:R4.

39 See http://www.intel.com/jobs/vietnam/ for more on Intel's role in these emerging global markets.

40 Duening TN, Click RL. *Essentials of business process outsourcing*. Hoboken, NJ: John Wiley & Sons Publishing; 2005.

Fundamentals of Economics in a Global Context

After studying this chapter, students should be able to:

- Understand the basic factors of every economy
- Recognize how price affects supply and demand
- Compare economic systems and evaluate entrepreneurial opportunities
- Identify barriers to conducting international business
- Leverage trading blocs and international trade agreements

TECH VENTURE INSIGHT

Safety Belt Manufacturer Finds Success in Exporting

Exports are among the few sectors of the U.S. economy having much success these days. More and more companies, especially small- and medium-sized ventures, are looking overseas to bolster their bottom line. Whether they're shipping consumer goods, industrial inputs, services, technology, or anything else, it is a great time to go global.

Even so, many entrepreneurial ventures hesitate to take that first step. Foreign markets can seem daunting and unreachable. There are a number of challenges, from linguistic and cultural differences to a host of complex regulatory systems. But getting started and being successful are relatively straightforward. In fact, most governments (state and federal) actively encourage exporting and have developed programs to assist small and start-up ventures in their quest to expand internationally.

Philip Clemmons is the owner of Elk River Safety Belt Inc., a manufacturer of safety belts and harnesses used in the construction, mining, and utility industries in the United States. Elk River's products offer protection for workers against falls and accidents. According to Clemmons, worker protection and safety standards in some countries are not what they are here in the states, creating a good market for his products.

Many of Elk River's international markets are growing faster than the U.S. market, proving the old adage that exports help smooth out fluctuations in sales from domestic economic cycles. "Construction is booming in Southeast Asia, a region we have targeted," Clemmons said. During its first year of exporting, Elk River's foreign sales climbed over 160%. Its initial success in the export market is no accident.

Elk River turned to the Alabama International Trade Center (AITC), a partnership program between the U.S. Small Business Administration (SBA) and the University of Alabama, for help. The AITC assisted Elk River in accessing several federal and state programs, including a U.S. SBA loan, the state's Linked Deposit Program for the construction of a new building, and an Alabama Export Council's grant for attending overseas trade shows. Clemmons also used the AITC's research and training programs to target markets and build skills within Elk River to deal with export procedures.

Clemmons also took the initiative to travel overseas, attend shows, and meet with prospective customers. While he faces stiff tariff barriers for his products in some markets, Clemmons has appointed several local representatives in Asia and is working with them and other organizations to overcome the obstacles and make sales.

Source: Adapted from Davis B. Small company finds safety in export markets. Alabama International Trade Center, Export Success Story; Travis T. Get in on the export boom. Entrepreneur *2008; November 20.*

INTRODUCTION

Understanding economics is essential to understanding business and entrepreneurship.[1] **Economics** is the study of how a society chooses to use scarce resources to produce goods and services and to distribute them to people for consumption. Economic theory is divided into two branches: microeconomics and macroeconomics. **Microeconomics** is the study of individual choice and how choice is influenced by economic forces. It considers economics from the viewpoint of individuals and firms. Microeconomics involves the study of household decision making on what to buy, business-pricing decisions, and how markets allocate resources among alternatives.

Macroeconomics is the study of inflation, unemployment, business cycles, and growth, focusing on the aggregate relationships in a society. Micro and macro analysis of an economy are related, and both need to be used to better understand how a country functions economically. Three issues are key to understanding economics: (1) resources, (2) goods and services, and (3) allocation of both resources and products.

2.1 ECONOMIC RESOURCES

A nation's resources consist of four broad areas: natural, capital, labor, and knowledge. **Natural resources** are provided by nature in limited amounts; they include crude oil, natural gas, minerals, timber, and water. Natural resources must be processed to become a product or to be used to produce other goods or services. For example, trees must be processed into lumber before they can be used to build homes, shopping malls, and schools.

Capital resources are goods produced to make other types of goods and services. Some capital resources, called "current assets," have a short life and are used up in the production process. These resources include fuel, raw materials, and paper. Long-lived capital resources, which can be used repeatedly in the production process, are called "fixed capital." Examples include factory buildings, personal computers, and railroad cars.

Labor resources represent the human talent of a nation. To have value in the labor force, individuals must be trained to perform either skilled or semiskilled work. For example, the job of a physicist requires extensive training, whereas only minimal training is needed to operate a service station's gas pumps. This collection of human talent is the most valuable national resource. Without human resources, no productive use of either natural or capital resources is possible.

Knowledge resources concern not only the accumulated knowledge residing within a society, but also the processes for creating new knowledge. Accumulated knowledge is contained within books, databases, and other sources. In today's highly interconnected world, accumulated knowledge is available to nearly everyone. Perhaps more important for long-term prosperity are the institutions and organizations that promote the development of new knowledge. The United States, for example, has an extensive and highly respected higher education system where public and private money is invested in basic research.

2.1.1 Goods and Services

A nation's resources are used to produce goods and services that will meet people's wants and needs. Wants are things people would like to have but do not absolutely need for survival. Such items as food, clothing, shelter, and medical care are needs; video recorders, cassettes, fashionable clothes, and luxury vacations are wants.[2]

A person's wants can be unlimited: As soon as one want is satisfied, another is created. Even wealthy people tend to have unlimited wants. Henry Ford was once asked how much money it would take before a person would stop wanting more. He reportedly answered, "Just a little bit more." Exhibit 2.1 demonstrates in very basic terms how households and businesses interact in an economy according to what is referred to as the "circular flow model." Essentially, households provide the basic factors of production, and businesses provide goods, services, and wages in return. Later, in Exhibit 2.5, we'll examine a more complicated model that integrates additional elements of a basic economy.

2.1.2 Resource Allocation

All countries face the age-old economic problem of limited resources and unlimited wants. We all know, for example, that the supply of oil and natural gas in the United

The basic circular flow model

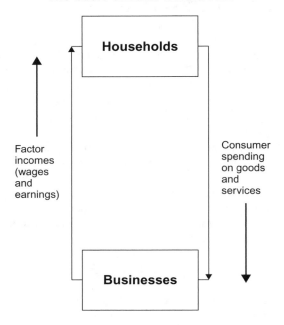

EXHIBIT 2.1

The Circular Flow of Goods and Services in an Economy.

States is a limited natural resource. Because we live in a world in which the quantity of all resources is limited, we must make choices about allocation—how these scarce resources are to be used. To make these choices, we have to answer three fundamental questions:

1. What goods and services will be produced, and in what quantities? What industrial goods and what consumer goods will be produced? Railroad cars or large trucks? Apartments or new houses?

2. How will goods and services be produced, and by whom? For instance, will energy be produced from coal, natural gas, nuclear, solar, or something else?

3. Who will use the goods and services? When the goods and services are divided, who is to benefit from their use? Rich or poor? Families or single people? Old or young?

Once these questions are answered, there is a basis for choosing how resources will be used (i.e., how they will be allocated to best satisfy consumers' wants and needs). In a market economy, allocation of resources also involves other issues. Should the need for business prosperity and success be a consideration? What priority should be given to government's need for resources? In the U.S. economy, allocating resources—especially scarce resources—involves all these questions. Allocation of scarce resources can be very complicated, indeed.

2.1.3 Product Distribution

The issue of allocation is not limited to scarce resources. It also involves the distribution of goods and services to the consumer. In this context, allocation involves an exchange (e.g., money, goods, time, services) between a business and a consumer (e.g., client, customer, another business). In an ideal pattern of distribution in a market economy, the business earns a profit and the customer is satisfied with the goods or service; the exchange provides mutual benefit. This is important in a market economy. A tailor able to earn a profit is likely to continue to work hard at the job. Likewise, a customer who likes the price and quality of the tailor's services will continue to use that tailor. When goods and services get to the customers who want or need them and mutual satisfaction occurs, both resources and products have been well allocated.

2.2 SUPPLY, DEMAND, AND PRICING

When there is little or no government intervention and control in an economy, the underlying law of **supply** and **demand** dominates the allocation of goods and services. Supply is the amount of goods and services sellers are willing to offer at various prices in a given time period and market condition. Demand is the amount of goods and services buyers are willing to purchase at various prices in the same time period and market conditions. The interaction of these two forces determines the market price as well as the amount of the product that will be produced and sold. Often buyers and sellers do not agree, and the equilibrium price does not correctly reflect the value of the quantity of the goods and services in the market. When this occurs, there is either a shortage (excess demand versus supply) or a surplus (excess supply versus demand).

In a competitive market economy, when a **surplus** or a **shortage** occurs, firms enact price changes until equilibrium is restored. In the case of a surplus, prices tend to be driven lower to meet the prevailing demand level in the short run. In the long run, either the supply will decrease, as the quantities offered are decreased by some firms making production cuts or by other firms going out of business, or the actual demand will change (increase) owing to buyers' willingness to purchase more of the product at the reduced prices.

When a shortage occurs, a company must carefully assess whether this condition is simply temporary or is actually a true market need not being satisfied. In these cases, prices rise to bring demand in line with current supply. Higher prices that can be charged in a strong customer demand environment may encourage new firms to enter the industry, increasing the available supply and putting downward pressure on prices. Technology entrepreneurs who have a relatively scarce yet high demand offering may be able to charge high prices in the short run. In the long run, competitors would respond to these pricing opportunities with offerings of their own, eventually pushing prices to a new equilibrium. One way to protect pricing advantage is to establish intellectual property that provides a competitive advantage or one could develop some other type of competitive advantage that can be sustained over time.

2.2.1 Price and Market Equilibrium

Often, one of the last considerations that the technology entrepreneur takes into account is the price that will be charged to customers. Yet, setting the price for products and services is a critical factor in the success of a venture. Setting a price too high might drive customers away and reduce the likelihood of success. Setting a price too low might attract customers, but it also might mean that the entrepreneur is not retaining enough profit to pay bills. Again, the likelihood of success is reduced. It's obvious that the "right price" is one that is neither too high nor too low.

One of the most straightforward business truths, yet one of the more difficult for the new entrepreneur to understand, is that the marketplace sets the price. The entrepreneur's marketplace is made up of customers and competitors. Competitor pricing strategy is an overriding consideration in setting price because the competition presents customers with an alternative to paying the entrepreneur's price.

Novice entrepreneurs often set the price of their products/services by adding up all of the costs involved in bringing the product or service to market, then adding a specific amount for profit (say, 15%). This is referred to as **cost-plus pricing**.[3] The problem with this approach is that customers don't care about an entrepreneur's costs—customers care about the price they have to pay for products and services and the value they receive for that price. If the entrepreneur uses the cost-plus pricing approach, one of two results is likely: The price is set too high and the customer does not buy, or the price is set too low and each sale returns less profit to the business than is possible.[4]

There are several common misconceptions among technology entrepreneurs about the reasons for setting a price. One common misconception is that the goal of pricing a product or service should be to increase sales volume. As most entrepreneurs learn, using price as a mechanism to drive sales volume can lead to problems. For example, sales of a product or service are likely to increase in proportion to price reduction. Following this logic, if the price is reduced to zero then sales volume will be at its peak. By using price to drive sales volume, the entrepreneur has reached or exceeded his sales goals, but he won't be in business long because there is no profit.

Nonetheless, many entrepreneurs use pricing strategy as a means of increasing sales volume. This is understandable as a price change in the market often has an immediate effect on sales volume. The effect of price on volume is well understood by basic economics. One of the most fundamental principles in economics is the law of supply and demand. Exhibit 2.2 demonstrates the standard effect of supply and demand on price.

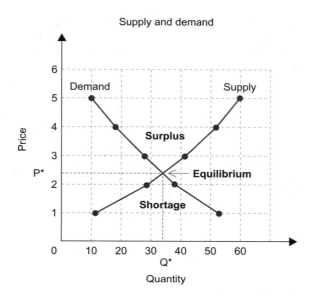

EXHIBIT 2.2

The Law of Supply and Demand.

This illustration assumes a free marketplace with willing buyers and willing sellers. Under these conditions, the law of supply and demand has four implications regarding price:

1. If supply is held static and demand increases, prices will rise.
2. If supply is static and demand falls, prices will fall.
3. If demand is static and supply increases, prices will fall.
4. If demand is static and supply decreases, prices will rise.

Note that this explanation deals with the supply and demand changes leading to price changes. Price strategy deals with the opposite side of the equation. Price strategy is an attempt to affect demand through alteration of prices. It is rare that a price strategy will be aimed at the supply variable in this basic economic situation.

An effective price strategy is one that is responsive to the type of demand present in the market. Economists recognize two basic types of demand in free markets: elastic and inelastic. An **elastic market** is one in which the overall demand in the market will expand if prices are lowered. An **inelastic market** is one in which demand will not respond to price changes. The most often cited example of an inelastic market is the market for salt. No matter the price of salt, people simply will not buy more than normal—the consumption rate of salt is based primarily on physiological need and is highly inelastic to price changes.[5]

A classic example of an elastic market is the passenger airline business.[6] As competition escalates among carriers and ticket prices drop, people tend to fly more frequently. The demand for air travel is highly elastic and very sensitive to price.

In an inelastic market, any increase in market share for one competitor has to be at the expense of one or more of the others. Therefore, any price-cutting strategy in such a market is likely to be matched by the competition. This competitive reaction leads to what is known as **price wars**. Price wars are destructive to markets and are generally won by the organization with the strongest financing.[7] It is illegal in the United States to use price-cutting as a method of putting competitors out of business. In an elastic market, the reduction of prices by a competitor has two potential consequences. First, it can lead to more buyers coming into the market, thereby expanding the market. Second, the reduction of prices can lead customers to buy the price cutter's product or service instead of the competitor's products. In an elastic market, it must be expected that competitors will react to any substantial loss of customers caused by a reduction in prices. When prices in a given market reach the level that the market is no longer able to expand, then the price reduction must stop or a price war will develop.

2.2.2 Price and Competition

A pricing strategy should never be executed without considering the reaction of competitors. The strength and speed of a competitor's reaction to a price change depends primarily on the effect the price has on the competitor's business. In an inelastic market, the reaction will be swifter and stronger than in an elastic market. In an inelastic market, any gain one participant makes must necessarily be at the expense of the others. Therefore, if a price strategy is effective and encroaches on the competitor's market share, there is likely to be a vigorous reaction.

In an elastic market, on the other hand, a price change may bring new customers into the market, in which case competitors may initially ignore the price change. However, if a price-reduction strategy results in increased profits for the firm executing the strategy, competitors usually will drop their prices as well. When the price reduction proves that the market is sufficiently expandable so that the price-reduction strategy is profitable for all participants, the lower price level tends to become established as the new market price.

The establishment of a lower price level in a marketplace is appropriate in industries where products are being manufactured for lower costs over time. In such situations the lowering of prices can be accomplished without reducing corporate profits whether or not the price reduction results in an increase in sales volume. When a price reduction is based on lower costs and the price move results in a larger number of buyers in the market, the strategy is a great success. Computers and other consumer electronic devices provide great examples of this cost-price effect. Entrepreneurs can take advantage of price reductions in these electronics components.

The reverse side of the success of price reductions based on lowered costs and an expanded market is also demonstrated in the computer and consumer electronics market. Technology advances have led to dramatic price reductions on computers and consumer electronics. When this pricing pattern continues for an extended period, consumers become **price sensitive**; i.e., they tend to shop for products based on low prices rather than brand recognition or other features. Discounters look for markets where consumers are price sensitive. They tend to drive prices down faster than the rate at which costs are reduced. The result is reduced margins for all participants in the marketplace and reduced profits.

A good example of a price-reduction cycle in an elastic market is the airline industry. It is well established in the United States that when prices for airline tickets drop, more people fly. In the 1960s and 1970s, before the advent of discount or low-cost airlines, the major air carriers were able to charge prices that created large margins and large profits. As the airlines became more profitable, however, labor began to demand its share of the earnings. This led to an increase in labor costs, especially costs associated with pilots and mechanics.

When low-cost airlines—such as Southwest—entered the market, they used price strategy as the principal method of competing with the major airlines. The major airlines, however, did not follow the price-reduction strategy employed by the low-cost carriers. Instead, they chose to rely on their better system of routes and established relations with business flyers, who represented the most profitable segment of their business. The major companies justified their higher prices by emphasizing safety and convenience. The market for air travel proved to be highly elastic, resulting in a greater tolerance for the price spread between the majors and the low-cost carriers. Over time, however, as the low-cost airlines gained better routing and more scheduled flights, the major airlines began to feel the competitive pressure. One outcome was that the financially weaker segment of the major airlines began to lose money, and many went out of business. All of these developments conspired to make air travel less expensive for the consumer. While this is good for the consumer, it is hard on the airlines. Nonetheless, this example clearly demonstrates how important competition is to price strategy.

Price increases, similar to price reductions, are equally subject to competitive pressures. When several carriers, for instance, attempted to add a modest fuel surcharge ($9–$10) to their ticket prices, the other airlines did not go along. The originators of the surcharge were forced to remove it from their pricing rather than allow their competitors to have even this small price advantage.

2.2.3 Price and Gross Margin

Technology entrepreneurs must also monitor the effect of their pricing strategy on the firm's **gross margin**. Gross margin is defined as the difference between the selling price of a product or service and the amount the entrepreneur has to pay for the raw materials that make up that product or service (the "cost of goods sold"). It represents the amount of money a venture has left to pay its selling costs, and the general and administrative expenses associated with operating the venture. After these latter two expense categories are covered, the remaining amount of money is the profit before tax for that business. Gross margin is expressed as a percentage of revenue, using the following equation:

$$\frac{\text{Revenue} - \text{Cost of Goods Sold}}{\text{Revenue}}$$

A price reduction usually results in a reduction in gross margin. This negative effect on gross margin is usually offset by an increase in sales volume that usually accompanies a price reduction. Increasing sales volume can lead to a reduction in the average **cost of goods sold** (COGS) since suppliers will often provide discounts to clients who buy in large volumes. Additionally, in some businesses an increase in sales volume leads to efficiencies in labor or equipment use that help ease the reduction of gross margin. Nonetheless, it is rare that these savings totally offset the negative effect of a price reduction on gross margin.

A price increase that increases sales volume—or, at least, does not reduce sales volume—is a powerful tool for profit. Since an increase in price does not entail a rise in costs, the added revenue falls directly to the venture's profit line (the "bottom line"). Businesses that are able to raise prices without cost increases and without sacrificing sales volume possess what is called **pricing power**. Pricing power is the ideal situation where a business sells goods into a willing market at a price it determines and, at least to some extent, controls.

2.2.4 Inflation

Inflation is defined as a rise in the general price level in a specific market. There are two primary causes of inflation: (1) Demand-pull inflation is caused by a market demand that is greater than the available product supply. (2) Cost-push inflation is caused by a shortage in the available supply of labor or materials that causes the cost of these items to increase.

A low inflation rate in the range of 3–4% is normal and even healthy in an economy that is expanding at a sustainable rate. Economic expansion is most often measured in terms of the **gross domestic product (GDP)**, which is defined as the total dollar value of all the goods and services produced annually in a given economy (such as that of the United States). Increases in GDP are made up of increases in the labor force, improvements in productivity, and inflation. The healthy and sustainable rate of growth in the GDP is comprised of increases in the employed labor force and its improved productivity. Some inflation in proper proportion to these healthy elements of GDP is desirable.

Price increases and cost increases are to be expected in periods when the economy is growing. The proper timing of any upward price adjustment for a particular company depends on its competitive position within the market. The major competitors are usually the leaders in price increases; the lesser competitors follow their lead. Inflationary price increases are usually modest and well spaced.

Price strategy difficulties occur more under **deflation** than inflation. In a deflationary economy, general price levels decrease. An entrepreneurial venture is forced to reduce its prices because competitors' prices have dropped. Failure to respond promptly to this market circumstance results in losing customers. However, if the entrepreneur does not or cannot cut costs simultaneously with decreasing prices, the profits of the venture will be reduced or even eliminated.

2.2.5 Price in a Compromised Market

A compromised market is one where the normal demand–supply–price equation (Exhibit 2.2) stops functioning. For example, compromised markets are common during wartime. An increase in military demand for goods over and above the civilian demand causes the supply side to become overwhelmed. Additionally, the transfer of labor from civilian jobs to military service handicaps the supply side. Finally, during wartime, manufacturing facilities and the civilian workforce are often converted from their normal activities to military applications. This realignment of labor and facilities causes a frictional loss of efficiency.

This change from "butter to guns" usually starts with the economy trying to service both objectives. The result is inflation, which is both demand-pull and cost-push simultaneously. Left unchecked, this inflation can become ruinous quickly. Furthermore, the usual method of checking inflation and monetary policy cannot be applied because of the government's wartime requirements for money and credit. Price and wage controls are usually instituted during wartime to stem runaway inflation. Under these conditions, the government dictates price strategy. Price and wage controls are never completely effective. Consequently, an excess profits tax is usually also instituted to prevent businesses from achieving overly large profits.

Price gouging is another strategy in a compromised market. Price gouging occurs when the supply of an item or items has been so severely restricted by special circumstances that the demand is completely out of balance with the supply. The circumstances that bring on price gouging are usually a catastrophic natural event—such as a flood—or an event arranged for profit reasons—such as the Super Bowl—where demand for access or accommodations near the event is excessive. Where health or safety of the public is involved there are government restrictions on price gouging. There are also regulations to control price gouging during events that are not critical to the health or safety of the public, although they are not always effective. Ticket "scalping," for example, is common near major sports events, even though the practice is illegal.

Dumping is another price strategy that is destructive to free markets. Dumping refers to disposing of goods in a foreign market, which are surplus in their native market, at prices below those in the home market or below the costs of manufacturing the goods. This practice undercuts the established market prices in the foreign market. The competitors in that market cannot compete with the dumping prices because they are below their costs. A company undertakes dumping as a price strategy in order to enter a market or to sell goods that are no longer wanted by consumers in the home market. Dumping is destructive to the market because it absorbs demand that is normally available to the other competitors at normal prices. Furthermore, dumping creates an erroneous perception among consumers that they are being overcharged at the normal prices. Dumping is prohibited in the United States and by U.S. companies.

2.2.6 Interest and the Price of Money

Interest is not often considered to be a part of price strategy; it is usually thought of as part of financial strategy. Yet, the term **interest rate** is defined as the price of money. The same supply–demand–price equation works with money as it does with products and services. When the demand for money is high and the supply of money is short then the price of money (the interest rate) goes up. When the reverse occurs, interest rates go down. The difference between product/service markets and the money market is that the money market is not completely free.

Money is such a fundamental requirement of all commerce that when its price is allowed to fluctuate freely with market forces some negative consequences can occur. When the price of money is high, virtually all commerce is restrained due to the cost of the money necessary to accommodate transactions. The capital-intensive sectors of the economy are the first to suffer from higher money costs. Real estate, steel, and automobile manufacturing are examples of capital-intensive industries. These industries represent such a large segment of the U.S. economy that higher prices in those industries spread to the rest of the economy fairly rapidly. These price increases can result in a high rate of inflation. If the inflation is left unchecked, it will ratchet itself up until the value of the monetary units of the country become virtually worthless. The ultimate result is a boom and bust cycle for the economy. A depression is the natural correction of such unchecked inflation.

The U.S. government regularly intercedes in the money market in an attempt to avoid excessive business cycles. The government agency that sets U.S. monetary policy is the **Federal Reserve Board**. The Fed, as it is commonly called, is charged with regulating money and credit in the United States. The Fed also has the power to print currency and strike coinage. As the country's commerce grows, the Fed prints more money and puts it into circulation to accommodate increased commercial transactions, avoiding the money supply shortage that would otherwise occur. As previously noted, if a shortage occurs, the price of money rises, acting as a restraint on the economy.

The Fed does not simply hand out the money it prints. Rather, it inserts the money into the economy by buying short-term government bonds. If the economy is growing too rapidly and the rate of inflation is rising, the Fed takes money out of the economy by selling government bonds, taking the cash it receives and placing it into Federal Reserve vaults. By slowing the rate of cash injected into the economy or by reducing the amount of currency in the economy, the Fed increases interest rates and regulates the growth of the economy.

It is important that technology entrepreneurs be aware of the price of money and how it can impact their individual ventures. When the money supply expands, it is easier to sustain a strategy of regular price increases. When the money supply contracts, a prudent company will consider the possible consequences of price-reduction strategies. While these are general observations, suffice it to say that a company in a capital-intensive industry needs to pay more attention to the money supply movements than the average business.

2.3 ECONOMIC SYSTEMS

An **economic system** is an accepted way of organizing production, establishing the rights and freedom of ownership, using productive resources, and governing business transactions in a society. There are three basic types of systems: (1) the government can produce almost all the goods and services (a planned economy); (2) private enterprise can produce almost everything (a pure market economy); and (3) there can be some government production and some private production (a mixed economy).

No economy is entirely privately owned, but a few (e.g., North Korea and Cuba) are almost entirely government owned. The United States is a mixed economy, with the majority of the production provided by the private sector.

2.3.1 Socialism

Socialism is an economic system that is based on government ownership of the factors of production. Government planning, rather than the market, is relied on to coordinate economic activity, including the production of goods and services, distribution of those goods and services, and, allocation of jobs and social benefits.

Prior to major changes in 1989 and 1990, countries such as East Germany (now part of Germany), Romania, Bulgaria, and the former Soviet Union had planned economies. The governments owned productive resources, financial enterprises, retail stores, and banks. In a planned economy, the government is the owner because planned economies believe that this is the best way to create prosperity and allocate wealth equally. Personal property, such as automobiles, clothing, and furniture, is owned by private citizens; however, the government owns almost all housing and the means of production. In a planned economy, politically appointed committees plan production, set prices, and manage the economy. Each factory receives detailed instructions on how many goods to produce.

Before 1990, socialist economic systems used administrative controls or central planning to answer the questions of what, how, and for whom. Recent reforms in countries such as China have reduced the central planner's power and role and increased the market's role.

Today, many European countries practice various forms of socialism. In fact, national economies continue to experiment with a variety of forms of private and public ownership of productive resources. In the United States, the election of Barack Obama in 2008 signaled a major change in government policies. In his first 100 days in office, President Obama enacted many reforms that some argue are a form of socialism. As we discuss below, this is not necessarily a bad thing as the United States has what is known as a "mixed economy."

2.3.2 Communism

Communism is a form of autocratic state socialism. A small group of nonelected officials decides what society's needs and goals are. In a pure communist economic system, the state owns most resources, including labor. Workers under pure communism have no claim to any of their work earnings. They provide their labor in service to the government.

Resentment of autocratic control in the former Soviet Union, East Germany, Hungary, and other countries led to demands for democracy in the late 1980s. The pure communist system has never actually existed, but it was the stated goal of many socialist countries until their demise. Today, many previously Communist Party–controlled countries hold elections and use multiparty political groups to change their political orientations.

The People's Republic of China was established as a communist state in 1949. China's famous five-year economic plans emphasized heavy industry, defense, and self-sufficiency. Wage differentials as a way to motivate workers were frowned on. In the late 1970s, China opened its doors to international trade, foreign investment, and communication. Worker incentives were improved through installing bonus systems, and private enterprise was encouraged in agriculture and consumer goods. The Chinese have even developed a stock market. The People's Republic of China is moving away ever so slowly from a planned communist-controlled state. There is still a high degree of central planning, but there are more market incentives.

Today, the only purely communist economy is North Korea. The nation's leadership determines most facets of daily life for the nation's millions of citizens. Although information about the productivity of the economy is difficult to obtain, it is clear that as a whole the nation underperforms world economic standards.

2.3.3 Mixed Market Economies

Capitalism is an economic system based on private ownership of property and the market in which individuals decide how, what, and for whom to produce. Under a pure capitalist system, individuals are encouraged to follow their own self-interest, while market forces of supply and demand are relied on to coordinate economic activity. Markets work through a system of rewards and payments. If you work, you get paid for that work; if you want a product or service, you pay the owner or provider for it. In a capitalist economy, individuals are free to do whatever they want as long as it's legal. The market is relied on to see that what people want to get and want to do is consistent with what is available. Under capitalism, fluctuations in prices coordinate individuals' wants.

Despite what many people think, the United States is not a pure capitalist nation. It is instead a mix of free markets plus government ownership and involvement. A pure capitalist nation would have no government control or ownership of markets. In a **mixed economy**, both the government and private business enterprises produce and distribute goods and services (see Exhibit 2.3). The government usually plays a role in supplying defense, roads, education, pensions, and some medical care.

A mixed market economy allows for the freedom to start a business.[8] Freedom of enterprise means that businesses and individuals with the capital may enter essentially any legal business venture they wish. This important feature of the market

System features	Socialism	Communism	Mixed economy
Ownership of enterprises	Basic industries by government, small-scale enterprises by private sector	Government owned, exception of small plots of land	Strong private sector, and larger than under socialism
What is produced	What planners believe is socially beneficial	Central planners decide what the citizens need	Consumer and profit oriented
Rights to profits	Officially exists in private sector only	Are not accepted	Are expected to generate profits
Management of enterprises	Government planning dominated	Centralized management system	Managed by owners
Rights of employees	Workers have right to choose their job, join unions, be influenced by government	Limited, against guarantee of no unemployment	Laborers have right to any job and freedom to join labor unions
Worker incentives	Limited in state owned, exist in private sector	Emerging currently	Exist in private sector, limited in public sector

EXHIBIT 2.3

Comparing Economic Systems.

Source: Ivancevich JM, Duening TN. Business: principles, guidelines, and practices. Cincinnati, OH: Atomic Dog Publishing; 2004.

economy permits individuals to seek out profit-making business opportunities. Under the market economy, any business or individual can earn a profit by producing a useful good or service. However, businesses and individuals do not have an automatic right to profit. Profit is a reward to a business for using scarce economic resources efficiently. Consumers must consider a good or service reasonable in price, quality, and value before a profit can be made.

Competition is yet another important part of a market economy. In general, competition refers to the rivalry among businesses for consumer dollars. Because of competition for consumer dollars, businesses have to be aware of what consumers want to buy. If they ignore consumer wishes, they are likely to lose sales, which directly affect the level of profit. A business that consistently loses money and makes no profit will fail. Consequently, competition among businesses generally provides

Company	Revenues	Profits
Wal-Mart Stores	$378,799	$12,731
Exxon Mobil	$372,824	$40,610
Royal Dutch Shell	$355,782	$31,331
BP	$291,438	$20,845
Toyota Motor	$230,201	$15,042
Chevron	$210,783	$18,688
ING Group	$201,516	$12,649
Total	$187,280	$18,042
General Motors	$182,347	-$38,732
ConocoPhillips	$178,558	$11,891
		All figures are in $millions

EXHIBIT 2.4

The Top 10 Largest Companies by Generated Revenue in the World in 2008.
Source: Fortune Magazine, July 2008.

consumers with lower prices, more services, and improved products. The continual fare wars in the airline industry illustrate how fierce competition among businesses can become.

Over time, companies can lose their stature or competitive position. Poor decision-making, improper business practices, lazy responses to competitors' actions, poor selection of top executives, inadequate training of the workforce, and ignoring signals of problems have all contributed to the fading of many business enterprises. Furthermore, within the past two decades, IBM, General Motors, and Sears Roebuck have lost their leadership positions on the world stage. These three seemingly invincible giants are today fighting for their lives against global competitors. In this short time period, Wal-Mart Stores has shot to the number one position in terms of revenues generated, as shown in Exhibit 2.4.

2.3.4 The U.S. Economic System

The United States has developed the world's largest economic system. The basic parts of the U.S. economy are illustrated in Exhibit 2.5. This model includes only the broadest parts of the economy; it does not include the government.

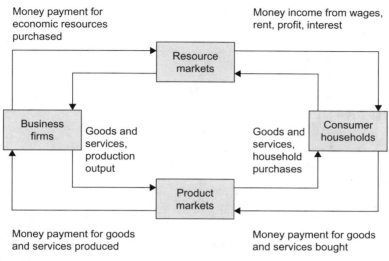

Money payment for
economic resources
purchased

Money income from wages,
rent, profit, interest

Money payment for goods
and services produced

Money payment for goods
and services bought

EXHIBIT 2.5

Flow of Products and Money in the U.S. Economy.

Source: Ivancevich JM, Duening TN. Business: principles, guidelines, and practices. Cincinnati, OH: Atomic Dog Publishing; 2004.

Note the differences between resource markets and product markets in the exhibit. **Resource markets** are places where economic resources—natural, labor, and capital—are bought and sold. The New York Stock Exchange, where money is invested in companies, is a financial capital resource market. The employment section in local newspapers is a resource market where labor is bought and sold. **Product markets** are the millions of markets around the world where business outputs (goods and services) are sold to consumers (the government can be viewed as a consumer in some cases). Consumers pay for goods and services with money. This type of consumer expenditure is called "retail sales." The money businesses receive from retail sales is "revenue."

Where do consumers get money to spend for goods and services? Exhibit 2.5 shows that consumer households supply economic resources to the resource markets. In return for money, people provide labor through work, invest in businesses (capital), and sell natural resources to businesses. The money received in payment for these economic resources is then used to purchase goods and services. Of course, businesses view money paid to suppliers of goods and services as an expense.

Two distinct types of economic resource flows are illustrated in Exhibit 2.5. The flow of economic resources and products is shown by the inner loop. It flows counterclockwise, showing that economic resources move from consumers to businesses and then return to consumers as finished goods and services. Money flow, on the other hand, is shown by the outer loop. This clockwise flow of money begins when firms pay consumers for the economic resource they purchase. Consumers use their money to purchase the goods and services produced by businesses. These two economic flows take place continuously and at the same time. As long as consumers spend all their money, the flow of money into and out of consumer households is equal. However, some money is diverted into savings, and the government needs to intervene to bring about a balance in the flows.

2.4 INTERNATIONAL TRADE

Scarcity of resources is perhaps the major reason why nations trade with each other.[9] A country with a surplus of some product may decide to sell this surplus to other nations. Such sales enable the country to purchase other products that it may not have the ability to produce.

No nation has every raw material or resource; no nation can produce everything it needs. Most nations specialize in producing particular goods or services. The United States, for example, has developed a specialty in producing agricultural products efficiently; thus, it sells food to many nations of the world. But the United States purchases much of its oil from other countries, such as Saudi Arabia, Venezuela, and Nigeria, that specialize in the production of crude oil.

A nation has an **absolute advantage** if it can produce a product more efficiently (using fewer resources) than any other nation. South Africa has an absolute advantage in the production of diamonds. Absolute advantages are rare because usually at least two countries can efficiently supply a specific product.

A nation has a **comparative advantage** if it can produce one product more efficiently than other products in comparison to other nations. For instance, countries with low labor costs, such as China and South American countries, have a comparative advantage in producing labor-intensive products such as shoes and clothing. Many nations become involved in international business because they have a comparative advantage. Firms sell those goods for which they have the greatest comparative advantage over other countries. The United States has its greatest comparative advantage in food products, engineering, medical technology, software, aircraft, and coal.

Comparative advantages shift frequently. For many years, the United States held a comparative advantage in manufacturing a variety of products, such as automobiles, television sets, and appliances. Today many of these products are made in Japan, Germany, and South Korea. The United States, for instance, experienced a dramatic loss of comparative advantage in manufacturing televisions. One reason for shifts in comparative advantages is competition.

Some nations want to become self-sufficient and thus do not specialize in the production and sale of particular products. The choice of specialization or self-sufficiency is generally a political and economic issue. For instance, communist nations traditionally have striven for self-sufficiency because they fear economic dependency on other countries. Some nations also view self-sufficiency as necessary for achieving military supremacy. Of course, no country is completely self-sufficient. To satisfy the needs of its citizens, every nation engages in some form of trade—exporting and importing—with other nations.

2.4.1 Exporting and Importing

The United States is one of the world's largest exporters and importers. In 2000, the United States exported about $1.80 trillion and imported about $2.5 trillion worth of goods and services.[10] Exporting is selling domestic-made goods and services to another country. Importing is purchasing goods and services made in another country. Most of the video recorders purchased in the United States are imported from Japan.

The United States imports a wide variety of products: food, oil, automobiles, electronic equipment, clothing and shoes, iron and steel, and paper products, to name a few. But U.S. exports are on the rise, up 121% since 1978. The major growth is in exports to Canada, Japan, Mexico, Taiwan, Korea, and Germany. The fastest growing U.S. exports are music, video, software, cigarettes and tobacco products, meat, pulp and wastepaper, and synthetic resins, rubber, and plastics.

2.4.2 Balance of Trade

A country's **balance of trade** is the result of importing and exporting. In 2008, the United States had a trade deficit (more imports than exports) of $700 billion.[10] The balance of trade is the difference (in monetary terms) between the amount a nation exports and the amount it imports. The United States has shown a deficit for many years. Some nations export more than they import and maintain a favorable balance of trade, or a trade surplus. Japan and China, for example, usually have positive balances of trade.

2.4.3 Balance of Payments

A country's **balance of payments** is the total flow of money into and out of the country. The balance of payments is determined by a country's balance of trade, foreign aid, foreign investments, military spending, and money spent by tourists in other countries. A country has a favorable balance of payments if more money is flowing in than is flowing out; an unfavorable balance of payments exists when more money is flowing out of the country than in. For many years, the United States enjoyed a favorable balance of payments. But in recent years, more money has been leaving the country than has been flowing in. Exhibit 2.6 shows the balance-of-payment deficit of the United States as a percentage of GDP since the 1950s. As you can see, the United States has become primarily an import economy since the early 1990s.

2.5 INTERNATIONAL MONETARY SYSTEM

In addition to the balance of payments, another area important for the technology entrepreneur to understand before engaging in the international business is the role of the International Monetary Fund (IMF) and the World Bank. Both the IMF and the World Bank were established in 1944 when representatives from 44 countries

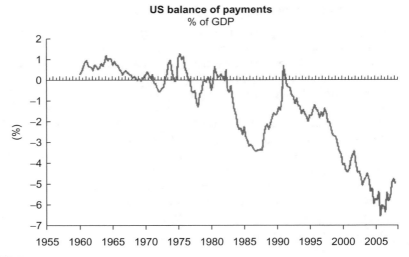

EXHIBIT 2.6

The U.S. Balance of Payments.

met at Bretton Woods, New Hampshire, to design a new international monetary system. The overall goal of the meeting was to design an economic order that would endure and facilitate the economic growth of the world following the end of World War II. While the IMF was established to maintain order in the international monetary system, the World Bank was established to promote general economic development. Each will be discussed in turn in terms of its importance for the technology entrepreneur.

2.5.1 International Monetary Fund

Due to the worldwide financial collapse, competitive currency devaluations, high unemployment, and hyperinflation occurring particularly in Germany in 1944, the IMF was established to avoid a repetition of these through discipline and flexibility. The discipline part of the equation was achieved through the establishment of a fixed exchange rate, thereby helping to control inflation and improving economic discipline in countries. This fixed exchange rate lasted until 1976 when a floating exchange rate was formalized and established.

To help minimize the rigidity of the original fixed exchange rate, limited flexibility was built into the system in the form of the IMF lending facilities and adjustable parities. Each member of the IMF provided gold and currencies to the IMF to lend to member countries to cover short-term periods of balance-of-payment deficits to avoid domestic unemployment in that country due to a tightening monetary or fiscal policy. The IMF funds could be lent to countries to bring down inflation rates and reduce the country's balance-of-payment deficits. Since a persistent balance-of-payment deficit would deplete a country's reserves of foreign currency, a loan to reduce this deficit would help a country avoid devaluing its currency. When extensive loans from the IMF fund occur in a country, the country and its macroeconomic policies are placed under increasingly stringent supervision by the IMF so that it can assist the country in addressing the issues within its economic policy.

The system of adjustable parities established by the IMF allows for the devaluation of a country's currency by more than 10% if the IMF feels that this will help achieve a balance-of-payment equilibrium in the country. The IMF felt that in these circumstances without devaluation, the member country would experience high unemployment and a persistent trade deficit.

2.5.2 The World Bank

The World Bank was established initially to help finance the rebuilding of Europe following World War II. Since this was successfully accomplished by the Marshall Plan of the United States, the World Bank focused on development in the form of lending money to Third World Countries. This general focus was power stations and transportation in the 1950s and the support of agriculture, education, population control, and economic development in the 1960s.

The World Bank makes loans for projects in developing economies through two schemes. The first one, the International Bank for Reconstruction and Development (IBRD) scheme, raises money for the project through the sale of bonds in the international capital market, previously discussed. The second scheme involves International Development Association (IDA) loans from money supplied by wealthier member nations. A technology entrepreneur may have the opportunity to be involved in one of these funded projects.

There are three other divisions within the World Bank: International Finance Corporation (IFC), Multilateral Investment Guarantee Agency (MIGA), and the International Center for Settlement of Investment Disputes (ICSID). MIGA, for example, provides companies with insurance to mitigate the risk associated with investing in developing countries.

2.6 ELEMENTS OF GLOBAL TRADE

There are varying attitudes throughout the world concerning trade. Starting around 1947 with the development of trade agreements and the reduction of tariffs and other trade barriers through international organizations, there has been an overall positive atmosphere concerning trade between countries. In saying that, the technology entrepreneur needs to be aware that risks and barriers also exist. Understanding each market and its specific environment will often determine success within those markets.

Global trade is facilitated by international agreements and treaties that articulate and enforce the rules of fair trade. Several organizations, including the World Trade

Organization (WTO) and others, have been established to influence global trade and to provide a context in which disputes can be adjudicated and remedied. In addition, global trading blocs have been set up to facilitate regional trading between nations whose fortunes are largely dependent upon one another's success. So, for example, the United States, Canada, and Mexico have crafted the North American Free Trade Agreement (NAFTA) to facilitate trade between the three North American nations whose futures and fortunes are highly intertwined. In addition to understanding how to leverage the rules and opportunities that are created by international and regional trade agreements, technology entrepreneurs must understand the nature of foreign currency exchange rates and how to manage the day-to-day fluctuations in these rates. Below, we explore some of these key elements of the global economy.

2.6.1 The WTO

One of the longest lasting agreements, and now the leading international organization on trade, is the WTO. Beginning in 1947 under U.S. leadership as the General Agreement on Tariffs and Trade (GATT), the WTO was officially established in January 1995 under the Uruguay Round (1986–1994) as a multilateral agreement among nations with the objective of liberalizing trade by eliminating or reducing tariffs, subsidies, and import quotas. WTO membership includes over 150 nations that have participated in eight rounds of tariff reductions. Mutual tariff reductions are typically negotiated between member nations through creating policies during each round. The ninth and current round, Doha (2001–present), have been in a stalemate due to disagreement on issues between developed and developing nations, but appear to be starting again.

The Dispute Settlement Board (DSB) was established in 1995 as the appellate body within the WTO to monitor member countrys' policies. Through the DSB member countries are able to bring disputes against other member countries if they feel that a violation in trade policy has occurred. Often these cases are brought against more than one nation by more than one country, forming something similar to a trade bloc. If the investigation uncovers a violation, violating countries are asked to change their policies and conform to the agreed-upon tariffs and agreements; or barriers in other sectors can be levied by prosecuting countries to compensate for lost revenues. With over 370 cases (as of April 2008) already brought before the DSB with successful trials, decisions, and according actions, many developing nations feel that a shift in power has occurred that gives them the power to bring justice against the major world powers.[11]

The case brought against the United States and its steel industry exemplifies the DSB's role and its unilateral actions against dumping. The U.S. Congress used an anti-dumping fine to aid U.S. steel companies under the name of the Byrd Amendment. Creating an environment solely beneficial to the U.S. steel companies, the U.S. Congress had allowed this "fine" to be directed solely into the coffers of U.S. companies affected as a result of dumping. With a myriad of countries claiming unfair trade practices, Japan, the European Union (EU), Mexico, South Korea, and a variety of others took the case to the DSB, and after appeal, the United States lost and the other countries were allowed to levy taxes against similar industries for recuperation.[12] Under the ruling, these nations could levy up to 72% of the money raised and distributed during the life of this amendment affecting other industries such as U.S. paper, farm goods, textiles, and machinery.[13]

Important from an entrepreneurial aspect, these types of cases can affect industries and sectors in which the technology entrepreneur's new venture is operating. The importance of understanding the implications for each business venture will be influential in deciding the direction and picking a strategy for each undertaking.

Although the support for the WTO varies and was relatively low in the 1970s, it increased in the 1980s due to the rise in protectionist pressures in many industrialized countries. The renewed support reflected three events. First, the world trading system was strained by the persistent trade deficit of the United States, the world's largest economy, a situation that caused adjustments in such industries as automobiles, semiconductors, steel, and textiles. Second, the economic success of a country perceived as not playing by the rules (e.g., Japan and then China) has also strained the world's trading system. Japan's and China's successes as the world's largest traders and the perception that their internal markets are, in effect, closed to imports and foreign investment have caused problems. Finally, in response to these pressures, many countries have established bilateral voluntary export restraints to circumvent the WTO. China did this during the economic prosperity of the 1990s due to the pressures from the world and particularly the United States.

2.6.2 Trade Blocs and Free Trade Areas

Around the world, groups of nations are banding together to increase trade and investment between nations within the group and excluding those nations outside the group. One little-known agreement between the United States and Israel, signed in 1985, establishes a Free Trade Agreement (FTA) between the two nations which phased out all tariffs and quotas, except on certain agricultural products, over a 10-year period. In 1989, an FTA went into effect between Canada and the United States that phased out tariffs and quotas between the two countries, which are each other's largest trading partners.

Many trading alliances have evolved in the Americas. In 1991, the United States signed a framework trade agreement with Argentina, Brazil, Paraguay, and Uruguay to support the development of more liberal trade relations. The United States has also signed bilateral trade agreements with Bolivia, Chile, Colombia, Costa Rica, Ecuador, El Salvador, Honduras, Peru, and Venezuela. The NAFTA among the United States, Canada, and Mexico is a much publicized agreement to reduce trade barriers and quotas and encourage investment among the three countries. Similarly, the Americas, Argentina, Brazil, Paraguay, and Uruguay operate under the Treaty of Asunción, which created the Mercosur trade zone, a free-trade zone among the countries.

Another important trading bloc has been developed by the EU. Unlike NAFTA or a similar FTA, the EU is founded on the principle of supranationality that prevents member nations from being able to enter into trade agreements on their own that are inconsistent with EU regulations. As nations are added, the EU trading bloc becomes an increasingly important factor for technology entrepreneurs doing international business as it prevents them from entering those markets.

2.6.3 Currency Exchange Rates

The price of a product is usually different in domestic and foreign markets. The costs of foreign trade, such as taxes, tariffs, and transportation, often result in higher prices in the foreign country. Exchange rates can also influence the price of foreign goods. When the technology entrepreneur and another company agree to execute a transaction and exchange currency immediately, the transaction is called a **spot exchange**. This is the exchange rate at which the currency of one country is converted into the currency of another at a particular time on a particular day. The spot rate also applies when someone is traveling in a foreign country and wants to convert, say, euros into British pounds at a bank in London. The spot exchange rate may not be the most favorable exchange rate as the value of the currency is determined by the interaction of demand and supply of each country's currency with respect to the currency of other countries on that particular day.

Currency swaps are used to protect against foreign exchange risk when there is a need to move in and out of a currency for a limited period of time. For example, suppose Alcoa buys and sells finished goods and parts with their wholly owned plant in Hungary and wants to ensure that the US $1 million it will need to pay bills in Hungary is available at a known exchange rate. It will also need to collect €20 million in 75 days (€ denotes the euro) when some accounts are due. If today's spot exchange rate is US$1 = €0.76 euro and the forward exchange rate is US$1 = €0.90, Alcoa can enter into a 75-day forward exchange currency swap for converting €20 million into dollars. Since the euro is today at a premium on the 75-day market, the company will receive $22.2 million (€20 million/0.90 = $22.2 million). Of course this could be reversed if the Euro was trading at less than 0.76 euro/US$1 in 75 days.

There are many other more sophisticated instruments and techniques for managing foreign exchange fluctuations. Technology entrepreneurs who venture into foreign markets should be aware of the potential for currency fluctuations to disrupt their business model. However, as a general rule, foreign exchange fluctuations should not be a profit driver for the venture. Companies that spend too much time trying to predict or speculate on changes in foreign exchange rates may end up losing focus on core products and lose out in the marketplace.

SUMMARY

This chapter examined a range of topics with the intention of exposing you to some fundamental concepts in economics and international business. Nearly any discussion of economic fundamentals will begin by highlighting the basic resources of natural resources, capital, labor, knowledge, and their scarcity. The challenge of any economic system is to allocate those scarce resources in a manner that maximizes the wealth and prosperity of everyone. This is no small undertaking and nations around the globe continue to experiment with various economic systems. From pure communism to pure market economies, most of the systems are a mix of government controls and market freedoms. The U.S. economy is best described as mixed and, depending on political fashions, swings back and forth from greater market freedoms to increased government intervention. For example, as this book is being written, the administration of President Barack Obama is said to be overturning much of the market-freeing de-regulation of the Reagan administration of the 1980s. Only time will tell if this experiment results in greater prosperity.

A second fundamental concept in economics is the basic law of supply and demand. In this chapter, we examined how supply and demand affect the price of goods and services. Technology entrepreneurs must be aware while setting the price of their venture's offerings that the marketplace sets the price. Attempting to set a price that ensures a particular profit margin may result in a price that is unacceptable to the market.

We also examined several economic phenomena that are related to price. In addition to consumers, competitors are also responsive to prices charged in the marketplace. In an inelastic market, lowering prices to increase market share can lead to destructive price wars. On the other hand, lowering prices in an elastic market can win new demand. Inflation is a marker of a general rise in prices throughout an economy. Occasionally, entrepreneurs may take advantage of compromised markets, where temporary disruption in supply or competition can support higher pricing.

Finally, this chapter introduced you to a number of issues pertaining to conducting business internationally. We reviewed the various strategies that technology ventures can use to enter foreign markets. These include exporting/importing, licensing, joint ventures, strategic alliances, counter trading, and direct investment. There are a host of issues technology entrepreneurs face when entering foreign markets, including fluctuating currencies, cultural differences, political barriers, and others. Overcoming these barriers takes patience, intelligence, and understanding. For most technology entrepreneurs, global markets and global competition will constitute at least part of their overall strategy.

STUDY QUESTIONS

1. Explain how a price is determined in the market using supply and demand. How do surpluses and shortages affect price?

2. How could a foreign exchange rate of a country either help or hinder a company looking to invest in that country?

3. Should foreign exchange be a profit driver for a venture? Why or why not?

4. Illustrate the law of supply and demand. What happens to the price of products in a market where demand stays the same while supply increases? Why?

5. Explain what is meant by the term "price sensitive." How should a technology entrepreneur adjust the venture's offerings when this occurs?

6. Identify three ways that technology entrepreneurs can enter foreign markets. For each of these, identify a technology venture that has used the approach.

7. Illustrate the circular flow of products and money. Your diagram should include households, businesses, product markets, and resource markets.

8. Explain how price wars occur in a market.

9. What is meant by the term "gross margin"? How can a technology venture use this ratio to manage the business?

10. Describe the three types of economic systems identified in this chapter. Which of these types does the U.S. economy closely resemble? Explain.

EXERCISES

1. The price of oil is something that most people track with some regularity. After all, the price of a barrel of oil purchased on the major commodity exchange markets will directly affect the price of a gallon of gas at the gas pump. For this exercise, students should track the price of a barrel of oil during a given trading day, and over the course of a week. During a single trading day, it's possible for the price of oil to vary by 5% or more. During the course of a week, variations in price can be even greater. Follow the price during these two suggested time periods, and try to determine the economic factors that are influencing the price and price changes. For example, the Organization of Petroleum Exporting Countries (OPEC) often sets supply quotas that directly affect price. Students should be encouraged to look for other factors that may be driving the price of a barrel of oil during the trading day and during a given week. Debate and discuss the factors that may affect oil prices. How does that affect the average American consumer? What can be done to raise or lower the price of oil and, thus, the price of a gallon of gas?

2. The People's Republic of North Korea (DPRK) is considered by most to be the least free economy in the world. For decades the nation has been run by Kim Jung Il or his father Kim Il Sung. The regime retains tight control over all of the factors of production, and it severely restricts travel by its citizens outside of the country or by foreigners inside the country. Students should familiarize themselves with this unique nation by visiting web sites, reading popular articles in the mainstream literature, and viewing documentaries about the nation. For example, these web sites will be helpful:

   - http://www.kcna.co.jp/index-e.htm
   - http://www.youtube.com/watch?v=FJ6E3cShcVU

Discuss what has been learned about this economy. How does it compare to the U.S. economy? What is the future of the DPRK? Should other countries emulate the DPRK's approach to organizing its economy? Why or why not?

KEY TERMS

Absolute advantage: A nation has an absolute advantage if it can produce a product more efficiently (using fewer resources) than any other nation.

Balance of payments: A country's balance of payments is the total flow of money into and out of the country.

Balance of trade: The balance of trade is the difference (in monetary terms) between the amount a nation exports and the amount it imports.

Capital resources: Goods produced to make other types of goods and services.

Capitalism: An economic system based on private ownership of property and the market in which individuals decide how, what, and for whom to produce.

Comparative advantage: A nation has a comparative advantage if it can produce one product more efficiently than other products in comparison to other nations.

Cost of goods sold: This refers to all the costs involved with raw materials and production of a product or service.

Cost-plus-pricing: A price-setting strategy whereby one adds up all the costs associated with a product or service, adds a preset profit margin, and derives the price to the customer.

Currency swaps: Financial instruments used to protect against foreign exchange risk due to fluctuations in relative currency prices.

Deflation: The decrease in general price level in a market.

Demand: The amounts of goods and services buyers are willing to purchase at various prices in the same time period and market conditions.

Dumping: Refers to disposing of goods in a foreign market, which are surplus in their native market, at prices below those in the home market or below the costs of manufacturing the goods.

Economics: The study of how a society chooses to use scarce resources to produce goods and services and to distribute them to people for consumption.

Economic system: An accepted way of organizing production, establishing the rights and freedom of ownership, using productive resources, and governing business transactions in a society.

Elastic market: A market in which the overall demand in the market will expand if prices are lowered.

Federal Reserve Board: An entity of the U.S. government charged with controlling inflation by regulating the supply of money, mostly via manipulation of basic banking interest rates.

Gross Domestic Product: The total value of all final goods and services produced in a particular economy.

Gross margin: The difference between the selling price of a product or service and the amount the entrepreneur had to pay for the raw materials that make up that product or service (the "cost of goods sold").

Knowledge resources: Knowledge resources concern not only the accumulated knowledge residing within a society, but also the processes for creating new knowledge.

Inelastic market: A market in which demand will not respond to price changes.

Inflation: The rise in general price level in a specific market.

Interest rate: The price of money in an economy.

Labor resources: Resources that represent the human talent of a nation.

Macroeconomics: The study of inflation, unemployment, business cycles, and growth, focusing on the aggregate relationships in a society.

Microeconomics: The study of individual choice and how choice is influenced by economic forces.

Mixed economy: Both the government and private business enterprises produce and distribute goods and services.

Natural resources: Resources provided by nature in limited amounts; they include crude oil, natural gas, minerals, timber, and water.

Price gouging: Occurs when the supply of an item or items has been so severely restricted by special circumstances that the demand is completely out of balance with the supply.

Pricing power: The ideal situation where a business sells goods into a willing market at a price it determines and, at least to some extent, controls.

Price sensitive: Consumers that have become price sensitive tend to shop for goods and services primarily based on low price.

Price wars: A condition that ensues in an inelastic market when one or more competitors begin to lower prices.

Product market: Markets where business outputs (goods and services) are sold to consumers.

Resource market: Places where economic resources—natural, capital, and labor—are bought and sold.

Shortage: A condition in a market where there is excessive demand and too little supply, which tends to lead to rising prices.

Spot exchange: When counterparties agree to execute a transaction and exchange currency immediately.

Supply: The amount of goods and services available in the market that sellers are willing to offer at various prices in a given time period and market condition.

Surplus: A condition in a market where there is excessive supply and too little demand, which tends to lead to falling prices.

WEB RESOURCES

The web sites below are intended as destinations for your further exploration of the concepts and topics discussed in this chapter:

1. http://www.eiu.com: This is the web site for the Economic Intelligence Unit. The EIU provides information, analysis, and forecasts for over 200 countries and eight key industries.

2. http://www.economist.com: This is the web site for the *Economist* magazine and provides information and commentary regarding current economic news.

3. http://www.federalreserve.gov: The web site for the U.S. Federal Reserve Bank that provides a safe, flexible, and stable monetary and financial system.

4. http://finance.yahoo.com/currency: Currency exchange rate converter provided by Yahoo.com that allows you to find out exchange rates throughout the world and compare exchange rates of various countries simultaneously.

5. http://www.bea.gov/international: This web site provides international economic account information provided by the Bureau of Economic Affairs (U.S. Department of Commerce).

6. http://www.worldbank.org: This is the official web site for the World Bank whose objective is to create a world free of poverty by providing financial and technical assistance to developing countries.

7. http://www.imf.org: This is the official web site for the International Monetary Fund, an international organization formed to provide stability, cooperation, and growth in the international monetary market as well as to provide temporary financial assistance to countries with balance-of-payment issues.

ENDNOTES

[1] Sowell T. *Basic economics: a citizen's guide to the economy*. New York: Basic Books; 2000.

[2] O'Connell B, Griffeth B. *CNBC creating wealth: an investor's guide to decoding the market*. Hoboken, NJ: John Wiley & Sons; 2001.

[3] Lucas MR. Pricing decisions and the neoclassical theory of the firm. *Manage Acc Res* 2003;**14**:201–17.

[4] Hanson WA, Kalyanam K. A cost-plus trap: pricing heuristics and demand identification. *Market Lett* 1994;**5**(3):199–209.

[5] Boote AS. Price inelasticity: not all that meets the eye. *J Cons Marketing* 1985;**2**(3):61–6.

[6] Eilers B, Schweiterman J. Strategies to stimulate discretionary air travel: an internal perspective. *J Transport Law Logist Pol* 2003;**Fall**:123–36.

[7] Desjardins D. Price wars bode ill for innovation. *DSN Retailing Today* 2003; **December 15**:28.

[8] Rothschild E. *Economic sentiments: Adam Smith, Condorcet, and the Enlightenment*. Cambridge, MA, and London: Harvard University Press; 2001.

[9] Krugman PR. *International economics: theory and policy*. Reading, MA: Addison-Wesley; 2000.

[10] U.S. Census, FT900: U.S. International Trade in Goods and Services. Released February 11, 2009.

[11] Ibid.

[12] Chronological List of Disputes Cases. World Trade Organization Web site. http://www.wto.org/english/tratop_e/dispu_status_e.htm. Accessed: April 22, 2008.

[13] United States–Definitive Safeguard Measures on Imports of Certain Steel Products. *World Trade Organization Dispute Settlement DS252*. http://www.wto.org/english/tratop_e/dispu_e/cases_e/ds252_e.htm. Accessed: May 5, 2008.

Technology Ventures in a Global Context

After studying this chapter, students should be able to:

- Identify major challenges technology ventures encounter in the global economy
- Evaluate venture opportunities in the global economy
- Understand factors involved in the creation, protection, and capture of market and enterprise value
- Explore how technology entrepreneurs can exploit the network effect
- Appreciate the distinction between market value and enterprise value
- Understand that the technology entrepreneur can pursue consumer markets (B2C) or business markets (B2B)
- Assess what it takes to create repeatable business systems

TECH VENTURE INSIGHT

Modern Technology Transforms Old Lands to Meet Rising Food Demands

Richard Spinks is something of a maverick entrepreneur. In November 2005, while at a bar in Warsaw, Poland, the 41-year-old Briton was chatting with friends about the biofuel boom occurring around the world. One friend suggested they buy land in Ukraine to grow rapeseed, a commodity used commonly in Europe to make biofuel for cars. Spinks and his friend liked the idea and, along with a third partner, they eventually invested nearly $1 million of their own money to form a company called Landkom International, PLC. Landkom began leasing land in Ukraine in early 2006, and by early 2008 had planted rapeseed, wheat, and barley over 25,000 acres of this newly acquired land.

Obtaining the land and farming it required not just innovative thinking and modern technology but also a fair share of resourcefulness on the part of Landkom Inc.

For example, Spinks and his partners learned that property laws in Ukraine prohibit private land sales, which meant that Landkom would have to sign leases with individual landowners. Ukrainian landowners are traditionally suspicious of the intentions of outsiders wanting to use their lands, a skepticism that goes back to the days of Stalin, when vast lands were confiscated in Ukraine to build collective farms, resulting in the starvation of millions.

Undaunted by these potential sticking points, Spinks set up Landkom's headquarters in the Ukrainian town of Bilyi Kamin. One evening, Spinks and his colleagues called a meeting of the town's 300 landowners, proposing to lease land at $14 an acre per year. Landkom also offered to pay the farmers a salary of $400 per month, twice the national average for manual labor. Although not everyone accepted the deal, the company was eventually able to lease more than 160,000 acres. Its ultimate goal was to lease 1 million acres by 2015. To achieve this, Landkom needed external funds, so in November 2007, the venture sold a 44% stake on the London Stock Exchange, raising $106 million. The company's market value has since then grown to $386 million, and Spinks's 5.8% stake is now worth over $20 million.

To manage profit margins amid rising costs for equipment, seed, and additional land, Landkom turned to modern farming techniques, investing $250,000 in facilities to install a command center that guides combines via satellite in Bilyi Kamin. The firm also invested in new combines from Germany. According to Vassil Stebnitski, Landkom's chief agronomist in the region, the farmers were in need of new technology. "All we had before was land and sky," he added.

Using new technologies in a part of the world that had been left behind by the global economy, Spinks was able to create and capture new value. Although there will be numerous challenges ahead, the actions of Landkom's creator represent a classic tale of a maverick with a vision and the courage to make it real. If Landkom succeeds in aggregating and making productive the scattered collective farms that fed the former Soviet Union, its venture would also have provided an important boost to global food supplies.

Source: Adapted from Miller JW. In Ukraine, mavericks gamble on scarce land. The Wall Street Journal 2008; May 12:A1; Landkom says on track with Ukrainian land bank. Reuter's News 2008; April 10.

INTRODUCTION

Twenty-first-century technology entrepreneurs face a vastly different world from what their counterparts faced in the twentieth century. This is not just idle talk based on a tendency for people to view their own historical period as unique; rather, the global economy and emerging new technologies have, in fact, radically changed both the opportunities available to entrepreneurs and the challenges they face. The Tech Venture Insight on Richard Spinks illustrates how one entrepreneur was able to leverage both the globalization of the economy and newly available technologies to carve out a thriving and unique business. Spinks founded Landkom International PLC only after he had worked in many different types of jobs all over Europe during the course of his career. In 2006, he seized upon the opportunity to lease land in Ukraine and to establish a modern technology-aided agribusiness that he felt was worth pursuing. Based on early results, Spinks's entrepreneurial adventures seem to be paying off.

In this chapter, we examine the challenges and opportunities for technology entrepreneurs in the global economy. The chapter is organized according to the main themes of this book: value creation, value protection, and value capture. Each will be discussed in the context of technology ventures in the global economy. We begin by discussing the technology entrepreneur's unique position in the context of globalization and trade.

3.1 GLOBALIZATION AND TECHNOLOGY VENTURES

The global economy can be described in many ways. One interesting perspective comes from *New York Times* columnist Thomas Friedman in his popular book, *The World Is Flat*.[1] Friedman seeks to convey the idea that while the world is not flat in the literal sense, today's economy has enabled a relatively unimpeded flow of money, talent, capital, and goods across national borders. At the same time, global trade policies are evolving constantly, with nations sharing the confidence that contracts will be upheld, goods sold to foreign buyers will be paid for, and disputes between trading partners will be managed through increasingly fair and predictable legal systems. This confidence also stems from the establishment of prominent governing bodies, such as the WTO, which sets rules for countries around the world on the principles of fair trade. The WTO also includes provisions for remedies in the case of trade disputes, as well as opportunities to propose and promulgate new rules as global economic conditions change.

Change is the normal order of business in today's modern global economy. For example, the past quarter century has produced enormous changes not only in technology, but also in *where* the technical work is performed and *who* is performing it. The trend of **business process outsourcing** or the movement of work to regions of the world with qualified and low-cost labor began in the late 1980s and accelerated throughout the 1990s. General Electric's former CEO Jack Welch is generally credited with being an early proponent of offshore outsourcing.[2] He identified the opportunity to shift some of GE's technical work to India. Today, many U.S., European, and other firms have followed suit by shifting their technical labor to countries such as India, China, the Philippines, Vietnam, and Mexico. Similarly, the U.S. aerospace industry has shifted much of its production work to Mexico to take advantage of that country's lower-cost yet technically skilled labor pool.[3] This migration of work to regions of the world with large pools of talented technical workers is likely to continue until labor rates become more homogeneous on a global scale.[4]

In addition to opportunities that exist with shifting work to regions of low-cost labor, technology ventures can also enjoy the benefits of global trade liberalization. As we discussed in Chapter 2, major regional trade agreements have opened the door for firms of all sizes to be able to sell goods and services while encountering fewer obstacles than in the past. One such agreement is the North American Free Trade Agreement or NAFTA, ratified in 1994 by the United States, Canada, and Mexico. NAFTA has been both hailed as a breakthrough agreement to unite North America into a single free-trade zone and vilified as an enabler for companies to shift work to the lowest cost labor market regardless of the consequences for domestic workers. Other regional trade agreements such as Latin America's MERCOSUR and Southeast Asia's ASEAN have faced similar difficulties.

The challenges that nations face attempting to iron out the details of regional trade agreements are amplified when they have to deal with various other global scale agreements. Despite these challenges, however, it is likely that the globalization trend is here to stay because the many benefits accrued by nations and peoples who participate in the global economy are far too numerous to think that our world can return to a pre-twenty-first-century economic state. In fact, technology entrepreneurs and the ventures they create are important drivers of global economic change and growth. Offshore outsourcing, for instance, would not have been possible were it not for the bold entrepreneurs who laid the fiber optic cables to create the "information superhighway." In the 1990s, new technology ventures such as Global Crossing, Qualcomm, and others were at the forefront of developing the telecommunications infrastructure that we take for granted today. While some of these pioneering firms experienced huge business disruptions during the dot-com bust in 2000, credit is due to them for having laid the foundations of the global communications revolution, linking workers, work, knowledge, and capital worldwide.

Technology entrepreneurs will undoubtedly continue to play an important role in the expansion of prosperity to people everywhere. In the comparatively wealthy nations, where students such as you are able to purchase textbooks such as this, it is easy to lose sight of the huge markets for technology goods and services that exist in the less wealthy regions of the world. One observer has called this untapped market "the fortune at the bottom of the pyramid."[5] Clearly, the technology entrepreneurs who can help solve the problems that are associated with population growth, emerging middle-class wealth in Asia and India, water shortages, depleting oil supplies, etc., will thrive in the coming years.

3.1.1 Technology Venture Drivers

The first two chapters introduced you to the fundamentals of business and economics as a starting point for your journey into technology entrepreneurship. Chapter 1 focused on business concepts, and presented the idea that a business is a going concern that consistently delivers value to an identified market. Chapter 2 was an introduction to fundamental economic issues. This chapter represents your first exposure to the *challenges* of technology entrepreneurship in the global economy.

Here, we begin the process of focusing on concepts, topics, challenges, and opportunities specific to technology ventures. We have identified four primary drivers of technology ventures in the global economy:

1. Capital intensity
2. Knowledge intensity
3. Accelerated pace of change
4. The network effect

These drivers are generic and may or may not apply to the specific technology venture you may be considering. Yet, if you are thinking of a specific type of technology venture opportunity, it is likely that most of these will apply to your venture.

3.1.2 Capital Intensity

In the previous chapter, you learned that the basic factors of production referred to in economics are **land**, **labor**, and **capital**. Launching a venture in some industries may require more of one factor of production than the others. A **capital-intensive** venture is one that requires a disproportionate amount of capital as compared to the other factors of production. For example, if you wanted to develop a new manufacturing facility in Asia, you would need a large amount of capital to get that started. You'd have to purchase the land for the factory, hire engineers and laborers, and set up a supply chain to your facility. Pharmaceuticals, shipping, semiconductors, and mining are other examples of capital-intensive technology industries.

By way of contrast, if you wanted to launch an Italian restaurant in downtown Beijing, much less capital would be required. Restaurants are **labor-intensive** ventures. They require a large number of people with a wide range of skill sets to run efficiently. A restaurant will require waitstaff, chefs and cooks, hosts, and table setup and cleanup staff. Individuals interested in starting a restaurant venture can usually do so with less capital than most technology ventures require. Other labor-intensive industries include garment production, medical transcription, data processing, and accounting.

Land-intensive ventures are those that either focus on growing things on the ground or extracting things from beneath the ground. Agriculture is an example of a land-intensive industry. Successful agriculture ventures are those that have learned systematically to till soil, plant seed, manage plant growth, and harvest crops over large tracts of land. Increasingly, agriculture has become corporate and systematic, with many corporations operating millions of acres of land.[6]

Technology ventures are generally more capital intensive than other types of start-up ventures. In fact, a vast venture capital network has developed in the United States and around the world to provide the needed capital for many types of technology ventures. In 2007, U.S. venture capital firms invested over $29 billion in technology and other start-up ventures.[7] Exhibit 3.1 shows how venture capital investing in the United States has varied over the first seven years of the twenty-first century, hitting a peak of over $40 billion in 2001.

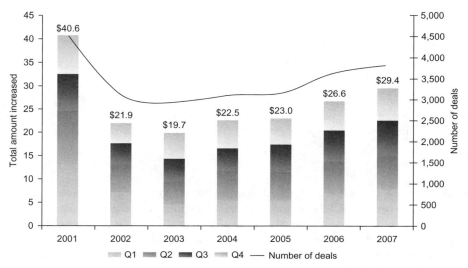

Investments maintain steady pace through 2007
US venture capital investments: 2001–2007

pricewaterhouseCoopers/National Venture Capital Association Money Tree™ Report
Based on data from Thomson Financial

EXHIBIT 3.1

The Money Tree™ Report.

Capital intensity varies dramatically across technology industries. Exhibit 3.2 shows the capital intensity of a variety of technology and other industries.[8] Here, you will see that pharmaceuticals, commodity chemicals, biotechnology, and semiconductors are all relatively more capital intensive than, for example, software ventures.

Capital intensity is not simply a measure of how much capital is required to start a venture in a particular industry; it is also potentially a **barrier to entry** in that industry. The greater the capital intensity, the harder it is to enter an industry. Consider the aspiring entrepreneur who is intent on launching his or her own semiconductor company. Typically, launching a semiconductor venture will require extensive capital investment. Most technology entrepreneurs do not have the capital to pursue their dream of running their own semiconductor company. As such, they are required to seek external capital to get started. The need for large amounts of external capital can be a daunting barrier to entry for technology entrepreneurs, especially those who are launching their first venture.

Seasoned entrepreneurs, such as Peter Wharton, the founder and CEO of the semiconductor start-up Adventiq, are more likely to find ways to finance a semiconductor start-up than first-time entrepreneurs. Wharton's company is in a new technology niche known as KVM. (The acronym stands for keyboard, video, and mouse.)

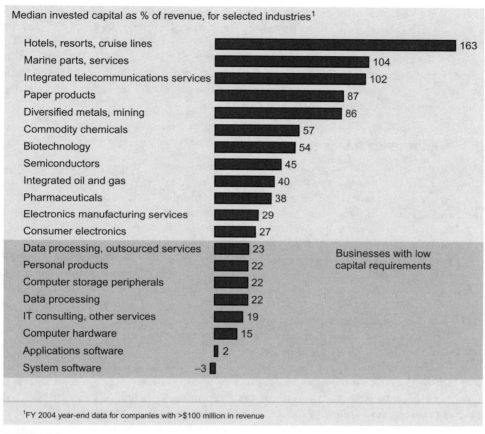

Median invested capital as % of revenue, for selected industries[1]

Hotels, resorts, cruise lines	163
Marine parts, services	104
Integrated telecommunications services	102
Paper products	87
Diversified metals, mining	86
Commodity chemicals	57
Biotechnology	54
Semiconductors	45
Integrated oil and gas	40
Pharmaceuticals	38
Electronics manufacturing services	29
Consumer electronics	27
Data processing, outsourced services	23
Personal products	22
Computer storage peripherals	22
Data processing	22
IT consulting, other services	19
Computer hardware	15
Applications software	2
System software	–3

Businesses with low capital requirements

[1]FY 2004 year-end data for companies with >$100 million in revenue

EXHIBIT 3.2

Capital Requirements Vary across Businesses.
Source: Thomson, McKinsey analysis.

Adventiq has developed a chip that sends KVM signals over IP networks. In addition to the technology, what makes Wharton's semiconductor undertaking unique is that it was able to deliver a working chip without taking any venture capital funding. Wharton's contacts in the technology community, including those who stand to gain from deployment of Adventiq's technology, stepped up to help fund the venture. Adventiq's chips have global customers in the United States, China, Taiwan, and Israel.[9]

It is important that technology entrepreneurs determine the capital intensity of starting a venture in their chosen industry. Not only will capital intensity provide an idea of the amount of capital required to become operational, it can also indicate how much **dilution** may occur in ownership over time. That is, most capital-intensive technology ventures must sell shares of stock to finance their growth. Every time a company sells shares of itself to investors, the ownership percentage of the current investors is diluted. Highly capital-intensive companies can require so many rounds of investing that the original owners have very little remaining equity. Peter Wharton preserved his equity in Adventiq by avoiding venture capital. He may not have been able to do so, however, if Adventiq was his first venture or if he lacked the industry contacts that provided the capital he needed.

3.1.3 Knowledge Intensity

Value creation in technology companies typically involves the application of some unique and proprietary knowledge. This knowledge often resides in the minds of the founders and leaders of the venture, but it also must be transferred to the products and services they create. Customers are interested in paying for products and services that help them solve problems, meet their aspirations, or demonstrate their own success (as in fashion items).

It is important for technology entrepreneurs to recognize that the knowledge they create is not valuable in and of itself. To become valuable, knowledge must be translated into a form that customers recognize, appreciate, and desire. For example, major pharmaceutical companies, such as Merck and Pfizer, spend hundreds of millions of dollars on basic research to discover new drugs. Undoubtedly, much of this investment leads to discoveries in the laboratory that are new to the world. This new knowledge is valuable to the small group of scientists who discovered it, but that doesn't mean that the discovery will lead to market value. In fact, the drug industry estimates that the cost of developing a single successful new drug is nearly $900 million, and can take more than 10 years to bring to market.[10] From these figures, it's easy to see that pharmaceutical companies must focus on research that leads to market value.

The **knowledge-intensive** nature of most technology companies requires that technology entrepreneurs hire highly intelligent people and create environments where they can thrive. In the global economy, the challenge to acquire and retain talented employees is greater than ever.[11] Technology venture leaders must recognize that highly talented knowledge workers will have what is called "divided loyalties."[12] Knowledge workers often are loyal primarily to their profession—the area of knowledge specialization that they have earned via formal and informal education. Technology entrepreneurs who do not recognize this may create a work environment that stifles the creative capacity of knowledge workers. The most likely outcome if that occurs is that knowledge workers will go somewhere else. Highly successful technology companies, such as Google, thrive on creating work environments that allow their knowledge workers to invent and create.[13]

The secondary loyalty of knowledge workers belongs to their company. If the technology venture is doing a good job of helping its knowledge workers build skills and capacities in their chosen profession, it is likely to create a satisfied workforce. It may seem paradoxical that the venture should help its knowledge workers build skills that, in fact, make them more mobile—i.e., more capable of finding employment elsewhere.

The resolution to the paradox is that most knowledge workers are more interested in creative technical challenges and personal growth than in maximizing personal income. Of course, knowledge workers want fair compensation for their work, but the technology venture that provides adequate compensation and appropriate technical challenges and personal growth can retain its vital pool of knowledge workers.

3.1.4 Accelerated Pace of Change

Key technologies that power the global economy have served to accelerate the pace of change by helping people from nearly anywhere in the world communicate and conduct transactions. For example, it's now possible to trade stocks on most global stock exchanges from the comfort of one's home. It's now also possible to chat and share documents with others around the world at zero cost via voice over IP (VOIP) technology. And it's now feasible for firms anywhere to set up web-based storefronts within minutes, including shopping carts and wish lists. The potential for competitors to emerge from nearly anywhere puts tremendous pressure on technology ventures constantly to engage in the innovation of core products, technologies, and services.

Technology ventures are particularly susceptible to the accelerated pace of change that characterizes the global economy. Not only is it possible for competitors to appear suddenly and challenge a venture's competitive position, but also technological innovation is so rapid that ventures that stand pat on their current technologies are at risk of becoming obsolete. For many technology ventures, speed to market is a critical factor in their potential to succeed. Getting to a market first, even though the offering may not be best-in-class, can establish customer loyalty and brand recognition that can be difficult for second movers to overcome. Of course, first mover products and/or services must meet and even exceed customer expectations.

The pace of change is difficult to measure. There is no single definition of "change" and no single way to measure its pace. Nonetheless, there are some indicators that point to an accelerating pace of change. For example, Exhibit 3.3 shows the rate of diffusion of key technologies in the United States. As you can see, the rate of diffusion of leading inventions to 30% of the population has accelerated dramatically between the advent of the automobile and the advent of the cellular phone.

Another indicator of the pace of change is the testimony of people who have to confront it on a daily basis. A McKinsey global survey of executives conducted in 2006 also pointed to an increasing pace of change.[14] The executives surveyed indicated that:

- Their companies are facing a more competitive environment than they were five years ago: 85% said "more" or "much more."

- Innovation and the free flow of information are the primary drivers of an accelerating pace of change in the global business environment.

- The second most important factor in the rate of change is "greater ease of obtaining information and developing knowledge."

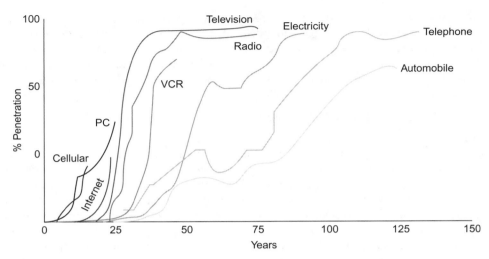

EXHIBIT 3.3

Innovation Is Accelerating.

(Measured by speed of market penetration.)

Technology ventures in the twenty-first century will need to pay heed to the accelerating pace of change and to the primary drivers of change in the global economy. Most analysts, pundits, and futurists don't anticipate that the pace of change will decelerate. Rather, they predict increased acceleration as emerging economies enter the global marketplace, and more people get an opportunity to participate as inventors, creators, and entrepreneurs.[15] Creating value amidst this pace of change requires constant innovation and an intense focus on customers.[16] A global survey of chief executive officers (CEOs) conducted by IBM indicates that the CEOs recognize that change has accelerated. An excerpt from the 2008 survey is provided in Exhibit 3.4.

3.1.5 The Network Effect

Many of you have probably heard of "Moore's Law." Moore's Law is attributed to Intel co-founder Gordon Moore, who famously predicted that the number of transistors that could be placed on a computer chip would double every two years.[17] Exhibit 3.5 illustrates how this law has held throughout history.

Moore's Law is part of the lore of technology ventures. It has achieved legendary status in large part because of its uncanny predictive success.

Another law that applies to technology ventures is less well known, but equally powerful. It is known as "Metcalfe's Law" after Robert Metcalfe, inventor of Ethernet, who noted that the value of a telecommunications network is proportional to the square of the number of users. This law has come to be known as the **network effect**. In general terms, it means that as the number of individual users of

FASTER, BROADER, MORE UNCERTAIN CHANGE

So what's causing this growing gap? Constant change is certainly not new. But companies are struggling with its accelerating pace. Everything around them seems to be changing faster than they can. As one U.S. CEO told us, "We are successful, but slow."

CEOs are also wrestling with a broader set of challenges, which introduces even greater risk and uncertainty. In 2004, market factors, such as customer trends, market shifts and competitors' actions, dominated the CEO agenda. Other external factors—socioeconomic, geopolitical and environmental issues—were seen as less critical, and rarely made it to the CEO's desk.

But in 2008, CEOs are no longer focused on a narrow priority list. People skills are now just as much in focus as market factors, and environmental issues demand twice as much attention as they did in the past. Suddenly everything is important. And change can come from anywhere. CEOs find themselves—as one CEO from Canada put it—in a "white-water world."

EXHIBIT 3.4

Faster, Broader, More Uncertain Change.
Source: IBM Survey of Chief Executive Offices, 2008.

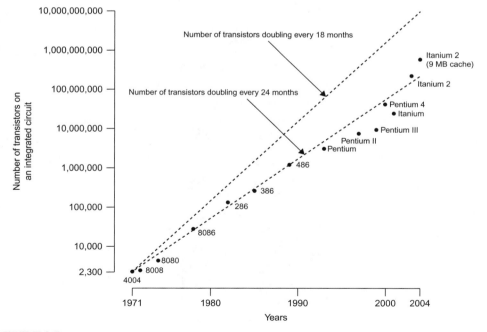

EXHIBIT 3.5

Moore's Law.

a new technology increases, its value accelerates exponentially. For example, imagine that you are the inventor of the world's first fax machine. The first photograph sent via fax occurred in 1924. Unfortunately, these early versions of the fax machine had limited market value because not many people had fax machines. It is fairly obvious that the more people who purchased and began to use fax machines, the greater was the value of a machine to an individual user. As more people acquired the technology, an individual could send faxes to more people, and the machine itself became more valuable to the user. It was not until the early 1980s that the fax machine became a regular part of the modern business office.[18] Exhibit 3.6 illustrates how the network effect exponentially multiplies connections and, thereby, the potential opportunity for value growth.

Many technologies are subject to the network effect. This is an important point to recognize for the technology entrepreneur. In founding Microsoft, Bill Gates and Paul Allen were driven by a vision of a computer on every desktop. With that guiding vision, they made strategic decisions that were not the most profitable, in the short term, at the time they were made. For example, when IBM was seeking a suitable operating system for its emerging personal computer (PC) line, Microsoft

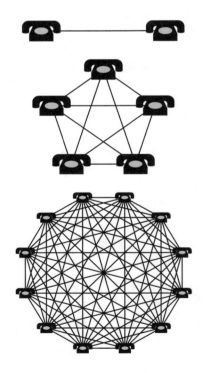

EXHIBIT 3.6

Metcalfe's Law.

had the option to sell its disk operating system (DOS) to IBM. However, Gates and Allen chose to license their technology instead, based on their vision that the network effect would lead to exponential sales and profit growth. Today, it is estimated that Microsoft provides the operating system for 90% of the world's computers and Microsoft is among the most valuable companies in the world.

The network effect has been studied extensively by scholars, analysts, and consultants. While there is little doubt that it exists, there are few quantitative tools to determine how it can be managed. Geoffrey Moore (no relation to Gordon Moore) noted that reaching the point where the network effect kicks in is similar to "crossing a chasm."[19] Moore noted that markets for new technologies often have a group of **early adopters**. These are individuals who are eager to try new technologies and who are not afraid of encountering potential bugs. According to Moore, technology entrepreneurs should pursue these early adopters to gain a critical beachhead into the chosen market. Then the chasm must be crossed.

Crossing the market chasm identified by Moore means developing a **value proposition** that appeals to a broader market than the risk-taking early adopters. It usually also means developing patience, persistence, and requisite capital to be able to stay in business long enough until the network effect is initiated. The phenomenon that occurs when a technology reaches a critical mass is called a **tipping point**, as coined by Malcolm Gladwell.[20] For a technology venture, the tipping point occurs when enough people own a new technology to the point that the value of the technology for an individual consumer rises rapidly. When that occurs, sales and marketing costs go down and profit margins go up. Tech Micro-case 3.1 explains how one software company used an early adopter to demonstrate the value proposition of their product.

TECH MICRO-CASE 3.1

Early Adopter Becomes Case Study for Software Company

wMobile is a software application developed by W-Systems Corporation of Ramsey, NJ. The application delivers full-featured and secure CRM to any mobile phone including BlackBerry, Palm, Windows Mobile, and iPhone. wMobile functionality includes full contact management, graphical team calendars, full email support with CRM integration, sales forecasting, graphical mapping functionality with door-to-door directions, and "List Closest" support to identify customers and prospects in proximity of the user.

In order to roll out its product, W-Systems enlisted an early adopter and tracked the success of the implementation. One early adopter was the A.L. Wilson Company. A.L. Wilson is a family-owned business that designs, manufactures, and markets textile stain removal products. Their products are sold in the United States and Canada. As a result of the wMobile deployment, A.L. Wilson greatly reduced the time sales managers spend on administrative tasks and reduced the time central office staff spends on recording and confirming orders.[21]

As we mentioned at the outset, there are no quantitative tools available to predict the size of market required to reach the tipping point where the network effect begins. However, technology entrepreneurs introducing disruptive technologies to a market must be prepared to persevere long enough until it does. This often requires sufficient capital to maintain sales and marketing programs that cost more than the revenue they produce in the early days. It also requires paying attention to customers and the feedback they are providing on the basic value proposition the company is articulating. Most technology companies will go through multiple iterations of their value proposition before they develop the messaging, pricing, and feature sets that appeal to the broader market.[16] Technology entrepreneurs must be careful to listen to the marketplace, and persevere until the network effect leads to greater demand for their products and services.

3.2 VALUE CREATION IN THE GLOBAL ECONOMY

In the global economy there are two types of value creation about which the technology entrepreneur should be concerned. The first type is **market value**. We define this as the value of the venture's products and/or services to customers. Value creation is fundamentally about serving a customer need. The second type of value creation is **enterprise value**. Enterprise value is defined as the value the technology venture as a whole has to investors or to other companies that may want to acquire it. The value of the venture's products and services to customers is part of the enterprise value, but not all. We discuss both types of value creation in the sections below.

3.2.1 Creating Market Value

Unlike business opportunities, such as restaurants, consulting services, and public relations, technology ventures are nearly always global. If they do not have global sales, they often face competition from global sources. Technology crosses international borders more easily than do personal services or culture-bound industries such as advertising or entertainment (although many ventures in these industries also are global in scope).

Value creation begins with evaluating a potential market, and then marketing and selling something of value to that market. It is very important for any type of venture to be aware of its customers and how its customers define value. Many entrepreneurs—and technology entrepreneurs in particular—often make the fatal mistake of thinking about value solely from their perspective. That is, many technology entrepreneurs are excited about the technology they have created and so think it is valuable. However, to customers, what constitutes a market "value" may be different from the technologist's definition. For example, the first hand-held computer was developed in the early 1990s by a company called GO Corporation. The inventors at GO believed they had created something of tremendous value—a computer that would fit in the palm of a person's hand. Judging by the ubiquity of today's diverse hand-held computing devices, one would guess that they were right. However, the first hand-held computer, for all the customer value it has created in markets today, was not valued in its day. After it burned through more than $75 million in venture funding, GO Corporation closed its doors for good in 1994. Had it been able to hang on another 4 to 5 years, it might have captured the wave of growth in the hand-held computer device marketplace. GO Corporation was simply ahead of its time. In other words, its definition of value was not shared by the marketplace.[22]

Technology ventures launched in the twenty-first century have an unprecedented opportunity to market and sell the value they create to global markets. In fact, the Internet and the World Wide Web have enabled anyone with a computer to build a web site that is viewable around the world. Of course, setting up a web site to announce your presence may be futile. Latest estimates indicate there are in excess of 100 million registered domain names, with more than 45 million active web sites in the world.[23] With all that noise, it is nearly impossible to be recognized simply by setting up a web site.

Technology start-ups can use the Internet to reach global markets, but most will also need to use the "old-fashioned" and well-established distribution and sales channels that have stood the test of time. For example, medical devices are often sold via distributorships. Technology entrepreneurs who invent a new medical device do not need to establish sales teams around the world to reach customers. Fortunately for them, there are existing distributor firms that are eager to add new products to their portfolios, and present them to customers they already are serving. Using distributors will cost the medical device company a commission on each sale, but that cost likely pales in comparison to what it would cost to set up a worldwide sales team.

Creating market value in the global economy means being flexible and nimble enough to be able to adjust products and prices, depending on a region's purchasing power and the unique tastes of customers. Cell phone makers, for example, have a wide range of products that are geared for different markets—mostly based on the varied purchasing power in those markets. While basic cell phone services are now nearly universal on our planet, people who use the basic service differ dramatically in their desire for advanced features. In Asian markets like Japan and Korea, cell phones are important to individual users, and usually include a wide range of advanced features. In parts of Africa and India, by way of contrast, some companies have been experimenting with different business models and phone types to reach markets with lower purchasing power.

3.2.2 Creating Enterprise Value

Enterprise value refers to the value of the enterprise as a whole. Enterprise value is easy to determine for companies that trade on the worldwide public stock markets. The value of an enterprise whose shares trade on public markets is simply the price per share times the number of outstanding shares. This is also referred to as the firm's **market capitalization** or **market cap**. In the case of private technology ventures, the efficiency and effectiveness of operations are the essence of its enterprise value. "Efficiency" refers to the cost-effectiveness of the venture's systems. "Effectiveness" refers to how well the systems produce value that customers demand.

Operating a technology venture consists of many things, but primarily it involves establishing repeatable systems that can deliver consistent value to customers. No matter if the technology venture is product-based, service-based, or some combination of both, growing a company requires setting up a variety of interlinked systems, including:

- *Production systems*: These are the systems that enable the production of a product or the delivery of a service to the customer. The systems that constitute production may include workflow, machinery, plant layout, and processes.

- *Customer relationship management (CRM) systems*: These systems track and record the venture's customers and each customer interaction. Sophisticated CRM systems keep track of every customer interaction, prompting employees to enter data about the interaction and its outcome. This helps the venture manage its customer relationships and prevents unwanted surprises or breakdowns in service.

- *Sales and marketing systems*: These systems are designed to help the venture sell to and communicate with customers and potential customers. They also provide the framework for articulating the venture's **value proposition**. The value proposition is a clear statement of the value the venture intends to provide to customers. Sales and marketing systems should send clear signals to customers and potential customers alike about the venture's value proposition. They should also make it easy for customers to conduct business with the venture.

- *Logistics systems*: Many technology ventures are product-oriented. Products are made from raw materials or assembled from components. Logistics systems concern the transportation of materials or components, including shipping, receiving, warehousing, tracking, and delivery. Logistics can be outsourced to third-party providers. That is, many ventures don't want to own and manage a fleet of trucks or warehouses. Instead, they contract with firms such as United Parcel Service (UPS) or FedEx who specialize in logistics.

- *Information technology systems*: Most technology ventures are highly knowledge intensive. Managing the flow of information and its conversion to knowledge in the workplace is increasingly challenging as the business grows and expands. Global technology ventures can use information systems to enable communications and collaboration with distributed workers and business partners.

- *Policy systems*: Most companies develop policies and procedure manuals to ensure that current and new employees know how to get things done.

Sometimes referred to as **standard operating procedures**, they describe the core processes of the venture. Policy manuals are required for firms that seek some types of certification, such as the **ISO Standard** certified by the **International Organization for Standards**. Some companies will require the technology venture to obtain ISO certification before they will purchase the venture's products or services.

■ *Human resource systems*: One of the more difficult systems for many technology entrepreneurs to put in place is the human resource system. This is because many technology entrepreneurs are highly self-motivated and cannot understand why an employee would not feel the same passion. However, employees in technology ventures often are not involved in venture ownership. They must become motivated through salary, working conditions, health benefits, and other incentives. Some of the top technology ventures have developed extensive and innovative human resource systems. Google, for example, is famous for its workplace benefits. These include 24-hour food service, assuring that employees are never hungry and can work long hours without worrying about going out to eat.

Each of these various systems is essential to an operating technology venture. Enterprise value is measured to a large extent by how efficiently these systems interact to create market value. One medical device entrepreneur said: "Run your company every day as if you intend to sell it tomorrow."[24] What this means is that putting operating systems in place early in the venture's life, with due attention to their efficiency and effectiveness, can lead to enhanced enterprise value at the time of exit.

3.2.3 Value Protection in Technology Ventures

Protecting the value created by the venture is a critically important element of successful technology ventures. Here again, we distinguish between a venture's market value and its enterprise value. Protecting market value is accomplished through a variety of means, most of which concern establishing and sustaining some type of competitive advantage.

The enterprise value of a venture, on the other hand, is reflected in the technologies that it has uniquely created and brought to market. This is generally referred to as the venture's **intellectual property (IP)**. IP refers to specific rights that inventors and creative individuals have over the works they create. These rights are usually granted via specific IP vehicles such as patents, copyrights, trademarks, and trade secrets. You will learn about each of these in much greater detail in Chapter 7. Here, we simply need to acknowledge the role that IP plays in the growth and vibrancy of technology ventures around the globe. Tech Tips 3.1 provides a few guidelines for protecting IP in your venture.

TECH TIPS 3.1

Tips for Protecting Your Venture's IP

Find your IP: Anything that sets your business apart can be protected, including your name, logo, inventions, designs, creative or artistic works.

Conduct an IP audit: Conducting an IP audit is essential in order to capitalize on your business. If you are thorough you will be able to make sure you are protecting all of your IP and not infringing anyone else's.

Get legal protection: The four most common ways to protect your IP is through trademarks, copyright, patents, and trade secrets.

Prevent infringement: You could lose your IP rights if someone else picks up your idea before you've had it protected. It is essential that all your business material including IP rights be kept safely and securely so you do not lose your claim to originality.

Respect others: It is important that you don't accidentally breach someone else's IP. If you do this, legal action could be taken out on you as well as orders to pay damages. To avoid this, carry out searches to ensure you are not breaching someone else's IP. In some cases you can get professional help with this from either a patent agent or trademark attorney.[25]

Another important measure of enterprise value in a technology venture stems from the venture's relative position within the industry that it competes in. The more rapid the growth of the industry and the more central the venture's position within that industry, the greater its enterprise value will be perceived. A discussion of principles of protecting market and enterprise value for the technology venture in the global economy follows.

Protecting market value

Market value is a function of customer demand. Protecting market value means that the firm protects its customer demand for its products and services, which is usually measured in terms of **market share**. However, protecting demand has become a tough challenge because customers in the global economy have an increasingly diverse array of choices.

Over the past quarter century, technology ventures were urged to protect the market value of their products and services by developing what is referred to as **sustainable competitive advantage**.[26] Sustainable competitive advantage is usually based on some sort of unique knowledge, property, or process that the company is able to protect as proprietary. However, as discussed earlier, the world shares knowledge at a breakneck speed, making it increasingly difficult for ventures to sustain a competitive edge based on standard IP mechanisms, such as patents and trade secrets. Today, competing ventures can find ways to avoid violating another venture's protected IP yet still offer value to its customers.

The process of obtaining, sustaining, and growing market share in the global economy requires that firms focus on continuous innovation. For technology ventures, it means developing **dynamic capabilities** that enable the venture to learn continuously and transfer that learning to value-added products and services in the marketplace.[27] Dynamic capabilities are centered on knowledge creation and knowledge absorption within the enterprise. That is to say, competitive advantage in the global economy depends more on the venture's ability to learn, adapt, innovate, and change, than on trying to protect IP. Companies around the world are focusing on continuous innovation as a means of staying one step ahead of competitors.

Protecting enterprise value

Despite the argument that competitive advantage in the marketplace stems more from continuous innovation than IP, IP remains important to enterprise value. Technology ventures are often based on unique technologies or processes that constitute the company's enterprise value. Frequently, the development of these technologies or processes has been possible only through massive investment of time and capital. Investors and potential acquirers of the enterprise, hence, will be very interested in the IP portfolio that the firm has created as a means of protecting enterprise value.

IP laws are well developed in the United States, and they are becoming increasingly robust in many other parts of the world. In the late twentieth century, China was a haven for IP law violations. Software piracy and other forms of IP law violations exacted a heavy toll on firms attempting to conduct business there. Estimates of lost business due to piracy in the software and entertainment industries alone range from $2.5 to $3.8 billion annually.[28] In recent years, the Chinese government has acted aggressively to curtail IP violations in an effort to become a more respected participant in the global economy.[29]

Establishing IP is a straightforward process, but it can be expensive. Costs associated with acquiring a **patent** in the United States, for example, include those associated with preparing and filing the patent application, various prosecution costs, and the patent issue fee. Once the patent is granted, there are additional patent maintenance fees. Filing fees for a U.S. patent range from $400 to $1,000. Patent applications covering technologies that are simple and easy to describe can cost as little as $2,000. More commonly, technology patents will cost in the range of $10,000–$15,000. It is best for the technology entrepreneur to use an experienced patent attorney since filing a readable and defensible patent application is imperative.

In addition to U.S. patent laws, the World Intellectual Property Organization has developed a **patent cooperation treaty** (PCT) that has been ratified by 139 countries around the world. These national governments have indicated a willingness to cooperate in the filing, searching, and examination of inventions and for protecting ownership rights to those inventions. The treaty allows for patent applications to be filed as international applications in any of the cooperating nations. The United States and Russia ratified the treaty in 1978, the year of its origin, while later entrants into the global economy, including China (1994) and India (1998), ratified the treaty in the 1990s.

Protecting the IP of technology ventures in the global economy is fraught with many challenges. While most nations abide by IP rights, treaties, and their obligations, the enforcement of IP laws remains inconsistent from country to country. For example, although technology entrepreneurs may apply for PCT patent protection and see their products distributed and sell well in China, they can anticipate a high likelihood that less expensive knock-offs will appear on the market. It's not that China lacks appropriate policies and regulations; rather, the country is simply so vast and growing so rapidly that the legal system does not have enough reach to enforce rules. Hence, technology entrepreneurs who experience a violation of IP policies must decide whether to pursue expensive and possibly fruitless legal action or to concentrate time and resources on the goal of gaining as much market share as possible. Depending on the circumstances, including the market size and the potential actually to win a legal proceeding and have it enforced, the entrepreneur may have differing strategic options. As such, it is not possible here to state that one option is better than another; however, with improving legal structures around the world, it is advisable for technology ventures to register and protect their IP claims as broadly as possible.

3.3 VALUE CAPTURE IN THE GLOBAL ECONOMY

As the title of this book indicates, successful technology ventures don't just create and protect value; they also *capture* the value they create. Value capture simply means that the venture has developed a **business model** that enables it to deliver value to a market that will pay fair price. Our definition of a technology venture requires that the venture deliver value, appropriate payment for that value, and do so repeatedly so as to constitute a going concern (see Chapter 1).

Value capture for a technology venture also involves capturing enterprise value. This refers to the *worth* of the technology venture to investors or other companies that may want to acquire it. For example, when Google purchased YouTube for $1.6 billion in 2006, the venture was less than 2 years old and had very little revenue. The value of YouTube as an enterprise was, in effect, the potential that it held to enable Google to make more money.

Next, we discuss approaches to capturing market value and enterprise value, respectively, in the global economy.

3.3.1 Capturing Market Value

Technology ventures can have many different types of customers. In general, however, markets for technology ventures are defined either as business-to-business (B2B) or as business-to-consumer (B2C). B2B businesses are those that provide products or services to other businesses. Many technology ventures serve only this market. Cisco, for example, provides high-end servers and other equipment to other businesses. Its business model requires that it market and sell to customers that need and can afford to pay for scalable, reliable, and expensive information technology systems.

Other technology companies use pure B2C business models. These ventures focus on consumer segments, and attempt to provide them with the technology products and/or services that enhance their lives and lifestyles. For example, video game maker THQ, Inc. publishes and distributes video games to a wide range of platforms. The challenge for a company like THQ is that it competes for consumer interests that evolve rapidly and are difficult to quantify. They are difficult to quantify because often consumers aren't sure about what they want. Thus, many technology companies must simply create the market opportunity rather than quantify and exploit it.

Capturing market value in the global economy is a challenge for technology companies, but the challenges for technology ventures can be less daunting than those for service ventures. Service companies often require subtle knowledge of local tastes and habits so that the services offered are tailored to local needs.

TECH MICRO-CASE 3.2

McDonald's Adapts to International Markets

McDonald's has made a fortune selling its brand of fast food to American consumers. McDonald's is also an international company that has restaurants in 118 countries around the world.[30] It has been successful in part because of its efficient and repeatable business systems that can be easily duplicated in nearly any part of the world. However, McDonald's learned early in its international expansion that it could not rely solely on the menu items that were top sellers in the United States. To thrive in international markets, the company had to expand its menu to include items that were unique to local tastes. To that end, McDonald's operates kosher restaurants in Jerusalem; it offers "veggie burgers" in nations such as India where beef is not eaten; and in some countries, such as Portugal, the company has opened McCafe's where it serves cakes, cookies, and sandwiches in addition to its regular menu.

Technology usually transfers more easily to international markets than services, but there are still country-specific business and cultural lessons to be learned. Many technology companies discover that setting up an office of some type in the countries where they want to do business is a helpful first step. For example, when Microsoft began selling software in China in 1992 it initially encountered many obstacles. The company's software was widely adopted in China, but it was also pirated openly and counterfeit copies were sold on the street for a fraction of the retail price. At first, Microsoft used aggressive legal tactics to try to stop the software pirates. Unfortunately, for every pirate that Microsoft shut down several more appeared. Finally, Microsoft executive Craig Mundie went to China to see what could be done. Mundie started by visiting China regularly, and he took a number of other Microsoft executives to the country for what he called "China Immersion Tours." The turning point came when Microsoft opened a research center in Beijing in 1998. The research center helped to improve Microsoft's image in China. Before long, the research center was considered to be the most desirable place in China for computer scientists to work. The improved image of the company reduced the likelihood that people would purchase illegal versions of its software. Now, Microsoft projects that China will one day become its largest market. The company even has a 5 year China plan that has been formulated to align with the 5 year growth plan the Chinese government uses to guide its policy decisions and actions.[31]

Capturing market value in the global economy normally begins at home. Technology ventures may find that their largest market opportunities lie with the massive population centers in China and India. However, most technology ventures are more likely to be successful internationally if they first discover how to be successful in their home markets. The only exception to that rule might be companies that are inhibited from developing local markets due to regulatory or other constraints. For example, medical device companies find that the approval process in the United States, which is governed by the Food and Drug Administration (FDA), is often slower than winning approval in other countries. In some cases, medical device companies are able to win the European CE Mark, the equivalent of FDA approval, sooner. In these circumstances, it may be advisable to begin marketing and selling in international markets prior to capturing market share at home.

3.3.2 Capturing Enterprise Value

Enterprise value can be captured in two primary ways: (1) a public offering of stock or (2) acquisition by a larger company. A public stock offering, usually referred to as an **initial public offering** (IPO), is where a venture places its stock for sale on one or more of the public stock markets around the world.

The United States has several public markets, including the New York Stock Exchange (NYSE) and the National Association of Securities Dealers Automated Quotation (NASDAQ) exchange. The NYSE lists the largest industrial companies in America, including multinational enterprises like GE, General Motors (GM), Procter & Gamble, and others. In all, nearly 3,000 companies from around the world list their shares for public sale on the NYSE. In 2007, the NYSE merged with Euronext, N.V., which was in itself a combination of the Amsterdam, Paris, and Brussels stock exchanges. The NYSE-Euronext exchange is by far the largest and most globally diverse stock exchange, listing more than 4,000 companies with an aggregate market capitalization in excess of $30 trillion.

NASDAQ has become known as the primary stock market for technology issues. The exchange lists more than 3,100 companies and handles more IPOs—over 1,400 since 2000—than any other exchange. Examples of companies that list on NASDAQ include eBay, JetBlue, Starbucks, Sirius Satellite, Google, and many others.

Internationally, the London Stock Exchange is an important exchange in Europe, listing both large companies on its main exchange and smaller companies through its Alternative Investment Market (AIM). The main exchange lists more than 2,800 companies, while the AIM exchange lists just over 1,100. AIM accounted for 65% of the IPOs in Western Europe. Other important exchanges around the world include the Japanese NIKKEI, the Shanghai Stock Exchange in China, the Bombay Stock Exchange in India, and the Frankfurt Stock Exchange in Germany. Each of these exchanges lists currently traded companies, and enables initial public offerings. Technology entrepreneurs may elect to go public in one or more of these exchanges, providing the shareholders of the company an opportunity to buy and sell their shares with other interested investors.

The other mechanism for capturing enterprise value is via acquisition from another, usually a larger company. Many technology entrepreneurs launch their ventures with this exit opportunity as a goal. In fact, many will have specific acquirers in mind as they build their technology, pursue markets, and gain international market share.

In the global economy, companies acquire other companies for a wide variety of reasons. Sometimes one company will acquire another to eliminate the threat of competition. Other times, a company will acquire another to gain access to its technologies, revenue, people, customer lists, or other unique feature. Whatever the reason, acquisition of one company by another is common in the global economy. In 2007, for example, the environmental blog Treehugger.com was acquired by the Discovery Network for $10 million. Why? The Discovery Network probably was interested in the unique visitors that the blog generates, and the advertising revenue that could be generated by these visitors.

Google is a company that has quite the track record in acquiring technology ventures. In fact, a common joke among the venture capital community is that every business plan they review has to "be acquired by Google" as its primary exit strategy. Google is active in acquisitions because it has cash on hand to execute transactions, and because it remains exceedingly aggressive about growth. Exhibit 3.7 shows Google's acquisition track record over the past few years.

Many technology entrepreneurs build their company from the beginning with an acquisition target in mind. While this is possible in some cases, in most cases the exit strategy will evolve based on changing markets and competitive conditions. Technology entrepreneurs should build their ventures every day with the goal in mind, someday, of participating in an exit that will convert equity into cash. The precise manner in which that takes place, whether through a public offering or acquisition, will depend to a large extent on the current state of the global economy.

3.4 GLOBAL MARKET ENTRY STRATEGIES

A technology venture that decides to enter an international market must select an approach. There are several alternative modes of entry to foreign markets, some requiring relatively low levels of commitment and others requiring much higher levels. Approaches to international business include exporting, licensing, joint ventures, strategic alliances, trading companies, countertrading, and direct ownership.

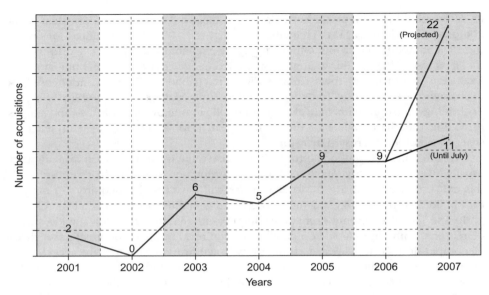

EXHIBIT 3.7

Google Acquisition Volume.

Source: Peter Cashmere. Google acquisitions by year, 2001–2007. Mashable, July 3, 2007. http://mashable.com/2007/07/03/google-acquisitions/.

3.4.1 Exporting

The simplest way to enter international business is through exporting—selling domestic goods to a foreign country. It requires the lowest level of resources and commitment. The U.S. government has created Export Assistance Centers in 17 U.S. cities to assist in conducting export business. More than half of the U.S. firms involved in global trade do so through exporting. In many cases, a firm can locate an exporting firm that can provide assistance in selling products to foreign countries, thereby making significant up-front investments.[32]

The United States is one of the world's largest exporters ($1.8 trillion in 2008) and importers ($2.5 trillion) in the world. Products (e.g., manufacturing, agriculture, computer chips) and services (e.g., transportation, consulting, license fees) have increased each year, especially since the 1970s, to a point where it is difficult to determine in which country products and services originated.

3.4.2 Licensing

In a licensing agreement, one firm (the licensor) agrees to allow another firm (the licensee) to sell the licensor's product and use its brand name. In return, the licensee pays the licensor a commission or royalty. For example, a beverage company such as

PepsiCo might enter into a licensing agreement with a firm in Taiwan. The Taiwanese firm would have the right to sell Pepsi products in Taiwan and would pay PepsiCo a specified percentage of the income from sales of Pepsi products.

Licensing offers advantages for both licensor and licensee. The licensor can become involved in international trade with little financial risk. The licensee gains products and technology that may otherwise be too costly to produce. For the licensor, the agreement usually results in a payoff of only about 5% of sales. Some American executives and managers believe that licensing agreements merely give away trade secrets for a meager 5% of sales; after the agreement expires (usually in less than 10 years), the licensee may continue to market the product without paying the licensor.

3.4.3 Joint Ventures

Firms may also conduct international business through a joint venture, in this case a partnership between a domestic firm and a firm in a foreign country. Because of government restrictions on foreign ownership of corporations, joint ventures are often the only way a firm can purchase facilities in another country. For instance, after years of negotiations, McDonald's of Canada opened a restaurant in the Soviet Union. The Moscow City Council's Food Service will own 51% of the enterprise, a requirement under Soviet joint venture laws.

In 2001, Walt Disney and its Japanese strategic partner, Oriental Land, opened the gates to DisneySea, a $2.8 billion marine-themed attraction built next door to Tokyo Disneyland, which opened in 1983. Disney and Oriental have joined together to attract customers and receipts. The partners are hoping that a combined 25 million visitors a year will visit both Japanese parks (about 17 million visit Tokyo Disneyland annually). The DisneySea park is estimated to generate $15 billion a year in related business in the area malls, gasoline stations, and support services.[33]

One major drawback to international joint ventures is that organizations may lose control of their operations. For example, because India does not allow foreign companies to own industries, Coca-Cola once entered into a joint venture with the Indian government. Despite the huge soft drink market in India, Coca-Cola pulled out of the agreement rather than risk giving up majority control and its secret formula.

3.4.4 Strategic Alliances

A recent approach to entering foreign markets is called a "strategic alliance," which occurs when two firms combine their resources in a partnership that goes beyond the limits of a joint venture. Strategic alliances have been growing in the highly competitive global marketplace at an annual rate of 20% since 1895.[34] IBM alone has established more than 400 strategic alliances with the U.S. and foreign firms. Trust is the major requirement for an effective partnership. If a firm can't trust its prospective partners, it shouldn't enter into a strategic alliance. Trust generally evolves over time, so firms must give strategic alliances adequate time to prosper. Ford and

Mazda formed a strategic alliance nearly 15 years ago to cooperate on new vehicles and exchange expertise. Ford is now the best-selling foreign auto in Japan; building trust over the years was a major factor in its success.[35]

3.4.5 Trading Companies

Another approach to international business is to use or form a trading company to provide a link between buyers and sellers in different countries. A trading company buys products in one country and sells them in another without being involved in manufacturing. Trading companies take title to products and move them from one country to another. Trading companies can simplify entrance into foreign markets because they are usually favored by the foreign governments. Many major corporations such as GE and Sears, Roebuck have developed trading companies.

Some countries—Brazil, for one—give trading companies tax advantages. In the United States, the 1982 Export Trading Company Act encourages the efficient operation of trading companies, helps finance international trade, and provides limited protection from antitrust laws when conducting export activities. After the act was passed, many major companies (such as GE, Kmart, and Sears, Roebuck) developed their own export trading companies.

3.4.6 Countertrading

Countertrading involves complex bartering agreements between two or more countries. **Bartering** refers to the exchange of merchandise between countries. Countertrading allows a nation with limited cash to participate in international trade. The country wishing to trade requires the exporting country to purchase products from it before allowing its products to be sold there. Agreements like this account for an estimated 20% of the U.S. exports, or $210 billion; the use of countertrade agreements is expected to grow. Countertrading provides an established trading vehicle for companies such as GE, Caterpillar, Pepsi, Boeing, Sears, and GM. For example, GM's trading section employs 25 people and does about $200 million annually in trade. Boeing has sold billions of dollars worth of 747 aircraft to Saudi Arabia in exchange for oil.

Countertrading has several drawbacks. First, it's often difficult to determine the true value of goods offered in a countertrade agreement. Second, it may be difficult to dispose of bartered goods after they're accepted. These problems can be reduced or eliminated through market analysis and negotiation. Companies have been developed to assist firms in handling countertrade agreements. Despite these drawbacks, companies that choose not to countertrade may miss significant opportunities.

3.4.7 Direct Ownership

A much more involved approach to international business is direct ownership—purchasing one or more business operations in a foreign country. Direct ownership requires a large investment in production facilities, research, personnel, and marketing activities. Many large companies, such as Ford, Polaroid, and 3M, own facilities outside the United States. Some well-known firms operating in the United States are actually subsidiaries owned by foreign firms. Magnavox, Pillsbury, Saks Fifth Avenue, and Baskin-Robbins are wholly owned subsidiaries of foreign multinational companies. Nonprofit organizations (such as the Red Cross) and the U.S. Army also own foreign subsidiaries or divisions.

Firms invest in foreign subsidiaries for a number of reasons. Direct ownership can reduce manufacturing costs because of lower labor and operating costs. Direct ownership also enables a firm to avoid paying tariffs and other costs associated with exporting. Additionally, by paying taxes in the host country and providing employment for local residents, a foreign country can build good relations with the host government. The greatest danger of direct ownership is that a firm may lose a sizable investment because of market failure or nationalization of its interest by a foreign government. When problems arise in a foreign country, it's often very difficult and expensive to move operations out of the country.

SUMMARY

This chapter introduced a number of factors affecting technology ventures in the twenty-first century. The global economy has changed some of the competitive rules, especially as they pertain to IP, factors of production, distribution and marketing, and the pace of change. A number of alternative techniques were also introduced that technology entrepreneurs can use to create, protect, and capture both market and enterprise value in the global economy. Finally, this chapter presented various modes of entry that are used to globalize a technology venture.

Each technology venture is different, and will have unique opportunities and obstacles to pursue and overcome. As you make your way through the remaining sections of this book, you should refer back often to this opening section and its fundamental lessons on business and economics. Technology ventures, by their very nature, almost always compete on a global stage. Technologies cross borders more readily than do personal services. As such, a technology venture has opportunities to market its products and services to global customers. On the flip side, technology ventures also have the challenge of facing competitors from nearly any place on the globe.

The journey to technology entrepreneurship success begins with an idea and culminates with a successful exit. While we can know with certainty these beginning and end points, the many possible variations to successful venture growth and development make it difficult to offer simple principles. Nonetheless, there has been enough research into technology entrepreneurship over the past quarter century to provide some guidelines. In the chapters that follow, you will be exposed to leading research, exemplar cases, and principles from expert entrepreneurs. Together, these various perspectives should provide you with the background necessary to decide which entrepreneurial path is right for you.

STUDY QUESTIONS

1. What is meant by the term "business process outsourcing"? Describe how a technology entrepreneur in the software development business might use this method to control operating costs.

2. One business analyst has asserted that there is a "fortune at the bottom of the pyramid." Explain what is meant by that term, and how it may lead to opportunities for technology entrepreneurs.

3. Explain why a "capital-intensive" industry can be difficult to enter for an entrepreneur. What might the entrepreneur do to overcome the entry barriers to a capital-intensive industry?

4. How can a technology company recruit and retain valued knowledge workers? Give an example of a company that excels in this regard. What does it do that makes it successful?

5. What are CEOs of global companies saying about the pace of change in the global economy? How can a technology venture deal with an accelerating pace of change?

6. How did Microsoft adjust its strategy in the Chinese market? What mistakes did it make in the early days in that market?

7. Explain what is meant by the term "sustainable competitive advantage." What should a technology venture do in the modern global economy to develop this?

8. Name three different modes of entry for a technology venture that is contemplating entering a foreign market. What factors should the entrepreneur consider when making this decision?

9. Under what operating system would a technology venture establish standard operating procedures? Why would a venture need to develop such procedures?

10. What is meant by the term "business model"?

EXERCISES

1. One of the things discussed in this chapter is the manner by which technology ventures protect the value of IP. A common technique is to develop patents for IP, which is key to the current and future value of the enterprise. For this exercise, students should familiarize themselves with the basics of conducting patent searches. A patent search is the first stage in determining whether a new invention or idea has or has not been invented by someone else.

Students should think of a new product concept, one that is not already readily available. For example, an inventor in Mesa, Arizona, named Joe Farinella had the idea of putting a timer on a bookmark to help his kids keep track of the amount of time they spend reading. He invented the "Mark-my-Time" digital bookmark. Students can look at Mr. Farinella's patent as a starting point and as a way to get acquainted with the major patent search engines. These are available at:
www.uspto.gov
www.delphion.com
Students should explore the patents that pertain to their own product ideas. Are there existing patents that overlap with the idea? How do they differ?

2. For this exercise, students should be divided into groups of three to investigate signs of the pace of change in the global economy. Each group should be assigned to dig into the primary and secondary literature, including popular business news articles, to find indications of the truth or falsity of the claim that the pace of change is accelerating. Each group should be prepared to discuss the information that was obtained and whether that information supports or disagrees with the claim of the accelerating pace of change.

KEY TERMS

Barrier to entry: The difficulties involved in a new venture entering an industry are referred to as barriers to entry.

Bartering: Bartering refers to the exchange of merchandise between countries.

Business model: A business model is simply a description of the way a venture makes money.

Business process outsourcing: This term refers to the process of shifting work (a business process) to a place in the world where it can be performed effectively and efficiently at the lowest cost.

Capital: This term is used to refer to the items to which a venture holds title, including its cash.

Capital intensity: The amount of capital required to launch a venture in an industry.

Dilution: When a company sells equity to investors the current investors are diluted in their ownership stakes proportionate to the sale of new equity.

Dynamic capabilities: This term refers to a venture's capacity to acquire, absorb, and use knowledge from the marketplace to create constant innovation.

Early adopters: Customers in a market who are most likely to buy a venture's products or services even before they have been thoroughly tested in a broader market setting.

Enterprise value: The value that the enterprise as a whole has to investors or other companies that may be interested in purchasing it.

Initial Public Offering (IPO): An IPO occurs when a venture that had previously been private offers its shares for sale on one of the major public stock exchanges around the world.

Intellectual property (IP): IP is the knowledge or processes that a venture claims and/or has registered to be proprietary.

International Organization of Standards (ISO): ISO is an international body that establishes, assesses, and certifies quality practices in organizations.

Knowledge intensity: This term refers to the amount of knowledge that is required to be successful in a particular industry.

Labor intensity: This term refers to the amount of labor that is required to be successful in a particular industry.

Land intensive: This term refers to the amount of land that is required to be successful in a particular industry.

Market capitalization: Market capitalization is a simple function of the number of outstanding shares of a publicly traded company times its price per share.

Market value: Market value is the term used to denote the value of a venture's products and services to a specific market.

Network effect: Also known as "Metcalfe's Law" the network effect asserts that the value of a technology increases exponentially with the number of instances of its use.

Patent: A patent is a vehicle for protecting IP.

Patent cooperation treaty: In 2008, 139 countries were signatories to the patent cooperation treaty, which specifies that patents granted in one nation are eligible for protection in all of the signatory nations.

Standard operating procedures: A set of rules that guide the manner in which a process or processes in a business should be executed.

Sustainable competitive advantage: The foundation of a venture's strategy is to develop and sustain an advantage over competitors.

Tipping point: A point where the network effect initiates and the value of a venture's technology experiences exponential growth.

Value proposition: The value of a venture's products and services to a market is articulated as the venture's value proposition.

WEB RESOURCES

The web sites below are intended as destinations for your further exploration of the concepts and topics discussed in this chapter:

1. http://landkomukraine.sharepointsite.net/home/default.aspx: This is the site for Landkom International, PLC, the Technology Ventures Insight featured at the beginning of this article.

2. http://mitworld.mit.edu/stream/264/: This web site is hosted by the Massachusetts Institute of Technology (MIT) and features a lecture by Thomas Friedman, author of *The World Is Flat*. Mr. Friedman's book was referenced in the opening paragraphs of this chapter. This lecture was presented to a class at MIT and is about 1 hour long.

3. http://en.wikipedia.org/wiki/Business_process_outsourcing_in_India: This Wikipedia site discusses the history of business process outsourcing in India.

4. http://news.morningstar.com/classroom2/course.asp?docId=2928&page=5&CN=com: This site provides a clear definition of the term "capital intensity."

5. http://www.intelproplaw.com/: This web site is the home of the Intellectual Property Law server. It provides a wealth of information and links to other IP-oriented web sites and information.

6. http://sprott.physics.wisc.edu/Pickover/pc/changing-pace.html: The pace of change in the world is difficult to quantify and articulate. The image on this web site is perhaps the most graphic example of how rapidly our world is changing. Join the discussion with people from around the world on how to interpret the image and its relevance to understanding the modern world.

7. http://www.globalinsight.com/Highlight/HighlightDetail10059.htm: This is the web site of a commercial venture focused on assisting technology companies in understanding global markets. In particular, students may want to download the white paper from this site. It contains some interesting perspectives on business information technology markets for the twenty-first century.

8. http://www.globalcompetitionpolicy.org/index.php?&action=900: This is the web site for the online magazine GCP, which stands for "Global Competition Policy." Articles in this online publication concern the rules of competition in various regions of the world.

ENDNOTES

[1] Friedman T. The world is flat: a brief history of the 21st century. *Picador* 2007;**July 24**.

[2] "The Evolution of BPO in India." St. Louis, MO: PricewaterhouseCoopers, 2005.

[3] Hawley C. Aerospace industry migrating to Mexico: companies follow automakers' lead to cut costs. *The Arizona Republic* 2008;**April 2**:A1, A12.

[4] Click R, Duening T. *Business process outsourcing: the competitive advantage*: John Wiley & Sons; 2004.

[5] Prahalad CK. *The fortune at the bottom of the pyramid: eradicating poverty through profits*: Wharton School Publishing; 2006.

[6] Ikerd J. "Corporate Agriculture and Family Farms." Presented at National Conference of Block and Bridle, National Collegiate Academic Organization, St. Louis, MO: January 20, 2001.

[7] PricewaterhouseCoopers, National Venture Capital Association. *2007 Venture capital investing hits six year high at $29.4 billion*. National MoneyTree Report; January 21, 2008.

[8] McKinsey and Company. Comparing performance when invested capital is low. *The McKinsey Quart* 2005;**November 15**.

[9] Business Weekly. Peter Wharton: CEO of Adventiq. *Business Weekly* 2007;**March 1**.

[10] Mudhar P. Pharma-biotech alliances: the answer to pharma's pipeline problems? *Clin Discovery* 2007;**March 1**.

[11] Ivancevich JM, Duening TN. *Managing Einsteins: leading high-tech workers in the Digital Age*. New York: McGraw-Hill; 2001.

[12] Cohen MD, March JG. *Leadership and ambiguity: The American college president*. New York: McGraw-Hill; 1974.

[13] "Google: What Makes it so Great?" *Fortune Magazine*, 100 Best Companies to Work for. http://money.cnn.com/magazines/fortune/bestcompanies/2008/snapshots/1.html. Accessed July 6, 2009.

[14] *An executive take on the top business trends: a McKinsey global survey*. McKinsey & Company technical report; April 2006.

[15] Malone MS. The next American frontier. *Wall Street Journal* 2008;**May 19**:A15.

[16] Carlson CR, Wilmot WW. *Innovation: the five disciplines for creating what customers want*. New York: Crown Business Publishing; 2006.

[17] Moore GE. Cramming more components onto integrated circuits. *Electron Mag* 1965;**April 19**.

[18] Taylor R. "Metcalf's Law and Network Marketing." *EzineArticles.com*. http://ezinarticles.com/?Metcalfes-Law-and-Network-Marketing&id=1244999. Accessed July 6, 2009.

[19] Moore G. *Crossing the chasm: marketing and selling disruptive products to mainstream customers*. New York: Collins Publishing; 2002.

[20] Gladwell M. *The tipping point: how little things can make a big difference*. Boston: Back Bay Books; 2002.

[21] WMobile affordable, feature-rich mobile CRM. *CRM Mag* 2008;**12**(9):8.

[22] Kaplan J. *Startup: a Silicon Valley adventure*: Replica Books; 2002.

[23] The size of the World Wide Web. *Pandia Search Engine News*; February 25, 2007.

[24] Colone B. Personal conversation with Thomas Duening. Mr. Colone launched, operated, and successfully sold his medical device company Endomed; February, 2009.

[25] *Guide to intellectual property (IP) and your business*. Retrieved from http://www.bytestart.co.uk/content/legal/35_2/intellectual-property-guide.shtml.

[26] Porter ME. *Competitive advantage: creating and sustaining superior performance*. New York: Free Press; 1998.

[27] Newey LR, Zahra SA. The evolving firm: how dynamic capabilities interact to enable entrepreneurship. *Br J Manage* 2009;**20**:S81–S100.

[28] Jade C. U.S. Pressures China on IP piracy. *Ars Technica* 2005;**October 27**.

[29] Gonzalez G. Top risk in China? Intellectual property theft. *Financial Week* 2007;**August 13**.

[30] McDonald's Corporation 2007 Annual Report. The report is available online at the following URL: http://www.mcdonalds.com/corp/invest/pub/2007_annual_report.html.

[31] Kirkpatrick D. How Microsoft conquered China. *Fortune* 2007;**July 17**.

[32] U.S. International Trade, Department of Commerce; June 21, 2001.

[33] Dawson C. Will Tokyo embrace another mouse? *Business Week* 2001;**September 21**:65.

[34] Rysbee E. *Developing strategic alliances*. New York: Crisp; 2000.

[35] Doz Y, Santos J, Williamson P. *From global to multinational: how companies win in the knowledge economy*. Boston, MA: Harvard Business School Press; 2001.

Legal Structure and Capital

Legal Structure and Equity Distribution

4

After studying this chapter, students should be able to:

- Examine a business as a distinct and separate entity from participants in the enterprise
- Understand the role of co-venturers in a business
- Define "limited liability"
- Differentiate between various forms of legal entity for structuring a business
- Describe the formalities of maintaining a limited liability status
- Characterize the nature of equity interest and its various forms
- Understand distribution of equity interest among founders of a venture
- Appreciate the role of employee stock options

effort) were required to take the project to the next stage. Two of the investors, Steele Elmi and Myron Barchas, opted not to risk the additional investment, so control passed to a subgroup of the partnership (two of the inventors, Melvin and Jeffrey Denholtz, and one of the investors, Bohm).

For whatever reason, the retiring partners in the Peroxydent Group were not bought out of their ownership interests. Instead, Elmi and Barchas retained their interest in the partnership, and the subgroup of the partnership formed a new entity, Evident Corporation, to which the partnership granted a license under the patent rights.

In 1997, Evident Corporation sought to enforce the patent rights, bringing a patent infringement suit against Church & Dwight (Arm & Hammer) and Colgate-Palmolive. The suit did not go well. The accused infringers brought a counterclaim asserting that the patent was unenforceable due to inequitable conduct on the part of the inventors. While the Peroxydent Group partnership was not involved in the initial infringement action, the counterclaim was brought against not only Evident Corporation, but also the Peroxydent Group partnership (as owner of the patent). The court held that "the inventors had been aware of three material references and withheld them from the patent and trademark office with intent to 'deceive,'" rendering the patent unenforceable. This led to the court finding that the case was exceptional, and Church & Dwight were awarded $1.3 million in attorney fees and costs.

To the consternation of the Peroxydent Group partners, the partnership was held to be liable along with Evident Corporation. Equity ownership in the Corporation did not make the shareholders' personal assets available to satisfy the judgment. However, as partners of the Peroxydent group, each of the individual partners, including investors Steele Elmi and Myron Barchas, who had previously opted not to take the risk of proceeding with the Evident Corporation, became personally liable for the entire $1.3 million to the extent not paid by Evident Corporation or the other partners. Elmi and Barchas may not have been personally involved in any of the activities that resulted in the liability, but as general partners of the Peroxydent Group, they were nonetheless each on the hook for the entire liability costs.

Source: Evident Corporation, Plaintiff/Counterclaim Defendant-Appellant, and Peroxydent Group, Counterclaim Defendant-Appellant, v. CHURCH & DWIGHT CO., INC., Defendant Counterclaimant-Appellee, and Colgate-Palmolive Co., Inc., Defendant Counterclaimant. United States Court of Appeals for the Federal Circuit. February 22, 2005. 399 F.3d 1310.

INTRODUCTION

Building a technology venture is a team sport, which means that at some time during the life of the typical venture, it will need to reach outside of itself for necessary resources. For many ventures, this occurs at inception. Consider that the genesis of a venture is often either an idea or recognition of an opportunity on the part of the individual (commonly referred to as the "founder" or, in technology ventures, the "inventor"). However, it is a rare individual who alone possesses the skills, expertise, and credibility necessary to develop a viable business around an idea or opportunity. At this stage, he or she must recruit other individuals or entities to supply the prerequisites needed to start a successful venture.

So, how are skills, expertise, and credibility transported into a venture? They can sometimes be brought in by forming a strategic alliance with another entity, or by hiring individuals with the needed skills, expertise, or necessary credibility. This, however, presupposes that the founder can actually *afford* to hire those employees. More typically, in start-ups, the individuals who can supply the needed skills, expertise, and/or credibility are brought into the venture by giving them the opportunity to share in the success of the venture. These individuals become **co-venturers** where they are provided an ownership interest or *equity* in the venture in lieu of all or part of the compensation that they might otherwise command in the open market. The relative ownership interests of the co-venturers, of course, depend upon their relative contributions to the venture.

The foundational choices made by an entrepreneur when setting up a venture can be critical to its ability to attract the interest of co-venturers or sources of capital. In addition, choosing the appropriate type of investor and strategically timing capital infusions are critical to obtaining the capital and/or resources necessary for the success of the venture. Finally, from the perspective of the founders, it is also critical that the cost of obtaining that capital is minimal in terms of interest paid and/or lost percentage of ownership.

In this chapter, we will explore some of those foundational issues: the various legal structures available for the business entity; tools for attracting the right talent and investors; and equity distribution in the embryonic venture. We begin by exploring some fundamental legal issues that technology venture founders must carefully consider when establishing a new venture.

4.1 OWNERSHIP AND LIABILITY ISSUES

A start-up venture is typically owned by one or more individuals or entities, each making a contribution to or an investment in the business in return for **equity interest** (percentage of ownership). The individuals or entities that initially form the venture are generally referred to as the **founders**. In general, the individuals or entities that own the venture are referred to as the **principals**, owners, or equity participants. Individuals or entities that make contributions to the business after it has been formed are referred to as **investors** or **lenders**.

In the United States, a business entity can take the form of any one of a number of different organic legal structures, including:

- Sole proprietorship
- General partnership
- Limited partnership

- Limited liability company (LLC)
- Corporation (C-corporation or S-corporation)

The choice of any one of these legal structures can affect the business in many ways, including the potential risk to participants, potential for business growth, availability of benefits, taxation of the business entity along with its individual principals, and the types of "exit" strategies available to principals. One of the primary considerations to make when selecting a particular form of business legal structure is the liability of the principals. Some legal structures limit the liability of the principals and some do not. We turn to that discussion next.

4.1.1 Limited versus Unlimited Liability

Certain forms of business entities, such as **sole proprietorships** and **general partnerships**, are considered to be the "alter ego" of the owners. These are further characterized as **unlimited liability companies** since each owner of the entity (sole proprietor or partner) can be held personally liable for the entire amount of the venture's debts, obligations, and liabilities. Here, liability is not limited to the *pro rata* share of ownership; rather each owner of an unlimited liability company is liable for all of the venture's obligations. All personal assets are at risk to satisfy the debt, obligation, or liability. In legal parlance, each partner is said to be *jointly* and *severally liable* for all obligations of the business.

Furthermore, the personal assets of each owner of an unlimited liability entity are at risk not only with respect to the debts, obligations, or liabilities that they create through their individual actions, but also those created by an employee or partner even without any personal involvement or knowledge on the owner's part. In other words, each owner is not only jointly and severally liable for all obligations of the business, but they are also **vicariously liable** for the acts of employees and partners in connection with the business.

This problem with unlimited liability entities has been duly recognized, and state governments in all 50 states have created other types of entities by statute. These include those that provide, from a liability and in some cases a tax perspective, a legal existence for ventures that is separate and apart from the individuals participating in the venture, and/or they limit the liability of the participants to their investment in the venture. These entities, referred to as **limited liability entities**, include **corporations**, **limited liability companies (LLCs)**, and **limited partnerships (LPs)**. Limited liability entities are also known as **statutory entities**. These legal forms limit the principals' liability exposure to the particular assets each contributed to the venture and shield their other, including personal, assets.

Why would the government want to shield the personal assets of an entity's owners from the debts and obligations of the entity? The reason for the existence of limited liability entities is to promote commerce. Certain types of ventures tend to require substantial infusions of capital well beyond the means of the typical entrepreneur. In order for those types of businesses to be created and sustained, the resources of investors must be brought to bear. The infusion of investor capital can take a number of different forms, including purchase of an equity position in the company. However, unless the venture takes the form of a limited liability entity, merely purchasing or otherwise acquiring an equity interest in the venture will make all of the investor's personal assets available to satisfy the debts and obligations of the venture.

Under unlimited liability conditions, it would be difficult to find investors, and if found they would be extremely expensive; the percentage of ownership per dollar invested would have to reflect the amount of risk assumed by virtue of the investment. This is particularly true with respect to wealthier investors since potentially exposing their entire wealth to satisfy the debts of the company would create a disproportionate risk as compared to less wealthy shareholders.

4.1.2 **The Extent of Limited Liability**

In general, the most important reason for establishing a venture as a limited liability entity is to protect the assets of principals and managers from the consequences of any financial or legal misfortune of the business. So long as these entities are properly formed, are compliant with all necessary formalities, and individuals observe a general fiduciary duty of loyalty and care to the venture in conducting its affairs, they are shielded from personal liability to third parties arising from those business affairs.

Suppose an employee orders goods on behalf of a venture on credit, but the business is unable to make the payments. Under normal circumstances, the employee would have no personal liability for that obligation—the obligation to pay is that of the business, not the employee. If the business was an unlimited liability entity, however, the owners of the business *would* be personally liable for that debt. On the other hand, if the business was set up as a limited liability entity, the owners would not be personally liable for the debt.

It is important to note, however, that individuals do not escape liability for their own improper conduct merely because a business entity was also involved. This personal liability arises not from the individual's equity interest in the business, but rather from the activities of the specific individual. For example, if there had been wrongdoing on the part of the individual placing the order for the goods (such as if the employee knew that payment would not be made), there may well be personal liability based on the individual's culpability. In one case, personal liability was imposed on the president of a limited liability entity when she personally made misrepresentations in order to obtain payment for equipment.[1]

Limited liability status, often referred to as the **corporate veil**, is almost always disregarded by courts for criminal acts of the officers, shareholders, or directors of a company. Further, federal and state tax laws generally impose personal liability on those individuals responsible for filing sales and income tax returns for the company. Owners, directors, and/or officers of limited liability entities have been held personally responsible for unpaid payroll taxes even if they had no personal involvement in the day-to-day function of the company and had no check-signing authority.

And a **single-member LLC** structure provides no liability protection to the owner for any taxes that have been unpaid by the company, regardless of the involvement or lack thereof.[2]

There are also circumstances under which limited liability entity status will be ignored, and liability imposed on the business owners, even in the absence of personal culpability, merely by virtue of their equity interests. The imposition of liability on the owners of a limited liability entity is often referred to as **piercing the corporate veil**. The issue typically comes up in the context of a lawsuit against the business, where it does not have sufficient assets to satisfy the damage claims. When this occurs, the plaintiff will attempt to convince the court that the business structure should not be respected and that the principals of the company, who may have personal assets sufficient to satisfy a judgment, should be personally liable. Piercing of limited liability entities typically occurs only under specific circumstances, such as:

- *Defective creation of limited liability entity*: If the entity was not properly established, the entity will typically be considered a sole proprietorship or general partnership, and personal liability will attach to the owners. Some jurisdictions, however, will forgive technical deficiencies as long as there was a good faith attempt to create a limited liability entity. The formalities entailed in creating and maintaining the various forms of limited liability entities vary from state to state.

- *Failure to observe procedural formalities*: As will be discussed, certain legal structures require that specific procedural formalities be observed. If those required formalities were not observed, the structure of the entity may be disregarded. For example, if the owners did not treat the business as an entity separate and apart from themselves and the business is nothing more than an alter ego for the owners, limited liability status is likely to be pierced.

- *Improper intent or fraud*: There are boundaries placed upon the extent of the limited liability entity protection on the basis of the intent and purpose of the business. For example, if a corporation or LLC was set up only to shield its owners from liability arising from a fraudulent technology development start-up deal and the owners siphon out the entity assets such that the entity is unable to compensate the victims of the fraud, a court is likely to set aside the limited liability entity and allow the victims to recover from the personal assets of the owners.

- *Statutory basis*: Some statutes, most notably the federal securities laws, specifically impose personal liability on persons in control of a limited liability entity.[3]

Examples of activities that may lead to disregarding limited liability entity status include:

- Knowingly incurring company debt or obligations when the company is already insolvent

- Removing unreasonable amounts of funds from the company, endangering its financial stability

- Personal use of company funds or assets by the principal owners

- Comingling company assets with the personal assets of the principal owners

- Hiring nonfunctioning employees who do nothing, yet draw a salary

- Assumption of the principal owner's personal debt by the company

- In the case of corporations, a pattern of consistent nonpayment of dividends or payment of excessive dividends

- Engaging in activities with the intent to defraud creditors

- Fraudulent misrepresentations by owners or management

- Unreasonable loans to company officials and extension of unwarranted credit

- Improper corporate guarantees of loans or contracts benefiting an owner or manager

- Little or no corporate business separations from the activities of the dominant owner

There are also instances where personal liability for corporate debt is imposed by statute, such as in cases involving employers' tax withholding obligations, wage and retirement benefits, environmental liability, and violations of the federal securities laws. Personal liability may also arise because of contractual obligations, such as when a shareholder signs a personal guaranty or fails to adequately identify that he or she is signing on behalf of the company, and as opposed to personally.

With these considerations in mind, the technology entrepreneur is ready to make an informed choice about the legal form for the new venture. We turn to that discussion next.

4.2 CHOICE OF LEGAL STRUCTURE

Each form of business entity has distinct characteristics that come with advantages and disadvantages, depending upon specific circumstances. In order to determine the optimum form of entity for a particular business venture, here are some factors to take into consideration:

- *The potential risks and liabilities entailed in the venture*: Limited liability entities limit the participant's exposure to the particular assets contributed to the venture.

- *Participants*: Certain forms of organization have limits on the number of participants, and the nature of those participants.

- *Capital growth needs and strategy*: Certain forms of organization are more amenable than others to raising high levels of capital from multiple sources.

- *Management structure*: Certain forms of organization are more flexible than others.

- *Tax implications*: Tax laws vary dramatically between the various types of entity structures.

- *Regulatory burden*: Securities laws have a greater impact on some forms of entity than others.

- *Administrative burden, formalities, and expenses involved in establishing and maintaining the various business structures*: Some forms of entity require that documents be prepared and filed with a designated state agency, and/or via written agreements between the participants.

- *Survivability*: Some entities terminate upon certain events.

- *Privacy*: Certain business and financial information regarding certain entities are required to be made public.

- *Exit strategy and liquidity needs*: Certain entity structures provide more flexibility for the participants as to liquidity and/or withdrawal.

As mentioned earlier, legal form choices for a new venture run from sole proprietorships to full-blown corporations. Let us explore each form in detail, and weigh in on the various advantages and disadvantages of each.

4.2.1 Sole Proprietorship

The most basic business legal form is the **sole proprietorship**. A sole proprietorship is created by default any time an individual owns a business without going through the specific formalities required to create a **statutory entity**. A sole proprietorship can operate under the individual owner's name or under a fictitious name. Most jurisdictions require that fictitious names be registered. A sole proprietorship does not have separate legal status from the owner.

For liability, as well as tax purposes, a business operated as a sole proprietorship is the alter ego of the proprietor. A proprietor is personally responsible for the debts and obligations of the business, as well as for the actions of employees. If the sole proprietor sells someone an interest of less than 100% of the business, the business by default becomes a general partnership (unless specific steps are taken to create a statutory entity).

A sole proprietorship is owned and managed by one person (or, for tax purposes, a husband and wife). From the perspective of the Internal Revenue Service (IRS), a sole proprietor and his or her business are one tax entity. Business profits are reported and taxed on the owner's personal tax return and taxed at the personal income tax rate.

Setting up a sole proprietorship is easy and inexpensive since no legal formation documents need be filed with any governmental agency. Once a fictitious name statement is filed in jurisdictions that require one, and other requisite basic tax permits and business licenses are obtained, the enterprise is ready for business.

When a business is operated as a sole proprietorship, essentially all income generated through the business is self-employment income and subject to the self-employment tax.[4] The primary and significant downside of a sole proprietorship is that it is an unlimited liability entity—its owner is personally liable for all the business debts. Given that fact, why would anyone operate a business as a sole proprietorship? The unfortunate answer is that most of the time they are simply ignorant of the unlimited liability. However, someone might choose to operate as a sole proprietorship because of the relative ease of, and lack of expenses entailed in, establishing the business. Exhibit 4.1 highlights the advantages and disadvantages of the sole proprietorship legal form.

4.2.2 General Partnership

When the profits and losses of a business are shared among more than one individual (or other entities), unless the formalities for creating a statutory entity have been followed, the business is by default a **general partnership**. For liability purposes, a general partnership is considered the alter ego of each of the partners. Any general partner can bind the business to an agreement, and each partner is liable for the acts of the others. A general partnership is an unlimited liability entity. A general partner's liability is not limited to that partner's percentage of ownership. In more legal terms, each general partner is jointly and severally liable for the debts and obligations of the partnership. This means that each partner could be held personally liable for the entire amount of the partnership debt and obligations, irrespective of their proportionate ownership in the venture, and that all of their personal assets are at risk to satisfy the obligations of the partnership.

While not required by law, partners typically enter into a written **partnership agreement** to define the internal rules under which the partnership will operate and how profits and losses will be allocated. In the absence of a partnership agreement, each partner has an equal say in management, and profits and losses are typically divided equally among the partners. Equity interests in a partnership are commonly referred to in terms of percentages of ownership or "units" corresponding to a percentage of ownership. And, there is no certificate or instrument (other than perhaps the partnership agreement) representing ownership interests.

From a tax perspective, a partner cannot also be an employee of the partnership. This does not mean that the partner cannot receive the equivalent of a salary. However, as in the case of a sole proprietorship, all monies received are considered self-employment income for tax purposes and subject to the self-employment tax. Since creation of the LLC legal form, most new businesses that choose a partnership form of entity do so primarily out of ignorance of the LLC form, which is discussed in Section 4.3.5. Unless the partnership operates without a formal partnership

Sole proprietorship	
Equity holder liability	Unlimited (Equity holder's personal assets are at risk to satisfy all debts, obligations, or liabilities of the business.) Vicariously liable for actions by employees
Equity interest types and terminology	100% owner/sole owner/sole proprietor
Management	Sole proprietor
Organic documents	None (If a name other than that of the sole proprietor is used for the business, some states require a registration of the "fictitious name" under which the entity does business.)
Formation procedure	None
Maintenance formalities	None
Limits on eligible owners	None (Multiple owners create a partnership.)
Limits on allocations of profit and loss	None
Fiscal year limitations	Must correspond to the tax year of the sole proprietor

EXHIBIT 4.1

Advantages and Disadvantages of Sole Proprietorship.

Taxation	
Income	No tax to the sole proprietorship as a business enterprise for enterprise income. All items of income, gain, or loss pass through and are taxed to the sole proprietor
Losses	All items of income, gain, or loss pass through and are taxed to the sole proprietor
Appreciated assets	The sole proprietor is taxed upon the sale of appreciated assets. Generally, there is no tax on the distribution of appreciated assets from business enterprise to sole proprietor
Entity upon liquidation	No tax to the sole proprietorship as business enterprise *per se* upon the sale or distribution of assets
Owners upon liquidation	Gain realized upon the liquidating sale of appreciated assets by the sole proprietorship passes to the sole proprietor. No gain is recognized upon distribution except to the extent that the money distributed exceeds the sole proprietor's basis in the business
Self-employment taxes	Sole proprietor is subject to self-employment taxes

EXHIBIT 4.1 Continued

agreement, in most states there is little cost savings in setting up a partnership over setting up an LLC.[5] Exhibit 4.2 highlights the advantages and disadvantages of the general partnership legal form.

4.2.3 Limited Partnership

A **limited partnership** (LP) is a legal entity that is created under the laws of a particular jurisdiction. The limited partnership is defined by certain documents—specifically, a **Certificate of Limited Partnership** that is filed with a designated state agency to register the new business and a limited partnership agreement, which defines the internal rules under which the limited partnership operates.

The LP has one or more general partners and one or more **limited partners**. The general partners are responsible for managing the partnership. The limited partners are, essentially, passive investors. Limited partners can have certain very limited veto and/or approval rights over certain actions, but any substantial involvement with the management or operation of the partnership will convert their status to a general partner.

General partnership	
Equity holder liability	Unlimited (Equity holders' personal assets at risk to satisfy all debts, obligations, or liabilities of the business.) Vicariously liable for actions by employees
Equity interest terminology	Partner Varies according to agreement: percentage, partnership units, etc. Different classes of partnership interests possible
Management	Per partnership agreement (In the absence of a partnership agreement, each general partner has equal say in the management of the partnership.)
Organic documents	None required Written partnership agreement preferred If a name other than that of the partners is used for the business, some states require a registration of the "fictitious name" under which the entity does business
Formation procedure	Formed by agreement (oral, or, preferably, written) between two or more prospective partners or by representing the business as a partnership to the public or connection with doing business The receipt of a share of profits (other than payment on a debt, interest, wages, rent, etc.) from a business is evidence of being a partner of that business
Maintenance formalities	None
Limits on eligible owners	Must be at least two partners; otherwise, no limits or requirements on the number or nature (e.g., individual, partnership, LLC, corporation, or trust) of partners A "partnership" with a single equity owner is a sole proprietorship
Limits on allocations of profit and loss	None. Allocations of profit and loss are determined by the partnership agreement. In the absence of an agreement, profit and losses are allocated equally among the partners
Fiscal year limitations	Must use the tax year of partners having a majority interest in the partnership, or the tax year of all principal partners if there is no majority interest

EXHIBIT 4.2

Advantages and Disadvantages of General Partnership.

Taxation	
Income	No tax to the partnership as a business enterprise for enterprise income. All items of income, gain, or loss pass through (per partnership agreement) and are taxed to the partners
Losses	All items of income, gain, or loss pass through (per partnership agreement) and are taxed to the partners
Appreciated assets	Generally, there is no tax to the partnership as a business enterprise upon the distribution of appreciated assets. The partners (as individuals) are taxed on gains upon the sale of appreciated assets
Entity upon liquidation	There is no tax to the partnership as business enterprise *per se* upon the sale or distribution of assets
Owners upon liquidation	Gain realized upon the liquidating sale of appreciated assets by the partnership passes to the partners (as in the case where there is no liquidation). No gain is recognized upon distribution except to the extent that the money distributed exceeds the partner's basis in his or her partnership units. Distributed assets will distribute at basis. Liabilities incurred by the partnership increase a partner's basis in his or her partnership interest
Self-employment taxes	General partners are typically subject to self-employment taxes

EXHIBIT 4.2 Continued

The general partners of the LP are jointly and severally liable with respect to debts of the LP. The limited partners are liable only to the extent of their ownership interest in the LP. In other words, if the limited partnership has a liability, all of the general partners' assets are at risk. In contrast, other than the limited partners' investment in the business, the limited partners' other assets are isolated from liability.

To relieve the burden of unlimited liability, the general partners of an LP often organize as a separate limited liability entity. For example, many venture funds are managed by general partners who raise capital from limited partners. The general partners organize as a separate LLC, which manages the limited partnership. By employing a limited liability entity as general partner, only the assets of the LP are at risk.

A limited partnership is, for tax purposes, treated as an extension of the partners (general and limited) and profits and losses **pass through** to the shareholders according to the partnership agreement. These profits or losses are then taxed at the individual income tax rate. It is very important to note that partners are taxed according to their ownership interests *whether or not* profits actually are distributed. The same thing applies to losses. That is, a partner can record losses on his or her personal income tax that have been allocated by the partnership venture.

Equity interests in a limited partnership are typically referred to in terms of percentages of ownership or units corresponding to a percentage of ownership. However, there typically is not any sort of certificate or instrument representing ownership interests. Exhibit 4.3 highlights the advantages and disadvantages of the limited partnership legal form.

4.2.4 Corporation

A **corporation** is a legal entity that can be created under the laws of a particular jurisdiction (typically under state law in the United States). The word "corporation" is derived from the Latin *corparæ*, which means to make corporal, or physically embody. In effect, the corporation embodies a business entity separate and apart from its owners.

The corporate entity is defined by certain documents, including **Articles of Incorporation**, which is also referred to as a "Charter" in some jurisdictions. These documents are filed with the designated state agency, usually through the Secretary of State office. A corporation does not have to incorporate in the state in which it is conducting business. In fact, many corporations elect to incorporate in Delaware or Nevada, two states that have governing laws that are favorable to the corporate legal form.

Structure of a corporation

A corporation is managed by a **board of directors** and various officers who are selected and serve at the behest of the board. Officers are responsible for the day-to-day operation of the venture, while the board of directors has fiduciary responsibility to the owners (shareholders) of the venture.

Equity interests in a corporation are managed and tracked via the distribution of stock or shares. There are typically written stock certificates representing the interests of the various owners. Not surprisingly, the owners of a corporation are referred to as shareholders, or stockholders. A corporation is a limited liability entity—shareholders are liable only to the extent of their contribution to the business.

A corporation is considered to be a separate entity from its owners, which is why a shareholder can be an employee of the corporation and can receive a reasonable salary for his or her services. That salary is subject to the Social Security and Medicare tax (FICA). Other monies paid out by the corporation to the shareholder, e.g., dividends or distributions, are not subject to either the self-employment tax or FICA.

Limited partnership	
Equity holder liability	General partner — Unlimited (see General Partnership) Limited partner — Potential liability limited to extent of investment
Equity interest terminology	General partner, limited partner Varies according to agreement: general partnership units, limited partnership units Different classes of partnership interests may exist
Management	Per partnership agreement In the absence of a partnership agreement, each general partner has an equal say in the management of the partnership Limited partners are precluded from taking an active role in management; if a partner takes an active role in management, the partner becomes a general partner with unlimited liability
Organic documents	Certificate of Limited Partnership filed with designated state agency Limited Partnership Agreement If a name other than that of the partners is used for the business, some states require a registration of the "fictitious name" under which the entity does business
Formation procedure	File a Certificate of Limited Partnership with a designated state agency Limited Partnership Agreement defines the internal rules under which the partnership operates
Maintenance formalities	None Limited partner may not have any substantial right to be actively involved with the management or operation of the partnership
Limits on eligible owners	Must be at least one general partner and one limited partner; otherwise, no limits or requirements on the number or nature (e.g., individual, partnership, LLC, corporation, or trust) of partners
Limits on allocations of profit and loss	None. Allocations of profit and loss are determined by the partnership agreement. In the absence of an agreement, profit and losses are allocated equally among the partners
Fiscal year limitations	Must use the tax year of partners having a majority interest in the partnership, or, if there is no majority interest, the tax year of all principal partners (defined as owning 5% or more), if no common year-end then that producing the least amount of change in income for partners

EXHIBIT 4.3

Advantages and Disadvantages of the Limited Partnership.

Taxation	
Income	There is no federal tax to the partnership as a business enterprise for enterprise income. All items of income, gain, or loss pass through and are taxed to the partners. May be special state taxes e.g., margin tax or B and O tax
Losses	All items of income, gain, or loss pass through and are taxed to the partners
Appreciated assets	Generally, no tax to the partnership as a business enterprise upon the distribution of appreciated assets. The partners (as individuals) are taxed on gains upon the sale of appreciated assets
Entity upon liquidation	No tax to the partnership as business enterprise *per se* upon the sale or distribution of assets
Owners upon liquidation	Gain realized upon the liquidating sale of appreciated assets by the partnership passes to the partners (as in the case where there is no liquidation). No gain is recognized upon distribution except to the extent that the money distributed exceeds the partner's basis in his or her partnership units. Distributed assets will distribute at basis. Liabilities incurred by the partnership increase a partner's basis in his or her partnership interest
Self-employment taxes	Limited partners are not subject to self-employment taxes except for guaranteed payments for services to the partnership. General partners may be subject to self-employment taxes

EXHIBIT 4.3 Continued

Tax laws and corporation types

In the United States, the tax laws have created two particular types of corporations: the subchapter C-corporation and the subchapter S-corporation. A C-corporation is taxed as an entity separate and apart from its shareholders.[6] A C-corporation's taxable income is subject to a graduated tax rate that increases with the amount of taxable income.[7] On the other hand, an S-corporation is, for tax purposes, treated as an extension of the shareholders and profits and losses pass through to the shareholders according to their ownership percentages. There are, however, certain limitations on the classes of stocks and number and types of shareholders permitted in an S-corporation (no more than 100 shareholders, no nonresident aliens, and only natural persons can be shareholders).

There are a number of noteworthy consequences that flow from the separate tax entity status of a C-corporation, and the choice of this form for a business has certain tax consequences:

- *Double taxation*: While salaries paid to corporate employees are deductible by the C-corporation, dividends paid to shareholders are not.[8] The dividends are thus taxed twice, first as income to the company (at the corporate tax rate) and then again as dividend income to the shareholder at the individual's tax rate.[9]

- *Retained earnings*: Retained earnings are corporate income (on which the corporate income tax is paid) that is retained by the company and not paid out to the shareholders. However, there must be good business reasons to retain earnings, such as research and development, expanding facilities, or acquiring another business entity. The IRS tends to view retained earnings as a potential attempt to avoid income tax on the shareholders. If the IRS deems the retained earnings excessive, they are subject to an accumulated earnings penalty tax. It is prudent for the board of directors and officers to document the reasons for retained earnings.[10]

In addition, if a C-corporation form is chosen for certain types of ventures, such as holding companies that receive royalties from technology licenses or consulting companies providing engineering services, the companies may be categorized as "personal holding companies" or "personal service corporations" and subjected to additional taxes and higher tax rates.[11]

On the other hand, U.S. tax laws also provide specific incentives for particular types of C-corporations that qualify as **Small Business Corporations**. A C-corporation is classified as a Small Business Corporation if the aggregate amount of money and other property received by the corporation for stock does not exceed $1,000,000. Under certain circumstances, individuals are given favorable treatment with respect to losses from investments in a Small Business Corporation. They are permitted to treat losses as **ordinary losses** that net out ordinary income as opposed to capital losses, which for the most part can be applied only against capital gains.[12]

C-corporations are typically the form of choice when the entity contemplates a large number of generally passive shareholders and/or a **public offering** of the shares of the business. Characteristics of a C-corporation and an S-corporation are summarized in Exhibits 4.4 and 4.5.

Maintaining corporate status

In order to maintain corporate status, certain practices and procedures must be observed. Since corporations are creatures of statute, unless the specific formalities required by the applicable statute are met (e.g., filing appropriate organizational papers with the designated state agency), the enterprise will be considered a sole proprietorship, or if more than one person has an equity interest, a general partnership. Filing the organizational papers and obtaining a charter, however, are merely the beginning. Failure to do so would result in the courts, tending to ignore the corporate structure and imposing liability on the individual owners and officers of the corporation.

If various corporate formalities are not consistently observed, the corporation can be easily disregarded and the individuals may be held personally liable. Within the court system, for instance, the corporate status is usually determined by whether or not the principals treat the corporation as a separate and distinct entity and hold it out to third parties as such. Similarly, a corporate structure will almost always be ignored and the corporate veil pierced when the management treats the corporation as its alter ego rather than as a separate entity, or the corporation is found to

S-Corporation	
Equity holder liability	Potential liability limited to extent of investment
Equity interest terminology	Shareholder, stockholder stock
Management	Board of Directors and Officers
Organic documents	Articles of Incorporation (also referred to as a "Charter" in some jurisdictions) "By-laws" — define the internal rules under which the corporation operates
Formation procedure	Articles of Incorporation are filed with the designated state agency to "give birth" to the new corporate entity
Maintenance formalities	Having and observing the by-laws, including meetings Minutes of shareholder and board of directors meetings and/or actions Separate uncommingled accounts and records Board resolutions for significant action/activities
Limits on eligible owners	May not have more than 100 shareholders Subject to certain exemptions, no non-individual shareholders (e.g., corporation, trust). No non-resident aliens Some states require more than one shareholder
Limits on allocations of profit and loss	Special allocations are not permitted. Dividends must be paid on stock ownership
Fiscal year limitations	Some state require S-corporations to use a calendar year, except under certain circumstances

EXHIBIT 4.4

Characteristics of the S-Corporation.

Taxation	
Income	There is no tax to the S-corporation except in two limited circumstances: (1) Recognized built-in gains, and (2) excess passive net income May be state taxes
Losses	Losses may be deducted by shareholders to the extent of their tax basis in their shares (not including any portion of the corporation's debt), subject to certain restrictions, including the basis, at-risk and passive loss limitations
Appreciated assets	Nontaxable at corporate level until gain is realized, distribution of assets to shareholders is done at current fair market value and can trigger a taxable event for the shareholder to the extent that fair market value exceeds basis
Entity upon liquidation	Generally nontaxable at corporate level but taxable at shareholder level through pass-through of corporate tax items
Owners upon liquidation	Shareholders taxed on gain Gain is recognized to the extent that fair market value of property distributed exceeds the shareholder's basis in his or her stock
Self-employment taxes	Self-employment tax does not apply to distributions paid to shareholders Shareholders pay self-employment tax on salary payments provided that compensation for their services is reasonable

EXHIBIT 4.4 Continued

be a "sham," and has been established only to facilitate fraud against third parties. In general, the following tend toward ensuring that corporate status is respected:

- Issuing of stocks
- Instituting and observing bylaws
- Filing annual reports with the state
- Holding annual shareholder and board of directors meetings
- Maintaining minutes of shareholder and board of directors meetings
- Promulgating formal written resolutions for significant actions
- Keeping up-to-date corporate records
- Separating and maintaining noncomingled funds and assets of the corporation among major shareholders

- Being adequately capitalized

- Avoiding dependency on the property or assets of a shareholder not technically owned or controlled by the corporation

- Avoiding personal use of corporate funds and assets by major shareholders except with full documentation

C-Corporation	
Equity holder liability	Potential liability limited to extent of investment
Equity interest terminology	Shareholder, stockholder stock. There may be different classes of stocks
Management	Board of Directors and Officers
Organic documents	Articles of Incorporation (also referred to as a "Charter" in some jurisdictions) "By-laws" define the internal rules under which the corporation operates
Formation procedure	Articles of Incorporation are filed with the designated state agency
Maintenance formalities	Having and observing the by-laws, including meetings Minutes of shareholder and board of directors meetings and/or actions Separate un-commingled accounts and records Board resolutions for significant action/activities
Limits on eligible owners	There are no restrictions on eligible owners
Limits on allocations of profit and loss	Special allocations are not permitted. Dividends must be paid on stock ownership
Fiscal year limitations	May use any fiscal year Personal Service Corporations are required to use a calendar year, under certain circumstances

EXHIBIT 4.5

Characteristics of the C-Corporation.

Taxation	
Income	Taxable income is, in general, subject to a graduated tax rate that increases with the amount of taxable income Also potentially subject to Retained Earnings Penalty Tax (Accumulated Earnings Tax), Personal Holding Company Tax, Personal Service Corporation Tax
Losses	Losses may not be passed through to or be deducted by shareholders
Appreciated assets	Potential double taxation. A tax is imposed at the corporate level upon the sale or distribution of appreciated assets. Additionally, there is a potential dividend or capital gains tax upon the distribution of sale proceeds to shareholders
Entity upon liquidation	Taxed on appreciation in assets upon the sale or distribution of assets. This may result in double taxation as these proceeds are distributed to shareholders in the form of liquidating dividends
Owners upon liquidation	Shareholders taxed on gain Gain is recognized to the extent that fair market value of property distributed exceeds the shareholder's basis in his or her stock
Self-employment taxes	Self-employment tax does not apply to dividends or distributions paid to shareholders Employees of the corporation are not made subject to self-employment tax by virtue of also being shareholders. The corporation pays the employer's portion of the withholding tax, which equals the self-employment tax

EXHIBIT 4.5 Continued

It is particularly important to maintain minutes of board and shareholder actions. Minutes can be the written record of meetings or can reflect the unanimous written actions of the directors or shareholders taken without a meeting. The secretary of the corporation typically prepares and signs the minutes of a meeting. To minimize the possibility of inaccuracies, those minutes are then approved by the board or the shareholders at the next meeting or in the next action. It is a good practice for the minutes to reflect that any requirements established by the by-laws with respect to meetings or actions were observed (e.g., proper notice was given or waived, a quorum was present, and so on). The minutes also typically identify who was present and who was absent, and note any abstentions or dissents on a vote. Any formal resolutions adopted by the board should also be included in the minutes. Many corporations have their minutes reviewed by legal counsel before completing them.

4.2.5 Limited Liability Company

A limited liability company, like a corporation and limited partnership, is a form of business entity that can be created under the laws of a particular jurisdiction. LLCs combine corporate and noncorporate features. The LLC is owned by "members," and is run either by the members or by one or more "managers" (who may or may not be a member).

The LLC as a form of entity is a relatively recent innovation in the United States, although a similar entity, the Giesellschaft mit beschrankter Haftung (GmbH), has been available in Germany and various other European nations for some time. The first LLC act in the United States was enacted by the state of Wyoming in 1977.[13] However, it was not until the 1990s, after the federal IRS broadcast a revenue ruling to the effect that an LLC formed under the Wyoming LLC Act would be classified as a partnership for federal income tax purposes even though all of its members had limited liability, that most states embraced the concept and enacted LLC statutes.[14] In 1997, the LLC became more popular still when the IRS established the **Check-the-Box Regulations**, permitting the LLC to choose whether it will be treated for tax purposes as a sole proprietorship, a partnership, a C-corporation, or an S-corporation.[15]

LLC characteristics

An LLC is created by filing a document referred to as **articles of organization** with an appropriate state agency, and entering into an **operating agreement** between the members. Equity interests in an LLC are typically referred to in terms of percentages of ownership or units corresponding to a percentage of ownership. There generally is no certificate or instrument (such as a share of stock) representing ownership interests.

The LLC is a limited liability entity and provides liability protection to its members in the same way that a corporation provides liability protection to its shareholders; members are liable only to the extent of their contribution to the LLC. LLCs are not subject to many of the formalities associated with corporations. LLCs are not required to hold annual meetings or prepare annual reports, although it may nonetheless be beneficial to hold regular meetings of the members, or to provide in the LLC operating agreement that meetings be held before certain actions are taken.

As noted above, an LLC can elect to be treated for tax purposes as a sole proprietorship, a partnership, a C-corporation, or an S-corporation. If no election is made, the default taxation for a single-member LLC is a sole proprietorship and a partnership for a multimember LLC. If either C-corporation or S-corporation treatment is elected, a member can be an employee of the LLC and can receive a reasonable salary for his or her services. That salary is subject to the Social Security and Medicare tax (FICA) deductions. Other monies paid out by the LLC to the shareholder, for example distributions, are not subject to either the self-employment tax or FICA. On the other hand, if either a sole proprietorship or partnership treatment is elected, then, from a tax perspective, a member of the LLC cannot also be an employee of the LLC. Essentially, all monies received from the LLC would then be subject to the self-employment tax.

In most respects, the LLC compares favorably with the other forms of limited liability entity. However, certain states, notably California, recognizing the popularity of the LLC as a form of entity, have begun charging relatively high fees to do business as an LLC.[16] Exhibit 4.6 highlights the various advantages and disadvantages of the LLC legal form.

Limited liability company	
Equity holder liability	Potential liability limited to extent of investment
Equity interest terminology	Member Membership interests, percentage interest, units. Different classes of membership interests Manager
Management	Managed either by all members, or by specifically designated managers. Members who participate in management are not personally liable
Organic documents	Entity is created by filing a document referred to as "Articles of Organization" with an appropriate agency, and by entering into an "Operating Agreement" between the members
Formation procedure	Entity is created by filing a document referred to as "Articles of Organization" with an appropriate agency, and by entering into an "Operating Agreement" between the members
Maintenance formalities	None; although annual meetings advisable
Limits on eligible owners	No restrictions on eligible owners Some states require more than one member
Limits on allocations of profit and loss	None. Allocations of profit and loss are determined by the operating agreement. In the absence of an agreement, profit and losses are allocated equally among the members
Fiscal year limitations	Some states require limited liability companies to use a calendar year, except under certain circumstances

EXHIBIT 4.6

Advantages and Disadvantages of the Limited Liability Company.

Taxation	
Income	The LLC can choose whether it will be treated for tax purposes as a sole proprietorship, a partnership, a C-corporation, or an S-corporation
Losses	The LLC can choose whether it will be treated for tax purposes as a sole proprietorship, a partnership, a C-corporation, or an S-corporation
Appreciated assets	Generally, no tax to the LLC as a business enterprise upon the distribution of appreciated assets. The members (as individuals) are taxed on gains upon the sale of appreciated assets
Entity upon liquidation	No tax to the LLC as business enterprise *per se* upon the sale or distribution of assets
Owners upon liquidation	Gain realized upon the liquidating sale of appreciated assets by the LLC passes to the members. No gain is recognized upon distribution except to the extent that the money distributed exceeds the member's basis in his or her membership units. Liabilities incurred by the LLC also increase a member's basis in his or her membership interest
Self-employment taxes	Depends upon choice of tax treatment Managers may be subject to self-employment tax on their distributed portion of income whether or not actually distributed Some tax practitioners have taken the position that a manager-managed LLC electing partnership taxation will be more similar to a limited partnership in that the manager will be subject to self-employment tax and the members would not be

EXHIBIT 4.6 Continued

4.3 LIMITED LIABILITY ENTITIES—A COMPARISON

Establishing the corporate form for an entity is never a black-and-white decision for the entrepreneur. Instead, many shades of gray intercede into the decision, and there are often good arguments to be made for establishing one form over another, and vice versa. Fortunately, the entrepreneur is not locked into a corporate form. It is easy enough to change a venture's legal form as its fortunes change, or as the ambitions of the founders change.

Still, making a good decision at the beginning is important to overall venture success. In the sections that follow, we look at some of the factors that may enter into decision making about the appropriate legal form of a technology venture.

4.3.1 Expense

Sole proprietorships and general partnerships are potentially easier and less expensive to set up than the alternatives, but they are unlimited liability entities. As such, they are rarely an appropriate choice for any technology venture unless it is substantially risk free. One such example may be a single-person technology consultancy that operates principally out of the home. However, even that simple type of business may be better organized as an LLC or other limited liability entity to protect against potential lawsuits and other types of claims.

4.3.2 Shareholder Options

Of the limited liability types, C-corporations typically are the legal form of choice when the entity contemplates a large number of passive shareholders or it anticipates an eventual public offering of stock. A C-corporation can sell shares of itself through private or public stock offerings. This makes it easier to attract investment capital and to hire and retain key employees by issuing employee stock options. An S-corporation can also sell shares of stock in private offerings, but it is limited to common stock only. In public offerings, a C-corporation legal form is the form most often used.

4.3.3 Taxation

While C-corporations pay taxes at the corporate tax rate on retained earnings, the corporate tax regime can be a benefit to some businesses. For example, owners of a corporation are not forced to pay personal income taxes on **phantom income** (profits that are not distributed). And, depending upon the relative corporate and personal tax rates applicable, it is conceivable that a corporation and its owners may have a lower combined tax bill than the owners of a business with a pass-through structure that earns the same amount of profit. An S-corporation does offer a unique advantage with respect to control of employment taxes (such as Social Security and the like). A shareholder of an S-corporation can be an employee of the business and can receive a reasonable salary for his or her services. Self-employment tax is paid only on salary payments, and provided the salary reflects reasonable compensation for services, monies can be paid by the corporation to that shareholder. Self-employment tax does not apply to dividends paid to shareholders. At the same time, the S-corporation is not plagued with the double taxation, retained earnings, personal holding company, and personal service corporation issues as a C-corporation.

4.3.4 Profits and Losses

The LLC tends to compare favorably with the other forms of limited liability entities, particularly for small businesses and start-ups. **Double taxation** and the other taxation issues endemic to the C-corporation can be avoided, and in contrast to the C-corporation, the LLC provides pass-through of profits and losses to members. An

S-corporation provides pass-through of profits and losses, but it is limited to apportioning profits and losses in accordance with ownership percentages. The LLC is more flexible. With the appropriate provisions in the operating agreement, profits and losses can be allocated in a flexible manner. For example, to entice an investor, an LLC owner may agree to allocate a disproportionate share of any losses to the investor. This would allow the investor to offset gains on personal income in other investments he or she may have made. The LLC also has the advantage of not being subject to the limitations on numbers and types of owners that are imposed on S-corporations. In addition, LLCs are not subject to the same formalities with respect to meetings, minutes, and the like that are required to maintain a corporation.

4.3.5 Partnerships

In comparison to a limited partnership, an LLC allows its members to elect the same pass-through tax treatment, without imposing a limitation on the extent to which the participants can exert management control without losing the benefit of limited liability. In addition, since limited partnerships require a general partner, and the general partner is very often a separate limited liability entity, the LLC structure provides the same advantage without the complexity.

4.4 EQUITY AND EQUITY TYPES

Equity is simply another term for an ownership interest in a business. Equity financing should be distinguished from debt financing. Debt financing, taking out a loan or selling bonds to raise capital, involves paying interest to a lender for the use of the lender's money. The cost of debt financing is typically finite, and the lenders will not share in any appreciation in the value of the business. However, in the short term, repayment of the loan can place constraints on the start-up venture's cash flow; both principal and interest must be repaid, and repayment is required whether the company is successful or not. The requirement to make periodic payments to service debt can make it difficult for the start-up venture to manage its cash flow.

Equity financing entails selling part of the ownership of the company. Equity financing is more flexible than debt financing with respect to cash flow concerns. There are typically no short-term repayment requirements, and equity investors share the risk. However, equity investors also share in any appreciation in the value of the business. And, while profit interests can be separated from governance and control, any time equity investors are added, there is some loss of independence on the part of the preinvestment ownership team.

4.4.1 Equity and Stocks

The particular terminology used to denominate equity varies depending upon the particular legal structure of the entity. As noted earlier, equity interests in partnerships and LLCs are typically referred to in terms of ownership percentages or **units** corresponding to a percentage of ownership. When a corporation is created, its articles of incorporation authorize it to issue up to a designated total number of **authorized shares** of stock. The original number of authorized shares is largely arbitrary, but it should be selected carefully on the basis of the venture's financial plan and future fund-raising objectives. The number of authorized shares can be changed, but only through a vote by the venture's shareholders. From the authorized pool of shares, the venture then can issue shares to founders and investors. It is the **issued shares** that constitute the ownership structure of the venture. That is, a shareholder's percentage of ownership of the venture is calculated as the proportion of shares held compared to shares that have been issued. Most ventures will reserve a portion of the authorized shares in the company's treasury for later use. The shares held in reserve are called "unissued shares."

It is also important to note that not all stock is created equal. A venture's articles of incorporation can be crafted to permit it to issue different classes of shares, each having a specific designation and unique preferences, limitations, and/or rights. These share classes are usually associated with financing rounds, and are often designated as Class A, Class B, and so on, or as Series A, Series B, and so on. The articles of incorporation either specify the particulars of each class of shares or delegate the authority to the board of directors. In general, the preferences and relative rights of classes of shares relate to voting, payment of dividends, and distribution of corporate assets on dissolution. Limitations of a class of shares typically relate to restrictions on transferability. Shareholders of private ventures normally do not want to leave it up to shareholders to decide to whom they may transfer their shares. Transferability is usually governed by **buy-sell** clauses in the shareholder's agreement.

Below, we discuss two types of stocks, common and preferred, that are commonly used to represent ownership interests and rights in a corporation.

4.4.2 Common Stock

Common stock refers to the baseline of ownership in a corporation and a right to a portion of profits from that company. Many ventures do not specifically define common stock. Instead, they refer to classes of stocks having different characteristics without ever specifically categorizing any particular class or classes as common stock. The designation for common stock tends to vary from company to company, often depending upon whether or not a special class of stock is issued to the founders. If only one class of stock is issued, it will be common stock. As used in this text, the term "common stock" means a class of stock that does not have any preference over another class of stock. Common stock is normally the last in line when dividends are paid and in the event of liquidation. In the event of liquidation, creditors, bondholders, and preferred stockholders typically are paid first and common stockholders are entitled only to what is left.

Owners of common stock usually have voting rights to elect the members of the board of directors who oversee the management and operations of the company.

Different classes of stocks may have different voting rights, as will be discussed. Profits of the company may be paid out in the form of dividends to the common stockholders. However, dividends on common stock are not guaranteed and usually only occur when the company is profitable. The amount and timing of dividends are determined by the board of directors.

The owners of common stock get the benefits of increases in value as the company grows. On the other hand, the value of their ownership interest decreases if the company is unsuccessful. So, ownership in common stock may provide the greatest upside potential to investors, but it may also provide the greatest risk of loss if the business does not succeed.

4.4.3 Preferred Stock

Technically, any stock that has any sort of advantageous characteristic over other classes of stocks is referred to as **preferred stock**. Preferred stock typically has provisions over and above common stock that:

- Provide a specific dividend on a periodic basis (e.g., monthly, quarterly, semi-annually, or annually)

- Classify the stock as senior to other classes of stocks in the order of distribution (i.e., payments are made to the preferred stockholder before dividends are paid to the other classes or when there is a distribution upon liquidation)

- Indicate that preferred shareholders have limited or no voting rights

Preferred stock tends to be a more conservative investment than common stock; it is not as volatile. Investors purchase preferred stock for the dividends. Unlike a dividend on common stock that is paid out to shareholders essentially at the discretion of the board of directors, dividends on preferred shares often are a fixed amount regardless of the earnings of the company. In some cases, the right to those dividends can accrue on a periodic basis. While payment of dividends on preferred classes of stock are not guaranteed *per se*, they are guaranteed to be paid before payment is made to any junior class of stock.

Preferred stock can include a wide range of advantages, also referred to as "sweeteners." A few of these advantages are discussed below.

4.4.4 Preferred Stock Distributions

A class of preferred stock can be specified to entitle the holders to distributions—dividends—and specify the manner in which the distributions are to be calculated. For example, dividends for a class of stock may be cumulative or noncumulative.

A cumulative class of stock has a provisional right to a dividend payment that accrues periodically, and all accrued dividends must be paid to the holders of that class of stock before any dividends can be paid to any junior class of stock. For example, assume a company issues a preferred class of stock that has a fixed dividend yield to be paid quarterly. However, because of financial problems the company is forced to suspend its dividend payments in order to pay expenses. If the company gets through the trouble and starts paying out dividends again, it will have to pay all of the dividends that have accrued to the holders of the preferred class of stocks before it can pay any dividends to the junior classes.

On the other hand, if the class of shares is noncumulative, dividends do not accumulate, and the shareholder is not entitled to receive dividend payments for the periods in which no dividend is paid. They are, however, typically still entitled to payment of dividends before payment is made to any junior class of stock when the venture is once again able to pay dividends.

4.4.5 Convertibility

A class of preferred stocks can also be redeemable or convertible into cash, debt, common stock, or other property. The conversion can be specified that it be triggered at the option of the corporation, the shareholder, or another person, and/ or on the occurrence of a designated event (e.g., a venture capital round of investment). The conversion can be a designated amount, or in accordance with a designated formula, or by reference to some prespecified extrinsic data or events.

4.4.6 Participating Preferred Stock

Preferred stock that pays a fixed dividend regardless of the profitability of the company is referred to as nonparticipating preferred stock. However, preferred stock can also be made to participate in the profits of the corporation. For example, a participating preferred stock may entitle the holder to the right to participate in any surplus profits in addition to a fixed dividend, and after payment of agreed levels of dividends to holders of common stock has been made.

In some instances, the participation may relate to dividends. For example, the preferred stockholder may receive the greater of the fixed dividend or the dividends paid to a share of common stock (or, if convertible, to the number of shares of common stock into which the preferred stock can be converted).

The participation can also relate to the distribution upon liquidation of the company. In the event that the company is liquidated, holders of nonparticipating preferred stock typically receive predetermined preferential return (e.g., an amount equal to the initial investment) plus any accrued and unpaid dividends prior to any distributions to junior classes of stocks. However, if the company is being sold at a sufficiently high valuation, holders of common stock may well receive more per share than holders of the preferred stock. If the preferred stock was convertible, then holders of preferred stock would have the option to convert their shares into common stock, giving up their preference in exchange for the right to share *pro rata* in the total liquidation proceeds. On the other hand, if the preferred stock was participating, the preferred stockholders could in essence have their cake and eat it too. Participating preferred stock might not only receive the predetermined preferential return, but also participate on an "as converted to common stock" basis with the common stock in the distribution of the remaining assets.

4.4.7 Voting Rights

While at least one of the classes of shares must have voting rights (and, at least in combination with the rights of other classes of shares, voting rights unlimited as to matters defined in the articles of incorporation that require votes), a preferred stock class may be given special, conditional, or limited voting rights or, for that matter, no right to vote at all.

One vote per share, typically, tends to be the norm for common stock, but a corporation can issue different "classes" of common stocks, distinguished by the specific voting rights associated with each share. The voting rights associated with common stock can range from no voting rights at all to multiple votes per share.

4.4.8 Founder's Stock

Founder's stock is used to refer to stock issued to the founders of the corporation. There is, however, no universal definition of founder's stock in terms of its characteristics. If the corporation issues a single class of stocks, founder's stock is simply common stock issued to the founders. However, founder's stock can sometimes be a separate class of stock with its own specific characteristics. For example, the company founders may issue two classes of stocks: Class A stock, which has one vote per share, and Class B stock, which has 10 votes per share. The founders would offer the Class A stock to investors and issue the Class B stock to themselves, with the intent that the super-voting rights would ensure that they retain decision-making control of the company. In addition to the super-voting privilege, there have been a number of characteristics proposed for founder's stock, including:

- The right to convert shares into any other class of stock when that class of stock is offered to investors

- Liquidation preferences at a designated amount such as actual monies invested by the founders, a specified percentage of the founder's holdings, or the valuation of the company prior to a successive round of financing

- Making the founder's stock preferred participating

Often, when part of the founder's contribution to the venture is **sweat equity**, founder's stock is subject to **vesting** (see page 127) in accordance with a predetermined, usually time-based, schedule or formula. Actual ownership of stock does not transfer until the vesting condition is met.

Of course, there tends to be a tension between having a special class of founder's stock with protective characteristics and the ability to attract future investors for subsequent financing rounds. In many instances, when a special class of founder's stock is in place, investors will attempt to negotiate away the founder's preferences as a precondition to investment.

It is important that founders keep in mind the tax consequences of receiving stock in return for services. As a general proposition, such stock is considered income to the founder. Under the U.S. tax law, the founder has the option to recognize income (the difference between fair market value of and the price paid for the stock) either when the stock is issued or when the stock vests.[17] However, if the founder/employee elects to recognize the value of the stock as income when the stock is issued, an **83(b) election** must be filed with the IRS no later than 30 days after the stock is issued. Since the value of the stock is likely to increase over time, it is generally beneficial for a founder/employee to make an 83(b) election. The 83(b) election starts the one-year capital gain holding period and phrases ordinary income (or alternative minimum tax) recognition to the issue date.

4.5 EQUITY DISTRIBUTION IN THE NASCENT VENTURE

We have discussed the various legal structures that a business can take and different forms of equity. We now return to the interrelational dynamics of the creation and growth of the business. When one or more founders form a business, each makes a contribution to the business in return for some consideration. Initial contributions by principals are typically made in return for, at least in part, an equity interest. The particular form of that equity interest depends upon the particular legal structure selected for the venture. The question we now address is: How should the equity be distributed among those individuals or entities?

When the business is founded by more than one person (or entity), the perceived relative values of the contributions that each makes (or will make) to the venture are typically reflected by the relative percentages of ownership. The issue is simplified if each of the founders makes a strictly monetary contribution to the venture in return for his or her equity interest. In that case, the relative percentage of ownership is easily determined. Each principal would hold an ownership interest equal to the ratio of the amount of money contributed by the individual to the total money contributed.

To the extent that founders make nonmonetary contributions to the venture, the venture can provide separate compensation, other than equity, for such contributions. For example, one or more of the founders will often be involved in the management and operation of the business. In that case, the individual will provide those services as an employee or independent contractor and will be separately compensated. Most often, the venture pays a salary or fee for those services. However, compensation can also be in the form of trade, e.g., the services are exchanged for an equivalent fair market value of services or product from the business. Likewise, a founder can also provide resources or IP to the entity for compensation other than equity. In that case, the founder provides the IP or resources in the role of an independent entity (e.g., a lessor or licensor) and is compensated with something other than equity, such as a lease or royalty payment.

Despite these alternative compensation options for nonmonetary contributions, such contributions often are compensated with equity. When nonmonetary contributions must be figured into the equity distribution equation, the determination of relative ownership percentage is more complicated. For example, when nonmonetary contributions are made for equity in the business, the relative value of that contribution must be determined. In many instances, this value can be difficult to determine. Nonmonetary contributions can take any number of other forms:

- Intellectual property
- An intangible right or relationship, such as a contractual right (e.g., the right to purchase property or a purchase order from a potential customer)
- Sweat equity (providing time and effort without drawing a "salary" or at a reduced "salary"; services provided at market rates are not a contribution to the venture)
- Resources (facilities, distribution, R&D, management)
- Credibility (e.g., credit, reputation) or access to sources of capital (network)

When the business is founded by more than one entity, the perceived relative values of the contributions that each makes or will make to the venture are typically determinative of the relative percentages of ownership. There are standard techniques for valuation of an existing business as will be discussed in Chapter 5. However, in many instances, those standard valuation techniques are not applicable to nonmonetary contributions. In fact, quite often with start-up ventures, the issue of relative value of contributions is simply a matter of the negotiation skills of the contributors. Factors tending to influence the valuation of a nonmonetary contribution include:

- The relevance of the contribution to the venture. For example, there is often a premium for contributing the idea or premise around which the venture is built.
- The intrinsic or demonstrable merit of the contribution. For example, consider the negotiating position of an inventor with an unproven idea as compared to that of an inventor who can provide a prototype.
- Favorable assessments by independent third parties.
- Legal exclusivity or protections. For example, consider the negotiating position of an inventor with an unproven idea as compared to an inventor with a patent.
- Uniqueness of the contribution.
- Demonstrable expertise in the technology or processes that form the basis of the venture.
- Relevant experience and track record.
- Relevant contacts and network.
- Reputation and goodwill.
- Efforts already taken that increase the likelihood of success of the venture.

In practice, pure force of personality and skills in negotiations often tend to play a significant and sometimes definitive role in determining the relative equity positions of the founders.

4.5.1 Employee Stock Options

Employee stock option (ESO) plans are a flexible tool used by many companies to reward employees for performance and attract and retain a motivated staff.[18] In effect, options preserve cash for the business while giving employees a stake in the success of the company. The popularity of ESO plans has seen remarkably steady growth since the 1980s. ESO plans became particularly popular during the days of the dot-com boom and indeed, for a period of time, were an expected component of a compensation package for employees of high-technology companies. Today, ESO plans continue to be a relatively popular method of compensating employees and garnering loyalty for the employer. One survey estimates that over 11.2 million employees are participants in such plans.[19]

In essence, an ESO plan is a contract between employer and employee that gives the employee the right to buy a specific number of shares in the company at a specified price, referred to as the **strike price**, over a specified period of time, referred to as the **exercise period**.[20] When the employee exercises the option, he or she will get the benefit of the difference between the fair market value of the stock at the time of exercise and the strike price. In a typical ESO plan, there are specific provisions regarding the employee's eligibility to earn options under the plan and various restrictions on how the option is exercised and transferability of the stock after the option is exercised. Several common provisions and restrictions are explored below.

Vesting

In the context of an employee stock option, **vesting** is a process or condition that must take place before an option can be exercised by the employee. For example, an ESO plan may require uninterrupted employment for a specified period of time (or some other length of service) before the employee is "vested." The vesting condition is most frequently the passage of time, but it may also be other things, such as the occurrence of an event or a milestone in the growth of the company. ESO plans may incorporate simple or elaborate vesting schedules. A typical vesting schedule for employees of a technology start-up is incremental over 4 years, with the employee having no exercisable rights until he or she has remained with the company for at least 12 months, at which point the right to exercise the option on the first 25% of the shares vests, and approximately 2% vest each month thereafter. No further vesting occurs after an employee leaves the company.

Restrictive clauses

ESO plans typically include restrictions or limitations on the transferability of the stock after the option is exercised to protect the company's interest. For example, an employee may be precluded from making any transfer of stock for a period of time after exercising the option. Or an employee who is leaving the company may be required by the plan to sell his or her stock back to the company. Some plans may also require that the option or the shares be offered back to company on a right of first refusal basis whenever a stock sale is contemplated to an outside or unrelated third party.

Tax issues

In the United States, the Internal Revenue Code creates two basic categories of employee stock option plans: incentive stock options (ISOs), also referred to as qualified stock options, and nonqualified stock options. In order to qualify as an ISO, the plan must meet relatively stringent requirements of the Internal Revenue Code. For example, the strike price cannot be discounted from fair market value, options cannot be available to nonemployee directors, and the exercise period cannot exceed 10 years.[21] In view of the restrictions of the qualification rules, most employee stock option plans are nonqualified.

In general, exercise of a nonqualified employee stock option has tax consequences for the employee. The spread at the time the option is exercised is considered ordinary income to the employee. Qualified employee stock option plans, however, qualify for special tax treatment. Neither the grant nor exercise of an ISO is a taxable event for the employee. Instead, the tax is deferred until the employee sells the underlying stock. At that point, any gain that the employee has made on the stock sale is taxed. If the sale occurs at least 2 years after the option is granted and at least 1 year after the option is exercised, the income is taxed at favorable long-term capital gains rates. If these holding-period requirements are not met, the spread at exercise is taxed as ordinary income and only the subsequent appreciation of the underlying shares is taxed as capital gain.

Nonqualifying stock options can create anomalous tax consequences for the employee in the event that the employee does not immediately sell the stock after exercising the option and the value of the stock goes down. Exercise of the option is a taxable event—that is the employee is taxed on the spread even though the actual gain (or loss) is less than the fair market value at the time the option was exercised. On the other hand, exercise of a qualified option is considered a preference item for purposes of calculating the alternative minimum tax.

SUMMARY

This chapter has examined the various legal forms that can be chosen when founding a new technology venture. As was pointed out, there are essentially two primary legal form classifications: limited liability and unlimited liability. It seems unwise for any business owner today to choose a legal form that includes the disadvantage of unlimited liability. Yet the sole proprietorship is far and away the most common business form in the United States. Unfortunately, the primary reason it is so common is that people who operate such entities are simply unaware of their alternatives. Of course, a sole proprietorship exists any time a person enters into a business transaction, but has not registered as a statutory entity with a state agency. In other words, a sole proprietorship will be automatically said to exist unless the business owner registers as another type of legal entity.

One of the primary considerations when choosing a legal form is the capital management plan for the venture over time. Technology ventures that intend to grow through large amounts of investor capital might as well establish themselves as C-corporations from the beginning. Such entities enable issuing both common and preferred stock, and they have well-established legal histories that allow predictability in the event of lawsuits or other legal issues. Of course, there are disadvantages with this legal form as well, including double taxation and the relatively greater expense associated with setting up a C-corporation form over other forms.

Some technology ventures are best established as limited liability companies (LLCs) or as S-corporations in the early days to take advantage of the pass-through taxation status of these entity types. In addition, these types are very easy to set up, and they provide limited liability to all shareholders.

Allocating shares to founders is often a difficult process. This is especially difficult when consideration is given to nonmonetary contributions, such as IP, real property, and even the so-called sweat equity. Nonetheless, founder's shares should be allocated early in the venture's life to ward off potential disputes as the venture begins to grow. All new ventures should establish appropriate governing documents, including operating agreements, to ensure compliance with prevailing laws and good operating procedures. Doing so, along with diligence to a host of other governance matters, can help the venture maintain its corporate veil and allow it to operate without risking the personal assets of its principals.

STUDY QUESTIONS

1. What are the risks of operating a business as a sole proprietorship?

2. What are the advantages associated with operating a technology venture as a limited liability company?

3. Define the significance of making a business a "limited liability entity."

4. Who is responsible for the wrongful acts of a businesses employee? Under what circumstances would be employee be personally liable? Under what circumstances would the owners of the business be personally liable?

5. How does an entrepreneur determine which form of entity to adopt for his or her business?

6. Under what circumstances is it most beneficial for a company to self-fund or obtain debt financing to reach a particular milestone?

7. Describe what is meant by the term "Employee Stock Option plan." How can an entrepreneur use such a plan to build a new venture?

8. How can a technology entrepreneur use a vesting schedule to motivate employees?

9. What are the responsibilities of a corporation's board of directors? How can a technology entrepreneur ensure compliance with statutes governing the corporate entity?

10. Explain the difference between common and preferred stock. What are some potential features that can be added to preferred stock offerings?

EXERCISES

1. *IN-CLASS EXERCISE*: This in-class exercise is designed to give the students a sense of what it feels like to distribute equity and decide legal structure. Groups of five students should be identified at random and asked to determine the equity structure for a start-up software venture. They are to allocate equity among the five "founders" according to their ability to contribute value to the growing venture. After 15–20 minutes, each team should report out on how they allocated their equity among themselves.

2. *OUT-OF-CLASS EXERCISE*: For this exercise, students should attend a local meeting of entrepreneurs. Most large metropolitan areas have many such gatherings. They go by names like "Enterprise Network" or "Technology Council" or simply "Entrepreneur Network." Students should attend one or more such events and listen to what the speakers have to say about equity distribution and management. Also, students should identify themselves as such to attendees and state that they are studying equity distribution and legal structuring. Students should learn from the attendees some of the "real world" challenges of deciding a legal structure, distributing equity, and setting up employee stock option programs. Students should report back to the class on their findings from attending these meetings.

KEY TERMS

83(b) election: A special tax status election that pertains to an Employee Stock Option plan whereby taxation of the value of the underlying shares can be deferred to a future date.

Articles of Incorporation: These are the formal documents that govern the operations of a business entity that is organized as a corporation, also referred to as a charter.

Articles of Organization: These are the formal documents that govern the operations of a business entity that is organized as an LLC.

Authorized shares: A designated maximum total number of equity shares that a corporation is authorized to issue in its articles of incorporation.

Board of directors: All corporations must establish a board of directors, which is charged with fiduciary responsibility for the business entity.

Buy-sell clause: This is a clause within a business entity's formal agreement that limits the conditions under which ownership interest can be bought and sold.

Certificate of limited partnership: These are formal documents that must be filed with a state agency specifying the limited partners in a limited partnership business entity.

Check-the-box regulations: This is an IRS provision that allows a limited liability company to specify whether it wants to be taxed as a partnership or as a corporation.

Common stock: Baseline ownership interest in a corporation and a right to a portion of profits from that company.

Co-venturers: Individuals or entities that make contributions to the business after it has been formed.

Corporate veil: The corporate veil is the limited liability protection enjoyed by principals of a statutory business entity that operates in good faith.

Corporation: The legal form of business entity that provides limited liability status to principals and also allows issuance of stock certificates. Corporations can be organized as subchapter S-corporations or as subchapter C-corporations.

Double taxation: A disadvantage of the C-corporation legal form is that profits may be taxed once as corporate profits and a second time as individual income when profits are distributed to owners as dividends.

Employee stock option: A contract between employer and employee that gives the employee the right (but not the obligation) to buy a specific number of shares in the company at a specified price (the grant, strike, or exercise price), over a specified period of time (the exercise period).

Equity interest: An ownership interest in a business enterprise.

Exercise period: The time period during which an employee may exercise an option to buy shares under an employee stock option plan.

Founders: The individuals or entities that initially form a business.

Founder's stock: Stock issued to the organizers of a business.

General partnership: The form of (unlimited liability) business entity created when the profits and losses of a business enterprise are shared between more than one individual, unless the formalities for creating a statutory entity have been followed.

Investors: Individuals or entities that provide capital to a venture in exchange for an ownership interest.

Issued shares: Shares authorized to a venture that have actually been distributed to principals (founders and/or investors).

Lenders: Individuals or entities that provide capital to a business entity and expect repayment and interest over a specified period of time.

Limited liability: Forms of business entities that limit liability of the protected principal to the particular assets that the principal contributed to the venture and shielding the principal's other assets. These entities include, for example, Corporations, Limited Liability Companies (LLCs), and Limited Partnerships (LPs).

Limited Liability Company (LLC): A statutory entity formed by filing Articles of Organization, owned by members, and managed either by the members or by one or more manager (who may or may not be a member).

Limited partner: A participant in a limited partnership who has limited involvement in the management and control of the business and has liability limited to the extent of his or her contribution to the venture.

Limited Partnership (LP): A statutory entity having one or more general partners and one or more limited partners. The general partners are responsible for managing the partnership. The limited partners are essentially passive investors and have limited liability.

Operating agreement: An agreement between members of an LLC that governs roles and responsibilities.

Ordinary losses: Losses due to investment in a Small Business Corporation that can be used to offset other personal income, as opposed to being allowed only to offset capital gains.

Pass-through taxation: Taxation on business entities that allocates profits and losses to principals to be taxed at the individual income tax rate.

Partnership agreement: An agreement among the partners in a general partnership.

Phantom income: Income that must be recognized on the basis of participation in a pass-through taxation entity, but that is not actually distributed to the principals.

Pierce the corporate veil: Circumstances under which limited liability entity status will be ignored, and liability imposed on the business owners, even in the absence of personal culpability, merely by virtue of their equity interests.

Preferred stock: Stock in a class that has any sort of advantageous characteristic over other classes of stocks, typically a liquidation preference and/or a dividend preference.

Principals: Also known as owners or equity participants, these are the individuals or entities that own a business (although there is specific terminology associated with the principals of different types of business entities).

Public offering: The sale of a business entity's shares on one of the major stock exchanges.

Single-member LLC: An LLC with only a single member.

Small business corporation: This is a special status for C-corporations that raises less than $1 million in external capital and contributions.

Sole proprietorship: The form of (unlimited liability) business entity created when a single individual owns a business without going through the specific formalities required to create a statutory entity.

Statutory entity: A business that is formally registered within a particular state is known as a statutory entity and will receive limited liability protection (such entities take either LLC, LP, or corporate form).

Strike price: The prespecified price at which shares allocated in an Employee Stock Option plan may be purchased from the company.

Sweat equity: Equity interest earned in a venture in exchange for labor contributions.

Units: Equity interests in a venture are sometimes referred to in terms of blocks of shares known as "units."

Unissued shares: Authorized shares held in reserve by Corporation.

Unlimited liability: The condition under which a business and its owner(s) are considered to be one and the same, and where the owner(s)'s nonbusiness assets are liable in the event of business failure or legal action against the business.

Vesting: A process or condition that must take place before a right can be exercised by, or actual ownership is transferred to, a beneficiary.

Vicarious liability: A business owner of an unlimited liability company is liable for the acts and commitments of employees and other owners.

WEB RESOURCES

The web sites below are intended as destinations for your further exploration of the concepts and topics discussed in this chapter:

1. http://www.sec.gov/: The Securities and Exchange Commission (SEC) web site.

2. http://en.wikipedia.org/wiki/Corporation: This is the Wikipedia section on corporations. It has a very thorough list of related topics that you can also link to

and explore in case any particular aspect of corporations is of more interest to you than others.

3. http://en.wikipedia.org/wiki/Limited_liability_company

4. http://en.wikipedia.org/wiki/Types_of_companies: Offers a list of the various types of business entities in various countries.

5. http://history.wisc.edu/dunlavy/i_corporations.htm: History of Capitalism: Corporations in History. Professor Colleen Dunlavy, Department of History, University of Wisconsin-Madison, is assembling a database of nearly 10,000 corporate charters granted between 1825 and 1870 by assembling a chronology of incorporation laws in the United States, Britain, France, and the German states.

6. http://www.nceo.org/index.html: The National Center for Employee Ownership site provides information on employee stock ownership plans and equity plans.

CASE STUDY

Legal Structure Complexity Blocks Funding Opportunity for Dignified Living, LLC

David Aitchison is having the time of his life operating his three companies. Each of his companies is built on the same vision: developing products to help people who are aging, or who have handicaps, to overcome everyday life challenges. Two of his companies are wholly owned by the third: a holding company that he calls "Dignified Living." Dignified Living, in turn, is 100% owned by Aitchison.

Dignified Living is currently serving its target markets through two subsidiary companies: Dignified Motors, LLC, and Dignified Products, LLC. Dignified Motors converts Honda elements into wheelchair-accessible vehicles. Technicians install various driving accessories depending upon the purchaser's needs. Dignified Motors will record nearly $1 million in revenue in 2008, with gross margins that average over 40%.

Dignified Products, LLC, developed its first product, a prescription label magnifier, in March 2007. Not long after the product was developed, Dignified Products gained Wal-Mart's agreement to carry the product at its in-store pharmacies. The product moved from concept drawing to manufactured product on the shelves of over 3,300 Wal-Mart stores in under 150 days. It is currently selling at a rate of approximately 0.6 units per store per week against a Wal-Mart target of 1.0 units per store per week. The product offers both Dignified Products and Wal-Mart a gross margin of approximately 50%. Dignified Products has approximately 300 additional concepts in its product development pipeline to fuel its growth through Wal-Mart and other planned distribution channels, including television shopping networks. Revenue for Dignified Products was expected to be less than $150,000 in 2008.

The company intends initially to concentrate on the U.S. market. According to the U.S. Department of Health and Human Services Administration on Aging, persons aged 65 and over numbered 37.9 million in 2007, the most recent year for which such a statistic is available—that is over 1 in 8 Americans. By 2030, it is estimated that such persons will number 71.5 million, an increase of 88.6%. These individuals, and even their younger counterparts who are part of the Baby Boomer generation, can be expected to suffer the challenges, deficiencies, and limitations regularly associated with aging—diminishing eyesight, diminishing hearing, diminishing muscle strength, etc.—for which the company will offer an array of products to add ease and dignity to their daily lives and those of their caretakers.

In 2009, Aitchison and his advisory team determined that a round of funding would enable the company to achieve several of its rapid growth goals. The company was poised for growth, Aitchison and his team reasoned, based on several unique features:

- Aitchison is a mechanical engineer and leader of the product development team. He has an intimate knowledge of and keen insight into the challenges and needs of the temporarily and permanently disabled and their caretakers inasmuch as he has been confined to a wheelchair for over 25 years due to spinal muscular atrophy.
- The company's product development and manufacturing team previously employed similar operational strategies to great success, conceiving, developing, and manufacturing products for a scrapbooking accessories company that propelled it from sales of approximately $2 million to $30 million in just 4 years, at which time it was acquired for $90 million.
- Its scope will be comprehensive, leveraging the multiple needs of its customers by offering mobility products and services through Dignified Motors, LLC, and consumer products through Dignified Products, LLC.
- Dignified Motors, LLC, is currently Arizona's only authorized dealer of state-of-the-art "drive-by-wire" technology, which enables even the most severely disabled individuals to drive a vehicle.

Aitchison and his team approached the Arizona Technology Investor Forum (ATIF) to inquire about its interests in reviewing the deal terms Dignified Living had put together. As one of 24 firms to apply for ATIF's June 5, 2009, investor meeting, Dignified Living was hopeful that it would have a chance to present to the 45-member angel investor group. Their hopes were further strengthened when Dignified Living was one of only six firms chosen to meet personally with the ATIF selection committee.

Aitchison had 25 minutes to present his vision, plan, and deal structure to the ATIF selection committee. During the presentation, he laid out his vision for the company's growth and how that was to be achieved. He mentioned that he was seeking up to $3 million in financing—money that would be used to develop the pipeline of products and get them into the already established distribution channel. The terms of the investment were such that investors would gain an equity interest in Dignified Living, LLC, the holding company. The structure of the investment opportunity is illustrated in the figure below.

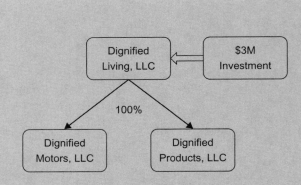

The meeting went well, but ultimately Dignified Living was not selected to present to the angel investors at the June 5 meeting. In a debrief meeting following its presentation to the ATIF selection committee, several factors were identified as contributing to Dignified Living's non-selection. The factor that stood out as a major barrier was the legal structure of the company and thus, the structure of the deal. The selection committee believed the angels would not be interested in investing in Dignified Living, but would be more interested in investing directly in Dignified Products, LLC. The products company was viewed as the high-growth opportunity, even though Dignified Motors was the source of most of the current revenue. Indeed, Aitchison had identified the products company as the major growth engine for the future.

Aitchison explained that the current structure was suited to his long-term vision. He wanted to build and own each of the companies, and sell each one at an appropriate future time. His thinking was that investors would share his desire to have multiple potential exit opportunities. As this book was going to press, Aitchison had not yet decided whether he wanted to adjust the terms of his investment offering to better suit the interests of the angel investors.

Source: Adapted from the Dignified Living, LLC, executive summary dated May 2009; personal conversations among David Aitchison, his advisory team, and Thomas Duening.

QUESTIONS FOR DISCUSSION

1. Why do you think the angel investors were reluctant to consider investing in Dignified Living, LLC?

2. How can David Aitchison better structure his deal terms to meet the needs of the angel investors? If you were going to invest $100K in this deal, how would you like to see it structured?

3. What should David Aitchison do with the Dignified Motors business if the Dignified Products business seems the better growth opportunity? Before you answer, consider that Dignified Motors currently provides a substantial portion of current revenue.

ENDNOTES

[1] Advantage Leasing Corp. v. NovaTech Solutions, No. 03-216 (Wis. Ct. App. March 24, 2005) (unpublished opinion).

[2] Littriello v. United States, 484 F.3d 372 (6th Cir. 2007).

[3] Securities Act of 1933 §§11,12,15.

[4] A federal "self-employment tax," analogous to the taxes on employee wages related to the Social Security and Medicare system, is imposed on the net "self-employment income" of anyone carrying on a trade or business as a sole proprietor, an independent contractor, a partner in a partnership, a member of a single-member LLC, or is otherwise self-employed. Under the Self-Employment Contributions Act Tax (SE Tax Act) 26 USC Ch 21. This tax is, however, not applicable to dividends and distributions.

[5] California is a notable exception because of fees charged with respect to LLCs. See endnote 16.

[6] The Internal Revenue Code, Title 26 United States Code (USC).

[7] 26 USC (IRC) §11. As of 2008, the graduated rate was:

   (A) 15 percent of so much of the taxable income as does not exceed $50,000

   (B) 25 percent of so much of the taxable income as exceeds $50,000 but does not exceed $75,000

   (C) 35 percent of so much of the taxable income as exceeds $75,000 but does not exceed $10,000,000

   (D) percent of so much of the taxable income as exceeds $10,000,000

[8] Because of the deductibility of salaries, the IRS scrutinizes the compensation paid to owner-executives. If the compensation is found to be excessive, the IRS can reclassify it as a dividend and increase the taxable income of the corporation.

[9] As of 2008, the tax on dividend income was 15%.

[10] 26 USC (IRC) §531 et seq. 26 CFR §1.531-1 et seq.

[11] 26 USC §541 et seq. 26 CFR §1.531-1 et seq. 26 USC §§541 et seq., 26 USC §11.

[12] 26 USC §1244.

[13] Wyoming Statute, Title17, Chapter 15.

[14] Rev. Rul. 88-76. 1988-2 C. B. 360, September 2, 1988.

[15] 26 C.F.R. §§301.7701-1-301.7701-3.

[16] In California, a Franchise Tax Board has been established to administer Personal Income Tax and the Corporation Tax. For the privilege of doing business in California, an LLC must pay not only an annual franchise tax (as of 2008, a minimum of $800 per year), but also, if it has total annual income above a specified amount (as of 2008, $250,000), it must pay a fee to the California Franchise Tax Board (as of 2008, ranging from $500 to $11,790, depending on the amount of the LLC's total income).

[17] 26 USC §83.

[18] Employee Stock Option Plan should not be confused with a type of retirement plan referred to as an Employee Stock Ownership Plan (ESOP).

[19] The National Center for Employee Ownership survey performed by the National Opinion Research Center (NORC) of the University of Chicago relying on survey information and statistics from the Department of Labor, Bureau of Labor Statistics survey. See details at http://www.nceo.org/library/eo_stat.html.

[20] An ESO plan can be part of a negotiated comprehensive employment agreement with an individual employee.

[21] The requirements include, among other things:

- The option may be granted only to an employee (grants to non-employee directors or independent contractors are not permitted)

- The option must be exercised (a) within 10 years of grant and (b) while still employed or no later than 3 months after termination of employment (unless the optionee is disabled, in which case this 3 month period is extended to 1 year)

- The option exercise price must be at least the fair market value of the underlying stock at the time of grant

- The ISO cannot be transferred by the employee (other than by will or by the laws of descent) and cannot be exercised by anyone other than the option holder

- No more than $100,000 in ISOs can become exercisable in any year

Capital and Deal Structuring

5

After studying this chapter, students should be able to:

- Develop a capital management plan for a technology venture
- Debate the relative merits of debt and equity financing
- Determine the value of a technology venture from multiple perspectives
- Evaluate how much capital a venture should raise during different stages of growth
- Identify alternative sources of financing besides debt and equity
- Structure a deal and develop materials needed for an investor presentation

---

TECH VENTURE INSIGHT

Angel Capital Helps Launch Flypaper Studios

Don Pierson was the CEO of an e-learning company based in Phoenix, Arizona. Pierson's company, Interactive Alchemy, was a successful provider of e-learning to a variety of large Fortune 1000 client companies. Interactive Alchemy provided a range of off-the-shelf online courses, such as sexual harassment, safety, leadership, supervising, and many other titles. Its specialty, however, was creating custom courses suited to the unique needs of clients. For example, many clients were business-to-business product companies.

To ensure that the sales team was always knowledgeable about existing and new products, Interactive Alchemy produced online training that highlighted the features and value elements of the products. Salespeople would have these courses at their disposal to freshen their knowledge before important sales calls. Other courses were available to a broader cross section of company personnel, providing them with just-in-time training.

While he was running Interactive Alchemy, Pierson and his team were also experimenting with software tools that targeted key corporate training needs. One software product the

---

team developed was a product that enables users to create Flash media presentations on web sites without intensive programming. Pierson and his team decided to create a new company around the software product, calling it "Freshbrew." They invested some of their own capital in the new venture, while bringing in additional talented people, including Greg Head as their marketing and business development leader. Head had previously served in a similar capacity in Internet companies, Act and SalesLogix. Following additional development work and conversion of the original software product into an Internet-based model, the team was ready to raise additional capital.

Given that the company still was not generating revenue, Pierson and his team decided to approach angel investors to raise $3 million. Over a period of approximately 6 months, they were able to gain commitments for nearly $2.5 million. In September 2007, Pierson contacted the Arizona Technology Investor Forum, an angel group associated with Arizona State University's Ira A. Fulton School of Engineering. Pierson made a presentation on a Friday, and by the middle of the following week, he had raised enough money to declare the investing round closed.

Subsequent to Freshbrew's successful fund-raising, Pat Sullivan was brought in to replace Pierson as CEO. Sullivan had more experience in building large Internet ventures than Pierson, and he was better connected to venture capital firms in Silicon Valley. When Sullivan came on board, the venture was renamed Flypaper Studios. Currently, the company is well on a path to success with a first-time fund-raising experience under its belt and a start-up venture that promises returns for both the founders and investors alike.

Source: Adapted from O'Grady P. Arizona technology investor forum seeks tech companies, angels. Phoenix Business Journal 2008:July 11; O'Grady P. Industry veterans, investors confident in Flypaper Studios' future growth. Phoenix Business Journal 2008:May 16.

INTRODUCTION

As was pointed out in Chapter 3, technology ventures often are capital intensive. Capital is required to conduct research, build prototypes, secure intellectual property, and many other activities related to venture growth and stability. Technology entrepreneurs have a number of options when it comes to how and where they obtain the capital they need to launch and operate their ventures. Capital needs change over time, and the potential sources of capital also change as the venture grows and matures.

This chapter explores the various sources of capital that are available to the technology entrepreneur. It delves into managing capital needs over time, the factors to be considered in developing a capital management plan for a technology venture, various fund-raising tools and techniques for raising capital, as well as alternatives to raising debt or equity capital.

5.1 ROLE OF CAPITAL IN TECHNOLOGY VENTURES

Entrepreneurs should raise enough capital to meet growth targets. At the same time, however, the entrepreneur should avoid raising too much capital. Although this may sound counterintuitive, it is possible to have too much capital in a venture. There are two primary problems associated with **overcapitalization**. First, with too much capital on hand, the entrepreneur may not develop disciplined internal cost controls and operational efficiencies. There is nothing that motivates the entrepreneur more to manage costs than the threat of running out of cash. Second, overcapitalization may mean the venture has sold more equity (ownership shares) in the company than it needed to, or it has taken on more debt than necessary.

The key to managing capital in a growing technology venture is to have just as much capital as necessary, and no more. Of course, that is not as easy as it sounds. Nonetheless, it is possible to develop a reasonable capital management plan that changes over time with the needs of the growing venture. Technology entrepreneurs also need a variety of tools to raise capital. For example, an up-to-date business plan is an essential component of the fund-raising process. Technology entrepreneurs are advised to develop and routinely update their business plan so that it is ready to present to investors or lenders. There are a number of other tools, and some tried and tested techniques for fund-raising that will be discussed later in this chapter.

Raising funds also requires that the entrepreneur develop what is referred to as a "deal." The term "deal" is used to refer to the conditions that the entrepreneur and the investors or lenders agree to in the transfer and use of funds. Investors and lenders have different expectations regarding the funds they provide to the technology venture. Investors are concerned with the returns they can earn on their invested capital, while lenders are concerned with whether the venture will be able to make the interest and principal payments on the loaned amount. The two different sources of capital require different conditions or terms in the deployment and use of the capital they provide. Technology entrepreneurs must learn how to structure deals that meet and exceed the interests of these various sources of capital.

Finally, technology entrepreneurs have opportunities to acquire capital from nontraditional sources through alternative forms of raising debt and equity capital. For example, the federal government has grant programs in key technology areas that have funded a large number of start-up ventures. The U.S. government's Small Business Innovation Research (SBIR) grant program has awarded more than $2 billion to a variety of technology entrepreneurs. This and other alternative funding opportunities will be discussed in this chapter.

TECH TIPS 5.1

Raising Capital Is a Primary Function of the Technology Entrepreneur

Raising capital is one of the primary functions of the technology entrepreneur. Most entrepreneurship training and textbooks remind the aspiring entrepreneur never to forget the mantra "cash is king." What that means is that a venture can operate without profits; it can operate without revenue; it can operate without intellectual property; but it cannot operate long without cash. Cash is required to purchase materials, pay bills, reimburse employees, and keep the lights on. Of course, every venture aspires to be able to operate solely on the cash that is generated from its own sales and operations. Until that day arrives, however, external funds are necessary.

5.2 CAPITAL SOURCES

The technology entrepreneur can pick from a wide range of capital sources that will vary depending on the venture's stage of development. All ventures can be described by virtue of their development stage, which in itself conveys useful information to lenders and investors, enabling them to sort through the deals they encounter. A common breakdown of the various stages of development is presented in Exhibit 5.1.

In the pre-start-up stage, most ventures are funded by the founders, and anyone else they can convince of the potential for their success. For most technology ventures in pre-start-up mode, this broader group of potential investors is limited to

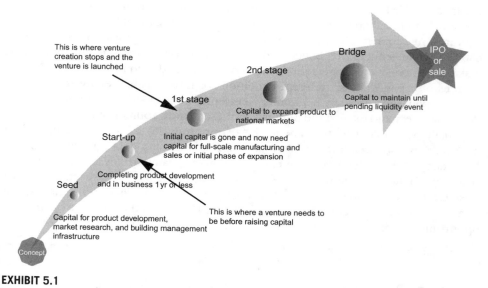

EXHIBIT 5.1

Company Funding Stages.

Source: StartupFlorida.com website: http://www.startupflorida.com/.

what is often referred to as "FF&F"—which stands for friends, family, and fools. This little memory device is only partly facetious. Scholarly research has demonstrated that most pre-start-up technology ventures are funded by the entrepreneur, family members, friends, and business partners.[1] The other "F"—fools—is used primarily as a reminder that start-up ventures are very risky investments.

Entrepreneurship guides and training programs often counsel the use of other sources of capital, such as bank loans that are available to start-up ventures. In reality, this is usually not the case. **Seed-stage** ventures, especially those that have not yet begun generating revenue from sales, are at a very high risk of failure. As such, investors and lenders have little incentive to put their capital to work at this stage. Research into the sources of funds for seed-stage ventures is very clear. In most cases, funds are derived from the assets of the founders. The good news is that most entrepreneurial ventures are launched with less than $20,000 in seed funds (Ref. 1, p. 79) The founders may tap into their personal savings, leverage a line of credit on a credit card, take out a home equity loan, and adopt other strategies to accumulate the needed start-up funds.

5.2.1 Angel Investors

As technology ventures launch and enter what is referred to as the **early stage**, most continue to be funded by founders and their network of family and friends. However, if the venture has matured beyond the prototyping stage and is preparing or actually beginning to sell into the target market, then **angel financing** may be available.

Angel investors are usually high-net-worth individuals who are interested in allocating a portion of their overall portfolio to high-risk and, potentially, high-return private ventures. Angel investing has become a significant source of capital for seed- and early-stage technology ventures in the United States. Angel investors provided over $26 billion in financing in 2007.[2] They often act alone, investing both their own and their family wealth. However, many angels aggregate in groups to review and discuss deals, and to coinvest with others in order to have a larger impact on the ventures they prefer.

According to the Angel Capital Association (ACA), there are more than 170 angel investor groups throughout the United States and Canada.[3] For example, there are three angel investor groups in the state of Arizona, one of which focuses exclusively on technology ventures. The Arizona Technology Investor Forum (ATIF) is an angel investor group that meets three times per year and reviews three or four ventures at each meeting.[4] Meeting formats are fairly standard across angel groups. Entrepreneurs generally are allotted time for a brief (20–30 minute) presentation, which is usually followed by 15–20 minutes of questions and answers. Some groups also perform extensive **due diligence** on the ventures that present. For groups that provide due diligence for members, the findings of the individual or team that performed the due diligence are also reported at the meeting.

Angel investment groups made an average of 7.3 investments in 2007, with average total funding of $1.94 million. Thus, the average size of an investment per venture was $262,000 in 2007. These numbers are similar to the numbers that angel groups reported in 2006, in which they invested an average total of $1.78 million in an average of 7.4 deals. Angel groups prefer to invest in seed-stage companies (81%) and early-stage firms (85%), than they do in growth-stage (38%) and later-stage companies (6%).[5] Angel groups expressed interest in a wide variety of industries in the survey. Five industry areas were of particular interest:

1. Software (83%)
2. Medical devices (75%)
3. Industrial/energy—including "clean tech" (64%)
4. Business products and services (64%)
5. IT services (63%)

Few angel groups specialize in any single industry, and most have interests and expertise among their members in a variety of areas. Five other industry areas were preferred by 50% or more of the responding angel groups, including biotechnology, consumer products and services, electronics and instrumentation, healthcare services, information technology services, and networking and equipment.

5.2.2 Venture Capital

For the technology venture that receives angel capital, the challenge is to grow the venture to attract additional capital to fuel still further growth. Ideally, the technology entrepreneur wants to grow the venture, increasing the enterprise value, which would result in a **return on investment** for the angel investors. If the growth is substantial and future growth projections even greater, the technology venture may become attractive to **venture capital**.

Venture capital is managed by venture capital (VC) firms. VC firms are typically built around a team of general partners who often are successful entrepreneurs or at least financially experienced individuals. VC firms raise money from high-networth individuals who participate as limited partners to the VC firm. These individuals pledge an amount of capital to the VC firm, and are required to put as much as 20% of it immediately into a fund that will be managed by the general partners. This capital is then invested in ventures according to the interests and expertise of the firm. As the fund is invested, the limited partners will be called upon to invest additional portions of their pledged amount.

Most VC firms have specific industries and venture types that they prefer. For example, one of the top VC firms in Silicon Valley, Benchmark Capital, invests primarily in high-growth Internet companies. The firm has invested in a wide range of well-known Internet ventures. Benchmark is composed of nine general partners. Matt Cohler became the ninth general partner when he was added to the firm in June 2008. Cohler, one of the first executives in the popular social networking company Facebook, joined the firm at the age of 31. Cohler has significant expertise in evaluating opportunities in the Internet industry, having also been one of the founders of LinkedIn.com.[6]

What do VC firms look for?

VC firms look for many specifics in technology ventures when deciding whether to make an investment. First and foremost, they seek out experience and talent in the management team. Venture capitalists often state that they invest in the management team, not the technology.[7] This is because no technology, no matter how revolutionary or disruptive, can be commercialized successfully without a solid management team. In addition, experienced VCs know that most technologies they invest in will change radically over time in response to changing market reactions and conditions. It is important to recognize that when VCs evaluate the management team, they do not require that the team be devoid of past venture failures. In fact, most VCs recognize the contingent nature of venture success and do not automatically consider a past failure in a negative light.[8]

Another primary area of consideration for VCs is the market potential for the technology.[9] The VC wants to know if the venture is "market ready," and they will attempt to forecast the likely marketplace demand. This will include forecasting the willingness of the customer to switch to the venture's products on the basis of the venture's stated value proposition. It will also include an analysis of the size of the market opportunity. Obviously, the VC will be more interested in large market opportunities than small ones.

Other items that VCs commonly consider to be important include the time to an **exit event**. An exit event is the manner in which shareholders can convert their stock into cash (see Chapter 6 for a thorough discussion of exit strategies). This is usually achieved via acquisition by a larger company or by going public on a major stock exchange. Either way, the capital invested by the VC is tied up until this event occurs. Obviously, the VC will be more interested in an earlier rather than later exit.[10]

In general, VCs are more risk averse than the founders of a technology venture seeking capital. In their respective evaluations of the investor readiness of the venture, the founders generally will have a more favorable perspective. VCs will evaluate a venture from both a risk and return perspective. Risk and return are evaluated in terms of the risk of venture failure and potential profitability.[11]

The technology entrepreneur tends to have a strong interest in innovation, and will often discount the factors that VCs find important. Most technology entrepreneurs place a higher importance on the technological aspects of the opportunity and consider the management team and marketability to be of lesser importance. In fact, research has identified certain cognitive biases, such as "overconfidence," that are common among technology entrepreneurs.[12] Recognizing these biases is an important part of raising VC funds. The entrepreneur who does not recognize the factors that VCs consider important may not receive an adequate hearing. On the other hand, technology entrepreneurs who tailor the VC presentation to address the issues VCs are most concerned with enhance their chances of receiving funds.[13]

TECH TIPS 5.2

Target the Right VC for Your Venture

In addition to recognizing what VCs are looking for, it's important for the technology entrepreneur to target the right VCs for a financing pitch. There are literally hundreds of VC firms in the United States, many of which invest in technology ventures. However, most VC firms have some areas of technology in which they specialize. For example, some will focus primarily on the Internet and web-based business models. Others will invest in biotechnology or semiconductor technologies. The technology entrepreneur who does not prequalify the VC firm before approaching for a potential financing pitch may waste a lot of time and money.[14]

5.3 EQUITY AND DEBT FINANCING

The technology entrepreneur has a variety of funding sources as described in the section above. Despite these choices, however, capital actually exists in the form of two types: equity or debt. The distinction is important, and presents another decision challenge to the technology entrepreneur in a growing venture.

There are advantages and disadvantages to each type of capital. The technology entrepreneur at each stage of the venture's development must weigh financing options in light of the venture's goals and objectives. During the **growth stage** of development, for instance, where market reaction to the venture's offerings is requiring it to grow, **debt financing** may become available to the venture. Debt financing is provided to ventures on the basis of criteria that differ significantly from angel or VC (equity) financing.

Equity financing is provided to a venture with no expectation of payments in the short term. Equity financing is provided with the anticipation that the long-term returns may be substantial. Debt financing, by way of contrast, is provided to generate a stream of interest payments. Usually, entrepreneurs who obtain debt financing are expected to begin paying back the principal, with interest, immediately. With no expectation of large future returns, the motivations behind debt financing differ significantly from those that underlie equity financing.

Accepting debt financing may subject the venture to difficult payment obligations and restrictive loan covenants. Accepting equity financing may dilute ownership percentages and remove incentives for working hard on building the venture. Let's consider the issues associated with each type of financing in turn, beginning with equity financing.

5.3.1 Equity Financing

Equity financing comes from individuals (angels) or institutions (venture capitalists). This type of financing is attractive to the entrepreneur because equity does not require that the venture pay back the capital invested. Unlike debt, equity financing does not require the venture to sign a **promissory note** specifying the terms under which the amount provided to the venture will be paid back. Equity financing is invested into a venture under terms specified in a **Private Placement Memorandum** (PPM). A PPM is a document that provides the equity investor with the full disclosure required by the U.S. Securities and Exchange Commission (SEC). The SEC governs all equity transactions in the United States. (See Section 5.6.1 for a lengthier discussion of the PPM.) Exhibit 5.2 shows the front page of a typical PPM.

Confidential Copy #_____

PRIVATE PLACEMENT MEMORANDUM

Common Stock for a total of $4,000,000

International Brand Management Consultants Corp. a Nevada corporation (the "Company," "we," or "us"), is offering (the "Offering") to sell shares of its Common Stock (the "Common Stock") for cash up to a total of $4,000,000.

The Offering will terminate upon the earlier of: (i) the completion of the sale of all of the shares; (ii) October 31, 2007. The Offering may be extended by the Company in its sole discretion (the "Offering Period"). The Offering may be closed from time to time, in tranches of any number of shares of Common Stock (collectively the "Closings"). The offering price of the shares of Common Stock has been determined by the Company.

Neither the Securities and Exchange Commission nor any other regulatory body has approved or disapproved these securities or passed upon the accuracy or adequacy of this Memorandum. Any representation to the contrary is a criminal offense.

	Price to Public(1)	Selling Commissions	Proceeds to the Company(2)
Per Share	$1.00	$ 0	$ 1.00

(1)   Please refer to the offering term sheet for pricing detail.
(2)   Before deducting the accountable expense allowed of the Company.

EXHIBIT 5.2

The Front Page of a Typical Private Placement Memorandum (PPM).

The PPM provides potential investors with details on the venture. It resembles the business plan in that it will describe the business, including its products and services, management team, financial projections, competitors, and the target market. In addition, the PPM goes far beyond the business plan in providing the investor with disclosure of the risks associated with an investment in the venture. Risks discussed in the PPM might include the potential for patent filings to be rejected by the patent office, misjudgment of the market potential of the venture's products and services, inability to raise sufficient capital in the future to operate, and many other contingencies. Most law firms that deal with start-up companies will have a PPM "boilerplate" that can be edited and tailored to a specific venture.

In the United States, the SEC governs all equity investments whether the venture associated with the equity transaction is public or private. Public companies are those listed on one of the major stock exchanges. They offer shares to investors in the public markets under the scrutiny of the SEC. Public companies are required to produce and distribute quarterly reports (called 10-K reports) and annual reports revealing their audited financial performance. These requirements ensure that the investing public is aware of how the company is performing, providing reasons to invest or not.

Security and Exchange Commission rules

The SEC has developed unique rules that enable private companies to sell shares to **accredited investors**. Private companies, such as a start-up technology venture seeking funds from friends, family, angels, or VCs, are not under the same comprehensive SEC rules. Nonetheless, private companies that desire to sell shares of their stock to others must comply with SEC rules allowing them to do so. Under the Securities Act of 1933, any offer to sell securities must either be registered with the SEC or meet an exemption. **Regulation D** contains three rules providing exemptions from the registration requirements, allowing private companies to offer and sell their securities without having to register the securities with the SEC. These rules are:

- *Rule 504 of Regulation D*: This rule provides an exemption from the registration requirements for some companies when they offer and sell up to $1,000,000 of their stock shares in any 12-month period. A company can use this exemption as long as it is not a **blank check company** and does not have to file reports under the Securities Exchange Act of 1934. Also, the exemption generally does not allow companies to solicit or advertise their securities to the public, and purchasers receive **restricted securities**, meaning that they may not sell the securities without registration or an applicable exemption. Rule 504 does allow companies to sell securities that are not restricted under some special circumstances.

- *Rule 505 of Regulation D*: This rule allows companies to decide what information to give to accredited investors, as long as it does not violate the antifraud prohibitions of the federal securities laws. But companies must give **nonaccredited investors** disclosure documents that generally are the same as those used in registered offerings. If a company provides information to accredited investors, it must make this information available to nonaccredited investors as well. The company must also be available to answer questions by prospective purchasers, and it must provide the following financial information:

- Can only offer and sell up to $5 million of its securities in any 12-month period;
- May sell to an unlimited number of accredited investors and up to 35 other persons who do not need to satisfy the sophistication or wealth standards associated with other exemptions;
- Must inform purchasers that they received restricted securities, meaning that the securities cannot be sold for at least a year without registering them;
- Cannot use general solicitation or advertising to sell the securities.
- Financial statements need to be certified by an independent public accountant.
- If a company other than a limited partnership cannot obtain audited financial statements without unreasonable effort or expense, only the company's balance sheet (to be dated within 120 days of the start of the offering) must be audited.
- Limited partnerships unable to obtain required financial statements without unreasonable effort or expense may furnish audited financial statements prepared under the federal income tax laws.

While companies using Rule 505 do not have to register their securities and usually do not have to file reports with the SEC, they must file a **Form D** after they sell their securities. Form D is a brief document that includes the names and addresses of the company's owners and stock promoters.

- *Rule 506 of Regulation D*: This rule is considered a safe harbor for the private offering exemption of Section 4(2) of the Securities Act. Companies using the Rule 506 exemption can raise an unlimited amount of money. A company can be assured it is within the Section 4(2) exemption by satisfying the following standards:

  - The company cannot use general solicitation or advertising to market the securities.
  - The company may sell its securities to an unlimited number of accredited investors and up to 35 other purchases. Unlike Rule 505, all nonaccredited investors, either alone or with a purchaser representative, must be accredited—that is, they must have sufficient knowledge and experience in financial and business matters to make them capable of evaluating the merits and risks of the prospective investment.
  - Companies must decide what information to give to accredited investors, so long as it does not violate the antifraud prohibitions of the federal securities laws. But companies must give nonaccredited investors disclosure documents

that are generally the same as those used in registered offerings. If a company provides information to accredited investors, it must make this information available to nonaccredited investors as well.

- ❑ The company must be available to answer questions by prospective purchasers.
- ❑ Purchasers receive restricted securities, meaning that the securities cannot be sold for at least a year without registering them.
- ❑ Companies using Rule 506 do not have to register their securities and usually do not have to file reports with the SEC, but they must file Form D.

Equity financing for a technology venture will be governed by any one of these Regulation D offering rules. There usually are some costs involved with preparing an offering, and the costs are usually greater as larger funding amounts are sought. This is so because larger amounts of funding usually require the technology entrepreneur to talk to and distribute documents to more people. In addition, larger transactions often are more complex and require more legal work to structure a deal that is attractive to both old and new investors.

5.3.2 Costs of Equity Financing

The costs associated with conducting a private equity offering are usually greater than applying for and receiving an equivalent amount of debt. The greater costs involved in an equity offering are part of the technology entrepreneur's deliberations when deciding on the type of financing to pursue.[15] Of even greater influence on this decision is the degree to which equity financing will **dilute** the ownership positions of the existing shareholders. Any time a venture sells shares to raise capital, the ownership positions of the existing shareholders will be affected. The degree to which dilution occurs is a function of the venture's **valuation** at the time of the financing.

A discussion of methods for determining a venture's valuation will follow later. However, here we will simply note the **effects** of different valuation levels on venture ownership. Suppose two individuals launch a technology venture, and they decide to split ownership equally between them, with each now owning 50% of the company. They have invested their own funds to launch the venture and so they have no other shareholders. Now, suppose they have developed their technology past the prototype stage and have received positive feedback from test markets. These two entrepreneurs are ready to expand the company, but they need capital to pursue that end. They decide that they will need to raise $1 million. What effect will that amount of capital raised have on their respective ownership positions?

The answer is: It depends. It depends on the valuation they persuasively can defend to investors. Let's assume the two entrepreneurs convince each other that their company—today, before they've raised any money—is worth $2 million. In the language of deal making, this is referred to as the **pre-money valuation** of the venture. That is, this is what the venture, arguably, is worth prior to any new money coming in. There are a number of factors to consider in establishing the value of a private company, but we will put off that discussion until later. For now, let's assume that the $2 million pre-money valuation is reasonable. A 50% stake in the company is thus worth $1 million (50% of $2 million).

If the technology entrepreneurs are successful in raising the $1 million, the venture would have a **post-money valuation** of $3 million. Note that the post-money valuation is simply the pre-money valuation plus the amount that is raised. What is the dilutive effect of this raise? It's plain to see that the investors who put in the $1 million will own 1/3 of a venture that has a post-money valuation of $3 million. The two entrepreneurs, who previously owned 50% each, have been diluted by 1/6 to a stake of 33.3%. Together, they maintain a controlling interest, 67%, in the company, but they have been diluted. Note, however, that despite the dilution in the percentage of their ownership, the dollar value of their shares remains $1 million (i.e., 1/3 of $3 million).

TECH TIPS 5.3

Equity Financing Requires a Growth Strategy

Technology entrepreneurs who want to raise equity financing should only do so with the intent of growing the venture. Equity investors, unlike loan providers, only get a return on the capital they provide if the venture's valuation grows. Consider the case discussed earlier where investors provided $1 million in financing to the venture. If the venture does not grow its valuation beyond $3 million, then the value of their investment is unchanged or worse, reduced. Notice also that as soon as the technology venture begins to use the invested capital (spend it), the venture's value will be lessened by the amount spent unless the spent amount leads to increased value via sales, greater intellectual property, or some other value-adding development.

5.3.3 Debt Financing

Debt financing is normally provided to technology ventures via an **institutional lender**, such as a bank. However, this may not always be the case. Some entrepreneurs structure deals where they take loans from private parties. While this is not a common practice, some estimates indicate that as many as 10 million people in the United States have accepted private loans from people they know.[16] Borrowing from family and friends can provide easier terms than an institutional lender, including a longer repayment period and a lower interest rate. However, borrowing from family and friends has some risks that are dissimilar from institutional lenders.[17] For example, if the entrepreneur does not repay the loan within a reasonable amount of time, the trust of the family and/or friend lenders could be lost. Worse, if the entrepreneur defaults on repaying the loan, relationships may become strained, family ties could be severed, and lawsuits could be filed.

It is important to carefully document the terms of the business loan between friends and relatives who lend money to a venture. This can help ensure that these individuals are less apprehensive about how their money is being used. It can also prevent messy or awkward situations in the event the venture has difficulties living up to the loan covenants. If details regarding remedies for late payments and even for complete default are discussed and agreed to in advance, the impact of such eventualities can be lessened.

TECH TIPS 5.4

Requesting a Loan from a Family Member

When requesting a loan from a family member to launch a venture, it is wise to follow formal procedures in establishing the terms of the loan. Here are some tips on how to proceed:

1. Establish a formal meeting when preparing to ask for a loan.
2. Present a carefully organized business plan or proposal.
3. If a loan is agreed to, make sure the parties document the terms in a promissory note. An attorney is not required but it might be a good idea to seek professional legal advice.
4. Items to address include payment schedule; what happens if the business is sold; how the lender will be kept abreast of the venture's progress, including its losses and profits; what new initiatives, marketing strategies, or products are being implemented to grow and expand the business.

5.3.4 Institutional Lender Requirements

Institutional lenders, such as banks, can be a source of financing for young technology ventures. However, banks will require more documentation than friends and family normally require; they will also require that the venture be able to secure the loan by pledging an equivalent amount of **collateral**. That is, most institutional lenders reduce the risk of their loans by gaining a **security interest** in property owned by the borrower that is adjudged to be equivalent in value to the loan principal amount. Collateral can be any property owned by the borrower, including personal property.

It is also not uncommon for technology entrepreneurs to provide—or to be required to provide—a **personal guarantee** on the amount of the loan. That is done when the venture has few assets to use as collateral, but the entrepreneur has personal assets that will suffice. For example, some entrepreneurs will use their personal savings, home equity, or real estate as collateral for loans from a bank for business purposes. In the case of default, the bank would then be able to exercise its claim on the collateral in an effort to recover whatever portion of the loan remains unpaid.

5.4 LOAN RATES, PAYMENT METHODS, AND LENDER TYPES

In addition to pledging collateral in the amount of the loan's principal, institutional lenders also normally require that the entrepreneur begin to pay back the loan immediately, with an added interest charge. Payments on a loan are normally made in monthly increments. The rate of interest charged to a small technology start-up for a loan will exceed the rate charged to large, well established, companies. The latter receive preferential rates from banks—usually called the **prime rate**—because they are less risky borrowers than a start-up venture. Start-up ventures will receive a rate that is termed "prime plus." That is, the rate will be some measure above the prime rate depending on the risk profile of the venture and, to some extent, its principals and the collateral they are able to pledge.

Most banks will have a similar prime rate, but they will vary significantly from one to another in the rates they charge to entrepreneur borrowers. For example, some banks specialize in originating real estate or construction loans. They will typically have loan officers who are familiar with the industry and will have experience in judging the risks involved in lending to this type of venture. A bank that specializes in real estate lending may not be able to evaluate the risk involved in lending to a technology start-up. It's important for the technology entrepreneur to conduct research prior to approaching a lender to determine whether it is capable of understanding the nature of the venture and its risks. Silicon Valley Bank, for example, specializes in loans to technology ventures. Exhibit 5.3 is an excerpt from the home page of Silicon Valley Bank.

Loans made by institutions also require payback to occur over a certain period of time referred to as the **term of the loan**. Short-term loans will carry slightly lower interest rates than long-term loans, but they will also require larger monthly payments. Long-term loans are generally considered to be those that have a term of 10 years or greater. Different lenders use different methods to calculate loan repayment schedules depending on their needs, borrowers' needs, the institution's interest rate policy, the length of the loan, and the purpose of the borrowed money. Normally, business loans are paid back on a monthly schedule in equal payments throughout the term of the loan.

Institutional lenders will also usually require that the start-up venture adhere to certain **restrictive covenants** in order to remain in good standing on the loan. Restrictive covenants enable the lender to maintain some control of the venture by specifying performance targets. In the event these performance targets are missed, the lender would have the option to **call the loan**. That is, it can demand complete

"For 25 years, Silicon Valley Bank has been dedicated to providing global financial services to some of the most innovative and entrepreneurial companies in the technology, life science, private equity and premium wine industries. Our experience with these industries affords us a deep understanding of our clients' business models and a high level of comfort with the business cycles inherent to these dynamic markets."

EXHIBIT 5.3

Statement from Home Page of Silicon Valley Bank.

payment of the outstanding principal before the term of the loan is completed. Restrictive covenants are of two types: **positive covenants** and **negative covenants**. Positive covenants specify performance targets that must be attained in order for the borrower to remain in good standing. For example, a bank and borrower may agree to establish positive covenants targeting total sales, cash flow, profitability, or others. Negative covenants establish performance floors below which the venture may not fall in order to remain in good standing. For example, the bank and the borrower may agree that the venture may not fall below target figures in sales, cash flow, the ratio of debt to equity capital in the venture, and others.

Restrictive covenants are usually determined through negotiations between the lender and the borrower, but many lenders have lending guidelines that establish in advance the covenants that they must put into any lending agreement. Technology entrepreneurs must be aware of the covenants that banks require, and determine whether they will impede the venture's ability to succeed. For example, many businesses are subject to fluctuations in sales based on business cycles and other factors. A restrictive covenant that did not account for these sales fluctuations may result in an unnecessarily premature decision by the lender to call the loan. In a time of decreased sales, such an action could be fatal to the start-up venture.

TECH MICRO-CASE 5.1

Microloan Helps Struggling Firm

A new type of institutional lender has emerged in recent years, providing microloans to small to mid-sized companies. Microlenders often are nonprofit organizations, providing loans in amounts ranging from $500 to $35,000. In the past, microlenders focused primarily on lower-income business owners, minorities, women, and entrepreneurs in developing countries. They tend to charge higher interest rates than banks, because their borrowers are often first-time entrepreneurs. They also can be more lenient than traditional banks and more willing to tailor repayment terms to meet the needs of the borrower.

Terry Bressier is the CEO of Skyline Trimmings, LLC, borrowed $22,000 from a Newark, NJ microlender in the summer of 2008. He used the money to upgrade his web site and to purchase new machinery for his 8-year-old company, which provides fabrics for window blinds. Bressier could not get the financing he needed from traditional banks, even though he had borrowed money from them in the past.[18]

Another type of loan facility that is exceedingly useful to the start-up venture is the **line of credit** or **revolving loan**. A line of credit is simply an amount of money that is set aside by a lender for a business to use as needed. The borrower can draw down the line of credit for business expenses without having to fill out a loan application each time funds are required. This saves the lender and the borrower a lot of time. Lenders provide lines of credit using the same risk calculations as any other loan, and they will also usually require collateral in the amount of the credit line. Borrowers benefit from a line of credit in that they only need to pay back the amount withdrawn, and pay interest only on the withdrawn capital. In addition to the interest collected, most banks will charge an annual maintenance fee to provide a line of credit to a company.

5.4.1 Small Business Administration Loans

Another type of loan commonly used by technology start-ups is the **Small Business Administration Loan**. The name of this loan type is confusing to some. The Small Business Administration (SBA) is an agency of the U.S. federal government. However, the SBA does **not** originate the loan made to the venture. A technology entrepreneur who wants to secure an SBA loan must do so through a commercial bank that provides such loans. The bank **originates** the loan, and the SBA **guarantees** payment on the loan up to a predefined percentage of the principal amount. In this way, the SBA provides a form of collateral and takes a good deal of risk out of the loan for the bank. The primary SBA loan programs available to technology entrepreneurs include:

- *Basic 7(a) loan guaranty*: This program serves as the SBA's primary business loan program to help qualified small businesses obtain financing when they might not be eligible for business loans through normal lending channels. Financing under this program can be guaranteed for a variety of general business purposes. Loan proceeds can be used for most business purposes including working capital, machinery and equipment, furniture and fixtures, land and building (including purchase, renovation, and new construction), leasehold improvements, and debt refinancing (under special conditions). Loan maturity is up to 10 years for working capital and generally up to 25 years for fixed assets.

- *Microloan 7(m) loan program*: This program provides short-term loans of up to $35,000 to small businesses for working capital or the purchase of inventory, supplies, furniture, fixtures, machinery, and/or equipment. Proceeds cannot be used to pay existing debts or to purchase real estate. The microloan program is available in selected locations in most states.

- *Loan prequalification*: This program allows business applicants to have their loan applications for $250,000 or less analyzed and potentially sanctioned by the SBA before they are taken to lenders for consideration. The program focuses on the applicant's character, credit, experience, and reliability rather than assets. An SBA-designated intermediary works with the business owner to review and strengthen the loan application. The review is based on key financial ratios, credit and business history, and the loan-request terms.[19]

While the SBA provides a credible and useful debt facility for many entrepreneurs, an SBA loan also has significant disadvantages. SBA loans generally require more documentation and disclosure on the part of the small business than does a traditional loan. SBA loans also frequently have higher interest rates than would a loan provided directly by a commercial lender. Still, there are many advantages to getting an SBA loan, not the least of which is the fact that many start-up ventures have no other options. The SBA has more loan programs than the primary ones listed above, including some that are designed specifically to support minority and women-owned enterprises.

TECH MICRO-CASE 5.2

An SBA Loan for a War Veteran

Upon retiring from the Air Force in 2006, Dave Brackett, 53, wanted to open an Italian-style trattoria in his hometown of Colorado Springs, CO one modeled after the rustic establishments he came to love during his many visits to Italy. Brackett was convinced that the trattoria concept would thrive in a "pizza wasteland" like Colorado Springs, where big-name corporate franchises dominate. The problem was funding. Banks were willing to loan him money, but only for a franchise pizza operation. "No way was I doing a franchise," he says. "I was going to do an independent restaurant or not do it at all."

Then he heard about Patriot Express, through which members of the military community can access loans backed by an SBA guaranty of up to 85%. Within weeks, Brackett had secured a $52,000 loan and begun the work of renovating a vintage brick building to house Pizzeria Rustica. Within a few months of opening, not only were lines forming to taste the pies produced in Brackett's wood-fired ovens, but the intimate, 40-seat restaurant had also earned a five-star review from the city's daily newspaper—a rare distinction for any restaurant, let alone a pizza joint. "The loan allowed me to do it the right way," he says.

5.5 FUND-RAISING TOOLS AND TECHNIQUES

Fund-raising for a technology venture is never a sure thing, but there are some tools and techniques that can improve the odds of acquiring needed capital. One of the primary tools of fund-raising is the business plan. It is rarely possible to raise money outside of the friends and family network without a business plan. Angel investors will almost always ask to see the venture's business plan before investing. Venture capitalists and institutional lenders will always require a business plan. Entrepreneurs should develop and maintain a business plan at all times because most start-up ventures are in near-constant fund-raising mode. A good business plan will describe the venture and its technologies in terms the investor or lender can understand and evaluate.

Another tool that should be part of the technology entrepreneur's fund-raising arsenal is the executive summary. The executive summary is an abridged version of the business plan, normally condensed down to a single page. Exhibit 5.4

(Company Name)

Mission: *i.e., A computer on every desktop*

Business Description: *Briefly describe the general nature of your company. From this section the stakeholders and reviewers must be convinced of the uniqueness of the company and gain a clear idea of the market in which the company will operate.*

Company Background: *Provide a short summary of the company's background.*

Product/Services: *Convey to the reviewers that the solution truly fills an unmet need in the marketplace. The characteristics that set the solution apart from the competition need to be identified (competitive advantage).*

Technologies: *In this section, highlight whatever aspects of your solution may be protected by current IP or patent law. Provide evidence of how your offerings are different and will be able to develop a barrier to entry for potential competitors. Also, identify any relevant dependencies.*

Markets: *Provide a clear description of your target market, and any market segments that may exist within that market. Include potential market size and growth rate. Also, mention your revenue model in this section.*

Distribution Channels: *Indicate which channels will be used to deliver your products/service to your target markets (i.e., systems integrators, independent software vendors, partner offerings, direct sales force, channel partners, etc. ...).*

Competition: *List any current or potential direct and indirect competition. Briefly describe the competitive outlook and dynamics of the relevant market in which you will operate.*

Financial Projections:

	FY 2009	FY 2010	FY 2011	FY 2012
Revenue				
Operating Income				

Solution Management:
CEO
CTO
Finance
Marketing
Product Development
Etc. ...

Industry: *i.e., Data Management*

Number of Employees:

Financing Sought: *i.e., $2M*

Use of Funds: *i.e., Product development, marketing/sales, distribution, etc. ...*

EXHIBIT 5.4

Template for a Technology Venture Executive Summary.

shows a common executive summary template that many start-ups use quickly to convey basic information about their ventures and funding needs. The Appendix to this text contains a more extensive overview of how to prepare a comprehensive business plan.

5.5.1 Private Placement Memorandum

As discussed above, private equity fund-raising requires that the venture prepare what is referred to as a Private Placement Memorandum (PPM). The PPM is a legal document that specifies all the risks associated with investing in the venture, including the potential for complete loss of all invested capital. The PPM also specifies the amount of capital that is to be raised, the type of security being offered, and the rights and privileges associated with investing. Most early-stage fund-raising will divide the amount to be raised into **units of investments**. For example, if a venture is raising $500,000 and selling its stock for $1.00/share, it does not want 500,000 individuals each purchasing one share. That would be an administrative nightmare. Instead, the venture would sell the $1.00/share stock in units of, say, $25,000. That means that anyone interested in investing would need to purchase at least one unit.

In addition to stating the amount that is going to be raised, the PPM also often specifies the minimum amount that needs to be raised in order for the venture to be able to use the funds. This is often referred to as the **min/max**. This concept can be understood most clearly from the perspective of the investor. If an investor puts money into a venture, he or she wants to be sure that the venture has a fair chance of success. However, if the venture raises only a small portion of what it needs, it may fail. The min/max specifies the minimum amount of money that is required for the venture to have a fair chance to succeed. The venture does not use any of the funds it raises until it achieves the minimum amount. Funds raised prior to reaching the minimum are held in trust in a bank account. Once the minimum has been raised, the venture is allowed to **break the bank** and begin to use the invested capital.

5.5.2 Subscription Agreement

The final document to include in equity fund-raising is the **subscription agreement**. A subscription agreement is a document that a potential investor signs, indicating an intent to invest at a certain amount. Even though a subscription document is not considered to be binding on the potential investor, it creates a psychological commitment on the part of the investor. For example, imagine that a technology entrepreneur completes a lengthy presentation to an angel investor in his or her office. The investor indicates an interest, but would prefer to look over the business plan and PPM before making a decision. If the entrepreneur walks out of the office with no signed commitment, the potential investor may not be interested enough to continue thinking about the deal. It is far more influential to have a signed agreement when conducting follow-up conversations, including the penultimate conversation where the entrepreneur asks for the check, than it is merely to have a promise.

5.5.3 "Elevator Pitch"

Technology entrepreneurs should also practice their investment "pitch," which is sometimes referred to as an **elevator pitch**. This term is used to conjure what it would be like to meet a potential investor in an elevator and, in the limited time available, describe the business in a manner that captures the investor's attention. An elevator pitch should articulate the venture's offering (product and/or service), its business model (how it will make money), and the size of the opportunity. In a full-investor presentation, the entrepreneur will need a slide deck (PowerPoint) that clearly articulates the business, the market, the value proposition, financial outlook, and the deal (including projected returns) for the investor.

5.6 ALTERNATIVES TO DEBT AND EQUITY FINANCING

While the primary sources of capital for the start-up venture are the sources of debt and equity discussed above, there are alternatives. Many technology ventures are able to get started and fund operations using government grants as their primary revenue source. Next, we look at two government grant programs available to the technology venture, and also at bootstrap financing, which relies on internal cash only to grow the venture.

5.6.1 SBIR Program

One particular type of government grant that is commonly used by technology ventures is the **Small Business Innovation Research (SBIR)** program. The U.S. Small Business Administration's Office of Technology administers the SBIR program. SBIR is a competitive program that encourages small businesses to explore their technological potential and provides the incentive to profit from its commercialization. Since its enactment in 1982, as part of the Small Business Innovation Development Act, SBIR has helped thousands of small businesses to compete for federal research and development awards. Small businesses must meet certain eligibility criteria to participate in the SBIR program:

- American-owned and independently operated
- For-profit
- Principal researcher employed by business
- Company size limited to 500 employees

Each year, 11 federal departments and agencies are required by SBIR to reserve a portion of their R&D funds to award to small business investments, including:

- Department of Agriculture
- Department of Commerce
- Department of Defense
- Department of Education
- Department of Energy
- Department of Health and Human Services
- Department of Homeland Security
- Department of Transportation
- Environmental Protection Agency
- National Aeronautics and Space Administration
- National Science Foundation

These agencies designate R&D topics and accept proposals. Following submission of proposals, agencies make SBIR awards on the basis of small business qualification, degree of innovation, technical merit, and future market potential. Small businesses that receive awards then begin a three-phase program.

- Phase I is the start-up phase. Awards of up to $100,000 for approximately 6 months support exploration of the technical merit or feasibility of an idea or technology.

- Phase II awards of up to $750,000, for as many as 2 years, expand Phase I results. During this time, the R&D work is performed and the developer evaluates commercialization potential. Only Phase I award winners are considered for Phase II.

- Phase III is the period during which Phase II innovation moves from the laboratory into the marketplace. No SBIR funds support this phase. The small business must find funding in the private sector or other non-SBIR federal agency funding.

5.6.2 Small Business Technology Transfer Program

Another grant program offered by the SBA is the **Small Business Technology Transfer Program (STTR)**. Central to this program is expansion of the public/private sector partnership to include joint venture opportunities for small business and nonprofit research institutions. The STTR program reserves a specific percentage of federal R&D funding to offer awards to small business and nonprofit research institution partners. The STTR program combines the strengths of both entities by introducing entrepreneurial skills to high-tech research efforts. The idea is that the small business partner is able to transfer from the laboratory to the marketplace the technologies and products developed within the nonprofit organization. As with the SBIR program, small businesses must meet certain eligibility criteria to participate in the STTR program:

- American-owned and independently operated
- For-profit
- Principal researcher need not be employed by small business
- Company size limited to 500 employees

The nonprofit research institution partner must also meet certain eligibility criteria:

- Located in the United States
- Meet one of three definitions
- Nonprofit college or university
- Domestic nonprofit research organization
- Federally funded R&D center (FFRDC)
- Each year, five federal departments and agencies are required by STTR to reserve a portion of their R&D funds to award to small business/nonprofit research institution partnerships:
  - Department of Defense
  - Department of Energy
  - Department of Health and Human Services
  - National Aeronautics and Space Administration
  - National Science Foundation

These agencies designate R&D topics and accept proposals. Following submission of proposals, agencies make STTR awards on the basis of small business/nonprofit research institution qualification, degree of innovation, and future market potential. Small businesses that receive awards then begin a three-phase program.

- Phase I is the start-up phase. Awards of up to $100,000 for approximately one year fund the exploration of the scientific, technical, and commercial feasibility of an idea or technology.

- Phase II awards of up to $750,000, for as long as two years expand Phase I results. During this period, the R&D work is performed and the developer begins to consider commercial potential. Only Phase I award winners are considered for Phase II.

- Phase III is the period during which Phase II innovation moves from the laboratory into the marketplace. No STTR funds support this phase. The small business must find funding in the private sector or other nonSTTR federal agency funding.

5.6.3 Bootstrap Financing

Another way that start-up technology ventures finance their growth is through what is often referred to as **bootstrap financing**. Here, the company uses its own sales and cash flows to invest in its growth. This type of internal growth is also referred to as **organic growth**. That is, the company grows only by virtue of its own ability to sell, control costs, and reinvest profits.

Bootstrap financing has the advantage of helping the firm steer clear of the dilutive effects of equity financing, and the debt burden effects of debt financing. The primary disadvantage of this type of financing is that it limits the venture's ability to grow rapidly. That could be a major disadvantage for technology ventures in highly competitive industries where acquiring market share is the key to long-term success.

SUMMARY

This chapter examined the sources of financing available to technology ventures throughout their lifetimes. During the early stages of the venture's life, friends and family are the primary sources of capital beyond the founders' own finances. As the firm begins to grow and develop sales, other avenues of financing may open. Angel financing becomes available when the firm is able to tell a more compelling story about its growth prospects, especially if it already has significant sales and well-known clients. In addition, firms that have developed collateral such as physical assets or highly regarded contracts may be eligible for debt financing. Commercial lenders, such as banks, are risk averse and lend only when there are assurances that the venture can pay back the principal with interest.

Raising capital is normally never ending for technology venture entrepreneurs. Maturing ventures seeking equity capital from angel investors must prepare legal documentation and sell shares in compliance with Securities and Exchange Commission rules. This chapter examined several rules that enable raising private equity from accredited investors. The most common capital raises for growing ventures are conducted under the SEC's Regulation D. Rules 504, 505, and 506 offer different frameworks for raising equity capital. Each rule shelters the venture from overbearing disclosure requirements that are required of public companies that sell shares to the investing public. The rules do require that ventures develop standard disclosure documents such as a PPM. The PPM will disclose all of the risks associated with investing in the venture, and detail the terms of the stock offering. In addition to the PPM, technology entrepreneurs typically also provide a business plan and executive summary to interested investors when raising capital.

Raising equity capital requires that the venture establish some reasonable valuation to determine ownership percentages. Valuation is an imprecise science, with several acceptable methods available. Each method is likely to produce a different valuation. Ultimately, the entrepreneur and investors negotiate a valuation they can agree to. The agreed valuation is referred to as the "pre-money valuation." After the investment is made, the post-money valuation is used to calculate the relative ownership shares of founders and investors. Founders who hold stock at the time of the investment will have their ownership percentages diluted on the basis of the amount of investment received.

The chapter also discussed options to debt and equity financing, focusing on several opportunities for nonrecourse government grants. The SBIR and STTR programs are excellent sources of financing for ventures that are in the research and development stage.

STUDY QUESTIONS

1. Explain what it means for a new technology venture to be "overcapitalized." What are the disadvantages of this?

2. Identify the type of investor who is most likely to invest in pre-start-up technology ventures. Explain how the technology entrepreneur should approach and work with these investors.

3. Explain what is meant by the term "angel investor." At what stage of development is the technology entrepreneur most likely to attract angel investors?

4. What does the average venture capitalist seek in the technology ventures they invest in?

5. What is the role of a private placement memorandum (PPM) in establishing and promoting a financing deal?

6. Explain how the U.S. Securities and Exchange Commission regulates investing in private technology ventures. What options are available to the technology entrepreneur to seek private equity financing?

7. How does investment in a private equity financing lead to dilution of a founder's overall percentage of ownership?

8. What would be an investor's equity ownership stake in a technology venture given the following information:

   - Pre-money valuation: $1.2 million
   - Investor share purchase: $300,000
   - What is the post-money valuation of this venture?

9. Explain the main differences between debt and equity financing. How does a lender, such as a bank, control the operations of a technology venture to which it lends money?

10. If a technology venture has net profits of $2 million, in an industry where other firms like it have been acquired for amounts equivalent to four times net profit, what is the value of the venture?

EXERCISES

1. *IN-CLASS EXERCISE*: For this exercise, students will practice the specialized art of asking for money. This chapter reviewed a number of elements that are important for different types of investors. Tech Tips 5.4 discussed tips for sitting down with a family member and discussing terms for a loan. Students should separate into groups of four, with two members representing the "family" and the other two representing the entrepreneurs seeking capital. The entrepreneurs should

seek a loan of $100,000 for a technology start-up venture they've created (any venture will do for this exercise). The two students playing the role of the "family" should assume they have as much as $250,000 to invest, and that the amount requested from the entrepreneurs is substantial, but not overwhelming.

Students should spend 10–15 minutes in the fund-raising discussion, and then reverse roles. Afterward, the instructor should lead a general discussion about the challenges of asking for money, and especially focus on the challenges of asking for money from family members. How does it feel to ask for money? How can a person get more comfortable with the overall fund-raising process?

2. *OUT-OF-CLASS EXERCISE*: For this exercise, students should go into the community and identify an event at which entrepreneurs request capital for their venture. Most large cities have angel investor groups or other events that feature new ventures. For example, the organization TIE (see www.tie.org) often features entrepreneurs giving one-minute pitches in rapid-fire fashion. These are fun events that often include exciting new venture ideas. Students should attend at least one such event and then discuss what they learned in a general classroom discussion.

KEY TERMS

Accredited investor: The Securities and Exchange Commission defines an accredited investor as someone who has over $1 million in net worth, or who has made an average income of $200,000 in the 2 most recent years.

Angel financing: Equity financing provided by an accredited investor to a private venture.

Angel investors: Individuals, usually of high net worth, who seek opportunities to invest in early-stage ventures.

Blank check company: A company with no specific business plan or purpose, other than an investment vehicle for the purposes of acquiring or investing in another venture.

Bootstrap financing: Financing that uses only funds provided by founders and which is internally generated by the venture.

Break the bank: This term refers to the point in time at which, after sufficient investor funds have been raised, the entrepreneur begins using the funds to operate the venture.

Call the loan: If a lender decides to recover loaned assets prior to the end of the term.

Collateral: Assets pledged to a lender that are of equivalent value to the size of a loan.

Comparable: A comparable is a venture that has achieved a definite valuation in an industry and is used to establish the valuation of another, similar venture.

Debt financing: Financing that is secured from a lender, that usually requires collateral, and where payback is established via terms that include interest.

Dilution: This is what happens to shareholders when additional shares of a venture are sold to new investors.

Discount rate: This is the rate that is used to discount a future cash flow stream to establish its present value.

Due diligence: The process used to examine the potential for a venture to succeed. Due diligence is usually conducted by investors or lenders prior to investing or lending.

Early stage: A venture development stage where the company is either close to or already selling into the target market.

Elevator pitch: A brief statement about the business and its prospects that could, theoretically, be delivered to a potential investor during a chance and brief encounter on an elevator.

Equity financing: Financing that is provided by an investor that results in a distribution of equity (shares) to the investor in an amount proportional to the investment, and which does not require payback.

Exit event: An exit event is the manner in which shareholders can convert their stock into cash.

Form D: Form D is a brief document that includes the names and addresses of the company's owners and stock promoters.

Growth stage: The stage of venture development where market reaction to the venture's offerings is requiring it to grow.

Institutional lender: An organization that is established to lend money as its primary mode of business.

Line of credit: A line of credit is available from an institutional lender to a borrower. Money is available to the borrower in the amount of the credit line, and can be borrowed over and over as long as the credit line is not exceeded.

Min/max: A technique used in a PPM whereby a venture specifies a minimum amount of capital required to allow the use of funds, and a maximum amount that is intended to be raised.

Negative covenants: Specifications in a loan agreement that preclude a venture from failing to achieve agreed upon milestones or objectives.

Nonaccredited investors: Investors who do not meet the qualifications of accredited investors.

Organic growth: Venture growth that is financed by reinvesting profits.

Overcapitalization: A venture that raises too much money is said to be overcapitalized.

Personal guaranty: Many lenders will require entrepreneurs to pledge personal assets as collateral for a business loan.

Positive covenants: Specifications in a loan agreement that require a venture to achieve agreed upon milestones or objectives.

Post-money valuation: The valuation of a venture immediately following a funding event.

Pre-money valuation: The valuation of a venture immediately prior to a funding event.

Present value: This term refers to what a future cash flow stream is worth in the present.

Prime rate: The interest rate that banks charge to their most favorable customers.

Private Placement Memorandum (PPM): A formal document that is required when selling private equity and that outlines both the promise and risks associated with a venture.

Promissory note: A loan agreement.

Regulation D: The Securities and Exchange Commission regulation that governs private equity offerings.

Restricted securities: A security (stock shares) that has not been registered with the SEC and that cannot be sold publicly.

Restrictive covenants: Covenants in a loan agreement, either positive or negative, that place restrictions on performance parameters a venture must achieve or avoid to remain in good standing with respect to a loan.

Return on investment: The returns to an investor that accrue from a successful exit from a venture.

Revolving loan: A specific type of line of credit that allows the borrower to use and repay institutional funds at will for a specific fee.

Security interest: A lender's property rights to collateral pledged to it for the purposes of securing a loan.

Seed stage: The earliest stage of a venture, where it is not selling and may not yet have a working prototype of its market offering.

Small Business Administration (SBA) loan: A loan originated by a bank and partially guaranteed by the U.S. Small Business Administration.

Small Business Innovation Research (SBIR) grant: A grant provided to technology ventures and issued by the U.S. federal government.

Small Business Technology Transfer (STTR): Similar to the SBIR grant, the STTR grant is narrower in scope and is also issued by the U.S. federal government.

Subscription agreement: A document that is used to get a signed commitment from an investor eventually to write a check.

Term of the loan: The length of time during which a loan, including principal and interest, is to be paid back to the lender.

Units of investment: The minimal amount an investor can invest in an equity offering.

Valuation: The process of establishing the value of a venture.

Venture capital: Private investors pool their money into a fund that is invested in growth- and later-stage companies, usually in amounts substantially larger than provided by angel or other individual investors.

WEB RESOURCES

The web sites below are intended as destinations for your further exploration of the concepts and topics discussed in this chapter:

1. http://www.flypaper.com: This is the site for Flypaper Studios, the Technology Ventures Insight featured at the beginning of this article.

2. http://www.angelcapitalassociation.org: This is the web site for the Angel Capital Association of America. It has some useful research and news items on angel investing in the United States.

3. http://atif.asu.edu: This is the web site for the Arizona Technology Investor Forum, a technology-focused angel investor group in Tempe, Arizona.

4. http://www.nvca.org/: This is the web site for the National Venture Capital Association. The web site is a rich source of news on where the venture capital money is going in the United States.

5. http://www.benchmarkcapital.com/: This is the web site for Benchmark Capital, a leading Silicon Valley–based venture capital firm.

6. http://www.sec.gov/: This is the web site for the U.S. Securities and Exchange Commission, the federal agency responsible for regulating both public and private securities in the United States.

7. http://www.sec.gov/answers/regd.htm: This is the site within the SEC web environment that discusses Regulation D private equity offerings.

8. http://www.svb.com/: This is the web site for Silicon Valley Bank, a debt financing organization that specializes in technology and other start-up ventures.

9. http://www.sba.gov/: This is the web site for the U.S. Small Business Administration.

10. http://www.sba.gov/services/financialassistance/sbaloantopics/index.html: This is the site within the SBA web environment that discusses SBA loans.

11. http://www.sba.gov/aboutsba/sbaprograms/sbir/index.html: This is the site within the SBA web environment that discusses SBIR and STTR grants.

CASE STUDY

TimeBridge Secures Series B Financing Following Introduction of New Software Tools

TimeBridge, Inc., a San Francisco–based software venture, announced in May 2009 that it closed a $5 million Series B round of financing led by Mayfield Fund and Norwest Venture Partners. The technology venture's B round of funding came after a year of significant user growth and ongoing expansion of capabilities and services that focus on helping people simplify the process of coordinating and executing business meetings. During the past year, TimeBridge has seen its user base increase tenfold from 30,000 to 350,000.

"This financing from some of the industry's most respected investors is a testament to the importance of simplifying the meeting process and the great momentum and leadership we have established," said TimeBridge CEO Yori Nelken. "The rapid growth we have experienced over the last year shows that people are in desperate need of a solution to streamline the experience of working with others, from the very beginning, when they are just trying to find a time to meet, through collaborating during the meeting itself."

In the year prior to the Series B funding, TimeBridge introduced a number of tools to facilitate productivity and further simplify and improve the meeting process, including:

- Free conference calling and an affordable Web conferencing service that is seamlessly integrated into the scheduling process, making it even easier for users to work together regardless of location

- TimeBridge Groups, which enables work teams to share their availability with each other as a group to further expedite scheduling; this new feature lets teams essentially create an *ad hoc*, cross-company Exchange server in the cloud for free with just a few clicks

- The Daily Brief, a log of the next day's appointments and activities sent to the user's inbox the evening before so that he or she is prepared and in control of their schedule for the day ahead

"In this economy, everybody is looking for ways to streamline efficiency and costs, and TimeBridge has a very practical solution which enables users to do both," said David Ladd, managing director, Mayfield Fund. "Workers at companies of all sizes across all industries are being asked to do more with less. Using TimeBridge, they can save time by offloading their scheduling for free, and also take advantage of the affordable Web conferencing services and continue to collaborate and be productive, no matter where they are or whether they can afford to travel to meetings in other locations."

Mayfield and Norwest had previously led the $6 million Series A financing, which closed in late 2006. Each of these firms has partners who serve as directors on the TimeBridge board. Nelken is the chairman of the TimeBridge board of directors.

Nelken founded TimeBridge in 2005 to make life easier for busy people by bridging the best of the desktop with the rich experience of web services. In 1997, Nelken founded Banter to help companies deal effectively with the explosion of e-mail communication. Banter built award-winning CRM and contact center solutions that automated processes to improve productivity, cut costs, and empower customers and information workers. Now part of IBM, Banter's products deliver efficiency improvements in many Fortune 100 contact centers today, including Wells Fargo and Bank of America.

TimeBridge's key idea is that it doesn't want to replace the calendar people already use; instead it wants to connect their calendar with everyone else's schedule, regardless of what software they're using. TimeBridge integrates with products like Google Calendar, Microsoft Outlook, and Apple iCal. That should mean more schedule sharing and automatic calendar updates, with less time manually fiddling with calendars and fewer meetings that slip through the cracks. On the other hand, the San Francisco company may also be vulnerable as calendars, themselves, make it increasingly easy to share schedules and as competitors like Tungle try to achieve similar goals.

"In TimeBridge, you have a solution that is helping people save time and money, which in today's economy is a service that cannot be undervalued," said Venkat Mohan, general partner, Norwest Venture Partners. "TimeBridge is a company with real momentum, solving a real problem for a broad and significant market."

TimeBridge's one-step scheduling is the most efficient way to schedule and conduct meetings with large groups or individuals across time zones, calendaring systems, and companies, drastically reducing the traditional back and forth of scheduling. TimeBridge facilitates the easy scheduling of meetings with availability sharing, automated meeting negotiation, calendar updates, meeting notifications, and reminders.

Questions for Discussion

1. Do you think it is a good idea for TimeBridge to have its venture capital investors serving on its board of directors? Explain your response. What are some potential advantages/disadvantages of this arrangement?

2. What role do you think TimeBridge's year of testing and software development played in the decision making for the B round financing?

3. Why do you think the same venture capital firms that provided the A round financing also provided the B round? Do you think TimeBridge should have sought other investors? Explain.

Source: Adapted from TimeBridge secures $5 million in Series B. Company Press Release retrieved from http://www.timebridge.com/press110509.php; Time Bridge receives a $5 million investment. Dealipedia. com 2009;May 18.; retrieved from http://www.dealipedia.com/deal_view_investment.php?r=14412; Ha A. Scheduling startup Timebridge adds cheap web meetings. VentureBeat 2009;March 9.

ENDNOTES

[1] Shane S. *The illusions of entrepreneurship: the costly myths that entrepreneurs, investors, and policy makers live by*. New Haven, CT: Yale University Press; 2008, 79.

[2] Sohl J. The angel investor market in 2007: mixed signs of growth. Center for Venture Research, University of New Hampshire.

[3] Wiltbank R, Boeker W. *Returns to angel investors in groups*. Kansas City, MO: Ewing Marion Kauffman Foundation; November 2007.

[4] For more information on this group, see http://atif.asu.edu.

[5] Angel Capital Association (ACA) Press Release. Angel groups cautiously optimistic about investing in startup businesses in 2008. Washington, DC February 21, 2008.

[6] Ha A. Facebook's Matt Cohler leaves for Benchmark Capital. *VentureBeat* 2008;**June 19**.

[7] Baeyens K, Vanacker T, Manigart S. Venture capitalists' selection process: the case of biotechnology proposals. *Int J Technol Manage* 2006;**34**:28–46.

[8] Cope J, Cave F, Eccles S. Attitudes of venture capital investors towards entrepreneurs with previous business failure. *Venture Cap* 2004;**April–September**:147–72.

[9] Eckhardt JT, Shane S, Delmar F. Multistage selection and the financing of new ventures. *Manage Sci* 2006;**52**(2):220–32.

[10] Stanley R. What a VC looks for in start-ups. *Drug Discov Dev* 2004;**7**(8):15.

[11] Roure JB, Keeley RH. Predictors of success in new technology based venture. *J Bus Venturing* 1990;**5**(4):201–20.

[12] Baron RA. Cognitive mechanisms in entrepreneurship: why and when entrepreneurs think differently than other people. *J Bus Venturing* 1970;**13**:275–94.

[13] Proimos A, Murray W. Entrepreneuring into venture capital. *J Private Equity* 2006; **Summer**:23–34.

[14] Roundtable Discussion. Not all VCs are created equal. *MIT Sloan Management Review* **42**(4); (Summer 2001): 88–92.

[15] Willoughby KW. How do entrepreneurial technology firms really get financed, and what difference does it make? *Int J Innov Technol Manage* 2008;**5**(1):1–28.

[16] Townes G. Financing a business with loans from family and friends. *The National Federation of Independent Business* 2005;**August 18**. http://www.nfib.com/object/IO_24228.html.

[17] Collins L. Finding funds. *Eng Manage* 2006;**16**(5):20–3.

[18] Cordero A. Microlenders widen their client base. *Wall St J* 2009;**March 31**:B4.

[19] Information on these loan programs was derived from the U.S. Small Business Administration website. The site is available at http://www.sba.gov.

Exit Strategies for Technology Ventures

6

After studying this chapter, students should be able to:

- Explain the various exit strategies options available
- Understand the importance of value creation and return on investment
- Execute the process of going public
- Identify aspects of a merger

TECH VENTURE INSIGHT

StumbleUpon Acquired by eBay for $75 Million

Internet search service StumbleUpon was acquired on May 30, 2007, by technology giant eBay. The venture was launched in 2002 by Geoff Smith, Garrett Camp, and Justin LaFrance. StumbleUpon gives people a way to discover relevant and entertaining content based on personal preferences and community recommendations. In its announcement of the deal, eBay said that it acquired StumbleUpon for an aggregate transaction value of $75 million. The acquisition provides eBay with exposure to a growing community-based service with approximately 2.3 million users at the time of acquisition.

StumbleUpon is a web site and content discovery service enabled by a browser toolbar. The site uses positive and negative user ratings to form collaborative opinions on web site quality. When users stumble, they will only see pages that friends and like-minded people have recommended. People who use StumbleUpon say they like it because of the surprise factor in what they see next. The StumbleUpon site says they have nearly 5.5 million users as of July 1, 2008, up from 1.7 million in December 2006. Over 10 million personalized recommendations ("stumbles") are delivered daily.

"StumbleUpon is a great fit within our goal of pioneering new communities based on commerce and sustained by trust," said eBay senior director Michael Buhr. "StumbleUpon's

---

downloadable toolbar provides an engaging and unique experience to its users, but it is the similarities in our approaches to the concept of community that make it such a compelling addition to eBay."

Garrett Camp, chief architect and one of the three StumbleUpon founders said, "We're excited about joining eBay, as we share the same values around community and we look forward to working with them to accelerate our growth." Camp came up with the idea for StumbleUpon as a means of solving his own problem. He needed a way to find photo sites on the Internet that just couldn't be done with Google or Yahoo. So he and a couple of his friends designed an algorithm and portal to fit their needs.

On completion of the acquisition, Buhr became general manager of StumbleUpon. The StumbleUpon founders and management team agreed to remain in their respective positions and will work with Buhr to enhance the user experience, evolve its unique product, and grow the community.

Source: StumbleUpon—the story behind the $45M buyout. JDSBlog.com April 19, 2007; Gonzalez N. eBay's StumbleUpon acquisition: confirmed at $75 million. TechCrunch May 30, 2007.

INTRODUCTION

In this chapter we explore the concept of an **exit strategy**. It may seem odd to think that an entrepreneur would launch a venture with an exit strategy in mind, but in fact that is quite often the case with technology entrepreneurs.[1] Technology entrepreneurs have several exit strategies available to them, the most well known of which is the **initial public offering** (IPO). An IPO is a process whereby a private company qualifies to sell its shares on a public stock exchange. The process is also referred to as "going public." The IPO was a preferred strategy for many of the dot-com ventures of the 1990s. Beginning with the unprecedented returns generated by the Netscape IPO, dot-com ventures found the IPO process to their liking prior to the bust that occurred in the early 2000s.[2]

The first purpose of a venture's exit strategy is to develop a road map by which early-stage investors can realize a tangible return on the capital they invested. The entrepreneur spells out a reasonable scenario—the exit strategy—by which investors will have an opportunity to convert their equity into cash. This later round should provide the original investors with an opportunity to sell their shares and realize a capital gain.[3] Second, the intent of an exit strategy is to suggest a proposed window in time that investors can tentatively target as their **investment horizon**, placing a limit on their involvement in the early stage funding deal. The founding team wants to assure investors that the venture will grow to a point whereby the initial investors will receive all of their original investment back, plus a sizable return for the risks they took by investing in a start-up venture. To investors and other shareholders of a private venture, the exit is a **liquidity event**. It is the event that enables them to convert their shares into cash.

The problem in projecting an exit strategy is that it's based on several major assumptions. The speed of market penetration, the ability to sell at expected price levels, the costs of doing business, the margins on sales, the management team's ability to arrange consistent deals, and the impact of competition and other economic factors collectively affect the new venture's projected market share and bottom line. Certainly, when these factors are all aligned in the most favorable way, the firm will experience significant market share, consistent high-growth sales rates, strong profit margins, and positive earnings. And such a scenario in the first 2–3 years is typically the story told by firms that eventually go public.[4]

There are three basic categories of exit strategies that are commonly pursued by technology entrepreneurs and their investors:

1. Acquisition
2. Merger
3. Public offering

These comprise the typical ways that investors use to cash out as the venture makes forward progress. Next, we discuss each of these, and then turn our attention to establishing a value for the enterprise at the time of exit.

6.1 ACQUISITION

Before the entrepreneur can build value attractive to a potential acquirer, he or she must understand the buyer's perspective.[5] Most acquisitions are undertaken because the acquiring company sees one or more strategic advantages that can be obtained by buying another business. These advantages or **value drivers** vary from industry to industry. They also change over time in specific industries because of evolving product life cycles and external developments that affect the industry. Generic categories of strategic value drivers include:

- *Broader product lines*: The buyer adds complementary products to increase revenues. This is a common strategy for both product and service companies.

- *Expanding the technology base*: The buyer adds technology skills or intellectual property that enhances or complements the company's current base.[6]

- *Adding markets and distribution channels*: The acquirer obtains channels it doesn't currently serve. Companies that start out with a **vertical strategy** can add new industry expertise by widening their distribution capabilities. A vertical strategy is one where a venture develops expertise in a given industry that can be expanded to other industries with only slight modifications.

- *Increasing the customer base*: The buyer adds a company that is similar in product offerings or in its business model yet focuses on a different customer segment. This strategy is enhanced if the target company has a good reputation or strong brand. It also can expand the acquirer's geographic coverage.

- *Creating economies of scale*: The combined company can offer a more efficient use of physical assets or overhead—a critical need in consolidating industries.

- *Extending internal skills*: The buyer can add new capabilities such as consulting or service offerings, international management skills, or various types of management and business skills. These skills can be offered as independent revenue producers or they can enhance a company's competitive edge.[7]

Many acquirers hope to leverage several value drivers in a single deal. The software company Microsoft, for example, has regularly used acquisitions as a strategic tool to bring in new skills, new technology, and, sometimes, new customers. Microsoft has developed expertise for integrating and delivering these newly acquired technologies in their complete product offering for Internet-enabled business.[8]

When the perspective of potential buyers is factored into the development of a new venture's strategy, the strategic planning process becomes very important. Internal characteristics of the company will influence the acquirer's final valuation.[9] These factors are related to fundamental business management, strong cash flow, and clean books and records. While they obviously are necessary to closing a deal, they must be augmented by significant value drivers to motivate the buyer to pay a premium price.[10]

TECH MICRO-CASE 6.1

Officenet Acquired by Staples

Officenet, founded in 1997 by two young entrepreneurs, Santiago Bilinkis and Andrés Freire, is an office solutions company. The largest office supply company in Latin America, it currently generates $57.4 million in revenues. After an initial investment of $7 million from investors, the company found success in the Argentine market and decided to expand its business in August 2000 to Brazil. Finding sustainable success there as well, the two founders sold their stake in the company, and it was acquired by Staples in 2004. Maintaining the Officenet name, Staples was able to expand and gain further presence in the region and the world.[11]

6.1.1 Due Diligence

The acquisition decision process involves additional considerations that must be taken into account. These decisions are usually made through a careful and often lengthy process of **due diligence** conducted by the firm that is interested in acquiring the technology venture.[12] Due diligence involves examining a number of key elements of the target venture, including its financial health, the status of the target firm's product line, the potential for synergy, the market position and future potential of the venture, the research and development history and road map for the venture, legal considerations, and plans for managing the acquired entity.[13]

Financial analysis

The due diligence process begins with a financial analysis—analyzing the profit and loss figures, operating statements, and balance sheets for the years of the company's operation, concentrating on the more recent years. Past operating results, particularly those occurring in the preceding 3 years, indicate the potential for future performance of the company. Key ratios and operating figures, concepts discussed in detail in Chapter 12, indicate whether the company is financially healthy and has been well managed. Areas of weakness, such as too much debt, too little financial control, dated and slow turning inventory, poor credit ratings, and bad debts are also carefully evaluated.

Product/service line

The past, present, and future of a firm's product lines will also be examined. The strengths and weaknesses of the firm's products are evaluated in terms of design features, quality, reliability, unique differential advantage, and proprietary position. The life cycle and present market share of each of the firm's products are verified. One method for evaluating the product line is to plot sales and margins for each product over time.[14] Known as S- or life-cycle curves, these indicate the life expectancy of the product and any developing gaps. The S-curve analysis could reveal that all products of this firm are at or near their period of peak profitability. Exhibit 6.1 is an example of the product life-cycle S-curve. It's important to know the life-cycle status of the target venture's product lines to understand whether there is a fit with the acquiring firm's product portfolio and operating expertise. For example, a venture that is good at introducing new products would not be a good fit with a company whose primary product line is mature and stable.

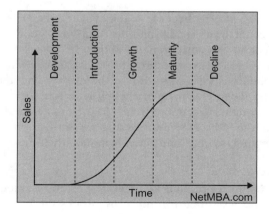

EXHIBIT 6.1

The Product Life-Cycle S-Curve.

Synergy

The phrase "the whole is greater than the sum of its parts" is a good definition of the concept of **synergy**. Technology entrepreneurs who seek to exit their companies via acquisition should consider whether the company provides any synergies to potential acquirers.[15] The synergy should occur in both the business concept—the acquisition functioning as a vehicle to move toward overall goals—and the financial performance. Acquisitions should positively impact the bottom line of the acquiring firm, affecting long-term gains and future growth. Lack of synergy is one of the most frequent causes of an acquisition failing to meet its objectives.

TECH TIPS 6.1

Warning Signs in Due Diligence

Evaluation of the potential for synergy in today's changing environment focuses not only on management and served markets, but also on the company's risks and vulnerability to changes in markets and technology. Some of the warning signs companies consider when evaluating an acquisition candidate include poor corporate communications, few management tools being used, poorly prepared financial statements, and a low number of new products and new markets being entered.

Market analysis

Acquiring firms carefully evaluate the entire marketing program and capabilities of the target venture. Although all areas of marketing are assessed, particular care is normally taken in evaluating the quality and capability of the established distribution system, sales force, and manufacturers' representatives. For example, one company may acquire a venture primarily because of the quality of its sales force. Another may acquire a venture to obtain its established distribution system, which allows access to new markets. Technology entrepreneurs should be aware of which of these elements of enterprise value will be of interest to potential acquirers.

Acquiring companies can gain insight into the market orientation and sensitivity of target ventures by looking at their marketing research efforts. Does the venture have facts about customer satisfaction, trends in the market, and the state of the art of the technology of the industry? Ventures that are setting up for an exit by acquisition should be collecting, storing, and actively mining their customer data.

Research and development

The future of the venture's products and market position is affected by its research and development. The acquiring firm normally probes the nature and depth of the candidate firm's research and development and engineering capability, assessing the strengths and weaknesses of each. Although the total number of dollars spent on research and development is usually examined, it is more important to determine if these expenditures and programs are directed by the firm's long-range plans.

Operations

The nature of the manufacturing process—the facilities and skills available—is also important in deciding whether to acquire a particular firm. Are the facilities obsolete? Are they flexible, and can they produce output at a quality and a price that will compete over the next 3 years? As we mentioned in Chapter 1, a going concern must be focused on repeatability and scalability. Acquiring firms do not want to build this infrastructure themselves. Typically, an acquiring company focuses on the growth prospects of the target venture, and looks carefully at its overall operation to determine if it is poised to deliver on its growth potential.[16]

Management and key personnel

Finally, acquiring companies in their due diligence process evaluate the management and key personnel of the candidate venture. The individuals who have contributed positively to past success in sales and profits of the firm should be identified. Will they stay once the acquisition occurs? Have they established good objectives and then implemented plans to successfully reach these objectives?

Additionally, acquiring companies examine whether any of the technology venture's personnel are indispensable. Generally, it is not good for the acquiring company if there are such individuals. There is too much risk associated with the key person leaving, becoming disabled, or worse, for the acquiring company to accept. As such, technology ventures should strive to ensure that knowledge and other key assets are not dependent on any single individual or groups of individuals.

6.1.2 The Acquisition Deal

Once the technology venture has been identified as a good candidate for acquisition, an appropriate deal must be structured. Many techniques are available for acquiring a firm, each having a distinct set of advantages to both the buyer and seller. The deal structure involves the parties, the assets, the payment form, and the timing of the payment. For example, all or part of the assets of one firm can be acquired by another for some combination of cash, notes, stock, and/or employment contract. This payment can be made at the time of acquisition, throughout the first year, or extended over several years.

Even though the number of mergers and acquisitions has significantly decreased between 2006 and 2007, their value has increased from about $450 million to $460 million during this same time period (see Exhibit 6.2). The hottest technology sectors for mergers and acquisitions include Internet, mobile software and technology services, systems and storage businesses, and recruiting firms. The acquisition targets companies exhibiting slow to moderate growth, strong cash flow, little debt, and businesses operating in saturated markets.

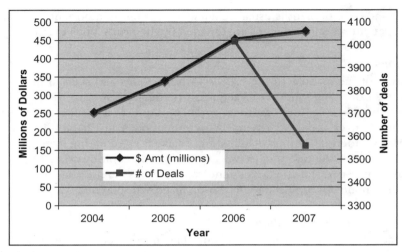

EXHIBIT 6.2

Technology Venture Merger and Acquisitions.

Source: 451 Group, Boston Technology Research Firm.

The two most common means of acquisition are the **direct purchase** of the target venture's entire stock or assets or the **bootstrap purchase** of these assets. In the direct purchase of the firm, the acquiring company often obtains funds from an outside lender or the seller of the company being purchased. The money is repaid over time from the cash flow generated from the operations. Although this is a relatively simple and clear transaction, it usually results in a long-term capital gain to the seller and double taxation on the funds used to repay the money borrowed to acquire the company.

In order to avoid these problems, the acquiring company can make a bootstrap purchase, acquiring a small amount of the firm, such as 20–30%, for cash. The acquiring company then purchases the remainder of the target venture by a long-term note that is paid off over time out of the acquired company's earnings. This type of deal often results in more favorable tax advantages to both the buyer and seller.

6.2 MERGERS

Another method for exiting a venture is via a **merger**—a transaction involving two (or more) companies in which only one company survives. Acquisitions are similar to mergers and, sometimes, the two terms are used interchangeably. In reality, they are quite different, with the primary differences between mergers and acquisitions centering on the relative size of the entities involved, and on who is in control of the combined entity. Mergers generally occur when the relative size of the ventures involved is equal—or at least they are perceived to bring equal value to the merged entity. Mergers are common among large companies. For example, when Hewlett

Packard wanted to enhance its line of personal computers, it merged with Compaq computer. Although Compaq was seen as a smaller company, its brand name was kept for computers offered by HP, and the Compaq executive team stayed on in the new venture for some time after the deal was closed.[17] HP was in charge of the merger, and the HP CEO was in charge of the merged entity.

When an entrepreneur decides to merge with another company, it's usually the case that the two entities are similar in size and offer similar value to the merged entity. For example, it would be quite unexpected for Google to merge with a five-person technology start-up. More likely, Google, with its tremendous assets and global reach, would be in control and simply acquire the smaller venture.

When two entrepreneurial ventures merge, the question about who will control the merged companies is part of the merger negotiations. An entrepreneur who wants to exit a venture may elect to merge with another company, but it may take some time for the entrepreneur to wriggle free of the new company. Often, when a merger occurs, the merged company requires that the top executives from each venture stay with the merged entity to ensure its success. Depending on the size of the new company, the entrepreneur may be able to negotiate a deal whereby he or she **earns out** of the company over a period of time.

An earn-out strategy is used for ventures that have begun to generate consistently strong positive cash flow.[18] The management team initiates a monthly or quarterly buyback of common stock from the owners of one of the merged entities. Typically, an earn-out can be accomplished over an agreed-upon period of time, and can provide the entrepreneur seeking to exit with a strong return as company sales expand and costs decline because of increased operating efficiencies. The purchase price can be scaled upward incrementally over time, as the entrepreneur seeking to exit provides the luxury of time for the management team to complete the deal.

Why should a technology entrepreneur merge? There are both defensive and offensive strategies for a merger. Merger motivations range from survival to protection to diversification to growth. When some technical obsolescence (loss of market or raw material) or deterioration of the capital structure has occurred in the technology entrepreneur's venture, a merger may be the only means for survival and exit. The merger can also protect against market encroachment, product innovation, or an unwarranted takeover. It can also provide a great deal of diversification as well as growth in market, technology, and financial and managerial strength.

A successful merger requires sound planning by the technology entrepreneur. The merger objectives, particularly those dealing with earnings, must be spelled out with the resulting gains for the owners of both companies delineated. Also, the technology entrepreneur must carefully evaluate the other company's management to ensure that it would be competent in developing the growth and future of the combined entity. The value and appropriateness of the existing resources should also be determined. In essence, this involves a careful analysis of both companies to ensure that the weaknesses of one do not compound those of the other. Finally, the technology entrepreneur should work toward establishing a climate of mutual trust to help minimize any possible management threat or turbulence.

6.3 VENTURE VALUATION

Investors expect technology entrepreneurs to use their money wisely and carefully to build enterprise value. As we noted in Chapter 5, only accredited investors are allowed to invest in private equity offerings. This is because the SEC assumes such investors are aware of and willing to accept the risks associated with investing in new ventures. Although investors in technology start-ups are willing to accept risk— including the risk of total loss of invested capital—they also expect to be rewarded, eventually, for the risks they take.

There exist a wide range of techniques that can be used to determine the value of a technology venture.[19] Some of these techniques employ sophisticated mathematical formulas and statistics; others use primarily qualitative judgments and educated hunches. We'll look at a subset of these techniques, including:

- *Multiples technique*: This is the least mathematical of the techniques for determining valuation, and perhaps the one most often used. It is a straightforward technique that relies on identifying a key metric within an industry, and on identifying industry comparables that have had a definitive valuation because of being acquired or having executed a recent equity sale. For example, in the Internet industry, a common key metric is "active members." Facebook, a popular Internet property, sold equity to Microsoft in 2007.[20] Microsoft purchased a 1.6% stake in Facebook for $240 million. Simple arithmetic calculates that that places the value of Facebook at $15 billion:

$$\frac{100}{1.6} \times \$240M = \$15B$$

  The number of active members that Facebook had at the time of this valuation, 50 million, can then be used to determine the value of each member. In other words, the $15B enterprise valuation places the value of each member at $300 ($15B/$50 million). Companies in similar industries can now use the same multiple (i.e., $300 per active member) to develop a reasonable valuation. Exhibit 6.3 lists some popular Internet sites and their valuation, using the Facebook valuation as one baseline, and Bebo and LinkedIn as others.

  Note that the valuations vary depending on which comparable a venture chooses to use in generating the value of each active member. In general, owners of a venture will want to choose the comparable that gives them the greatest value, while potential investors will argue for the comparable that gives the venture the least current value (which means share prices are cheaper). There is no such thing as an "absolute" or "true" venture value. It all depends on the argument that can be made for one or another metric and comparable company to serve as the baseline for calculating valuation.

- *Discounted cash flow*: To the start-up technology venture, the old mantra clearly applies: "Cash is king." What that simple phrase means is that start-up

Site	Bebo Deal	Value Based On LinkedIn Deal	Facebook Deal
Myspace.com	$ 3,279,184,220	$ 18,045,014,284	$ 19,981,067,260
Facebook.com	$ 2,461,718,519	$ 13,546,584,411	$ 15,000,000,000
Bebo.com	$ 850,000,000	$ 4,677,462,781	$ 5,179,308,642
Hi5.com	$ 322,144,914	$ 1,772,730,405	$ 1,962,926,983
Ameblo.jp	$ 318,796,110	$ 1,754,302,281	$ 1,942,521,702
Buzznet	$ 275,843,872	$ 1,517,940,523	$ 1,680,800,648
Skyrock.com	$ 270,563,270	$ 1,488,881,911	$ 1,648,624,331
Mixi.jp	$ 270,469,365	$ 1,488,365,160	$ 1,648,052,138
Poczo.com	$ 228,179,520	$ 1,255,648,482	$ 1,390,367,244
Studivz.net	$ 204,513,483	$ 1,125,416,713	$ 1,246,162,884
Linkedin.com	$ 181,722,451	$ 1,000,000,000	$ 1,107,290,188
Tagged.com	$ 171,021,351	$ 941,112,948	$ 1,042,085,134
Netlog.com	$ 162,716,039	$ 895,409,664	$ 991,478,335
Orkut.com	$ 150,388,212	$ 827,570,901	$ 916,361,138
Hyves.nl	$ 138,543,637	$ 762,391,416	$ 844,188,535
Friendstar.com	$ 108,105,350	$ 594,892,648	$ 658,718,792
Perfspot.com	$ 106,593,717	$ 586,574,288	$ 649,507,953
Dada.net`	$ 102,789,834	$ 565,641,910	$ 629,329,737
Nasza-Klasa.pl	$ 90,801,993	$ 499,674,050	$ 553,284,172
Multiply.com	$ 75,472,812	$ 415,319,140	$ 459,878,809
Odnoklssniki.ru	$ 64,882,232	$ 357,040,264	$ 395,347,181
Badoo.com	$ 55,315,218	$ 304,393,371	$ 337,052,457
Metroflog.com	$ 50,608,401	$ 278,492,837	$ 308,372,386
Vkontakte.ru	$ 44,002,291	$ 242,140,090	$ 268,119,346
Sonico.com	$ 33,431,396	$ 183,972,514	$ 203,710,960

EXHIBIT 6.3

Popular Internet Sites and Their Valuations.

Source: Arrington M. Modeling the real market value of social networks. TechCrunch June 23, 2008.

ventures must organize their growth and operations to ensure an ample supply of cash on hand to meet current obligations, including operating expenses and debt liabilities. One of the primary ways that entrepreneurs track and manage cash in the venture is via the cash flow statement. Unlike the income statement, the cash flow statement does not track operating performance according to rules of accounting. The cash flow statement tracks the actual movement of cash into and out of the firm. The technical term "discounted cash flow" is a refined measure of cash, which subtracts one-time capital expenses and dividend obligations from the projected cash position at some future date.

To determine valuation using the discounted cash flow method requires understanding the concept of *present value*. Present value is simply defined as the value in the present of some future cash flow. A simple technique has been developed to determine the present value of a future cash flow. Let's say that you are promised some cash right now, or $100 in the future. How much cash would you need to get right now to decide to forego the $100 future cash flow? Important to this consideration is how much a person could have earned by investing cash in hand over the 1 year period. If the investment would return more than the $100, then you should take the cash in the present and invest it. If it would return less than the $100, then you should take the money in the future. The interest rate used to determine the return on current cash is called the **discount rate**. Here's how the calculation works:

$$PV = \frac{FV}{1 + r}$$

In this equation, PV is the present value of the future value (FV) divided by the discount rate, r. If the discount rate was 8% (0.08) with an FV of $100, someone would need to give you at least $92.59 in the present to make it worth your while to forego the future cash. Applying this logic to a venture requires identifying a future cash flow that will be put into the calculation above. Most ventures will provide interested investors with cash projections for at least 3 years into the future. A simple way to generate the FV variable in this equation is by using the projected cash position of the venture at the end of the 3 year period. This would be the last line of the cash flow statement in month 36 (for more on this, see Chapter 12).

Investors' expectations of future returns is based on their estimate of risk. One element of that risk is the alternative investments they could have made. One alternative investment would be to purchase fixed-income securities, such as U.S. Treasury bills. Such notes are backed by the U.S. government and are assumed to be risk free. Thus, the prevailing interest rate paid on U.S. Treasuries is known as the risk-free rate of return. At minimum, investors in technology ventures would expect to meet and exceed this risk-free rate. If the technology entrepreneur could not demonstrate how that was possible, investors would purchase the bonds instead.

In addition to the alternative investments that comprise one element of the risk associated with investing in any particular venture is the risk of the venture itself. The risk-free rate of return is relatively straightforward, and can be known with a high degree of accuracy. However, the risks that are associated with any particular technology venture are more difficult to quantify. One technique that is used to arrive at a discount rate that includes both the alternative investment risk and the venture risk is known as the "risk-adjusted discount rate method," or RADR. This approach is summarized in the equation below.

$$r_{vt} = r_{ft} + RP_{vt}$$

This equation states that for a particular venture, v, that yields an uncertain cash flow at some future time, t, the discount rate is expressed as r_{vt}.

In the equation above, r_{ft} is the so-called risk-free rate of return, and RP_{vt} is the risk premium that is associated with the venture. The discount rate that is used to determine the present value of future cash flows, then, is a combination of the risk-free rate and the risk premium. This is the minimum rate of return that investors would expect to receive on their exit from the venture.[21]

The important point to remember as a technology entrepreneur is that valuation is based on the likely future cash flows that will be generated by the venture. The various assets of the organization, such as intellectual property, products and services, customer lists, brand value, and many other things, are meaningless to investors if they can't be translated into future cash returns. Many technology entrepreneurs don't have a lot of experience in predicting those future cash flows, and are at a disadvantage when the time comes to place a value on the venture. This lack of experience can be offset by consulting with independent firms that specialize in placing a value on ventures. Paying for such a service prior to seeking capital, especially from experienced venture capital firms who are usually very good at determining the value of a venture, can be a worthy expense. Arming oneself with rational and defensible arguments about the future value of a venture can help when the time comes to negotiate with others.

6.4 GOING PUBLIC

Going public via what is called an initial public offering (IPO) occurs when the technology entrepreneur and other equity owners of the venture offer and sell some part of the company to the public through a registration statement filed with the SEC pursuant to the Securities Act of 1933. The resulting capital infusion to the company from the increased number of stockholders and outstanding shares of stock provides the company with financial resources and a relatively liquid investment vehicle. Consequently, the company will have greater access to capital markets in the future and a more objective picture of the public's perception of the value of the business. However, given the reporting requirements, the increased number of stockholders, and the costs involved, the technology entrepreneur must carefully evaluate the advantages and disadvantages of going public before initiating the process.

6.4.1 Advantages of Going Public

There are three primary advantages of going public: obtaining new equity capital, obtaining value and transferability of the organization's assets, and enhancing the company's ability to obtain future funds. Whether it is first-stage, second-stage, or third-stage financing, a venture is in constant need of capital. The new capital provides the needed working capital, plant and equipment, or inventories and supplies necessary for the venture's growth and survival. Going public is often the best way to obtain this needed capital on the best possible terms.

Going public also provides a mechanism for valuing the company and allowing this value to be easily transferred among parties. Many family-owned or other privately held companies may need to go public so that the value of the company can be disseminated among the second and third generations. Venture capitalists view going public as the most beneficial way to attain the liquidity necessary to exit a company with the best possible return on their earlier-stage funding. Other investors as well can more easily liquidate their investment when the company's stock takes on value and transferability. Because of this liquidity, the value of a publicly traded security sometimes is higher than shares of one that is not publicly traded. In addition, publicly traded companies often find it easier to acquire other companies by using their securities in the transactions.

The third primary advantage is that publicly traded companies usually find it easier to raise additional capital, particularly debt. Money can be borrowed more easily and on more favorable terms when there is value attached to a company and that value is more easily transferred. Not only debt financing, but also future equity capital is more easily obtained when a company establishes a track record of increasing stock value.

6.4.2 Disadvantages of Going Public

Although going public presents significant advantages for a new venture, they must also carefully weigh the numerous disadvantages. Some technology entrepreneurs want to keep their companies private, even in times of a hot stock market. Why do technology entrepreneurs avoid the supposed gold rush of an IPO?

One of the major reasons is the public exposure and potential loss of control that can occur in a publicly traded company. To stay on the cutting edge of technology, companies frequently need to sacrifice short-term profits for long-term innovation. This can require reinvesting in technology, which in itself may not produce any bottom-line results, particularly in the short run.

Some of the most troublesome aspects of being public are the resulting loss of flexibility and increased administrative burdens. The company must make decisions in light of the fiduciary duties owed to the public shareholder, and it is obliged to disclose to the public all material information regarding the company, its operations, and its management. One publicly traded company had to retain a more expensive investment banker than would have been required by a privately held company in order to obtain an "appropriate" fairness opinion in a desired acquisition. The investment banker increased the expenses of the merger by $150,000, in addition to causing a 3 month delay in the acquisition proceedings. Management of a publicly traded company also spends a significant amount of additional time addressing queries from shareholders, press, and financial analysts.

If all these disadvantages have not caused the technology entrepreneur to look for alternative financing other than an IPO, the expenses involved may. The major expenses of going public include accounting fees, legal fees, underwriter's fees, registration and blue sky filing fees, and printing costs ("blue sky" refers to laws protecting against fraud). The accounting fees involved in going public vary greatly, depending in part on the size of the company, the availability of previously audited financial statements, and the complexity of the company's operations. Generally, the costs of going public are around $300,000–$600,000, although they can be much greater when significant complexities are involved. Additional reporting, accounting, legal, and printing expenses can run anywhere from $50,000 to $250,000 per year, depending on the company's past practices in the areas of accounting and shareholder communications.

The underwriters' fees include a cash discount (on commission), which usually ranges from 7% to 10% of the public offering price of the new issue. In some IPOs, the underwriters can also require some compensation, such as warrants to purchase stock, reimbursement for some expenses—most typically legal fees—and the right of first refusal on any future offerings. The NASD regulates the maximum amount of the underwriter's compensation and reviews the actual amount for fairness before the offering can take place. Similarly, any underwriter's compensation is also reviewed in blue sky filings.

There are also other expenses in the form of SEC, NASD, and state blue sky registration fees. The final major expense—printing costs—typically ranges from $50,000 to $200,000. The registration statement and prospectus discussed later in this chapter account for the largest portion of these expenses. The exact amount of expenses varies, depending on the length of the prospectus, the use of color or black-and-white photographs, the number of proofs and corrections, and the number printed. It is important for the company to use a good printer because accuracy and speed are required in the printing of the prospectus and other offering documents.

TECH MICRO-CASE 6.2

Silicon Storage Technology Goes Public

Not only can going public be a costly event, the process of getting ready to go public can be exasperating as well. Just ask Bing Yeh, who went through some trying circumstances when he decided to go public in July 1995 until his company—Silicon Storage Technology (SST)—issued their IPO on November 22. While the exact process varies with each company, the goal is the same as that for SST—make over the company so that it is seen in the best possible light and is well received by Wall Street. The many changes in a company that occur usually take place over a 6 month to 1 year period of time, not the 100 days it took for SST.

Regardless of how much preparation occurs, almost every technology entrepreneur, like Bing Yeh, is unprepared and wants to halt it at some time during the makeover process. Yet, for a successful IPO, each technology entrepreneur must follow Yeh's example and listen to the advice being given to make the recommended changes swiftly.

6.4.3 Timing

Is this a good time for the venture to initiate an IPO? This is the critical question that technology entrepreneurs must ask themselves before launching this effort.[22] Some critical questions in making this decision follow.

First, is the company large enough? While it is not possible to establish rigid minimum-size standards that must be met before a technology entrepreneur can go public, New York investment banking firms prefer at least a 500,000 share offering at a minimum of $10 per share. This means that the company would have to have a past offering value of at least $12.5 million in order to support this $5 million offering, given that the company is willing to sell shares representing not more than 40% of the total number of shares outstanding after the offering is completed. This size offering will only occur with past significant sales and earnings performance or a solid prospect for future growth and earnings.

Second, what is the amount of the company's earnings and how strong is its financial performance? Not only is this performance the basis of the company valuation, but it also determines if a company can successfully go public and the type of firm willing to underwrite the offering.

Third, are the market conditions favorable for an IPO? Underlying the sales and earnings, as well as the size of the offering, is the prevailing general market condition.[23] Market conditions affect both the initial price that the technology entrepreneur will receive for the stock and the aftermarket—the price performance of the stock after its initial sale. Some market conditions are more favorable for IPOs than others.

Fourth, how urgently is the money needed? The technology entrepreneur must carefully appraise both the urgency of the need for new money and the availability of outside capital from other sources. Since the sale of common stock decreases the ownership position of the technology entrepreneur and other equity owners, the longer the time before going public, given that profits and sales growth occur, the less percentage of equity the technology entrepreneur will have to give up per dollar invested.

Finally, what are the needs and desires of the present owners? Sometimes, the present owners lack confidence in the future viability and growth prospects of the business or they have a need for liquidity. Going public is frequently the only method for present stockholders to obtain the cash needed.

6.4.4 Underwriter Selection

Once the technology entrepreneur has determined that the timing for going public is favorable, he or she must carefully select a managing **underwriter**, an investment bank that will take the lead in forming an **underwriting syndicate**.[24] For example, Goldman Sachs is an investment bank with broad and deep experience as an IPO underwriter. The firm selected to perform as the lead investment bank is of critical importance in establishing the initial price for the stock of the company, supporting the stock in the aftermarket, and creating a strong following among security analysts. A syndicate is a group of investment banks and other, usually institutional, investors who subscribe to the IPO. That means they designate in advance how many shares they will purchase at the IPO. This is important as it could be disastrous for a venture to have an IPO where no one purchased the stock.

Although most public offerings are conducted by a syndicate of underwriters, the technology entrepreneur needs to select the lead or managing underwriter(s). The managing underwriters will then develop the strongest possible syndicate of underwriters for the IPO. A technology entrepreneur should ideally develop a relationship with several potential managing underwriters (investment bankers) at least one year before going public. Frequently, this occurs during the first- or second-round financing, where the advice of an investment banker helps structure the initial financial arrangements to position the company to go public later.

TECH TIPS 6.2

Selecting the Investment Bank

Since selecting the investment banker is a major factor in the success of the public offering, the technology entrepreneur should approach one through a mutual contact. Commercial banks, attorneys specializing in securities work, major accounting firms, providers of the initial financing, or prominent members of the company's board of directors can usually provide the needed suggestions and introductions. Since the relationship will be ongoing, not ending with the completion of the offering, the technology entrepreneur should employ several criteria in the selection process, such as reputation, distribution capability, advisory services, experience, and cost.

The success of the offering also depends on the underwriter's distribution capability. A technology entrepreneur wants the stock of his or her company distributed to as wide and varied a base as possible. Since each investment banking firm has a different client base, the technology entrepreneur should compare client bases of possible managing underwriters. Does the client base consist of predominant institutional investors, individual investors, or balanced between the two? Is the base more internationally or domestically oriented? Are the investors long term or speculators? What is the geographic distribution—local, regional, or nationwide? A strong managing underwriter and syndicate with a quality client base will help the stock sell well and perform well in the **aftermarket**. The aftermarket is the term used to refer to the performance of a stock after the excitement of the IPO has subsided. How a stock performs in the long run is often dependent on the ability of the underwriters to gain interest from *their* investors.[25]

The final factor to be considered in the choice of a managing underwriter is cost. Going public is a very costly proposition and costs *do* vary greatly among underwriters. The average gross spread as a percentage of the offering between underwriters can be as high as 10%. Costs associated with various possible managing underwriters must be carefully weighed against the other four factors. The key is to obtain the best possible underwriter and not try to cut corners, given the stakes involved in a successful IPO.

6.4.5 Registration Statement and Timetable

Once the managing underwriter has been selected, a planning meeting should be held of company officials responsible for preparing the **registration statement**, the company's independent accountants and lawyers, and the underwriters and their counsel. At this important meeting, frequently called the "all hands" meeting, a timetable is prepared, indicating dates for each step in the registration process. This timetable establishes the effective date of the registration, which determines the date of the final financial statements to be included. The company's end of the year, when regular audited financial statements are routinely prepared, is taken into account to avoid any possible extra accounting and legal work. The timetable should indicate the individual responsible for preparing the various parts of the registration and offering statement. Problems often arise in an IPO owing to the timetable not being carefully developed and agreed to by all parties involved.

After the completion of the preliminary preparation, the first public offering normally requires 6–8 weeks to prepare, print, and file the registration statement with the SEC. Once the registration statement has been filed, the SEC generally takes 4–8 weeks to declare the registration effective. Delays frequently occur in this process, such as (1) during heavy periods of market activity, (2) during peak seasons, such as March, when the SEC is reviewing a large number of proxy statements, (3) when the company's attorney is not familiar with federal or state regulations, (4) when a complete and full disclosure is resisted by the company, or (5) when the managing underwriter is inexperienced.

In reviewing the registration statement, the SEC attempts to ensure that the document makes a full and fair disclosure of the material reported. The SEC has no authority to withhold approval of or require any changes in the terms of an offering that it deems unfair or inequitable, so long as all material information concerning the company and the offering is fully disclosed. The National Association of Securities Dealers (NASD) will review each offering, principally to determine the fairness of the underwriting compensation and its compliance with NASD bylaw requirements.

6.4.6 The Prospectus

The prospectus portion of the registration statement is almost always written in a highly stylized narrative form, as it is the selling document of the company. While the exact format is decided by the company, the information must be presented in an organized, logical sequence and in an easy-to-read, understandable manner in order to obtain SEC approval. Some of the most common sections of a prospectus include the cover page, prospectus summary, the company, risk factors, use of proceeds, dividend policy, capitalization, dilution, selected financial data, the business, management and owners, type of stock, underwriter information, and the actual financial statements.

The cover page includes such information as company name, type and number of shares to be sold, a distribution table, date of prospectus, managing underwriter(s), and syndicate of underwriters involved. There is a preliminary prospectus and then a final prospectus once approved by the SEC.

The preliminary prospectus is used by the underwriters to solicit investor interest in the offering while the registration is pending. The final prospectus contains all of the changes and additions required by the SEC and blue sky examiners and the information concerning the price at which the securities will be sold. The final prospectus must be delivered with or prior to the written confirmation of purchase orders from investors participating in the offering.

- *Part II*: This section of Form S-1 contains specific documentation of the issue in an answer format and exhibits such things as the articles of incorporation, the underwriting agreements, company bylaws, stock option and pension plans, and contracts. Other items presented include indemnification of directors and officers, any sale of unregistered securities within the past 3 years, and expenses related to going public.

- *Form S-18*: In April 1979, the SEC adopted a simplified form of the registration statement—Form S-18—for companies planning to register no more than $7.5 million of securities. This form was designed to make going public easier and less expensive by having fewer reporting requirements. Form S-18 requires less detailed description of the business, officers, directors, and legal proceedings; requires no industry segment information; allows financial statements to be prepared in accordance with generally accepted accounting practices rather than under the guidelines of Regulation S-X; and requires an audited balance sheet at the end of the past fiscal year (rather than the past 2 years) and audited change in financial positions and stockholder equity for the past 2 years (rather than the past 3 years). Although Form S-18 can be filed for review with the SEC's Division of Corporation Finance in Washington, DC, as are all S-1 forms, it can also be filed with the SEC's regional office.

6.4.7 The Red Herring

Once the preliminary prospectus is filed, it can be distributed to the underwriting group. This preliminary prospectus is called a **red herring** because a statement printed in red ink appears on the front cover, which states that the issuing company is not attempting to sell its shares. The first page of the red herring for UTI Asset Management Company, Limited, an Indian company, is shown in Exhibit 6.4. The registration statements are then reviewed by the SEC to determine if adequate disclosures have been made. Some deficiencies are almost always found and are communicated to the company either by telephone or a **deficiency letter.** This

Please read section 60B of Companies Act
(This Draft Red Herring Prospectus will be updated upon ROC filling)
DRAFT RED HERRING PROSPECTUS
Dated January 9, 2008
100% Book Built Offer

UTI Mutual Fund

UTI ASSET MANAGEMENT COMPANY LIMITED

(We were incorporated as UTI Asset Management Company Private Limited under the Companies Act, 1956 on November 14, 2002. Our status was subsequently changed to a public limited company and the name of our Company was accordingly changed to UTI Asset Management Company Limited by a special resolution passed at the Annual General Meeting on September 18, 2007. A fresh certificate of incorporation consequent to the change of the name was granted on November 14, 2007 by the Registrar of Companies, Maharashtra, Mumbai. The registered office of the Company was at 13, Sir Vithaldas Thackersay Marg, New Marine Lines, Mumbai 400 020. Pursuant to a Board resolution dated May 20, 2003, the registered office was shifted to UTI Tower, Gn Block, Bandra Kurla Complex, Bandra (East), Mumbai 400 051, which is the current Registered Office. For details in the changes of name and registered office, see the section titled "History and Corporate Structure" on page 88 of this Draft Red Herring Prospectus.)

Registered and Corporate Office: UTI Tower, Gn Block, Bandra Kurla Complex, Bandra (East), Mumbai - 400 051.
Compliance Officer: Mr. Kiran N. Vohra
Tel: (91 22) 6678 6666 **Fax:** (91 22) 2652 8991, **Email:** investor@uti.co.in, **Website:** www.utimf.com

PUBLIC OFFER OF 48,500,000 EQUITY SHARES OF Rs. 10 EACH ("EQUITY SHARES") OF UTI ASSET MANAGEMENT COMPANY LIMITED ("UTI AMC" OR THE "COMPANY") THROUGH AN OFFER FOR SALE BY THE SELLING SHAREHOLDERS FOR CASH AT A PRICE OF Rs. [●] PER EQUITY SHARE (INCLUDING A SHARE PREMIUM OF Rs. [●] PER EQUITY SHARE) AGGREGATIG TO Rs. [●] (THE "OFFER"). THE OFFER ALSO COMPRISES A RESERVATION OF NOT LESS THAN 485,000 EQUITY SHARES FOR SUBSCRIPTION BY ELIGIBLE EMPLOYEES (AS DEFINED HEREIN) (THE "EMPLOYEE RESERVATION PORTION") AND THE OFFER TO THE PUBLIC OF 48,015,000 EQUITY SHARES (THE "NET OFFER"). THE NET OFFER WILL CONSTITUTE 48,015,000 EQUITY SHARES AND [●]% OF THE FULLY DILUTED POST OFFER PAID-UP CAPITAL OF THE COMPANY.

PRICE BAND: RS. [●] TO RS. [●] PER EQUITY SHARE OF FACE VALUE RS. 10 EACH

THE FLOOR PRICE IS [●] TIMES THE FACE VALUE AND THE CAP PRICE IS [●] TIMES THE FACE VALUE

In case of revision in the Price Band, the Bidding/Offer Period will be extended by three additional days after revision of the Price Band subject to the Bidding /Offer Period not exceeding 10 working days. Any revision in the Price Band and the Bidding/Offer Period, if applicable, will be widely disseminated by notification to the National Stock Exchange of India Limited ("NSE") and the Bombay Stock Exchange Limited ("BSE"), by issuing a press release, and also by indicating the change on the websites of the Book Running Lead Managers and at the terminals of the Syndicate.

The Offer is being made through the 100% Book Building Process wherein not less than 50% of the Net Offer shall be available for allocation on a proportionate basis to Qualified Institutional Buyers ("QIBs"), out of which 5% shall be available for allocation on a proportionate basis to Mutual Funds only. The remainder shall be available for allocation on a proportionate basis to QIBs and Mutual Funds, subject to valid Bids being received from them at or above the Offer Price. Further, not less than 15% of the Net Offer will be available for allocation on a proportionate basis to Non-Institutional Bidders and not less than 35% of the Net Offer will be available for allocation on a proportionate basis to Retail Individual Bidders, subject to valid Bids being received at or above the Offer Price. Further up to [●] Equity Shares shall be available for allocation on a proportionate basis to the Eligible Employees, subject to valid Bids being received, at or above the Offer Price.

RISK IN RELATION TO THE FIRST OFFER

This being the first public Offer of Equity Shares of our Company, there has been no formal market for the Equity Shares of our Company. The face value of the Equity Shares is Rs.10 per Equity Share and the Offer Price is [●] times of the face value. The Offer Price (as determined by our Company in consultation with the Book Running Lead Managers and the Selling Shareholders on the basis of assessment of market demand for the Equity Shares offered by way of the Book Building Process and as stated in the section titled "Basis for Offer Price" on page 28 of this Draft Red Herring Prospectus) should not be taken to be indicative of the market price of the Equity Shares after the Equity Shares are listed. No assurance can be given regarding an active and/or sustained trading in the Equity Shares of our Company or regarding the price at which the Equity Shares will be traded after listing.

This Offer has been rated by [●] as [●] (pronounced [●]) indicating [●]. For details see the section titled "General Information" on page 12 of this Draft Red Herring Prospectus.

GENERAL RISKS

Investments in equity and equity-related securities involve a degree of risk and investors should not invest any funds in this Offer unless they can afford to take the risk of losing their investment. Investors are advised to read the risk factors carefully before taking an investment decision in this Offer. For taking an investment decision, investors must rely on their own examination of the Offerer and the Offer, including the risks involved. The Equity Shares offered in the Offer have not been recommended or approved by the Securities and Exchange Board of India ("SEBI"), nor does SEBI guarantee the accuracy or adequacy of this Draft Red Herring Prospectus. Specific attention of the investors is drawn to the section titled "Risk Factors" on page (xiii) of this Draft Red Herring Prospectus.

COMPANY'S ABSOLUTE RESPONSIBILITY

The Company, having made all reasonable inquiries, accepts responsibility for and confirms that this Draft Red Herring Prospectus contains all information with regard to the Company and the Offer that is material in the context of the Offer, that the information contained in this Draft Red Herring Prospectus is true and correct in all material aspects and is not misleading in any material respect, that the opinions and intentions expressed herein are honestly held and that there are no other facts, the omission of which makes this Draft Red Herring Prospectus as a whole, or any information or the expression of any opinions or intentions, misleading in any material respect.

LISTING ARRANGEMENT

The Equity Shares offered through this Draft Red Herring Prospectus are proposed to be listed on the NSE and the BSE. We have received in-principle approval from NSE and BSE for the listing of our Equity Shares pursuant to letters dated [●] and [●], respectively. For purposes of the Offer, the Designated Stock Exchange is [●].

GLOBAL COORDINATORS AND BOOK RUNNING LEAD MANAGERS			REGISTRAR TO THE OFFER
JM FINANCIAL **JM Financial Consultants Private Limited** 141, Maker Chambers III Nariman Point Mumbai 400 021 Tel: (91 22) 6630 3030 Fax: (91 22) 2204 7185 Email: grievance.ibd@jmfinancial.in Contact Person: Ms. Poonam Karande Website: www.jmfinancial.in	**citi** **Citigroup Global Markets India Private Limited** 12th Floor, Bakhtawar Nariman Point Mumbai 400 021 Tel: (91 22) 6631 9999 Fax: (91 22) 6631 9803 Email: utiamc.ipo@citi.com Contact Person: Mr. Anuj Mithani Website: www.citibank.co.in	**ENAM** **Enam Securities Private Limited** 801, Dalamal Towers Nariman Point, Mumbai 400 021 Tel: (91 22) 6638 1800 Fax: (91 22) 2284 6824 Email: utiamcipo@enam.com Contact Person: Mr. Pranav Mahajani Website: www.enam.com	**KARVY** **Karvy Computershare Private Limited** Plot No 17 to 24, Vittalrao Nagar, Madhapur, Hyderabad 500 081 Tel: (91 40) 2343 1553 Fax: (91 40) 2343 1551 Email: utiamcipo@karvy.com Contact Person: Mr. Murali Krishna Website: www.karvy.com

BOOK RUNNING LEAD MANAGERS				
Goldman Sachs **Goldman Sachs (India) Securities Private Limited** 951A, Rational House Appasaheb Marathe Marg Prabhadevi, Mumbai 400 025 Tel: (91 22) 6616 9000 Fax: (91 22) 6616 9090 Email: uti_amc_ipo@gs.com Contact Person: Mr. Sachin Dua Website: www.gs.com/ country_pages/india	**UBS** **UBS Securities India Private Limited** 2/ F, Hoechst House Nariman Point, Mumbai 400 021 Tel: (91 22) 2286 2000 Fax: (91 22) 2281 4676 Email: utiamc@ubs.com Contact Person: Mr. Venkat Ramakrishnan Website: www.ibb.ubs.com/ Corporates/indianipo	**ICICI Securities** **ICICI Securities Limited** ICICI Centre H.T. Parekh Marg Churchgate Mumbai 400 020 Tel: (91 22) 2288 2460/70 Fax: (91 22) 2282 6580 Email: utiamc_ipo@isecltd.com Contact Person: Mr. Rajiv Poddar Website: www.icicisecurities.com	**SBI Capital Markets Limited** 202, Maker Towers 'E' Cuffe Parade Mumbai 400 005 Tel: (91 22) 2218 9166 Fax: (91 22) 2218 8332 Email: tecpro.ipo@sbicaps.com Contact Person: Mr. Rohan Talwar Website: www.sbicaps.com	**CLSA** ASIA - PACIFIC MARKETS **CLSA India Limited** 8/F Dalamal House Nariman Point Mumbai 400 021 Tel: (91 22) 6650 5050 Fax: (91 22) 2285 6524 E-mail: utiamc.ipo@clsa.com. Contact Person: Mr. Anurag Agarwal Website: www.india.clsa.com

BID/OFFER PROGRAMME			
BID/ OFFER OPENS ON	[●]	**BID/ OFFER CLOSES ON**	[●]

EXHIBIT 6.4

Red Herring for UTI Asset Management Company Limited.

preliminary prospectus contains all the information contained in the final prospectus except that which is not known until shortly before the effective date: offering price, underwriters' commission, and amount of proceeds.

6.4.8 Reporting Requirements

Going public requires a complex set of reporting requirements:

- The first requirement is the filing of a Form SR sales report, which the company must do within 10 days after the end of the first 3 month period following the effective date of the registration. This report includes information on the number of securities sold and still to be sold, and the proceeds obtained by the company and their use. A final Form SR sales report must be filed within 10 days of the completion or termination of the offering.

- The company must file annual reports on Form 10-K, quarterly reports on Form 10-Q, and specific transaction reports on Form 8-K. The information in Form 10-K on the business, management, and company assets is similar to that in Form S-1 of the registration statement. Of course, audited financial statements are required.

- The quarterly report on Form 10-Q contains primarily the unaudited financial information for the most recently completed fiscal quarter. No 10-Q is required for the fourth fiscal quarter.

- A Form 8-K report must be filed within 15 days of such events as the acquisition or disposition of significant assets by the company outside the ordinary course of the business, the resignation or dismissal of the company's independent public accountants, or a change in control of the company.

- The company must follow the proxy solicitation requirements regarding holding a meeting or obtain the written consent of security holders. The timing and types of materials involved are detailed in the Securities and Exchange Act of 1933.

These are but a few of the reporting requirements of public companies. All the requirements must be carefully observed, as even inadvertent mistakes can have negative consequences on the company. The reports required must be filed on time.

SUMMARY

This chapter focused on the exit strategies most commonly used by technology entrepreneurs. An exit strategy is often selected early on in a venture's life. Ventures that aim for an exit by acquisition, for example, should establish a strategy and operations that are conducive to achieving that objective. In contrast, ventures that target an IPO as an exit strategy are likely to select other business objectives and strategies.

Exiting a venture provides a liquidity event for the entrepreneurs who founded the venture and the investors who supported it financially along the way. The shares they have been holding onto suddenly can be converted into cash. The amount of cash that is available to the founders and investors depends on the valuation of the enterprise at the time of the exit. Technology entrepreneurs must be mindful of the factors that acquiring firms use to evaluate the value of company they intend to acquire. Such factors as the financial health of the venture, the position within the product life cycle of its portfolio of products and services, the operational efficiencies, and other factors are all part of the due diligence process used by acquiring firms. Technology entrepreneurs should structure their company in a way that easily conveys the value of what they have created.

Merger is another exit option for technology entrepreneurs. Mergers are similar to acquisitions, but a merger often results in more negotiations regarding control of the merged venture. Technology entrepreneurs in merged ventures may take longer to exit the merged venture than they would from an acquisition. Many technology entrepreneurs negotiate earn-out deals, whereby their shares gradually are purchased by the merged entity until they are bought out.

Valuing a technology venture is complex, and many different metrics are available. The easiest metrics to understand and apply in a rough way are known as "multiples." Companies in an industry are acquired for a sum that can be calculated as a multiple of some key datum, such as revenue, profit, or cash flow. This multiple can then be applied to the same datum in any company in the industry to determine its relative value.

In the end, valuation is a function of the future cash flow that can be expected to be generated by a venture. More sophisticated valuation techniques attempt to estimate the future cash flows likely to be generated by a venture, and then discount that future cash flow to a corresponding present value. The discount rate used to determine present value is a combination of the risk-free rate of return available to any investor and the risk premium that is associated with the particular venture being valued.

Finally, this chapter examines the processes and challenges associated with an IPO. This is the most expensive exit strategy for a technology venture, but also potentially the most lucrative. Technology ventures that choose the IPO route will need to retain the services of an investment banking firm to handle the underwriting chores. The bank chosen to lead the IPO will attempt to syndicate the deal to its investors and other investment banking firms. The success of this syndication will have a large impact on the aftermarket performance of the stock as well.

Whether via acquisition, merger, or IPO, the exit strategy for the technology entrepreneur is always an important consideration. Although the specific exit strategy may not be fully known in advance, the technology entrepreneur should build the venture each and every day with the objective of selling all founder and investor shares at some future date. With that in mind, the technology entrepreneur will take care of those factors that are attractive to others who may find the technology venture a tempting investment of their own.

STUDY QUESTIONS

1. What is the difference between an acquisition and a merger?

2. How does a technology entrepreneur estimate the value of the venture using multiples? Give some examples.

3. What is the most important element in valuing a technology venture? Explain.

4. What is meant by the term "due diligence"?

5. Explain why the selection of an investment bank to be the lead underwriter is an important decision for a technology entrepreneur.

6. What are some of the costs associated with a company going public?

7. What are the various sections of a registration statement and what purpose does each of them serve?

8. Why does an acquiring firm want to understand the position in the product life cycle of the products and services in the target venture's portfolio? Explain.

9. What are some of the disadvantages of an IPO? Advantages?

10. Explain how an exit strategy can influence the day-to-day decision making of a technology entrepreneur.

EXERCISES

1. *IN-CLASS EXERCISE*: For this exercise, the instructor will guide the students to the web site IPO Central (http://www.hoovers.com/global/ipoc/index.xhtml). Look at some recent IPO deals and how they performed in the marketplace. Students should discuss why some deals did well, and why others did not perform so well. What are the fundamental differences? How could a venture that did not do well in its IPO have done better? What IPO deals are scheduled to occur in the near future? Is the timing right for these deals? Track them and see.

2. *OUT-OF-CLASS EXERCISE*: Since it was launched late in 1996, BizBuySell has become the Internet's largest business-for-sale site. Of the more than 10,000 listings in its databases, some 15% are listed by personal owners, and the remainder comprises about 700 business brokers. Manufacturing businesses account for 15% of the businesses listed; wholesale and retail for 40%; and services make up roughly 30%. Of the thousands of people who visit the site, most are looking for a business to buy. Not only may buyers search the database at no charge, they may also register to be notified about new business listings that match their interests.

    Students should visit the BizBuySell web site at http://www.bizbuysell.com and explore the variety of businesses listed there for sale. Each student should select a category (e.g., Pet Shops & Supplies) to explore and pick at least five small businesses that are listed for sale on the site. For each business selected, take note of the following information:

- Asking price
- Gross sales
- Cash flow
- Inventory
- Financing options

Compare the five businesses selected and consider the following questions:

1. Which one seems to offer the "best deal"?
2. Which one seems to have the highest valuation? Why?
3. Among the various industries, which ones seem to have the highest multiples of gross sales? Why?

KEY TERMS

Acquisition: A transaction involving two companies in which one company purchases the other. The company that is purchased no longer exists, but is incorporated into the other company.

Aftermarket: The term used to refer to the performance of a stock after the excitement of the IPO has subsided. How a stock performs in the long run is often dependent on the ability of the underwriters to gain interest from *their* investors.

Bootstrap purchase: The acquiring company can acquire a small amount of a firm, such as 20–30%, for cash. The acquiring company then purchases the remainder of the target venture by a long-term note that is paid off over time out of the acquired company's earnings.

Direct purchase: In the direct purchase of the firm, the acquiring company often obtains funds from an outside lender or the seller of the company being purchased.

Discount rate: In conducting a present value analysis, a future cash flow must be discounted by the rate of interest that otherwise could be earned by present dollars.

Due diligence: Due diligence involves examining a number of key elements of a target venture, including its financial health, the potential for synergy, the market position and future potential of the venture, the research and development history and road map for the venture, legal considerations, and plans for managing the acquired entity.

Earn out: An earn-out strategy is used for ventures that have begun to generate consistently strong positive cash flow. The management team initiates a monthly or quarterly buyback of common stock from the owners of one of the merged entities.

Exit strategy: The technology entrepreneur's strategic withdrawal from ownership or operation of their venture.

Initial Public Offering (IPO): An IPO is a process whereby a private company qualifies to sell its shares on a public stock exchange.

Investment horizon: This refers to the time an investor's money will be tied up in a technology venture until the time of exit.

Liquidity event: The exit from a private venture for shareholders represents their opportunity to convert shares to cash.

Merger: A transaction involving two or more companies in which the companies join to form a new bigger company.

Present value: This is the value in today's dollars of some future cash flow. Because money today can earn interest, future cash flows must be discounted by at least as much as that interest amount to convert to its present value.

Red herring: The preliminary prospectus document has been given this name owing to the red ink used in the printing of the front cover.

Registration statement: Document filed with the SEC before a company can go public. It consists of two parts: the prospectus and the registration statement.

Synergy: The phrase "the whole is greater than the sum of its parts" is a good definition of the concept of synergy. Technology entrepreneurs who seek to exit their companies via acquisition should consider whether the company provides any synergies to potential acquirers.

Underwriter: A company that administers the public issuance and distribution of securities. A company will utilize one when filing an IPO.

Underwriting syndicate: A syndicate is a group of investment banks and other, usually institutional, investors who subscribe to an IPO.

Value drivers: Strategic advantages that lead to value creation in an industry and that are attractive elements of an acquisition candidate in an industry.

Vertical strategy: A vertical strategy is one where a venture develops expertise in a given industry that can be expanded to other industries with only slight modifications.

WEB RESOURCES

The web sites below are intended as destinations for your further exploration of the concepts and topics discussed in this chapter:

1. http://money.cnn.com/news/: A service of CNN, the leader in reporting breaking news around the world. This site reports the most current business and financial news.

2. http://www.investopedia.com/university/mergers/: This site is run by Forbes Media Company and provides an introduction to mergers and acquisitions.

3. http://www.thebodyshop.com/bodyshop/: The official web site for the Body Shop International PLC, the manufacturer and retail store discussed in the case analysis for acquisitions.

4. http://www.loreal.com/dispatch.aspx: The official web site for L'Oréal, the cosmetics company discussed in the case analysis for acquisitions.

5. http://www.marketwatch.com/: A division of the Wall Street Journal, this site provides news and commentary on business markets, technology, and finances.

6. http://www.sec.gov/: The official web site for the SEC. Companies must register with the SEC in order to issue a public offering.

7. http://www.finra.org/index.htm: The largest nongovernmental regulator for all U.S. security firms.

CASE STUDY

Omnisio, an Easy and Practical Acquisition for Google

Only a couple of months after the Atherton, CA, start-up Omnisio publicly launched, Google made moves to acquire the company. Omnisio's web service allows users of the popular YouTube to slice and edit clips of audio and video to customize and create interactive videos. Their application for online videos has introduced social interaction where commentary can be embedded with the video. Additionally, users are able to take video clips and merge them together to form one video. The acquisition was expected to be one of the steps toward helping YouTube maintain its position as the market leader of online video.

In mid-2008, it was estimated that Google offered $15 million for the fresh start-up. While not a normal occurrence for the majority of start-ups, it appears that the founders of Omnisio have more than plain luck to be an acquisition target so quickly after launch. Tracing back the steps that got this start-up moving reveals a story of a determined and skilled group of young Australian engineers who came to form a team in Silicon Valley after working together at the venture-backed start-up Sensory Networks.

Omnisio was founded by Ryan Junee, Julian Frumar, and Simon Ratner. Ryan is known as the Omnisio visionary and was the key to providing the driving force behind the company's strategic direction. Ryan's preventure educational track record exemplifies his internal drive to learn and grow. Earning a Bachelor's degree in Computer Engineering, Finance and Commercial Law, a Master's in Electrical Engineering, and the start of a PhD in Electrical Engineering demonstrates his passion for technology and personal growth.

At some point into his PhD program, he chose to redirect his energy and "passion for high-tech start-ups." Along with school, Ryan experienced building up and guiding Sensory Networks; he had the opportunity to wear many hats for the company. With his strong technical background, work experience, mixed with the desire to grow, it is no surprise he contributed to being able to so quickly develop a successful start-up a company such as Omnisio.

The second founder, Julian Frumar, was responsible for the overall user experience of Omnisio. A user interface is particularly important to software applications as it is what makes statements about the quality and brand image of the service. Omnisio gained a reputation of having a slick interface, with fantastic and easy usability where people can simply drag and drop to use the tool. This was likely derived from Julian's experience and expertise. Additionally, Omnisio gained advantages from his previous experiences with developing the image and online brands of other start-ups such as Vquence and Sensory Networks.

The final founder, as well as technical lead, of Omnisio was Simon Ratner. Simon's strong engineering background in software and electrical engineering contributed to the backbone of the company. He too was one of the first employees at Sensory Networks where he gained the skills and confidence in his ability to build up a successful product from the early stages. The experience surely contributed to the technical know-how in building Omnisio. The technical knowledge of the team helped them make smart decisions, such as ensuring that the online tool utilizes YouTube's data servers, while not actually needing to host anything. This model minimizes Omnisio's costs. While each member of Omnisio's original team had a technical background, they each had unique skills and different experiences to bring to the table and ensure the success of the company.

Pulling together their knowledge, the Australian founders Ryan, Julian, and Simon started Omnisio with the support of Y Combinator. Y Combinator is a company that places small investments in premature-stage companies, usually at the idea stage. For a small investment in the companies that Y Combinator funds, they usually expect a 2–10% stake. Y Combinator is driven to get these start-ups to the point where larger investors or acquirers will have an interest in the company. Y Combinator has helped provide the seed funding to other successful companies. For example, the Y Combinator start-up Auctomatic was acquired by a Canadian company, Communicate.com.

Y Combinator showcased Omnisio as one of the promising start-ups at their Spring 2008 Y Combinator Demo Day. They create an opportunity for their start-ups to draw in investors and acquirers. Omnisio formally launched shortly after Demo Day, and only a few months later Google proposed an offer for an acquisition of the company.

So why would Omnisio be such a key element to the YouTube division of giant corporation Google? Omnisio is a company whose early success appears to align with the timing of the rise and demand of socializing and expanding the interactions of users on the

Internet. The demand and desire to personalize, customize, and share opinions online have resulted from the flurry and growth of social networking web sites. Omnisio's features, which include speech bubbles, spotlights, video pauses, and notes, are key elements to help YouTube keep up with the demand of web social interaction.

Not only are the features from Omnisio important to help keep YouTube fresh, Omnisio lined themselves up to be a natural and easy acquisition for the company. Although it is not known if the original intent of the founders was to create this company in order to be snatched up by Google, they certainly managed to align their features and services to meet a need of YouTube. Omnisio already utilized the video from YouTube and Google video; the fact that they made use of Google's assets makes the transition phase into Google's corporation and environment much simpler.

While it could be argued that the founders of Omnisio should feel lucky that they just joined the leading online video business, they have certainly much to offer the giant market leader, Google. The founders have a strong and key technical expertise that will be key in the future enhancement of YouTube's viewer and user experience. Their vision for change in the online video experience is necessary to help YouTube maintain their market share from strong, promising competitors such as Hulu.

Questions for Discussion

1. Assuming that the founders of Omnisio created the company with the intention of being acquired by Google, what aspects of the business plan and technology would Omnisio have structured around that intention?
2. What type of analysis do you think Google did before acquiring Omnisio? What are the key factors for Google to make sure this acquisition of Omnisio is a success?
3. Y Combinator showcased 19 companies as promising start-ups at their Spring 2008 Y Combinator Demo Day including Omisio. Investing in the ideas of 19 people seems quite risky. What strategies do you think Y Combinator has to minimize the risk in these companies?

Source: Adapted from Communicate Acquires Y Combinator Startup Auctomatic, Unveils New Business Strategy. http://www.techcrunch.com/2008/03/26/communicate-acquires-y-combinator-startup-auctomatic-unveils-new-business-strategy/; March 2008; Google Acquires Omnisio to Spice Up YouTube. http://www.techcrunch.com/2008/07/30/google-acquires-omnisio-to-spice-up-youtube/; July 2008; Getting Started: Creating or editing annotations, You Tube Help Center. http://www.google.com/support/youtube/bin/answer.py?answer=92710; 2008; Ten Y Combinator companies that want to revolutionize social networking, search, databases and anything else you can think of. http://venturebeat.com/2008/03/19/ten-y-combinator-companies-that-want-to-revolutionize-social-networking-search-databases-and-anything-else-you-can-think-of/; March 2008; Y Combinator Alum Omnisio joins Google's YouTube. http://www.xconomy.com/boston/2008/07/30/y-combinator-alum-omnisio-joins-googles-youtube/; July 2008.

ENDNOTES

[1] Bernard S. Finish line. *Entrepreneur* 2007;**35**(8):36.

[2] Rozens A. IPO drought to linger into 2010. *Investment Dealers' Digest* 2009;**75**(3):6–8.

[3] Capon A. Exit strategies for entrepreneurs. *Global Investor* 1997;**November**:16–19.

[4] Newton D. Enticing investors. *Entrepreneur* 2001;**January**.

[5] Nolop B. Rules to acquire by. *Harv Bus Rev* 2007;**85**(9):129–39.

[6] Andriole SJ. Mining for digital gold: technology due diligence for CIOs. *Commun AIS* 2007;**2007**(20):371–81.

[7] Mero J. People are his bottom line. *Fortune* 2007;**155**(7):30.

[8] Hamerman J. Microsoft deal pro sounds off. *Mergers Acquisit Rep* 2005;**October**:1–12.

[9] Dalziel M. The seller's perspective on acquisition success: empirical evidence from the communications equipment industry. *J Eng Technol Manage* 2008;**25**(3):168–83.

[10] Adams M. Exit pay-offs for the entrepreneur. *Mergers Acquisit: Dealmaker's J* 2004;**March**:24–8.

[11] Historia de la Empresa, Quienes somos? Officenet web site. http://www.officenet.com.ar/ (accessed April 6, 2008).

[12] Quinlan R. Ideal acquisitions need effective evaluation. *Financ Exec* 2008;**24**(5):17.

[13] Watson DG. Acquisitions: how to avoid implementation pitfalls and improve the odds of success. *Global Bus Org Excellence* 2007;**27**(1):7–19.

[14] Kvesic DZ. Product lifecycle management: marketing strategies for the pharmaceutical industry. *J Med Mark* 2008;**8**(4):293–301.

[15] Wang C, Xie F. Corporate governance transfer and synergistic gains from mergers and acquisitions. *Rev Financ Stud* 2009;**22**(2):829–58.

[16] Morrison NJ, Kinley G, Ficery KL. Merger deal breakers: when operational due diligence exposes risks. *J Bus Strategy* 2008;**29**(3):23–8.

[17] Roy P, Roy P. The Hewlett Packard-Compaq computers merger: insight from the resource based view and the dynamic capabilities perspective. *J Am Acad Bus* 2004;**5**(1/2):7–14.

[18] Vaughan B. Earn outs come back into fashion. *Buyouts* 2008;**21**(17):24.

[19] Hering T, Olbrich M. Valuation of startup Internet companies. *Int J Technol Manage* 2006;**33**(4):409–19.

[20] Stone B. Microsoft buys stake in Facebook. *N Y Times* 2007;**October 25**.

[21] Smith JK, Smith RL. *Entrepreneurial finance 2E*. Hoboken, NJ: John Wiley and Sons; 2004.

[22] Yoon-Jun L. Strategy of startups for IPO timing across high technology industries. *Appl Econ Lett* 2008;**15**(11):869–77.

[23] Beales R, Cox R, Silva L. Master of market timing. *Wall Street J, East Ed* 2008;**251**(55):C12.

[24] Kulkarni K, Sabarwal T. To what extent are investment bank differentiating factors relevant for firms floating moderate sized IPOs? *Ann Finance* 2007;**3**(3):297–327.

[25] Gleason K, Johnston J, Madura J. What factors drive IPO aftermarket risk? *Appl Financ Econ* 2008;**18**(13):1099–110.

Intellectual Property and Contracts

Intellectual Property Management and Protection

After studying this chapter, students should be able to:

- Differentiate between types of intellectual property (IP)
- Understand basic mechanisms for protecting rights in technology and other IP
- Recall steps involved in obtaining, maintaining, and exploiting IP rights
- Describe basic record-keeping procedures
- Explain the prerequisites for trade secret status
- Be familiar with patent, trade secret, copyright, and trademark requirements

TECH VENTURE INSIGHT

VisiCalc Pays for Decision *Not* to Patent First Spreadsheet Program

Many people believe that one of the key contributors to the growth of personal computers was a spreadsheet program called "VisiCalc." VisiCalc was the brainchild of Dan Bricklin while he was studying for his MBA at Harvard Business School in 1978. He observed that an error in a single cell of a common (paper) spreadsheet required that the value in every other cell be changed. As a way to solve this problem, Bricklin envisioned electronic spreadsheets where a cell entry could be changed, and all the other cells would change automatically according to specified formulae.

To implement this new idea, Bricklin enlisted the expertise of a software programmer named Bob Frankston. Their partnership led to the creation of Software Arts, Inc. Soon, Bricklin and Frankston's fledgling company garnered the attention of Personal Software, a software publisher that loaned Software Arts an Apple II computer on which to perform a demo of VisiCalc.

Personal Software eventually funded the development of VisiCalc by licensing it from Software Arts and becoming its sole distributor. What once began as an innovative idea in

Bricklin's apartment building in mid-1978 went on to become, in 1979, the first spreadsheet program ever released on the market. VisiCalc was considered revolutionary in many circles.

However, VisiCalc's story does not end there. No sooner had the company completed its initial investment and established a firm market for its product, other companies began to place conceptually identical spreadsheet programs on the marketplace. These newcomers not only profited from publicly available information of the original spreadsheet program, including its established market acceptance, but arguably they were able to muster greater resources to overcome VisiCalc's market leadership.

According to Bricklin, a conscious decision was taken not to pursue patent protection of VisiCalc. He stated that Personal Software (later renamed VisiCorp) had retained a patent attorney. The patent attorney explained that there were many obstacles involved in obtaining a patent on the software, estimating only a 10% chance of success. Based on this advice, and the potentially high costs involved, Bricklin decided not to pursue a patent. Instead, copyright and trademark protection were vigorously pursued to keep others from replicating their work.

"The enormous importance and value of the spreadsheet, and of protections in addition to copyright to keep others from copying our work, did not become apparent for at least two years, too late to file for patent protection," said Bricklin. "If I invented the spreadsheet today, of course, I would file for a patent," he added.

Source: Hormby, VisiCalc and the Rise of the Apple II, http://lowendmac.com/orchard/06/visicalc-origin-bricklin.html. US Patent and Trademark Office; Dan Bricklin, Patenting VisiCalc, http://www.bricklin.com/patenting.htm.

INTRODUCTION

Intellectual property, the intangible asset that results from creativity, innovation, know-how, and reputation, is fast becoming the primary asset of modern business. Only a few decades ago, the bulk of U.S. corporate assets were tangible in nature. Intangibles such as intellectual property comprised only 20% of corporate assets. However, by 1998, the ratio of intangible to tangible corporate assets had essentially reversed; the market value of the S&P 500 was approximately 80% intangible assets.[1]

Creating and maintaining rights in intellectual property and avoiding infringement of the intellectual property rights of others are becoming increasingly important aspects of any business. This is particularly true with small and emerging technology ventures. In this chapter, we will examine the various types of intellectual property and how to protect them. We begin by discussing how to recognize intellectual property assets and then will review the characteristics of the individual types of intellectual property and protection mechanisms.

7.1 INTELLECTUAL PROPERTY AND TECHNOLOGY VENTURES

In 2005, then Federal Reserve Chairman Alan Greenspan proposed that as much as 75% of the value of all publicly traded companies in the United States could be attributed to intangible assets.[2] The annual revenue realized from just the licensing of patented technology had grown from less than $10 billion in 1990 to nearly $120 billion by 2002.[3] Even so, a 2007 survey of technology executives indicated that fully 85% believe that intellectual property would increase in importance for their organizations over the next 3–5 years.[4]

Demonstrable rights to intellectual property are also a bellwether for the perceived likelihood of success of a technology venture. A study of venture capital fundings between 1995 and 2002 showed that businesses that had intellectual property were 34% more likely to be successful in obtaining subsequent rounds of venture capital funding.[5]

From the perspective of an emerging technology venture, intellectual property rights can be a great equalizer in the marketplace. Faced with established competitors, an emerging venture must have some form of sustainable competitive advantage if it is to succeed. A potent sustainable competitive advantage is possession of a legally protectable, exclusive right to intellectual property. However, unless the appropriate legal foundation is laid to protect intellectual property rights, competitors will be able to legitimately appropriate or copy the feature, and the competitive advantage will be lost.

Recently, so-called open models of doing business have gained notoriety. Examples include various consortia and open source licensing of software under various general public licenses. Examples of such consortia include SEMATECH (semiconductor devices), Semiconductor Test Consortium (STC), RFID Industry Patent Consortium (RF identification), and the Symbian Foundation (platform for converged mobile devices). Open source software licensing regimes include the Berkeley Software Distribution license, the GNU General Public License, and others.[6] Even these business models rely on intellectual property protection.[7]

The laws relating to intellectual property, however, are anything but intuitive. Valuable rights in intellectual property can be unwittingly lost by seemingly innocent courses of action, and failing to consider third-party rights can lead to disaster.

7.2 INTELLECTUAL PROPERTY PROTECTION

Intellectual property can take many forms. As will be discussed, the terms "trade secret," "utility patent," "design patent," "copyright," "mask work," and "trademark" have come to denote both an intellectual property asset and legal mechanism for protecting the underlying assets. These mechanisms provide a framework for establishing and maintaining rights in intellectual property. A strategy employing combinations of the various legal mechanisms should be developed to maximize protection for facilitating the venture's goals. Exhibit 7.1 highlights some of the characteristics of the respective legal mechanisms.

Protection Mechanism	Term	Source of law	Protectable subject matter	Scope of protection	Compatible concurrent forms of protection
Trade secret	Potentially infinite (as long as kept secret)	State law (Uniform Trade Secret Act)	Anything that can be kept secret	Precludes unauthorized copying or use. No protection against independent development	Trademark
Trademark	Potentially infinite (as long as in use and capable of identifying source) Registrations renewable 10-year terms	State Common Law; State Statutes; Federal Statute (15 USC §1051 et seq)	Anything nonutilitarian that is capable of identifying the source of goods or services	Precludes unauthorized use of trademark in context that creates any likelihood of confusion as to source, sponsorship, or affiliation	Trade Secret, Design Patent, Utility Patent, Copyright
Copyright	70 years after death of the last surviving author, or if work for hire, 95 years from publication	Federal Statute (17 USC §101 et seq)	Nonutilitarian works of authorship	Precludes unauthorized copying of copyrighted aspects of a work of authorship	Trademark, Design Patent, Utility Patent
Mask work	Until end of 10th calendar year after earlier of registration or first commercial exploitation	Federal Statute (17 USC §901 et seq)	Mask works (images, representing three-dimensional patterns in the layers of a semiconductor chip)	Precludes unauthorized use of reproductions of masks in production of competing chips	Trade Secret, Trademark, Utility Patent
Design patent	14 years from issue date	Federal Statute (35 USC §§1 et seq, §171)	Nonfunctional aspects of ornamental designs	Precludes unauthorized application of patented design to any article of manufacture, or sale or exposing for sale such article of manufacture	Utility Patent, Copyright, Trademark
Utility patent	20 years from filing date of application	Federal Statute (35 USC §1 et seq)	New and useful process, machine, manufacture, or composition of matter, or new and useful improvement thereof	Unauthorized making, using, offering to sell, selling, or importing any patented invention	Copyright, Design Patent, Maskwork Trademark

EXHIBIT 7.1

Comparison of U.S. Intellectual Property Protection Mechanisms.

7.2.1 **Recognizing Intellectual Property**

A starting point for developing intellectual property assets is, simply, recognizing their existence. When a venture is built around a particular product or idea, intellectual property embodied in that product or idea is readily identified as a candidate for protection. However, valuable intellectual property also may reside in other aspects of the business.

A potential intellectual property asset is created every time a problem is solved. When a problem is encountered in connection with a business, it is likely that its competitors will encounter the same problem. If the business develops a solution to the problem that is better than that of its competitors, obtaining the exclusive rights to use that solution can give the business an advantage. If the problem is significant and the solution is notably better than the competition's, that advantage can be significant too.

Technology ventures should continuously and systematically analyze their products and operations to identify potential intellectual property and seek opportunities to develop protectable intellectual property assets. The venture then should obtain and maintain exclusive rights to the intellectual property to the extent available and exploit it as part of an overall strategy. The starting point in recognizing intellectual property is to identify everything that gives the business an advantage, including:

- Identifying the reasons that customers are attracted to the business instead of its competitors

- Dissecting the operations and systems, products, services, and communications and analyzing each component and feature to determine whether it contributes to the competitive advantage, and whether there is anything about it that is unique, better, or distinctive

- Developing a strategy to exploit the intellectual property by securing exclusive rights through application of the appropriate legal mechanisms and/or promoting the use by others to drive sales of other products or services of the company

An effort should also be made to identify technologies or processes that are not in use by the venture, but that might be of use to others. Exclusive rights to those developments may be used to generate income through licensing or to broaden a competitive advantage.

By properly identifying and protecting as many intellectual property assets as possible, a venture will be more attractive to investors. Ventures with intellectual property holdings are more likely to be successful in obtaining multiple rounds of venture capital funding. In analyzing a company, investors evaluate not only its intellectual property, but also the systems and processes the venture has in place to protect intellectual property assets.

7.2.2 **Record Keeping**

In many instances, the basic foundation for intellectual property protection and protecting against charges of intellectual property infringement by third parties is a complete and accurate evidentiary record. There are a number of scenarios where it would be necessary to prove the date and nature of technical innovations and the project with which the innovations are associated. For example, such records often are vitally important in:

- Disputes regarding ownership of technology—whether a certain technology was first made under a particular development contract or government contract

- Disputes regarding whether a particular technology is covered by a particular license agreement

- Disputes regarding whether a certain technology is subject to a confidentiality or nonuse agreement

- Proving that intellectual property was independently developed and not derived from third-party proprietary material

- Proving that an invention was previously developed, not abandoned, suppressed, or concealed, as a defense to patent infringement[8]

- Interference proceedings before the U.S. Patent and Trademark Office or courts to determine priority of invention

IP ownership is often governed by a variety of contract and agreement types, including development agreements, license agreements, and confidentiality and non-use agreements. Good records are often necessary to retain ownership under those agreements.

Development agreements

Development agreements often specify that a particular party will obtain rights (e.g., title or a license to make, use, and sell) to all technology arising from work done under the contract or first conceived or first made under the contract. For the developer under such contracts, it is imperative to be able to show that:

- Certain technology was created prior to the contract

- Certain technology that was developed on other projects during the term of the agreement was, in fact, developed on those other projects

License agreements

A license agreement grants a licensee someone else's invention. In general, it is desirable that license agreements clearly and concisely specify the licensed technology, the term of the agreement, and the geographical area it covers. Certain license agreements sometimes relate to all technology made prior to the date of the agreement. It then becomes imperative to be able to prove, after the fact, when technology was made *vis-à-vis* the date of the agreement. The ability to do this, however, typically relies upon the sufficiency of the records that were generated contemporaneously with the technology.

Confidentiality and nonuse agreements

Confidentiality and nonuse agreements are often entered into in connection with various business negotiations. Such agreements often include provisions to the effect that technology that:

- Was already in the possession of the company or
- Is independently developed by the company
- Is not subject to the confidentiality and nonuse provisions

It is generally incumbent on the recipient of information, however, to prove prior possession or independent development. Thus, records that show the history of the development of the technology, when specific acts occurred, and the particular individuals involved in those acts should be maintained contemporaneously with the development.

Prior development defense

In some instances, the ability to prove precisely when an invention was made can be critical to support a defense against a patent infringement allegation, or to prevail in a contest for a patent. In the United States, if an accused infringer can prove he or she was the first to have made the invention in the United States, and did not abandon, suppress, or conceal the invention, the earlier invention is a complete defense against the charge of infringement and will invalidate the patent.[9] The term "suppress or conceal" can mean, for example, maintaining an invention as a trade secret for an extended period of time. If the accused patent infringer treated the invention as a trade secret, a U.S. patent would not be invalidated by the earlier invention.[10]

Similar issues arise if more than one party files patent applications claiming the same invention. When this occurs, the relative priority of the inventors is determined by an **interference proceeding** conducted by the USPTO,[11] or, if more than one patent is issued on the same invention, the federal courts. In an interference proceeding, the location of the activities related to making the invention is irrelevant. However, an earlier inventor defense requires that the invention be made in the United States.

Making an invention is a two-step process that involves *conceiving* the invention and then *reducing* the invention to practice. **Conception** is basically the mental portion of the inventive act. **Reducing the invention to practice** is building the invention and proving that it works for its intended purpose. The filing of a patent application is a constructive reduction to practice.

In a contest between inventor A and inventor B, where inventor A was the first to conceive an invention, if inventor A was also the first to reduce it to practice, he or she will be deemed the first to have made the invention. However, if inventor B, while conceiving the invention after inventor A, was the first to reduce it to practice, inventor A will be deemed to be first only if he can prove reasonable diligence in reducing the invention to practice from a point in time prior to when inventor B conceived the invention. Reasonable diligence means that there was reasonably continuous activity to reduce the invention to practice. If inventor A cannot prove that, then inventor B will be deemed the first to have made the invention.

While the law and evidentiary requirements vary from country to country, in order to prevail with an earlier inventor defense, relevant dates must be typically proven on the basis of documentary evidence. And, in the United States, each of the elements of the two-step process of making an invention must not only be established but also corroborated by evidence independent of the inventors.

7.2.3 Record-Keeping Procedures

Given the variety of situations where it becomes necessary to prove the precise nature and timeframe of development activities, it is best practice to establish procedures that facilitate complete and accurate records, keeping in mind the requirement for independent **corroboration**. They are discussed below.

Notebooks

Many companies employ some form of **invention disclosure forms** to establish positive proof of a conception date in the United States. As a practical matter, however, invention disclosure forms typically do not contain the details necessary to prove actual reduction to practice of an invention or diligence to do so. For this reason, detailed, contemporaneous **laboratory notebooks** should be maintained.

The value of an entry in a laboratory notebook as proof of conception, diligence, and/or reduction to practice of an invention is directly proportional to the specificity of the entry and the care taken to date and sign each entry and have each entry read, signed, and dated by a witness. The following are guidelines for developing laboratory notebooks:

- The context of the entry in a laboratory notebook can sometimes be used to prove a date. For example, if an entry showing conception is found in a *bound* notebook between entries dated the 3rd of January and the 5th of January, then that is relevant proof the invention was conceived sometime between the 3rd and 5th of January. It would not be as relevant, however, if a loose-leaf binder had been used instead of a bound notebook as loose-leaf binders allow for inserting pages.

- Every notebook entry should be signed and dated by the project leader, indicating the particular project with which the entry is associated. Entries should also be signed and dated by a nonparticipating witness, if possible.

- All computations, circuit diagrams, and test results should be contemporaneously entered into the notebook. It is as easy to do calculations in the notebook as on scratch paper. So long as the entry is legible, there are no particular format or neatness requirements.

- It is imperative that each notebook entry identifies the subject of the work with particularity and contains all relevant details. An entry such as "worked on new sharpener" sheds little light on whether the "new sharpener" included a specific feature.

- All persons involved in the work should be identified in the corresponding notebook entries. Unless participants are identified, it is often difficult to establish, long after the fact, those involved in or who witnessed particular activities.

- It is important that all loose papers, such as blueprints, schematics, flow charts, oscillographs, photographs of models, etc., be signed and dated, cross-referenced to a particular notebook entry, and, preferably, mounted in the body of the appropriate notebook entry. Similarly, physical results of tests, such as samples, models, prototypes, etc., should be carefully labeled with the date, cross-referenced to notebook entries, and retained.

- Notebooks should be maintained with the idea of proving not only the dates of conception and reduction to practice, but also reasonable diligence in reducing to practice. To this end, it is important to have the documentary evidence and dated notebook entries describe all testing performed, the particular types of equipment used, and the results of the testing.

- Procedures should be implemented to ensure that significant tests or demonstrations showing or comprising reduction to practice are witnessed by noninventors, or that such tests or demonstrations are repeated before noninventor witnesses.

- Audio and/or video recordings of significant tests showing and identifying all participants, preferably including noninventor witnesses, should be considered.

In all, a documentary record should be maintained that is capable of establishing the dates and activities comprising each of the elements of making an invention, identifying individuals involved in the work who can provide testimonial proof, and identifying the particular project with which technical work is associated.

7.3 TRADE SECRETS

Secrecy is probably the most ancient form of intellectual property protection. Any proprietary information that can effectively be kept secret can be a **trade secret**. It is protected in the sense that if the competition doesn't know it, it can't be copied. Trade secrecy is the primary protection mechanism for information, data, *know-how*, and expertise, but it can also be used to protect inventions.

In the United States, trade secret rights that are enforceable against others are provided under the laws of the various individual states. In order to qualify for protection, the subject matter of the trade secret right must meet certain prerequisites:

- It must derive some value from being kept secret.
- It must be confidential.
- It must be subject to reasonable efforts to maintain its secrecy.

Information that is generally known in an industry and basic skills or practices employed in an industry while unquestionably valuable to a business do not qualify for trade secret protection. A typical example of such **nonproprietary know-how** is the skills and knowledge acquired by an employee who is trained in the operation of a commercially available machine. A venture's interest in nonproprietary know-how is best served by ensuring that the know-how is possessed by a number of people within the business and ensuring that the know-how is well documented or recorded.

7.3.1 Trade Secret Protection

Trade secret protection has a potentially infinite duration. The information is protected as long as it is not accessible to competitors. However, a trade secret can be very fragile. Once a trade secret becomes generally known, irrespective of how it becomes known, the protection is lost. A trade secret also provides no protection against independent development of the technology by others. Moreover, it is conceivable that an individual who later independently develops the technology could obtain a patent on it and foreclose others from further use.

It should also be noted that some types of technology are unsuited for trade secret protection. For example, any technology embodied in a product that is sold to the public and can be reverse engineered cannot be maintained as a trade secret. In general, in the absence of an express or implied contractual obligation, any unpatented technology that comes into an entity's possession legally can be freely used and copied.

7.3.2 Maintaining Trade Secrets

In theory, the procedure for maintaining a trade secret is simple. Implementation of the procedure, however, requires discipline. All technology considered proprietary should be clearly treated and marked as such. For example, all printouts, flow charts, schematics, layouts, blueprints, technical data, and test results that contain confidential information should be marked as such. However, it is important to be aware of two points with respect to use of markings to indicate proprietary information:

- Merely marking something as "confidential" does not make it so. Providing someone with a document marked "confidential" does not in and of itself create any obligation to keep the secret.

- Indiscriminate use of proprietary markings can dilute the significance of the marking when used on things that are, in fact, proprietary.

Tight security should be maintained and access restricted to the area in which a trade secret is practiced or kept. Access to and knowledge of a trade secret should be permitted only on a need-to-know basis. Records should be kept about all persons given access to any portion of the trade secret. All copies of trade secret documents should be accounted for.

It is imperative that **confidentiality agreements** be executed with all parties who are given access to any part of the trade secret technology. Confidentiality agreements are the primary mechanism for creating the obligation of secrecy necessary for trade secret status. As a general proposition, permitting any entity to have access to the trade secret information without first imposing an obligation of confidentiality will preclude or destroy trade secret status. Exhibit 7.2 illustrates a generic confidentiality agreement.

While employees are typically obligated by law to keep their employer's trade secrets confidential, there is great room for dispute regarding the scope of that obligation and precisely what should be considered a trade secret. Accordingly, each principal and/or employee of a company who has access to company technology or sensitive business information should sign a confidentiality agreement that defines and creates presumptions with respect to things deemed confidential. Generally, such agreements should be executed at or prior to the time of hiring the employee.

TECH MICRO-CASE 7.1

Coke Employees Try to Sell Secrets to Pepsi

In 2006, three employees of Coca-Cola, Inc. were arrested for attempting to sell trade secrets to rival Pepsi. A mysterious informant writing from the Bronx, NY, and identifying himself only by the name "Dirk" wrote to PepsiCo, claiming to be a high-level employee with Coca-Cola. He asked for $10,000 for the trade secrets and $75,000 for the recipe for Coke. PepsiCo called the police, and an undercover agent offered Dirk $1.5 million for the trade secrets. The perpetrator fell for the ruse and was apprehended. The leak apparently came from an executive assistant for a high-level Coke executive, who supplied "Dirk" with his documents and information.[12]

When employees who have access to trade secret information leave the company to take a similar position with a competitor, disclosure or use of the former employer's **confidential information** is sometimes inevitable, notwithstanding any agreement to the contrary. In those circumstances, a noncompetition agreement with the employee should be considered as a mechanism for ensuring that the confidential information is not disclosed. However, such agreements are disfavored and strictly construed by the courts.[13] Under any circumstances, the noncompetition agreement should be carefully crafted so that geographic scope, duration, and scope of prohibited employment are restricted to the minimum necessary to protect the proprietary rights of the former employer. If any of those aspects of the agreement is deemed to be overly broad or overreaching, a court is likely to find the agreement unenforceable.[14]

AGREEMENT REGARDING CONFIDENTIAL INFORMATION AND TECHNOLOGY

This is an Agreement between _____ and _____.

The parties each have particular expertise and has developed or otherwise obtained certain confidential information.

The parties wish to enter into a business relationship which is likely to require that certain confidential information of one party be disclosed to the other. However, the parties do not wish to diminish their rights in their confidential information, and require assurances that their rights therein will not be adversely affected by the business relationship.

Accordingly, the parties agree as follows:

1. **"Technology"** as used in this Agreement means, processes, machines, manufactures, compositions of matter, improvements, technological developments, methods, techniques, systems, mask works, software, documentation, data and information (irrespective of whether in human or machine-readable form), works of authorship, and products, whether or not patentable, copyrightable, or susceptible to any other form of protection and whether or not reduced to practice.

2. **"Providing Party"** means, with respect to an item of information or technology, that party which provides, reveals or discloses that item to the other party.

3. **"Recipient Party"** means, with respect to an item of information or technology, that party to which that item is provided, revealed or disclosed.

4. **"Confidential Information"** of a Providing Party means any and all Technology and/or information which: (a) is created, developed, or other wise generated by or in behalf of the Providing Party; or (b) is otherwise marked Confidential by the Providing Party; AND (c) is provided to the Recipient Party during the term of this Agreement; EXCEPT such information which: (a) at the time of this Agreement is clearly publicly and openly known and in the public domain; (b) after the date of this Agreement becomes publicly and openly known and in the public domain through no fault of the Recipient Party; (c) comes into the Recipient Party's possession and lawfully obtained by the Recipient Party from a source other than from the Providing Party, and not subject to any obligation of confidentiality or restrictions on use; or is approved for release by written authorization of the Providing Party.

5. The Providing Party reveals Confidential Information to the Recipient Party in strict confidence, and solely for the purpose of performing in connection with the business relationship with the Providing Party. The Recipient Party shall not use, or induce others to use, such Confidential Information for any other purpose whatsoever, and shall take all reasonable measures to maintain the confidentiality of such Confidential Information. At a minimum, the Recipient Party shall not: (a) directly or indirectly, print, copy or otherwise reproduce, in whole or in part, any such Confidential Information, without prior consent of the Providing Party; (b) disclose or reveal any such Confidential Information to anyone except those of the Recipient Party's employees who have a specific need to know, are directly involved in the business relationship, and have agreed to be bound by the terms of this Agreement.

6. Upon the Providing Party's request, the Recipient Party will deliver over to the Providing Party all of the Providing Party's Confidential Information, as well as all documents, media, items and comprising, or embodying, such Confidential Information as well as any other documents or things belonging to the Providing Party that may be in the Recipient Party's possession. The Recipient Party shall not retain any copies.

7. **This Agreement shall be governed by and construed in accordance with the laws of the State of ___ without reference to principles of conflicts of laws.** It may be amended only in a writing signed by the parties, and there are no other understandings, agreements, or representations, express or implied. If any clause or provision is or becomes illegal, invalid, or unenforceable, the remaining provisions shall be unimpaired, and the illegal, invalid or unenforceable provision shall be replaced by a provision, which, being legal, valid and enforceable, comes closest to the intent of the parties underlying the illegal, invalid or unenforceable provision.

8. This Agreement shall become effective upon execution by the parties and shall remain in effect until terminated. The minimum term of this Agreement shall be _____. Thereafter, either party may terminate this agreement upon 30 days written notice. The parties obligations with respect to Confidential Information disclosed hereunder, however, shall survive the termination of the Agreement.

Date: Date:

By: _____   By: _____
Name: Name:
Title: Title:

EXHIBIT 7.2

Example of a Generic Confidentiality Agreement.

7.4 PATENTS

The founders of the United States decided to give Congress the power to enact laws relating to patents. Article I, section 8, of the U.S. Constitution reads: "Congress shall have power to promote the progress of science and useful arts, by securing for limited times to authors and inventors the exclusive right to their respective writings and discoveries." Under this power, Congress has enacted various laws relating to patents. The first patent law was enacted in 1790. The patent laws underwent a general revision on July 19, 1952, that came into effect January 1, 1953. On November 29, 1999, Congress enacted the American Inventors Protection Act of 1999 (AIPA), which further revised the patent laws.

A patent for an invention is the grant of a property right to the inventor, issued by the USPTO. Generally, the term of a new patent is 20 years from the date on which the application for the patent was filed. U.S. patent grants are effective only within the United States, U.S. territories, and U.S. possessions.

The patent law specifies the general field of subject matter that can be patented and the conditions under which a patent may be obtained. According to the statute, any person who "invents or discovers any new and useful process, machine, manufacture, or composition of matter, or any new and useful improvement thereof, may obtain a patent," subject to the conditions and requirements of patent law. The word "process" is defined as a process, act, or method, and primarily includes industrial or technical processes. The term "machine" used in the statute needs no explanation. The term "manufacture" refers to articles that are made, and includes all manufactured articles. The term "composition of matter" relates to chemical compositions and may include mixtures of ingredients as well as new chemical compounds. These classes of subject matter taken together include practically everything that is made by humans and the processes for making the products.

There are three basic patent types:

- *Utility patent*: Issued for the invention of a new and useful process, machine, manufacture, or composition of matter, or a new and useful improvement thereof, the utility patent generally permits its owner to exclude others from making, using, or selling the invention for a period of up to 20 years from the date of patent application filing. Approximately 90% of the patent documents issued by the USPTO in recent years have been utility patents, also referred to as "patents for invention."

- *Design patent*: Issued for a new, original, and **ornamental** design for an article of manufacture, the design patent permits its owner to exclude others from making, using, or selling the design for a period of 14 years from the date of patent grant.

- *Plant patent*: Issued for a new and distinct, invented or discovered asexually reproduced plant including cultivated sports, mutants, hybrids, and newly found seedlings, other than a tuber propagated plant or a plant found in an uncultivated state. The plant patent permits its owner to exclude others from making, using, or selling the plant for a period of up to 20 years from the date of patent application filing.

Generally, when people refer to a "patent," they are referring to a utility patent.

7.4.1 Patentability

Threshold criteria must be met before an inventor is awarded exclusivity to an invention. In order to be patentable, an invention must be (a) within certain broad categories of subject matter, (b) **novel**, and (c) **nonobvious**.

The categories of subject matter that are eligible for utility patent protection encompass essentially "anything under the sun made by humans," and exclude only those things that logically could not be subject to exclusivity, or with respect to which exclusivity could not realistically be enforced.[15] For example, laws of nature, natural phenomena, and abstract ideas are not eligible for patent protection.[16] Likewise, mental processes, processes of human thinking, and systems that depend for their operation on human intelligence alone are not patent-eligible subject matter.[17]

There has been considerable controversy as to whether computer software or methods of doing business are eligible for patent protection. It is now settled that computer programs or methods of doing business can be patented. However, the part of the patent document that defines the invention protected by the patent is drafted so broadly that it preempts substantially all uses of a fundamental principle or mathematical algorithm, it is not patent-eligible. The same is true if the claim covers purely mental processes. With these principles in mind, a claimed process is patent-eligible if:

- It is tied to a particular machine or apparatus.
- It transforms a particular article into a different state or thing.[18]

It is important to note that patent claims intended to cover an invention can be drafted in many different ways, and in some cases, form is exalted over substance. The specific manner in which the claim is drafted can determine whether or not the claim is found to meet the above criteria.

7.4.2 Requirements for Novelty and Nonobviousness

Novelty and nonobviousness are measured against what is referred to as **prior art**. Novelty is defined in the patent law by exception; an invention is considered novel unless specific circumstances, referred to as **statutory bars**, have occurred.[19] In effect, when a statutory bar occurs, the subject matter of the statutory bar becomes prior art against which the patentability of an invention is measured. There are two basic triggers for statutory bars: the filing date of the patent application and the date of invention:

- Prior art by virtue of relative timing with the application filing date includes things that were:
  1. Commercialized by the applicant for the applicant's business more than 1 year before the application for patent was filed

2. Ascertainable from publicly available information more than 1 year before the application for patent was filed
3. Claimed in a foreign patent granted prior to the U.S. filing of an application filed more than 1 year before the application for patent was filed

- Prior art by virtue of relative timing with the invention includes things that:

    1. The inventor knew from third-party sources when he or she conceived the invention
    2. Were ascertainable from publicly available information before the invention was conceived
    3. Invented by someone else (and not maintained as a trade secret) before the invention was made

Under the novelty and nonobviousness sections of the law, a claimed invention is not patentable if the prior art fully anticipates or renders the claim obvious. A claim is **anticipated** if a single prior art reference teaches each and every element of the claim. If all of the elements of the claims are taught by a combination of references, the claim may be rendered obvious.

If the invention as claimed is obvious to a **person of ordinary skill in the art** after considering the pertinent prior art, the claim is not patentable. To assess the nonobviousness of an invention, a number of factors must be reviewed:

- The scope and content of related prior patents and publications
- The level of "ordinary skill in the art"
- The differences between the invention as claimed and the prior art
- Whether the invention provides unexpected results, fulfills a long-felt need, and/or is commercially significant

Nonobviousness is measured not against what was subjectively obvious to the inventor, but rather against the general knowledge of practitioners in the pertinent area of technology at the time of the invention.

If the claim calls for no more than the predictable use of prior art elements according to their established functions, then it is obvious. However, the invention must be considered in total context and without hindsight. For example, the solution to a problem may be patentable even though the solution, with the benefit of hindsight, is very simple; the identification of the problem can constitute a patentable invention.

Rights can also be lost if (a) filing a U.S. application is unduly delayed (i) after making the invention (abandoned) or (ii) relative to filing a foreign patent application or (b) by intentionally misnaming the inventors.

7.4.3 Exclusive Right

As noted above, a patent provides an **exclusive right** to the inventor. An exclusive right must be distinguished from a right of monopoly. A patentee has the right to exclude others from practicing the invention.[20] However, the grant of a patent does not necessarily give the patentee the right to practice the patented invention. It is this distinction that makes the U.S. patent system so effective in advancing industry. Through this mechanism, patent protection can be provided for improvements without degrading the protection provided for basic inventions.

An unauthorized item infringes a patent if it includes elements corresponding to each and every element in any claim in the patent. It is irrelevant that the item includes additional elements, even if the additional elements (or combination) are patentable in their own right.

To illustrate the "exclusive" nature of patent protection, assume that when the "stool" and "chair" were first invented, there was a patent system in place. Inventor A obtains a patent on the "stool," claiming: "Apparatus comprising a seat and a support structure for maintaining the seat at a predetermined level from the ground." Inventor B purchases a stool and determines that it can be improved by incorporating a back support. Ultimately, he or she invents the "chair." He or she obtains a patent on the "chair," claiming: "Apparatus comprising a seat, a support structure for maintaining the seat at a predetermined level from the ground, and a back extending above the seat."

Both inventor A and inventor B have patents. However, notwithstanding the addition of the back, the chair still includes elements corresponding to each and every element of the claim of, and thus infringes on, the basic patent on the stool. Similarly, while inventor A is free to make, use, and sell the stool, he or she cannot put a back on the stool without infringing the patent on the chair.

As a practical matter, the result is that inventors A and B each obtain a license from the other under the respective patents, and, where before there was only one stool manufacturer, there are suddenly two chair manufacturers.

7.4.4 The Patent Application Process

A patent application is, in effect, a proposed patent, submitted to the USPTO for its approval. A patent is divided into two major sections: the written description (typically including a drawing), which teaches the public how to use the invention, and the claims, which define the particular intellectual property to which the inventor obtains rights. Exhibit 7.3 is the first page of a patent for a compact high power alternator invented by Mssrs. Lafontaine and Scott, and owned by Magnetic Applications, Inc. It issued in 2006 from an application filed in 2004.

Written description

The patent statute requires that an application for a patent include a written description of the invention, a drawing where necessary for understanding of the invention, with at least one claim, and an oath or declaration by the applicant. The written description is the primary vehicle for teaching the public how to use the invention. Each element of the invention is described, with specific reference to the drawing. Each element shown in the drawing is designated in the drawing by a numeral. The designation used in the drawing is specifies the element when mentioned in the specification. The patent law requires that the description be in sufficient detail to enable a person skilled in the art to make and use the invention. As a basic proposition,

US007122923B2

(12) **United States Patent**
Lafontaine et al.

(10) **Patent No.:**   **US 7,122,923 B2**
(45) **Date of Patent:**   **Oct. 17, 2006**

(54) **COMPACT HIGH POWER ALTERNATOR**

(75) Inventors: **Charles Y. Lafontaine**, Berthoud, CO (US); **Harold C. Scott**, Lafayette, CO (US)

(73) Assignee: **Magnetic Applications Inc.**, Lafayette, CO (US)

(*) Notice: Subject to any disclaimer, the term of this patent is extended or adjusted under 35 U.S.C. 154(b) by 0 days.

(21) Appl. No.: **10/889,980**

(22) Filed: **Jul. 12, 2004**

(65) **Prior Publication Data**

US 2005/0035673 A1   Feb. 17, 2005

Related U.S. Application Data

(60) Provisional application No. 60/486,831, filed on Jul. 10, 2003.

(51) **Int. Cl.**
H02K 9/00   (2006.01)
(52) **U.S. Cl.** .... **310/58**; 310/61; 310/60 A
(58) **Field of Classification Search** .... 310/52–64, 310/153, 112, 113; 290/1 B
See application file for complete search history.

(56) **References Cited**

U.S. PATENT DOCUMENTS

765,078	A *	7/1904	Jigouzo	310/112
4,467,229	A *	8/1984	Ogita	310/60 A
4,900,965	A *	2/1990	Fisher	310/216
4,931,683	A *	6/1990	Gleixner et al.	310/89
5,625,276	A	4/1997	Scott et al.	322/24
5,705,917	A	1/1998	Scott et al.	322/46
5,886,504	A	3/1999	Scott et al.	322/15
5,929,611	A	7/1999	Scott et al.	322/46
6,018,200	A	1/2000	Anderson et al.	240/40
6,034,511	A	3/2000	Scott et al.	322/46
6,384,494	B1 *	5/2002	Avidano et al.	310/58
6,441,522	B1	8/2002	Scott	340/156.23
6,744,157	B1 *	6/2004	Choi et al.	310/62
2002/0053838	A1	5/2002	Okuda	310/59

FOREIGN PATENT DOCUMENTS

DE	33 29 720	2/1984
DE	3329720 A1	2/1984
DE	195 13 134	10/1996
DE	19513134 A1	10/1996
FR	2 536 222	5/1984
FR	2536222 A	5/1984
JP	60118036	6/1985
JP	60118036 A	6/1985
JP	08322199	12/1996
JP	08322199 A	12/1996

* cited by examiner

Primary Examiner—Dang Le
(74) *Attorney, Agent, or Firm*—Michael A. Lechter; David E. Rogers; Squire, Sanders & Dempsey, L.L.P.

(57) **ABSTRACT**

An apparatus for converting between mechanical and electrical energy, particularly suited for use as a compact high power alternator for automotive use and "remove and replace" retrofitting of existing vehicles. The apparatus comprises a rotor with permanent magnets, a stator with a winding, and a cooling system. Mechanisms to prevent the rotor magnets from clashing with the stator by minimizing rotor displacement, and absorbing unacceptable rotor displacement are disclosed. The cooling system directs coolant flow into thermal contact with at least one of the winding and magnets, and includes at least one passageway through the stator core. Various open and closed cooling systems are described. Cooling is facilitated by, for example, loosely wrapping the winding end turns, use of an asynchronous airflow source, and/or directing coolant through conduits extending through the stator into thermal contact with the windings.

129 Claims, 43 Drawing Sheets

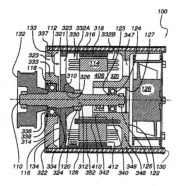

EXHIBIT 7.3

Patent for Compact High Power Alternator.

it is desirable to include as much detail as possible in the description and to be exceedingly careful to fully describe each and every feature that is to be protected. In addition, the application must describe the best mode contemplated by the inventor of carrying out the invention.

Claims

The claims define the scope of protection provided by the patent. An accused device or process infringes a claim if it includes elements corresponding to each and every element of the patent claim. The broader and less specific the terms of the claim, the broader the protection afforded by the patent but also the more vulnerable it is to invalidation based on prior art. The language of the **patent claims** must, therefore, be drafted with the utmost precision. It is permissible to have a number of different claims in the patent application. As a matter of practice, claims of varying scope, ranging from the most general to the most specific, are submitted. In this way, if it appears after the fact that some relevant piece of prior art exists that invalidates the broad claims, the other, more specific claims are not necessarily invalidated. In this manner, the inventor not only can obtain protection on the broad aspects of his invention, but also on the specifics of the particular product that is put on the market.

Duty of disclosure

A patent applicant has a strict duty of candor in dealing with the USPTO. Any information known to be material to patentability must be disclosed to the USPTO. This duty is typically observed through an **information disclosure statement**, filed within 3 months of the original application. It is very important that the Examiner be apprised of all relevant facts and prior art. If the most pertinent art is before the Examiner, the patent is accorded a strong presumption of validity when it is enforced. On the other hand, if information is withheld from the Examiner, the applicant may have committed a fraud on the USPTO. Not only would any patent obtained be invalid, but also, if an attempt was made to enforce the patent, the patentee could be assessed for attorneys' fees and be subject to a counterclaim of violating antitrust laws. When in doubt as to the materiality of information, the safest course of action is to disclose the information to the USPTO.

Types of applications

In practice, there are two types of complete patent applications, in addition to a standard application patent, continuation applications and continuation-in-part applications.

- A **continuation application** is an application filed while a parent application is still pending, which is substantively identical (except perhaps as to the claims) and contains a specific reference to the parent application.[21] The continuation is treated as if it was filed on the date that the parent application was originally filed. This is particularly significant with respect to defining the prior art against which patentability is measured. However, the term of the patent issuing from the continuation application is measured against the filing date of the parent application.[22]

- Once a formal application has been filed with the USPTO, no "new matter" can be added to the application. In this context, new matter refers to new embodiments or details not described or shown in the originally filed application. In the event that further details or embodiments of the invention which, in and of themselves, warrant protection are developed after an application has been filed, a **continuation-in-part (CIP) application** may be filed. The CIP application includes an additional description of the further details or embodiments of the invention and claims relating to those new features. The claims relating to the material described in the parent application are given the benefit of the original filing date. The claims covering the additional details are given the filing date of the CIP application. However, the term of the patent issuing from the CIP application is measured against the filing date of the parent application, even with respect to the claims relating to the new matter. A CIP application can be filed at any time before the original application issues as a patent or is abandoned.

In general, the patent statute requires that an application for a patent include a written description of the invention, a drawing where necessary for understanding of the invention, at least one claim, and an oath or declaration by the applicant.[23] The statute also provides for what is referred to as a **provisional application**, which does not require a claim or declaration. A provisional application can establish the filing date (and thus the prior art) with respect to inventions that are described in it with sufficient detail. It, however, is not intended to provide any enforceable rights. It is not examined and is automatically abandoned 12 months after filing. It does not itself ever mature into a patent. For a patent to issue on the subject matter described in the provisional, a regular continuation application claiming priority on the provisional application must be filed within a year of the provisional. Once a provisional application has been filed, however, it is permissible to mark products including the invention described in the application with **Patent Pending**.

7.4.5 The Patent Examination Process

After the patent application is prepared, it is filed with the USPTO. The USPTO gives the application a serial number and filing date, and assigns it to a Patent Examiner having expertise in the particular technological area of the invention. The Patent Examiner then conducts an investigation, and searches the USPTO files to determine if there is any relevant prior art in addition to that supplied by the applicant. The Patent Examiner then issues what is known as an "Office Action." In brief, the Office Action lists all of the references considered by the Examiner and indicates whether he or she considers the claims to be of proper form, and whether he or she considers the claims to be anticipated by the prior art or rendered obvious by the prior art.

If necessary, a response to the Office Action should be filed within 3 months of the mailing date of the Office Action. The response must answer each and every issue raised by the Examiner by arguing against the Examiner's positions, amending the claims, or canceling the claims. In effect, there is a negotiation with the Patent

Examiner to determine the exact scope of the claims to which the inventor is entitled. If the Examiner agrees with respect to the exact scope of the claims, the application issues as a patent.

7.4.6 Patent Pending

Filing a patent application does not, in itself, provide any enforceable rights. That is, no enforceable rights are provided until the patent is actually granted. However, during the period the patent application is pending before the USPTO, a patent pending notice can be placed on products including the invention described in the application. A patent pending notice can sometimes scare off certain types of potential infringers. Care must be taken, however, not to mismark; marking a product with a patent pending notice when it is not actually described in pending application can be a violation of the false marking statute.

7.5 PATENT OWNERSHIP

The rules with respect to ownership of patents are far from intuitive. In the United States, unless there is an express or implied contractual agreement to the contrary, the actual inventor owns the invention and any patent on the invention. Where there are joint inventors, each owns an equal undivided interest in the whole of the invention and any patent on the invention. Absent an agreement to the contrary, all joint inventors are entitled to make, use, and sell the invention, without accounting to the other coinventors.[24] Of course, these are default rules and may be varied by agreement. In reality, the default rules are typically superseded by written agreements.

Under certain circumstances (e.g., where an employee is hired to invent) there is an implied agreement that the employer will own all rights to that invention. Similarly, where an invention is made on company time and/or using the business's facilities, the business may acquire **shop rights**—a royalty-free license—to the invention. However, in order to avoid any disputes, it is a best practice for a venture to require each of its employees to execute an employee's **invention assignment agreement** as a condition of employment—explicitly obligating the employee to assign all rights in relevant inventions to the business.

7.6 INTERNATIONAL PATENT PROTECTION

The rights granted under a U.S. patent extend only throughout the territory of the United States and have no effect in other countries. Almost every country has its own patent law, and a technology entrepreneur desiring a patent in a particular country must make an application for patent in that country, in accordance with that country's requirements. Fortunately, there are several international treaties that aid in applying for international patent protection. The most notable are the Paris convention and the **Patent Cooperation Treaty**, otherwise known as the PCT.

Under the Paris Convention, a patent application filed in a first signatory country on a particular date, is treated as if filed on that date in another signatory country as long as a corresponding application is actually filed in the other country within one year of the original filing.

The PCT facilitates the filing of applications for patents on the same invention in countries that have ratified the Treaty. It provides for centralized filing procedures where a single application filed in a member country governmental receiving office constitutes an application for patent in one or more member countries. Once the application is filed, one of the eligible PCT governmental searching offices will perform a patent search. When the search is completed, the applicant may then elect to enter what is known as Chapter II of the process. At this point, a PCT governmental office evaluates the patentability of the application pursuant to standards set forth in the Treaty. Eventually, the applicant will be required to have the PCT application entered into the national patent office of each of the countries from which patent protection is desired.

There are several advantages to the PCT process. First, the applicant can file a single patent application rather than filing one for each country. The single application is much less expensive than individual national filings. Although the applicant will eventually be required to incur a cost similar to the national filings when the PCT application is entered in each national patent office, the PCT procedure allows these costs to be delayed for up to 18 months. This period allows the venture to analyze the patentability and profitability of the invention, and therefore to make an informed decision about where the patent application should be filed.

A second advantage of the PCT process is that the evaluation of patentability made by the PCT governing body handling the examination should lead to more uniform results in connection with the patentability of the invention in each country. Although individual countries are not bound by the determination made during the PCT process, a positive PCT decision on patentability is often persuasive in member country patent offices.

Patent laws of many countries differ in various respects from the patent law of the United States. In most foreign countries, publication of the invention before the effective date of the application will bar the right to a patent. It is thus advisable to file the U.S. application prior to public disclosure of the patent even though the U.S. patent law provides for a 1 year grace period. As long as corresponding foreign applications are filed within one year of the U.S. filing, under the Paris Convention, the effective date of the foreign application will be that of the U.S. application, and the public disclosure will not bar the foreign application.

Patent maintenance fees are required in most countries, but they will differ in timing and amount from U.S. fees. In addition, most countries require that a patented invention be manufactured in that country after a certain period, usually 3 years. This requirement is generally referred to as the "working requirement." If there is no manufacture within this period, the patent may be void in some countries, although in most countries the patent may be subject to the grant of compulsory licenses to any person who may apply for a license.[25]

7.7 COPYRIGHTS

Copyrights protect artistic expression. A copyright arises automatically as soon as an original work of authorship becomes fixed in a tangible form of expression. In other words, copyright protection is secured automatically as soon as the work is fixed in a form that can be read or visually perceived either directly or with the aid of a machine or device. Copyright protection is expressly limited by the statute to nonutilitarian expression:

> *In no case does copyright protection for an original work of authorship extend to any idea, procedure, process, system, method of operation, concept, principle, or discovery, regardless of the form in which it is described, explained, illustrated, or embodied in such work.*

Neither publication nor registration is necessary to secure copyright protection. Exemplary categories of original works of authorship include literary works; musical works; dramatic works; pantomimes and choreographed works; pictorial, graphic, and sculptural works; motion pictures and other audio/visual works; and architectural works. Certain aspects of computer programs are also considered literary works.[26] Copyright protection is also available for original compilations (creative selection or arrangement of preexisting materials or data).[27]

To be "original," a work must be independently created by the author and possess a minimal degree of creativity.[28] Those aspects of a work that are not created by the author are not copyrightable subject matter. For example, facts are said to be discoveries, as opposed to those created by an act of authorship. The same is true with respect to those elements of a work that are in the public domain and those elements of the work dictated by function. In addition, some types of works are simply not sufficiently creative to warrant copyright protection. These include, for example, numbering schemes for parts, command codes, and fragmentary words and phrases.[29]

A work may include expressions or descriptions of ideas and concepts, facts, and utilitarian elements. In this context, the term "utilitarian element" means a "useful article" or anything that (a) has an intrinsic function other than just portraying appearance or conveying information or (b) is required to be in a particular form because of some external factor. Examples of utilitarian elements include logical sequences,

---

TECH MICRO-CASE 7.2

A Classic Example of Copyright Law

A classic example is found in the case of Baker v. Selden.[30] Mr. Baker, an accountant, wrote a book describing a double entry accounting system that he had developed. Mr. Selden, also an accountant, bought the book and began using the double entry accounting system developed by Mr. Baker. Mr. Baker contended that the use of his accounting system by Mr. Selden constituted copyright infringement. The Supreme Court, however, disagreed. The use of the ideas described in a copyrighted work does not constitute infringement of the copyright.[31]

procedures, processes, systems, and methods of operation. The copyright gives the author exclusive rights with respect to the way in which the ideas, facts, and utilitarian elements are expressed or described. It does *not* give any sort of exclusivity with respect to the use of the ideas, facts, and utilitarian elements themselves.

Expression which is inseparable from an idea, either because (a) protecting the expression would, in effect, give exclusive rights to the underlying idea or (b) because the expression serves a functional purpose or is dictated by external factors is likewise not protectable under the copyright law. However, an original compilation comprising an arrangement or selection of those unprotectable elements that is itself nonutilitarian and sufficiently original is protected by copyright.

7.7.1 Considerations with Respect to Software

Certain aspects of computer programs are also considered literary works.[32] However, it must be kept in mind that ideas, facts, and utilitarian elements embodied in the software are not protected by copyright.

Historically, there has been much debate as to whether software is a "work of authorship," and thus qualifies for copyright protection. It is well settled that copyright protection is applicable to human-readable software, such as flow charts, documents, machine-readable source code representations of a program, and audio/visual displays in game programs.[33] However, it is not universally accepted that a copyright registration of the underlying program code that generates the screen displays extends to the displays; arguably, the screens must be subject to a separate copyright. It is prudent to consider copyright protection of the various screens separately from the copyright protection of the underlying program code. It may be advantageous to register copyrights on the respective screens as audio/visual works separate and apart from the program code.[34]

The most important limitation on the scope of copyright protection, as it applies to software, however, is the nature of the copyright itself. As previously discussed, copyright protection is limited to original nonutilitarian works of authorship. A copyright clearly protects against actual copying of substantial portions of copyrighted program code. However, categorizing aspects of a computer program other than literal code as protectable expression as opposed to idea or utilitarian is often extremely difficult. Applying an "abstraction-filtration-comparison" process, the program is dissected into varying levels of generality (aspects). For example, a computer program can often be parsed into at least six levels of generally declining abstraction:[35]

1. The main purpose
2. The program structure or architecture
3. Modules
4. Algorithms and data structures
5. Source code
6. Object code

Each level of abstraction (aspect) is then examined to filter out non-protectable elements. For example, elements of a computer program that are necessary to effect a specific logical process or algorithm, comply with hardware standards and mechanical specifications, software standards and compatibility requirements, computer manufacturer design standards, industry practices and demands, and computer industry programming practices are not copyrightable subject matter.

The remaining protectable elements are then compared with the allegedly infringing program to determine whether those protectable portions of the original work that have been copied constitute a substantial part of the original work.[36] Depending upon the nature of the elements at issue, the courts require either "substantial identity" or "substantial similarity" between the elements in order for there to be an infringement.[37] In general, a substantial similarity test is applied to literal elements of the software and a virtual identity test is applied to nonliteral elements and compilations.

In any event, a copyright does not in any way protect the owner from independent creation of a similar program by another, even if the other is generally aware of the copyrighted program. A competitor can, in general, study a copyrighted program, determine the central concept and basic methodology of the program, and then write its own program to accomplish the same results. In practice this is often done using the so-called "clean rooms."

7.7.2 Notice

It is advantageous that all published copies of a work bear a notice of copyright. Basically, the notice of copyright includes three elements: the copyright symbol ©, the word "copyright," or the abbreviation "Copr"; the named owner of the copyright; and the year of first publication of the work. There are various requirements as to where the copyright notice must be placed on the work.[38] However, in general, the placement of the notice should be sufficient if it is placed in a prominent position on the work, in a manner and location as to "give reasonable notice of the claim of copyright."

7.7.3 The Term

The term of the copyright for all works created after January 1, 1978, is the author's life plus an additional 70 years after the (last surviving) author's death.[39] The term for a work for hire is 95 years from publication or 120 years from creation, whichever is shorter.[40]

7.7.4 Copyright Registration

Copyright registration is not a prerequisite for copyright protection. However, registration is significant in several respects. A registration is normally necessary before the copyright on works originating in the United States can be enforced.[41] If the registration is made before publication, or within 5 years after publication, the mere fact of registration establishes the validity of the copyright and of the facts stated in the copyright certificate in court.[42] Also, the copyright statute provides for "statutory damages and attorneys' fees," which may range from $750 (in the case of an innocent infringer) to up to $150,000 (in the case of the willful infringer).[43] If a registration is made within

3 months after publication of the work or prior to the infringement of the work, the copyright owner has the option to elect to take statutory damages instead of the actual damages and profits that he can prove, as well as attorneys' fees.[44] Unless made within 3 months after the publication of the work, a registration not made until after the infringement only entitles the copyright owner to be awarded the damages and profits that can actually be proven.[45]

A registration is obtained by filing the appropriate completed application form, a specified fee for each application, and two complete copies of the work. In general, two complete copies of the "best" edition must be filed. However, special provisions are made in the Copyright Office regulations for deposit of "identifying portions" of machine-readable works, in lieu of complete copies.[46]

7.7.5 Copyright Ownership

Copyright ownership is one of the areas of IP law that tends to turn the unwary into casualties. It is counter-intuitive. One would think that the person who paid for a work to be created would own it. This is not necessarily so under the copyright law. Unless there is a written assignment or the work qualifies as a "work for hire," the creator/originator of a work owns the copyright.[47] If the work qualifies as a "work for hire," then the employer of the creator, or the entity that commissioned the work, is considered to be the author and holder of the copyright.[48]

A work prepared by two or more authors with the intention that the respective contributions be merged into inseparable or interdependent parts of a unitary work

---

TECH TIPS 7.1

The Requirements of a Work-for-Hire Designation

To be a work for hire, the work must either (1) have been prepared by an employee within the scope of the employee's duties or (2) fall within one of certain specified categories of works, be specially ordered or commissioned, and be the subject of an express written agreement specifying that it will be a work for hire.[49]

The requirements to qualify as a work for hire are very strictly construed. Unless the creator qualifies as an employee, and the work is created within the scope of employment, the existence of an express written agreement is imperative.[50] While no single factor is determinative, as a practical matter, unless the creator is treated as an employee for tax and social security purposes, the person is likely to be deemed an independent contractor and the owner of the copyright in the absence of a written agreement.

---

is referred to as a **joint work** by **coauthors**. Absent agreement to the contrary, coauthors are co-owners of the copyright. Each coauthor owns a proportionate share of the copyright and, in the absence of an agreement, is entitled to a share of any royalties received from licensing. A joint owner may generally use or license the use of the work without the consent of co-owners, but must account to them for their shares of profits derived from any license to a third party.

7.8 MASK WORKS

Under the Semiconductor Chip Protection Act, **mask works** are defined as a "series of related images, however fixed or encoded, that represent three-dimensional patterns in the layers of a semiconductor chip."[51] Protection is available for any mask work unless:

- The mask work is not original

- The mask work consists of designs that are "staple, commonplace, or familiar in the semiconductor industry, or variations of such design, combined in such a way that, considered as a whole, is not original"[52]

- The mask work was first commercially exploited more than 2 years before the mask work was registered with the Copyright Office[53]

- The mask work has no nexus to the United States[54]

In essence, the Chip Protection Act protects against the use of reproductions of registered mask works in the manufacture of competing chips. It provides a number of very powerful remedies against infringers, including injunctive relief, actual damages, any profits made by the infringer that are not accounted for in the actual damages, and, at the court's discretion, all of the costs of the suit, including attorneys' fees.[55] However, the Act makes it absolutely clear that competitors are not precluded from reverse engineering the chip for purposes of analysis or from using any unpatented idea, procedure, process, system, method of operation, concept, principle, or discovery embodied in the mask work.[56]

Mask work protection commences upon the earlier of the first commercial exploitation of the chip anywhere in the world or upon registration of the mask work, whichever occurs first. The protection then runs for a term of 10 years, expiring at the end of the tenth calendar year.[57] However, in order to maintain mask work protection, the mask work must be registered with the Copyright Office within 2 years of the first commercial exploitation.[58] Registering a mask work involves the filing of a Copyright Office form together with particular identifying material, and a fee.

Typically, the owner of a mask work is the person(s) who created the mask work. Presumably, in analogy to the copyright statute, cocreators of the mask work would be co-owners of the mask work protection. However, where the mask work is made within the scope of the creator's employment, the employer is considered the owner of the mask work.[59]

7.9 TRADEMARKS

A **trademark** or **service mark** is used to identify the source or origin of a product or service. That is, a trademark distinguishes goods or services of one company from those of another. It is through a trademark that a customer connects the goodwill and reputation of the company to its products. Under the law, a competitor is prevented from capitalizing on another venture's reputation and goodwill by passing off possibly inferior goods as those of the venture. Thus, proper use of a trademark can protect the sales value, the venture's reputation, and that of the product, as well as its investments in advertising and other promotional activities used to develop goodwill. However, trademark protection does not prevent the competition from copying or reverse engineering a product.

7.9.1 Acquiring Trademark Rights

In the United States, trademark rights are acquired through use of the mark in legal commercial transactions. That is, a venture simply adopts a proper mark, begins to use it commercially, and through that use acquires proprietary rights in the mark. Once the mark is used in interstate commerce (or a good faith *bona fide* intent to use the mark in interstate commerce is formed), a federal registration may be obtained.

Under common law, trademark rights are acquired through use of the mark. The first to use a given mark in connection with particular goods or services in a given geographical area obtains the exclusive rights in the mark for use with those particular goods or services in that particular geographical area. However, if someone else adopts the mark somewhere outside of that geographical area without knowledge of the prior use of the mark, that person would acquire valid common law rights to the mark in the remote area. Federal registration provides constructive knowledge of the mark and can prevent subsequent remote users from obtaining rights.

Use of a trademark requires physical association of the mark on or in connection with the product or service. With a trademark, it is sufficient to apply the mark to labels or tags affixed to the product or to the containers for the product, displays associated with the product, or the like. Trademark usage cannot be established just through use of the mark in advertising or product brochures. However, if it is not practicable to place the mark on the product, labels, or tags, then the mark may be placed on documents associated with the goods or their sale.[60] On the other hand, if services are involved rather than a physical product, use of a mark in advertising is a proper usage for a **service mark**. A service mark is a mark used in sales, advertising, or services to identify the source of the services.

7.9.2 Registering a Trademark

Once a *bona fide*, good faith intention to use a mark in interstate commerce can be alleged, an application for registration of the mark may be filed. The USPTO maintains two separate trademark registers: the Principal Register and the Supplemental Register. Registration on the Principal Register provides a number of procedural and substantive advantages. However, registration on either register provides a number of very valuable rights.

Registration on either register provides notice of the mark to would-be users. In the first place, marks on both the Principal and Supplemental Registers are available to the Trademark Examiners and can prevent someone else from obtaining a registration on a confusingly similar mark. In addition, the mark may show up in any investigations made by a company prior to adopting a mark, causing the company to reconsider adopting the mark.

After the application is filed, a Trademark Examiner reviews the application to determine whether the mark is registerable. That is, whether it is capable of distinguishing the applicant's goods from the goods of others. The Examiner also reviews both the Principal and Supplemental trademark registration files maintained at the USPTO to determine if the mark is confusingly similar to any mark already being used. Thus, registering a mark and making it available to the Trademark Examiner tends to prevent registration of confusingly similar marks.

If the application is based on an *"Intent to Use"*, the applicant must file a declaration of commencement of use, together with specimens of the use within a specified period. Once the mark is actually used and the registration granted, the registrant obtains a nationwide constructive use priority effective as of the date of the application. The constructive use priority provides rights to the mark against all other persons except those who (a) used or applied to register the mark prior to the date of the application or (b) obtained an earlier "effective filing date" by applying to register the mark under Section 44 of the Lanham Act (obtained a priority on the basis of a filing in a foreign country).[61]

Once a federal registration on either the Principal or Supplemental register has been obtained, the registrant is entitled to use a registration notice such as ®, which is typically placed as a superscript to the registered mark. The use of a registration notice is not mandatory. However, the notice does provide constructive notice of the registration and sets off the mark. Setting off the mark signifies that the term or symbol is intended to be a mark that indicates the source or origin of the goods as opposed to a descriptive term for the goods.

A registration notice is appropriate only when used with the specific term or symbol shown in the registration, and only when the mark is used in connection with goods that are within, or are natural extensions of, the definition of the goods set forth in the registration. Accordingly, when a registered trademark is used in conjunction with a new product, a determination should be made as to whether or not the new product falls within the scope of the definition of goods of the registration. If not, a registration notice should not be used and consideration should be given to filing an application for a new registration.

TECH TIPS 7.2

Using the ™ Symbol in Lieu of Registration

In circumstances where there is no applicable registration, a ™ symbol is often used to set off a mark. The use of the ™ symbol has no legal significance other than to signify that the term or symbol is intended to be a source-identifying trademark.

7.9.3 Principal Register

Registration on the Principal Register provides various substantive and procedural advantages. In addition to providing actual notice of the mark, registration of a mark on the Principal Register also provides constructive notice of the registrant's claim of ownership of the mark.[62] This prevents someone else in a remote geographical area from subsequently adopting and obtaining rights in the mark.

Registration of the mark on the Principal Register is evidence of the validity of the registration, the registrant's ownership of the mark, and the registrant's exclusive right to use the mark in commerce in connection with the goods or services specified in the certificate.[63] In addition, if an affidavit is filed with the USPTO to the effect that the mark has been used continuously for 5 consecutive years subsequent to the date of registration and is still being used in interstate commerce, the registered mark is incontestable and can be attacked only on certain limited bases.[64] Once a mark is registered on the Principal Register, it can be filed with the U.S. Customs Service to prevent any article of imported merchandise that copies or simulates the trademark from entering the country.[65]

TECH TIPS 7.3

Trademark Registration Restrictions

Any mark capable of distinguishing the goods of one company from those of another can be registered on the Principal Register unless the mark:

1. Includes immoral, deceptive, or scandalous matter or matter that may disparage or falsely suggest a connection with persons living or dead, institutions, beliefs, or national symbols, or bring them into contempt or disrepute
2. Includes the flag or coat of arms or other insignia of the United States, or of any State or municipality, or of any foreign nation, or any simulation thereof
3. Includes the name, portrait, or signature of a particular living individual or a deceased president during the life of his widow, unless by consent
4. Is confusingly similar to the mark of another
5. Consists of a mark that, when applied to the goods of the applicant, is merely descriptive or deceptively misdescriptive of such goods
6. Consists of a mark that, when applied to the goods of the applicant, is primarily geographically descriptive or deceptively misdescriptive of the goods
7. Consists of a mark that is primarily merely a surname[66]

7.9.4 Supplemental Register

Any mark in actual use that is capable of distinguishing the applicant's goods or services, but does not qualify for the Principal Register because it is merely descriptive of the goods, a geographical origin of the goods, or a surname, can be registered on the

Supplemental Register.[67] The Supplemental Register can be used for any trademark, symbol, label, package, configuration of goods (such as the shape of a bottle), name, word, slogan, phrase, surname, geographical name, device, or a combination thereof, as long as the mark is capable of distinguishing the applicant's goods or services from those of others and is not utilitarian or directed by the function of the goods.

Registration on the Supplemental Register does not provide all of the rights accruing to registration on the Principal Register. A supplemental registration is not constructive notice of the mark, is not evidence of the validity of the registration, is not evidence of the registrant's ownership of the mark or the registrant's exclusive right to use the mark, and cannot be the basis for stopping importations through customs.[68] The primary difference in an infringement action concerning a mark on the Supplemental Register, as compared to a suit concerning a mark on the Principal Register, is that the registrant must actually prove that the mark is distinctive and identifies that the goods originate with a single source (e.g., the registrant). On the other hand, registration on the Supplemental Register does not preclude subsequent registration on the Principal Register, and is expressly not an admission that the mark is not distinctive.[69]

7.9.5 Application Based on "Intent to Use"

Rights in a mark can also be created by filing an application for registration based upon intent to use the mark.[70] Once a proper mark has actually been used in inter-state commerce, it is eligible for registration with the USPTO. Once a registration based upon intent to use is obtained, the registrant is accorded a constructive use priority, effectively equivalent to actual use of the mark on the date of the application for registration. To obtain a registration, however, actual use of the mark must commence within a predetermined period of a notice of allowance from the USPTO indicating that subject to a proper showing of the use in commerce, the mark is entitled to registration.[71] An application based on intent to use establishes constructive use of the mark as of the filing date of the application.[72]

7.9.6 The Strength of a Mark

A symbol or word that cannot effectively identify the source of the goods cannot be used as a trademark. The protection afforded by a particular trademark is a direct function of the distinctiveness of the mark and how closely associated the mark is with the source of the product as opposed to the product itself. Marks can be cat-egorized with respect to the degree of protection accorded by the mark as generic, descriptive, suggestive, and fanciful.

Generic marks

A **generic mark** uses a term that refers to the genus of the particular product. For example, using the mark "oscilloscope" for an oscilloscope apparatus would be con-sidered a generic mark. A generic term does not identify the source of the goods

EXHIBIT 7.4

Kleenex: Trademark or Generic Term for tissue?

and cannot be utilized as a trademark. Examples of terms that began as trademarks, but became generic include: Asprin, Escalator, Murphy Bed, and Thermos.[73] One example of a mark struggling against becoming generic is provided in Exhibit 7.4. The term "Kleenex" is a registered mark for a Kimberly-Clark product. However, it is widely used generically to describe tissue paper, and even defined as a generic term in some dictionaries. It is possible that Kimberly-Clark will continue to hold trademark rights in the stylized Kleenex logo shown in Exhibit 7.4, but will lose any exclusive rights to the term Kleenex per se.

Descriptive marks

A **descriptive mark** is a term that conveys an immediate idea of the ingredients, qualities, or characteristics of the goods. It describes the intended function, purpose or use of the goods, size of the goods, class of user of the goods, effect upon the user, and the like. A descriptive term is just one step removed from a generic term. It can be a word, picture, or symbol that describes the size, color, purpose, class of users, or the effect on the users of the product it is associated with. To be registered, or for legal protection from imitations, it needs to have acquired distinctiveness (a secondary meaning) through association with a particular manufacturer or provider, and thus identify the source of goods or services. Even then a competitor cannot be stopped from using the term in its normal descriptive sense.

Suggestive marks

A **suggestive mark** requires imagination, thought, and perception to conclude the nature of the goods from the term used as the mark. Suggestive marks can be a word, picture, or symbol that only suggests but does not describe some aspect of an associated good or service. For example, "Diehard" for a car battery hints at but does not explicitly describe its expected longevity. Suggestive marks are considered distinctive in their own right in trademark law, and do not require evidence of a secondary meaning for legal protection as a trademark.

Fanciful and arbitrary marks

The highest degree of trademark protection is provided through the use of a **fanciful** or **arbitrary mark**. A fanciful term is a word created strictly and entirely for use as a trademark. It has no meaning before being used as a trademark. Well known

EXHIBIT 7.5

The Apple Computer Logo: An Example of a Fanciful Trademark.

examples include Kodak, Starbucks, Verizon, Exxon, and Cingular. (Acronyms, abbreviations, phonetic variations, misspellings, and foreign words, however, all afford the same protection as the corresponding correctly spelled English word). An arbitrary mark is a common English word with a normal meaning that bears no relationship to the goods or services to which they are applied. The word "Apple" is arbitrary as applied to computers. The graphic logo for Apple Computers is a well-known example of a fanciful mark (Exhibit 7.5).

It should be apparent that a given term may be generic in one market and arbitrary in another market (e.g., the mark "oscilloscope" for an oscilloscope apparatus is generic). However, the mark "oscilloscope" for chewing gum is arbitrary. Terms can, however, change from one category to another. At one time, "escalator," "cellophane," "aspirin," and "shredded wheat" were valid trademarks. However, through improper use, these marks became associated with the goods in general (i.e., became a generic term for the goods). This is the reason that some companies are fighting a continual war against people who are improperly using their trademarks. In the past, some goods that have found themselves in a generic status have, in effect, made a comeback. For example, the marks "Goodyear" and "Singer" at one time were considered to be generic, but eventually reacquired the status of protectable trademarks.

7.9.7 Choosing a Mark

There are certain basic guidelines with respect to choosing a mark. The strongest word mark is a relatively euphonious, easily pronounced, coined word that does not include components commonly used in marks. It is desirable that the mark be simple—a simple mark is more easily protected. Where a mark includes a large number of elements, there is the possibility that another could adopt some, but not all, of the elements of the mark. If both a word mark and a symbol mark are adopted, they should be completely separate in at least some of the instances where they are used to ensure that each can be protected separately. It should also be noted that purchasers more easily remember a simple mark. Also, it has been found that a mark that can be depicted and put into words is more easily remembered than a mark that can only be depicted or verbalized but not both.

Before a mark is adopted for a product or service, it is prudent to undertake an investigation to ensure that no one else is using it. Basically, this involves examining trademark registration files maintained by the USPTO (USPTO.gov) to see whether any similar mark is already registered to another, or an application has been made for registration of a similar mark for similar goods and/or services.

Potential trademark problems arise when the proposed trademark is confusingly similar to another mark. Three rule-of-thumb criteria for determining whether or not a mark resembles another are:

1. Do the marks look alike?
2. Do the marks sound alike?
3. Do the marks have the same meaning or suggest the same thing?

The similarities and dissimilarities of the goods themselves must also be considered. In this regard, the manner of marketing the goods is relevant. Do the respective goods move in the same channels of trade? Are they sold in the same type of store? Are they bought by the same people? What degree of care is likely to be exercised by the purchasers?

Trademarks when used on products purchased by relatively sophisticated purchasers are less likely to cause confusion as to the source of goods, sponsorship, or affiliation than when used on goods typically sold to unsophisticated purchasers. The ultimate answer as to whether marks are "confusingly similar" is the cumulative effect of the differences and similarities in the marks and in the goods or services.

7.9.8 Term of the Registration

While trademark protection is of potentially infinite duration, the registration itself must be periodically renewed after each 10-year period. There is no limit on the number of times that the registration can be renewed. The registration is renewed for the next 10-year period by filing the appropriate papers and fees within 6 months before the expiration of the registration. There is no necessity to file a Section 8 affidavit during the second or further terms of registration. However, when the registration is renewed, it is necessary to indicate that the mark is being used in interstate commerce. In addition, the renewal application must recite only the goods on which the mark is in actual use at that time.

7.9.9 Maintaining Trademark Rights

After rights in a mark are acquired, it is important to prevent the mark from losing its trademark significance. If the mark ceases to distinguish the company's goods or services from those of others, the mark becomes public property and can be used by anyone. In other words, if the primary significance of the mark to the consuming public becomes a descriptor for the nature or characteristics of the goods rather than an indication of the source of the goods, the mark becomes a common descriptive term and any trademark rights in the mark are lost.

SUMMARY

Intellectual property is the great equalizer in the world of business. It is often the primary factor that enables an emerging business to compete successfully against larger established competitors with vastly more marketing power. Intellectual property assets can not only be leveraged to sustain competitive advantage, but also create credibility in the industry and with investors. Credibility is based not only on a venture's intellectual property assets, but also on the venture's systems and processes to develop and protect them. IP-based strategic alliances and/or licensing can also be an alternative to raising capital, providing the benefit of the co-venturer's resources without the venture having to spend the time and money to develop them on its own.

Technology ventures should continuously and systematically analyze their products and operations to identify potential intellectual property and seek opportunities to develop protectable intellectual property assets. It should develop a comprehensive strategic plan to acquire, maintain, and reap maximum benefit from those assets. All of the various protection mechanisms should be employed, singly and in combination, as part of that plan. At the same time, the venture must be mindful of third-party intellectual property rights. A few relatively simple procedures and precautions can mean the difference between the success and failure of a technology venture.

In this chapter, we examined the various types of intellectual property that a technology venture can develop and protect. Of course, in order to develop intellectual property, the venture first must learn to recognize what it is and when it is created. The importance of record keeping and developing an engineering notebook was emphasized. Good record keeping can support the venture's claims to priority over others, and help protect against damages that may occur from infringement lawsuits.

Trade secrets are one form of intellectual property. They are elements of a venture that provide it with competitive advantage, and which are not generally known or practiced. Trade secrets must be treated as secrets in order to be protected by courts.

This chapter also discussed the various forms of U.S. patents, including how to apply for a patent and how to protect those that are granted. Patents are granted for inventions that are novel and nonobvious. Patents provide the inventor with exclusive rights to practice the invention. The chapter also discussed international patent protection and the Patent Cooperation Treaty (PCT). Most technology ventures are global in nature, and it may be in their interest to explore and utilize PCT member country protection.

The chapter examined copyright law, noting that an original work does not need to be registered to receive copyright protection. However, it is always advisable to use the copyright mark © for items that an author wants to protect.

Mask works are a form of intellectual property, pertaining the design of semiconductor chips. Mask work protection is based on the Semiconductor Chip Protection Act, and applies only to those products that were developed or distributed in the United States.

Trademarks are those marks that distinguish the goods or services of a company from others in the industry. It is through a trademark that a customer associates the goodwill and reputation of a company with its products. When a trademark is registered with the USPTO, the venture is able to use the ® symbol to indicate its property rights. Prior to registration, the ™ symbol may be used to indicate an intent to register a mark.

STUDY QUESTIONS

1. Explain how intellectual property can be the "great equalizer" for an emerging company competing with a larger and better financed competitor.

2. The chapter identified six legal mechanisms for protecting intellectual property. Name each and briefly explain the type(s) of intellectual property it is intended to protect.

3. Explain how obtaining exclusive rights to intellectual property can sustain a competitive advantage. How can a company identify potentially protectable intellectual property?

4. How can a company sustain a competitive advantage with intellectual property rights and at the same time generate income by licensing others to use those rights?

5. When analyzing a company, why would sophisticated potential investors and/or co-venturers evaluate not only its intellectual property, but also the systems/processes that the enterprise has in place to identify and protect potential intellectual property assets?

6. Why is it important to keep complete and accurate records of development efforts? Why is it important to have those records witnessed? Why is independent corroboration important?

7. What is an interference proceeding? Identify each point the participants need to prove.

8. Anxious to receive credit for developing a new device, the inventor immediately publishes a technical paper. What are the ramifications with respect to obtaining a U.S. patent on a device? What are the ramifications with respect to obtaining patents on the device in other countries?

9. John, employed as a quality control engineer by XA Company, had an idea for a new product that would fill out XA's product line. He stayed after hours, without pay, to develop and test a new product. What rights, if any, does XA have to the product?

10. XA Company engaged an independent contractor to develop software to implement a new process. XA's standard nondisclosure agreement was modified to include a detailed description of the expected deliverable from the contractor and to specify a date by which it was to be delivered and payment terms. What rights, if any, does the independent contractor have to the software?

EXERCISES

1. *IN-CLASS EXERCISE*: For this exercise, students should be asked to distinguish between the various types of trademarks and identify examples under each. The types of trademarks are generic, descriptive, suggestive, and fanciful. List at least three examples under each mark.

2. *OUTSIDE-OF-CLASS EXERCISE*: For this exercise, students should be asked to conduct a patent search for an idea identified by the class. Brainstorm on products that people would like to see invented, but that are not currently in the market (to the best of everyone's knowledge). Students should visit either the USPTO web site or Delphion.com (or another site) to conduct their searches. Find patents that are close to or identical to the idea developed in class. Have any of these ideas been commercialized? If not, why not?

KEY TERMS

Conceiving an invention; conception: The mental portion of the inventive act.

Confidentiality agreement: The primary mechanism for creating the obligation of secrecy necessary for trade secret status.

Confidential information: Information, data, and know-how that is not readily ascertainable from publicly available information.

Continuation-in-part (CIP) application: An application filed while a parent application is still pending and that includes an additional description of details or new embodiments of the invention and claims relating to those new features.

Continuation application: An application filed while a parent application is still pending and that is substantively identical (except perhaps as to the claims) to an earlier filed parent application claiming the benefit of the parent application.

Corroboration: Proof by evidence independent of the inventors.

Descriptive mark: A mark that conveys an immediate idea of the ingredients, qualities, or characteristics of the goods.

Design patent: A government grant of exclusive rights with respect to an ornamental design.

Development agreements: This type of agreement often specifies that a particular party will obtain rights to all technology arising from work done under a contract or first conceived or first made under a contract.

Exclusive right: The right to prevent unauthorized making, using, importing, offering for sale, or selling the patented invention for a period of up to 20 years from the date that the application for patent is filed.

Fanciful or arbitrary marks: Marks that are made-up words, or existing words that have normal meanings totally unrelated to the goods or services with which they are used.

Generic mark: Uses a term that refers to the genus of the particular product.

Intellectual property: The intangible assets that result from creativity, innovation, know-how, and good relationships with others.

Interference proceeding: A proceeding performed by the USPTO to determine priority of patent claims by two or more contesting parties.

Invention assignment agreement: An agreement, typically written, that transfers all rights (ownership) of an invention (and related patent applications and patents) to an assignee. Assignment provisions are often incorporated into employee agreements requiring the employee to assign all inventions to the employer.

Invention disclosure forms: Employed by many companies to establish positive proof of a conception date in the United States.

Know-how: Accumulated practical skill, expertise, data, and information relating to a business and its operations, or performing any form of industrial procedure or process.

Laboratory notebook: A contemporaneous record of the conception and reduction to practice processes involved in making an invention. The laboratory notebook can be a more powerful proof of invention priority than the simple disclosure form.

License agreements: An agreement that grants use rights to a licensee for someone else's invention.

Making an invention: A two-step process consisting of conception and reducing to practice.

Mask works: Under the Semiconductor Chip Protection Act, a "series of related images, however fixed or encoded, that represent three-dimensional patterns in the layers of a semiconductor chip."

Nonproprietary know-how: Information that is generally known in an industry and basic skills or practices employed in an industry beneficially employed by a venture.

Novelty: In patent law, not *anticipated* by the *prior art*. No applicable *statutory bar*. An invention is considered novel unless specific circumstances (statutory bars) have occurred.

Nonobvious: In patent law, means the invention would not be obvious to a person of ordinary skill in the art. If all of the elements of a patent claim are taught by a combination of prior art references, the claim may be rendered obvious.

Ornamental: Nonutilitarian, nonfunctional.

Patent claims: The portion of the patent claims defining the invention protected by the patent. An accused device or process infringes a claim if it includes elements corresponding to each and every element of the patent claim.

Patent Cooperation Treaty (PCT): The PCT facilitates the filing of applications for patents on the same invention in countries that have ratified the Treaty.

Patent specification: The written description and claims of a patent. The patent specification is required to provide a description of the best mode of the invention in sufficient detail to enable a person of ordinary skill in the art to make and use the invention.

Person of ordinary skill in the art: The typical person practicing in the field of the invention—e.g., the average engineer, technician, scientist, or worker in the particular area of technology of the invention.

Plant patent: Issued for a new and distinct invented or discovered asexually reproduced plant including cultivated sports, mutants, hybrids, and newly found seedlings, other than a tuber-propagated plant or a plant found in an uncultivated state.

Prior art: Refers to the body of information against which patentability is measured.

Provisional patent application: An application that can be filed to provide a priority filing date for a corresponding regular patent application.

Reducing the invention to practice; reduction to practice: Building the invention and proving that it works for its intended purpose. The filing of a patent application is considered to be a constructive reduction to practice.

Service mark: A mark used to identify the source or origin of a service.

Shop rights: A royalty-free license to an invention created by an employee of a company on company time and using company resources.

Statutory bar: Specific circumstances defined in the patent law (35 U.S.C. §102) that render an invention unpatentable. In effect, when a statutory bar occurs, the subject matter of the statutory bar becomes prior art against which the patentability of an invention is measured.

Suggestive mark: A mark that requires imagination, thought, and perception to conclude the nature of the goods from the term used as the mark.

Trade secret: Information, data, and know-how that is not readily ascertainable from publicly available information, and has been subject to reasonable measures to maintain secrecy.

Trademark: A mark used to identify the source or origin of a product or service.

USPTO: United States Patent and Trademark Office.

Utility patent: A government grant of exclusive rights with respect to an invention.

WEB RESOURCES

The web sites below are intended as destinations for your further exploration of the concepts and topics discussed in this chapter:

1. http://www.USPTO.gov: This is the web site for the U.S. Patent and Trademark Office. It provides search capabilities and has other useful information about obtaining patents and trademarks in the United States.

2. http://www.epo.org/patents/patent-information/free/espacenet.html: This web site offers the general public free access to worldwide patent information. It has the following main aims:

    - To offer basic patent information to individuals, small and medium-sized enterprises, students, etc.
    - To increase awareness and use of patent information at the national and European levels
    - To support innovation and reduce wastage in the innovation cycle
    - To supplement existing channels for the dissemination of patent information

3. http://www.delphion.com: This is another site that enables people to conduct their own patent searches. The first page of every patent is available for free download, and other pages are available for a small fee.

4. http://www.wipo.int/portal/index.html.en: This is the web site for the World Intellectual Property Organization (WIPO). For technology entrepreneurs interested in building a global business, this site has useful information about developing IP in international markets.

CASE STUDY

Microsoft Attempts New Approach to Intellectual Property Protection in China

On Friday, May 15, 2009, Microsoft announced a partnership aimed at helping make the eastern Chinese city of Hangzhou a model for innovation and protection of intellectual property, in the company's latest attempt to combat rampant software piracy. A 3 year agreement executed between the company and Hangzhou calls for Microsoft to establish a center in Hangzhou aimed at developing new applications and business models for cloud computing. Microsoft will also launch a second technology center in the city that will work with local companies to put their systems on Microsoft platforms, and it will expand an existing technology center operated in conjunction with the Hangzhou government. The company expects to invest more than $1 billion in the new technology development centers in Hangzhou.

Hangzhou is attempting to position itself as a hub for high-tech industries such as software outsourcing. It is headquarters to Alibaba Group, one of China's largest Internet companies with operations in search, online retailing, and the business-to-business transaction platform Alibaba.com. As part of the partnership program, Microsoft plans to offer heavily discounted software to start-up companies in Hangzhou, and provide technology and training to a local university. Hangzhou is the capital of eastern Zhejiang province and home to many small, export-oriented companies. In return, Hangzhou has set targets for enforcement of intellectual property rights, including ridding shops of pirated software and encouraging local enterprises to use legitimate programs.

Software piracy is still rampant around the world despite individual countries' attempts at cracking down. Research commissioned by the Business Software Alliance, an industry trade group, found that 82% of the software used in China in 2007 was not legitimately purchased, more than double the worldwide piracy rate of 38%. In addition, a study by research firm IDC estimates that 80% of software installations in China last year were done without proper authorization, down from 90% in 2004, thanks to measures such as a requirement that computer makers sell PCs with legitimate software preinstalled. In January 2009, a Chinese court convicted 11 people for manufacturing and distributing counterfeit Microsoft software, in a case hailed by the company as a milestone in IPR enforcement here.

But Hangzhou, one of China's wealthiest cities, is seeking to build up its technology industries as it shifts away from textile making and other traditional manufacturing. The aim of the partnership initiative is to establish a model city where intellectual property rights have greater protections than elsewhere in China. Microsoft executives say they may adopt a similar model elsewhere in the country to reward those that clamp down on piracy with greater investment.

Technology companies around the world have tried various measures to stop the drain on their revenue that piracy represents. The move marks a new approach for Microsoft, which in the past has mainly prodded the central government to step up enforcement actions. With several Chinese cities competing to become high-tech hubs, Microsoft is hoping that more can be prompted to follow Hangzhou's lead. Last month, the western city of Chongqing also declared itself to be an "Intellectual property rights (IPR) protection model city," and said it will "robustly protect international IPR rights throughout China for those companies which decide to locate in the city."

The deal with Microsoft came after Hangzhou pledged to improve its enforcement of antipiracy laws and promote the use of legitimate, nonpirated software by individuals, government offices, and companies based in the city, which is west of Shanghai. The deal also calls for the two sides to set up a working team that will hold regular meetings to assess progress in that area.

The partnership will focus on educating local people and businesses on the importance of fighting piracy of software and other intellectual property to their own economic future. Raising consumer awareness was the motivation behind Microsoft's Windows Genuine Advantage program, which turns the wallpaper of computers using pirated Windows software black and notifies users, urging them to get a legitimate copy. That effort continues despite complaints from some Chinese computer users.

QUESTIONS FOR DISCUSSION

1. What is your impression of the strategy that Microsoft has taken to protect its intellectual property in China? Do you think the $1 billion in will spend in Hangzhou is worth it?

2. Do you think that other software companies conducting business in China will benefit from Microsoft's efforts there?

3. If China as a nation is really interested in developing a stronger IPR reputation, what steps will it need to take? Explain how intellectual property law plays a role in a nation's overall ability to engage in global commerce.

Sources: Adapted from Elaine Kurtenbach, "Microsoft, China's Hangzhou set 'model city' pact," Associated Press, May 15, 2009; Aaron Back, "Microsoft to invest in Hangzhou following antipiracy pledge," The Wall Street Journal, May 15, 2009; "Microsoft sued for antipiracy measures," Shanghai Daily, October 29, 2008.

ENDNOTES

[1] Brookings Institute. See also: The challenge of valuing intellectual property assets, Jody C. Bishop, 1 Nw. J. Tech. & Intell. Prop. 4 at http://www.law.northwestern.edu/journals/njtip/v1/n1/4, Spring 2003.

[2] ASME, Strategic issues and trends, November 2005.

[3] Mun J. *Real options analysis: tools and techniques for valuing strategic investments & decisions*: Wiley & Sons; 2002.

[4] PricewaterhouseCoopers, "Technology executive connections: exploiting intellectual property in a complex world," June 2007.

[5] Ocean Tomo, Historical impact of IP on VC, 2007.

[6] See http://www.gnu.org & http://www.fsf.org.

[7] In those models, intellectual property is not dedicated to public. Trade secret protection of the technology involved is typically abandoned, it is either provided subject to a license to a select group of entities, e.g., a pool or consortium, requiring a cross license of their intellectual property, or subject to some form of general public license (again relying on underlying intellectual property rights, typically copyright and occasionally patent) making certain requirements of the users. See Jacobsen v. Katzer, 87 USPQ2d 1836, 535 F.3d 1373 (Fed. Cir. 2008), Wallace v. International Business Machines Corp., 80 USPQ2d 1956, 467 F.3d 1104 (7th Cir. 2006).

[8] 35 USC 102(g).

[9] 35 USC 102(g)(2).

[10] 35 USC §273.

[11] 35 USC §135.

[12] "Pepsi Snitches on Coca-Cola Trade Secrets Thief." *The consumerist*, July 6, 2006.

[13] California go so far as to prohibit non-competition agreements with employees. Bus. & Prof. Code, §16,600.

[14] *See, e.g.*, Cambridge Engineering v. Mercury Partners, 27 IER Cases 68 (Ill. App. Ct. 2007); H&R Block Eastern Enters. v. Swenson, 26 IER Cases 1848 (Wis. Ct. App. 2007); Whirlpool Corp. v. Burns, 457 F. Supp. 2d 806 (W.D. Mich. 2006); Mohanty v. St. John Heart Clinic S.C., 866 N.E. 2d 85, 225 Ill.2d 52 (Ill. 2006); Coventry First LLC v. Ingrassia, No. 05-2802, 23 IER Cases 249 (E.D.Pa. July 11, 2005); Scott v. Snelling & Snelling, Inc., 732 F. Supp. 1034, 1043 (N.D. Cal. 1990); The Estee Lauder Co. v. Batra, 430 F. Supp.2d 158 (S.D.N.Y. 2006); MacGinnitie v. Hobbs Group, LLC, 420 F. 3d 1234 (11th Cir. 2005).

[15] Diamond v. Chakrabarty, 447 U.S. 303, 309 (1980).

[16] Diamond v. Diehr, 450 U.S. 175, 101 S. Ct. 1048, 209 USPQ 1 (1981); In re Bilski, 88 USPQ2d 1385, 1389 (Fed. Cir. 2008).

[17] In re Bilski, 88 USPQ2d 1385, 1389 (Fed. Cir. 2008), In re Comiskey, 84 USPQ2d 1670, 499 F.3d 1365, 1371 (Fed. Cir. 2007).

[18] In re Bilski, 88 USPQ2d 1385, 1391 (Fed. Cir. 2008), citing Gottschalk v. Benson, 409 U.S. 63, 70 [175 USPQ 673] (1972) and Diamond v. Diehr, 450 U.S. 175, 192 [209 USPQ 1] (1981).

[19] 35 USC §102.

[20] Vaupel Textilmaschinen RG v. Meccanica Euro Italia S.P.A., 944 F.2d 870 (Fed. Cir. 1991).

[21] 35 USC §120.

[22] 35 USC §154 (a) (2).

[23] 37 CFR §154 (f).

[24] See 35 USC §262; Willingham v. Star Cutter Co., 555 F.2d 1340, 1344 (6th Cir. 1977); Lemelson v. Synergistics Res. Co., 669 F.Supp. 642, 645 (S.D.N.Y. 1987); Intel Corp. v. ULSI System Technology Inc. USPQ2d 1136 (Fed. Cir. 1993); Schering Corp. v. Roussel-UCLAF SA 41 USPQ2d 1359 (Fed. Cir. 1997) ; Ethicon Inc. v. United States Surgical Corp. 45 USPQ2d 1545 (Fed. Cir. 1998).

[25] More information on the Patent Cooperation Treaty (PCT) and international patent law can be found at the U.S. Patent and Trademark Office (USPTO.gov).

[26] 17 USC §101 (a "computer program" is a set of statements or instructions to be used directly or indirectly computed in order to bring about a certain result; "Literary works" are works …

expressed in words, numbers, or other verbal or numerical symbols or indicia, regardless of the nature of the material objects, such as … film, tapes, disks, or cards, in which they are embodied.); 17 USC §117; Gates Rubber Co. v. Bando Chemical Indus., Ltd., 9 F.3d 823, 28 USPQ2d 1503, 1513 (10th Cir. 1993).

[27] 17 USC §101 (a "compilation" is a work formed by the collection and assembling of preexisting materials or of data that are selected, coordinated, or arranged in such a way that the resulting work as a whole constitutes an original work of authorship); 17 USC §103.

[28] *Feist* Publications, *Inc. v. Rural Telephone Services Co.*, 111 S. Ct. 1282, 1296, 18 USPQ2d 1275 (1991).

[29] Hutchins v. Zoll Medical Corp., 492 F.3d 1377, 83 USPQ2d 1264 (Fed. Cir. 2007), *CMM Cable Rep, Inc.* v. *Ocean Coast Properties, Inc* 97 F.3d 1504, USPQ2d 1065, 1077-78 (1st Cir. 1996) (copyright law denies protection to "fragmentary words and phrases"… on the grounds that these materials do not exhibit the minimal level of creativity necessary to warrant copyright protection). See also *Arica Inst., Inc. v. Palmer*, 970 F.2d 1067, 1072-73, 23 USPQ2d 1593 (2d Cir. 1992) (noting that single words and short phrases in copyrighted text are not copyrightable); *Alberto-Culver Co. v. Andrea Dumon, Inc.*, 466 F.2d 705, 711, 175 USPQ 194 (7th Cir. 1972) (holding that "most personal sort of deodorant" is short phrase or expression, not an "appreciable amount of text," and thus not protectable); National Nonwovens, Inc. v. Consumer Products Enterprises, Inc., 78 USPQ2d 1526, 397 F.Supp.2d 245, 256 (D. Mass. 2005) (There are no stylistic flourishes or any other forms of creative expression that somehow transcend the functional core of the directions), *Perma Greetings, Inc. v. Russ Berrie & Co., Inc.*, 598 F.Supp. 445, 448 [223 USPQ 670] (E.D.Mo. 1984) ("Cliched language, phrases and expressions conveying an idea that is typically expressed in a limited number of stereotypic fashions, [sic] are not subject to copyright protection.").

[30] *Baker v. Selden*, 101 U.S. 99 (1979).

[31] *CMM Cable Rep, Inc.* v. Ocean *Coast Properties, Inc.* 97 F.3d 1504, 1516 41 USPQ2d 1065 (1st Cir. 1996).

[32] 17 USC §101 (a "computer program" is a set of statements or instructions to be used directly or indirectly computed in order to bring about a certain result; "Literary works" are works … expressed in words, numbers, or other verbal or numerical symbols or indicia, regardless of the nature of the material objects, such as … film, tapes, disks, or cards, in which they are embodied.); 17 USC §117; *Gates Rubber Co. v. Bando Chemical Indus., Ltd.*, 9 F.3d 823, 28 USPQ2d 1503, 1513 (10th Cir. 1993).

[33] *See, e.g.*, Asset Marketing Systems Inc. v. Gagnon, 542 F.3d 748, 88 USPQ2d 1343 (9th Cir. 2008), *Sega Enterprises, Ltd. v. Accolade, Inc.*, 977 F.2d 1510 (9th Cir. 1993); *Atari Games Corp. v. Nintendo of America, Inc.*, 975 F.2d 832 (Fed. Cir. 1992); *Computer Associates, Int'l, Inc. v. Altai, Inc.*, 23 USPQ2d 1241 (2d Cir. 1992); *Johnson Controls v. Phoenix Control Sys., Inc.*, 886 F.2d 1173 (9th Cir. 1989). *Stenograph L.L.C. v. Bossard Assocs., Inc.*, 144 F.3d 96, 100, 46 USPQ2d 1936 (D.C. Cir. 1998). See also *Triad Systems Corp. v. Southeastern Express Co.*, 64 F.3d 1330, 1335, 36 USPQ2d 1028 (9th Cir. 1995) (where defendant's conduct "involved copying entire programs, there is no doubt that protected elements of the software were copied"). See, e.g., *Closed* Development *Corp. v. Paperback Software Int'l*, 740 F. Supp. 37 (D. Mass. 1990), *MiTek Holdings*, Inc. *v. Arce Eng'g Co.*, 89 F.3d 1548, 39 USPQ2d 1609, 1617 (11th Cir. 1996); *Digital Communications Assocs., Inc. v. Softklone Distrib. Corp.*, 659 F.Supp. 449, 463, 2 USPQ2d 1385 (N.D.Ga. 1987).

[34] *Digital* Communications *Assoc. v. Softklone Distrib. Corp.*, 659 F. Supp. 449 (N.D. Ga. 1987).

[35] *Gates Rubber Co. v. Bando Chemical Indus., Ltd.*, 9 F.3d 823, 28 USPQ2d 1503, 1509 (10th Cir. 1993).

[36] *Gates Rubber Co. v. Bando Chemical Indus., Ltd.*, 9 F.3d 823, 28 USPQ2d 1503, 1512-13 (10th Cir. 1993).

[37] *MiTek Holdings, Inc. v. Arce Eng'g Co.*, 89 F.3d 1548, 39 USPQ2d 1609, 1616-17 (11th Cir. 1996). See *Apple Computer, Inc. v. Microsoft Corp.*, 35 USPQ2d 1435, 1446, 32 USPQ2d 1086 (9th Cir. 1994) (as to a work as a whole [i.e., a compilation], "there can be no infringement unless the works are virtually identical"), *cert. denied*, _____ U.S. _____, 115 S.Ct. 1176, 130 L.Ed.2d 1129

(1995); *Harper House, Inc. v. Thomas Nelson, Inc.*, 889 F.2d 197, 205 (9th Cir. 1989) ("as with factual compilations, copyright infringement of compilations consisting of largely uncopyright-able elements should not be found in the absence of 'bodily appropriation of expression'").

[38] 37 CFR §201.20.

[39] 17 USC §§302(a)-(b).

[40] 17 USC §§302(c).

[41] 17 USC §411.

[42] 17 USC §410.

[43] 17 USC §504.

[44] 17 USC §505.

[45] 45 17 USC §412.

[46] 37 CFR §202.20-21.

[47] See 17 USC §201.

[48] 17 USC §§101, 201(b).

[49] USC §101.

[50] See, e.g., Community for Creative Nonviolence v. Reid, 490 U.S. 730 (1989).

[51] The Semiconductor Chip Protection Act of 1984, 17 U.S.C. §901(a)(2).

[52] 17 USC §902(b).

[53] 17 USC §908(a).

[54] 17 USC §902(a).

[55] 17 USC §911(f).

[56] 17 USC §§902(c), 906(a)(2).

[57] 17 USC §904.

[58] 17 USC §908(a).

[59] 17 USC §901(a)(6).

[60] 15 USC §1127 (definition of "use in commerce").

[61] 15 USC §1057(c).

[62] 15 USC §1072.

[63] 15 USC §1057(a), (b).

[64] 15 USC §1115(b).

[65] 15 USC §§1124, 1125(b), 19 C.F.R. §133.

[66] 15 USC §1052.

[67] 15 USC §1091.

[68] 15 USC §1096.

[69] 15 USC §1095.

[70] 15 USC §1057(c).

[71] 15 USC §1051(d) (1988).

[72] 15 USC §1126 (1988).

[73] Bayer Co. v. United Drug Co., 272 F. 505, 509 (S.D.N.Y. 1921); Haughton Elevator Company v. Seeberger (Otis Elevator Company substituted), 85 USPQ 80 (Comm'r Pat. 1950); The Murphy Door Bed Co. Inc. v. Interior Sleep Systems Inc., 10 USPQ2d 1748 (2d Cir. 1989); King-Seely Thermos Co. v. Aladdin Indus., Inc., 321 F.2d 577, 579 (2d Cir. 1963).

Contracts

8

After studying this chapter, students should be able to:

- Describe the sources of contract laws and requirements for contract formation
- Understand the concepts of contract offer, acceptance, consideration, and defenses
- Differentiate between substantial performance and a material breach
- Classify alternative remedies in the event of a breach
- Explain the basic structure of a contract
- Identify types of contracts encountered by a technology entrepreneur

TECH VENTURE INSIGHT

Informal Agreements Can Lead to Intractable Arguments

A litigation involving Facebook.com founder Mark Zuckerberg, described by one judge as a "blood feud," illustrates the need for definitive written agreements. In December 2002, Harvard student Divya Narendra went to his fellow students Cameron and Tyler Winklevoss with an idea for an online social network for college students. They founded a venture, originally called "Harvard Connection" (later called ConnectU), to develop the idea.

Harvard Connection needed a programmer. Mark Zuckerberg, who later founded Facebook, had gained notoriety when he was placed on probation by Harvard University for hacking into the school's servers. Harvard Connection hired Zuckerberg and provided him their business plan and code. It wasn't long after that a dispute arose.

It is alleged that Zuckerberger understood that the information provided was proprietary. There was, however, no *written* agreement. Supposedly, Zuckerberg stalled Harvard Connection's launch while working on Facebook. In January 2004, three days after telling the Harvard Connection team that he had their coding done, Zuckerberger registered the domain name "TheFacebook.com." E-mail correspondence indicates that the Facebook site was complete prior to a final meeting between Zuckerberg and Harvard Connection.

On September 2, 2004, ConnectU sued Zuckerberg for breach of contract, misappropriation of trade secrets, and copyright infringement. Zuckerberg countered that there was no contract, that he had never promised Harvard Connection anything, and that he was simply helping them out. The litigation lasted more than 3 years before it was ultimately settled prior to trial.

By February 2008, ConnectU had still not been provided with access to the relevant documents from Facebook. Nonetheless, the parties agreed to suspend proceedings and submit to mediation. After a 2 day mediation, a one-and-a-half page "Term Sheet & Settlement Agreement" was signed. Facebook agreed to give the ConnectU principals a certain amount of cash and Facebook stock in return for their stock in ConnectU. However, ConnectU learned that its understanding of the value of the Facebook stock was in error, and the parties could not agree on the formal documents. ConnectU took the position that the mediation document lacked crucial terms and was insufficient to establish a meeting of the minds. The court disagreed. It held that the document was sufficient to create a contract and that ConnectU could have conditioned the agreement on a specific Facebook valuation or done its own due diligence. It had done none of those things.

The moral of the story is that complete and definitive agreements can prevent much heartache. Oral agreements and incomplete, imprecise written agreements are grist for the litigation mill.

Source: ConnectU, Inc. v. Facebook, Inc. et al., District of Massachusetts, case number 1:2007cv10593; The Facebook, Inc. v. ConnectU, LLC et al., Northern District of California, case number 5:2007cv01389; O'Brien. Poking Facebook. November/December 2007.

INTRODUCTION

A contract is an agreement between two or more parties that is legally binding and enforceable in a court of law. Performance in accordance with the agreement is an obligation or duty under the law. A failure to perform is a **breach** for which the law offers a remedy.

Businesses are free, within broad legal limits, to agree upon whatever terms they see fit. Those terms define respective rights and obligations of the parties for the duration of the contract and the rules under which they operate. A contract serves, in effect, as the private law of these businesses. It is important, therefore, to understand when a contract is created and how it can be enforced. A failure to form a valid contract when intended, or inadvertently committing a business to obligations, can be catastrophic.

In this chapter, we will consider the basic sources and concepts of contract law, the anatomy of a typical agreement, and then introduce various types of agreements often encountered by technology entrepreneurs.

8.1 SOURCES OF CONTRACT LAW

When an issue of contract law arises, it is generally in connection with one of the following questions: Has a contract been formed? If so, what are the terms? And, if there is a breach, is there an applicable defense? If not, what are the remedies?

Answering these questions, however, is complicated by the fact that a different body of contract law applies depending upon the subject matter of the transaction, the location of the parties to the transaction, and the terms of a written agreement.

Two primary sources of law govern contracts between parties located within the United States. Historically, the **common law** of a state (judicial decisions) was applied to contracts formed or performed within its boundaries.[1] Consequently, the law varied from state to state. As the frequency of interstate transactions increased, a **Uniform Commercial Code** (UCC) was developed for, among other things, the sale of goods.[2] The UCC, consisting of 10 articles, covers the rights of buyers and sellers in transactions. Drafted in 1952, the UCC has been adopted in its entirety by every state except Louisiana (which has adopted about half of the code). Each state has unique adaptations of the code to fit its own common law findings. Generally, business owners can access a state's UCC through that state's Secretary of State Office. The UCC defines "goods" as all things that are movable at the time of identification to a contract for sale.[3]

The UCC is significantly different from the common law in a number of respects. Most notably, actions that would not create an enforceable contract under the common law may do so under the UCC, and the effective terms of an agreement may be different when interpreted under the UCC rather than under the common law.

In addition, since 1988, the United States has been a signatory to the 1980 United Nations Convention on Contracts for the International Sale of Goods (CISG). It applies to the sale of goods between a party in the United States and a party in another CISG country.[4] Effective August 2009, 71 countries are signatories of the CISG including Canada, China, Mexico, Japan, Germany, South Korea, France, Venezuela, the Netherlands, Italy, and Belgium, all of the top 10 trading partners with the United States, excepting the United Kingdom and Saudi Arabia.

The CISG covers agreements that would otherwise be covered by the UCC as shown in Exhibit 8.1. It is, however, significantly different from the UCC in a number of respects, notably whether or not a contract is formed, the requirement for a written agreement,[5] and the availability of certain remedies.

However, the parties to a transaction can opt out of the CISG by an express and explicit statement in a written agreement, such as:

> *The rights and obligations of the parties shall not be governed by the United Nations Convention on Contracts for the International Sale of Goods.*

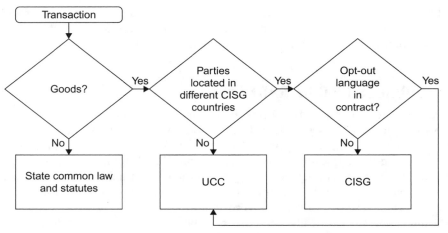

EXHIBIT 8.1

The CISG and UCC Relationship.

It is important to be able to determine the particular body of law that governs a transaction. Whether or not there is a contract, and if so, the terms of the agreement may be different depending upon the applicable law.

8.2 CONTRACT FORMATION

To begin to understand the concept of contract formation, the first question to ask is: Has the interaction between two parties created an enforceable contract? Traditionally, in order for an agreement between respective parties to create a legally enforceable obligation:

- The parties must have the legal capacity to enter into a contract.
- There must be an offer and an acceptance of that offer.
- There must be **consideration** to support the contract.
- The subject matter of the contract must be legal and not against public policy.
- There must not be any viable defense, such as mutual mistake, fraud, duress, etc.

In addition, under certain circumstances, the contract must be reflected by a written instrument.

8.2.1 The Offer

The basic foundation of a contract is **mutual assent** or a meeting of the minds. In order for there to be a meeting of the minds, an offer must be made, and that offer accepted. An offer is a proposal meeting certain requirements, made by an offeror to an offeree. In order for a proposal to constitute an offer, it must:

- Identify the person or class of persons (the offeree) to whom it is made

- Be unconditional

- Not require anything from the offeree other than an indication of acceptance

- Indicate an intent to create a contract with the offeree, and not a mere invitation for offers (like an advertisement or request for proposals)

- Be objectively reasonable to believe that the person making it is prepared to be bound by the terms if it is accepted—i.e., a reasonable person would not consider it a joke or facetious statement

- Provide a basis for determining the essential terms of the proposed agreement. Nonessential terms can be filled in by the court on a "reasonableness" standard[6]

The offeree's response to the offer determines whether or not a contract is formed. It can either reject or accept the offer. Rejection can be express or by lapse of time. Once rejected, the offer need not be honored or renewed. Generally, only the person to whom an offer was directed may accept an offer. Under common law and the UCC, an acceptance is deemed to be effective when it was mailed, as opposed to when it was received.[7] This is sometimes referred to as the **mailbox rule**. Under the CISG, however, acceptance is not effective until it reaches the offeror. The point at which an acceptance is effective becomes relevant when there is an attempt to revoke an offer (or an acceptance).[8]

8.2.2 The Counteroffer

A response to an offer that proposes the parties go forward on different terms from those proposed in the offer is a **counteroffer**. Whether or not a counteroffer constitutes acceptance and forms a contract is a function of the applicable body of law. Unless under the UCC, an acceptance must unequivocally meet and correlate with the offer in every respect in order to be effective. This is referred to as the **mirror image rule**.[9] Even if a response is labeled an acceptance, if it proposes any deviation from the terms of the original offer, it is considered a counteroffer and a rejection of the original offer.[10] If the counteroffer is not acceptable, the original offeree cannot fall back on the original offer. The original offer need not be honored or renewed.

The mirror image rule does not apply to the sale of goods covered by the UCC.[11] The result is the sometimes difficult and confusing **battle of the forms**. When purchasing goods, a company often sends a preprinted purchase order form containing contract terms and conditions to a vendor. The vendor often responds with its preprinted order-acknowledgment form, containing terms and conditions, almost always different from those contained in the order/offer.

Under the common law, the vendor's order-acknowledgment form would be a counteroffer, and the original offer can be rejected. No contract would be formed. If, as is often the case, the parties fail to notice or ignore the inconsistencies in preprinted forms, and proceed with the transaction, and there is a dispute, contract law remedies would not be applicable. The parties would have to rely on the judicial doctrine of quasi-contract (see Section 8.4.4) to prevent unjust enrichment.

Under the UCC, however, any definite and timely expression of acceptance will form a contract regardless of whether it contains different terms—unless acceptance is expressly made conditional on assent to the new terms.[12] Between merchants, the terms of the counteroffer become part of the contract unless:

- The offer expressly limits acceptance to the terms of the offer

- They materially alter the offer

- A notification of objection has already been given or is given within a reasonable time after notice of them is received[13]

A number of issues tend to arise: Which writing between the parties constitutes an offer and which constitutes the acceptance? Does a term proposed in a responsive communication "materially alter" the offer? And, even if the forms are so completely at odds that no written contract is formed, conduct may still form a contract.[14] The issue then becomes: What are the contract terms?

The UCC uniquely assists courts in determining provisions of contracts that are not explicitly spelled out between parties. If parties to a contract go to court to dispute something that was not explicitly spelled out in a written agreement, the court will rely upon the UCC to fill in the missing details. The UCC provides the court with what is known as **gap fillers**. These are provisions that are derived from the UCC and that help complete a contract. For example, if a party agrees to provide labor for $100, it may not be spelled out in the contract that U.S. currency is preferred. If a dispute arises because the second party wants to pay in Turkish lira (a currency that devalues quickly), the first party may go to court, where the court will use gap fillers to determine the *intent* of the offer, acceptance, and consideration. The UCC also provides "supplementary" terms (e.g., implied warranties) that become applicable "unless otherwise agreed" or conspicuously disclaimed. Accordingly, the effective terms of the contract between the parties become the terms common to the "offer" and "acceptance," together with the supplementary terms supplied by the UCC[14a]. This can result in contract terms very different from those envisioned by either party.

8.2.3 Acceptance

The **acceptance** is the assent on the part of the receiving party to the terms of the offer. For example, if the offer to provide labor services was suitable, the party to receive the services would accept the terms. Signing of a contract is one way a party may demonstrate assent. Alternatively, an offer consisting of a promise to pay someone if that person performs certain acts that he or she would not otherwise do (such as repair a vehicle) may be accepted by the requested conduct instead of a promise to do the act. The performance of the requested act indicates objectively the party's assent to the terms of the offer.

The essential requirement is that there must be evidence that the parties had each, from an objective perspective, engaged in conduct indicating their assent. This requirement of an objective perspective is important in cases where a party claims that an offer was not accepted, taking advantage of the performance of the other party. In such cases, courts can apply the test of whether a reasonable bystander would have perceived that the party has impliedly accepted the offer by conduct.

8.2.4 **Revocation of Offer or Acceptance**

Occasionally, a party will attempt to revoke or retract an offer or an acceptance. If an offer is retracted before it is accepted, or acceptance retracted before it is effective, there is no contract. A revocation is effective when received.[15] An acceptance, under the mailbox rule of the common law and UCC, is effective on dispatch.[16] Under the CISG, it is effective when it reaches the offeror.[17] If the revocation of an offer reaches the offeree after acceptance is dispatched, but before it is received by the offeror, under the UCC (and common law), the revocation is too late, and a contract is formed. Under the CISG, however, there is no contract.

8.2.5 **Consideration**

Consideration is the bargained-for exchange of an agreement. The adequacy of consideration is irrelevant to the formation of a contract. However, in the absence of consideration, there is generally no enforceable contract. Nevertheless, in some cases, it is reasonable for a person to rely upon a promise. And, if they do so to their detriment, a legal theory known as **promissory estoppel** provides a substitute for consideration, and the promise is enforceable.

8.3 **DEFENSES AGAINST CONTRACT ENFORCEMENT**

In order for there to be an enforceable contract, there must be mutual assent as to the terms of the agreement. Assent, however, can be negated by a mutual mistake as to the subject matter of the agreement.[18] If both parties are mistaken as to a basic assumption regarding the contract, the contract may be avoided. On the other hand, a mistake on the part of only one party to a contract, i.e., a **unilateral mistake**, is generally not a defense to enforcement of the contract. However, if the nonmistaken party knew or should have known of the mistake, or if it was the result of a misstatement or misleading statement by the nonmistaken party regarding a material element of the contract, the contract may be **voidable**.

A contract is unenforceable if the subject matter of the agreement is against public policy or illegal. Examples include:

- Agreements in restraint of trade. For example, an overreaching employee non-competition agreement may be unenforceable as against public policy
- Gambling contracts (in some states)

- Usurious agreements

- Agreements to commit crimes or torts

An enforceable contract also requires true consent to the terms. If the agreement is entered into as a result of undue duress or coercion, there is no true consent. If the agreement is induced by fraud (e.g., the victim was not aware that the contract was being formed) such **fraud in the inducement** negates true consent and renders the contract voidable.

Under the **Statute of Frauds**, in the absence of a signed writing, certain contracts are voidable. This concept is incorporated into both common law and the UCC. Significantly, there is no statute of frauds under the CISG.[19] Under the statute of frauds, the following subjects must be in writing:

- Surety contracts that are agreements to pay the debt of another

- Promises in consideration of marriage (e.g., "I'll marry you if you'll pay for that surgery I want"—where still enforceable)

- Long-term interests in real property such as the purchase of real estate, mortgages, easements, and leases with the term in excess of 1 year

- Contracts that by their terms cannot possibly be performed within 1 year (e.g., 5 year employment contract)

- Promises by the executor of an estate to pay the estate debts out of the executor's private funds

- Sale of goods having a value of more than a specified amount, typically $500[20]

- Certain consumer purchases

The writing need not be a formal contract, or even a single document, but must reflect the existence of a contract. The common law requires that the writing includes the identity of parties, subject matter of the agreement, essential terms of the agreement, consideration, and signature of the party against which the contract is to be enforced. Under the UCC, the writing need only be signed by the party to be charged, and specify a quantity term.[21]

8.4 PERFORMANCE AND BREACH

Contracts typically spell out performance expectations for the various parties. For example, a contract to purchase a vehicle from a car dealer specifies that the car will be delivered to the buyer in good quality and in working order. The dealer must perform to these standards to meet its obligations. The buyer is obligated to meet payment terms, including a down payment and future monthly payments if the vehicle is financed. There are different types of performance on a contract:

- **Complete performance** means completely fulfilling all obligations under the contract in accordance with its terms.

- **Substantial performance** means essential fulfillment of the obligations but with slight variances from the exact terms, unimportant omissions, and/or minor defects that do not defeat the purpose of the contract.
- **Inadequate performance** or **nonperformance** means a significant variance from the terms of the contract.

Complete performance **discharges** a contract; anything less is a **breach**. A failure to provide at least substantial performance is a **material breach**.[22] Exhibit 8.2 illustrates the alternative remedies available to an aggrieved party in the event of a breach.

Performance can also be made conditional on specified events or metrics. Satisfaction of a **condition** can be a prerequisite for a duty to arise, terminate an existing duty, or change the nature or extent of a duty. Failure to meet a condition is not a breach. Rather, the contract contemplates the possibility of the condition not being met and is carried out or terminated according to its terms. In the absence of complete performance, the parties to a contract also can generally agree to discharge each other's performance through:

- A mutual **rescission**, whereby both parties are restored to the positions they were in before the contract was entered
- A **substituted** or **amended contract**, whereby the original duties are discharged and new duties imposed
- An **accord and satisfaction**, where a party to the contract agrees to accept a different performance in satisfaction of the existing duty

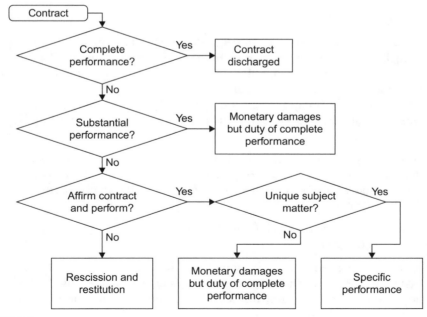

EXHIBIT 8.2

Remedies for Unexcused Breach of Contract.

A **novation** is an amended contract that substitutes a third party for one of the original parties to the contract.

Occasionally, after an agreement has been entered into, something will occur that renders performance either improper or impossible. The law recognizes these extraordinary situations, and in those cases excuses the duty to perform. For instance, if the subject matter of an agreement was legal when the contract was formed, but becomes illegal at a later time, failure to perform is excused.

Circumstances that make performance more difficult or burdensome, however, will typically not excuse contractual duties. However, under a commercial impracticability test adopted by many states, and the UCC and CISG, performance by a party is excused if it is rendered impracticable by an event that is:

1. Beyond the control of the party
2. Not reasonably foreseeable by the party
3. Contradicts a mutual assumption on which the contract was based.

Commercial impracticability requires more than just unexpected hardship or increased cost. It requires extreme and unreasonable difficulty or expense caused by an unforeseeable supervening event. The concept of commercial impracticability is often explicitly written into contracts. Such a provision is often referred to as a **force majeure clause**. Alternatively, an agreement can explicitly call for a party to assume the risk that a certain event may occur, in which case commercial impracticability would not excuse nonperformance.

8.4.1 Damages

In general, the aggrieved party to a breach is entitled to all damages that it can show were actually caused by the breach of the contract, as long as the damages were reasonably foreseeable at the time the parties entered into the contract, and are sufficiently certain so as not to be "speculative."[23]

Compensatory damages place the aggrieved party in the position it would have been in had there been complete performance under the contract.[24] This includes:

- The loss of value (the difference between the value of the benefits that would have been received and those actually received) less any savings experienced from being excused from any obligations under the contract (the benefit of the bargain).

- **Incidental damages** directly resulting from the breach (e.g., costs of obtaining substitute performance).

- Consequential damages, also sometimes referred to as "special damages," which include any other damages or losses actually caused by the breach, so long as they are reasonably foreseeable. Agreement terms can make clear that certain types of losses are foreseeable.

- Reliance damages are an alternative to compensatory damages, intended to place the nonbreaching party in the position it would have been in had the contract never been formed. Typical elements include expenses incurred in preparing for and performing under the agreement and lost opportunities.

In general, there is a duty to **mitigate** damages. Damages that could have been avoided with reasonable effort and without undue risk or burden cannot be recovered.[25] The particular acts required to mitigate damages depend upon the circumstances. For example, an aggrieved buyer of goods must cover by purchasing substitute goods at a reasonable rate and without unreasonable delay or is foreclosed from receiving any consequential damages that would have been prevented.[26]

Commercial contracts often include provisions purporting to limit liability. Contractual limitation of liability can allocate risk between the parties and be reflected in the consideration paid. If the party limiting liability assumes more risk, the price paid by the other party would be proportionately increased. The provision thus permits lower charges. This is, of course, totally inapplicable to contracts of adhesion. In general, limitations on liability are not favored by the courts.

8.4.2 Rescission and Restitution

Rescission means, in effect, to undo a contract. **Restitution** means a return of the benefits the aggrieved party unjustly conferred upon another. The aggrieved party is placed in the position it would have been in had the transaction never taken place. Rescission and/or restitution are available:

- As an alternative remedy for a party injured by material breach[27]

- To a party that breaches after partial performance for any benefit conferred in excess of the loss caused by the breach[28]

- When there has been a fault in the formation of a contract, or it is otherwise unenforceable or voidable[29]

Restitution may also be pursued in situations where damages and specific property are difficult or impossible to prove and recover and can sometimes also result in a greater overall award.

8.4.3 Specific Performance

When monetary damages are inadequate to compensate an aggrieved party, a court may order the breaching party to completely perform under the contract-specific performance.[30] For example, money damages are inadequate where the subject matter of the contract is rare or unique (e.g., land or fine art). Under common law and the UCC, specific performance is not available unless monetary damages are inadequate. However, the parties in international sales transactions under the CISG may obtain specific performance anytime there is a material (fundamental) breach, so long as they have not already resorted to an inconsistent remedy.[31]

8.4.4 **Quasi-contract**

Even in the absence of an enforceable agreement, a court may construct a fictional contract, a **quasi-contract**, to prevent unjust enrichment. Quasi-contractual recovery is applicable when a party has partially performed in the mistaken belief that there was a contract or under a contract that is unenforceable. It permits the party to recover the reasonable value of the partial performance.

8.4.5 **Reformation**

In some cases, a written agreement will not accurately reflect the intentions of the parties. This most often occurs when there has been an error in documentation, a fraud, misrepresentation, or mutual mistake. In those cases, the equitable remedy of reformation is employed by the court to effectively rewrite the contract to reflect the actual intentions of the parties.

In other instances, a particular provision of an agreement may be void or unenforceable as **unconscionable**, against public policy, or illegal (due to a change in the law) if interpreted literally. When that occurs, some courts will hold the entire agreement unenforceable. Others will effectively rewrite the offending agreement so that it is valid and enforceable.[32]

8.5 **ANATOMY OF A CONTRACT**

Written contracts can take many forms and adopt many formats. As discussed earlier, an enforceable contract can be found even in the absence of a written agreement. However, a comprehensive written agreement can be a tool to ensure good relationships between participants. In this section, we will examine the general structure of a typical negotiated written agreement and consider the significance of some of the typical provisions.

TECH TIPS 8.1

Writing an Effective Written Contract

Some pointers to remember in crafting a contract include:

- Carefully and precisely define all terms.
- Explicitly, unambiguously, and completely state all items of importance to the parties.
- Attempt to contemplate all contingencies and clearly set forth the rights and obligations of the parties if these should occur.
- Addressing potential future problems does not imply any particular expectation that they will in fact arise, nor does a complete and unambiguous contract imply any lack of trust between the parties. To the contrary, it assures that the parties have a complete understanding of each others' expectations. A minimalist agreement, on the other hand, leaves too much room for misunderstanding.

8.5.1 Preamble

The preamble is a vehicle for identifying the parties; in many instances, for denominating the agreement for identification purposes; and for establishing the effective date of the agreement.

8.5.2 Recitals

Recitals are a vehicle for ensuring a complete understanding of the party's respective expectations, and memorializing any relevant background information and extrinsic considerations. Recitals preferably should identify a range of considerations, including:

- Related agreements
- A third party that may play a role in the performance or that may be a beneficiary
- The relationship between the parties
- The basic premise of the agreement and expectations of the parties
- Any facts or circumstances relevant to the possibility of particular damages (and establishing that such damages were foreseeable)

8.5.3 Definitions

A well-written agreement will include a definition for each term used that could conceivably be misunderstood. In complex agreements, these definitions are typically included as a separate section either at the beginning or end of the agreement. However, in shorter, simpler agreements, the definitions may be provided in the body of the agreement, as the terms are introduced in context. By convention, defined terms are typically designated within the text of the agreement by capitalization. Explicit definitions are particularly important in international transactions where there are cultural and language differences.

8.5.4 Performance

The provisions of an agreement relating to performance by the parties vary greatly depending upon the subject matter of the agreement. In any event, it is extremely important that all expectations be explicitly reflected in the provisions. A simple rule of thumb: if a detail is not explicitly recited in the agreement, do not expect it to happen.

Conditions and contingencies are frequently built into the performance provisions. Use of relative timing (e.g., "Developer will deliver a prototype within 30 days of receiving the specification") rather than absolute dates can be used to expressly make performance conditional on the performance of the other party.

There are myriad ways to specify the performance of the parties. Each performance event can be described with particularity in successive provisions. All deliverables should be precisely and completely defined, including all details that are significant to the parties. Deliverables can be described integrally with the performance events, or in separate provisions. Such descriptions are often provided by reference to a specification or documentation.

In situations where there is considerable interaction between the parties, a **performance matrix** can be a useful organizational tool, particularly when used in conjunction with separate detailed descriptions of deliverables. Exhibit 8.3 illustrates a skeletal

Time reference	Developer action	CP action
Phase I		
Upon execution of the AGREEMENT		Provide initial draw-down retainer of $xx Provide detailed specification for the existing CP product
Within 10 days of receipt of CP product specification	Present proposed development schedule	
Within 5 working days of receipt of proposed development schedule		Provide a written notice of acceptance or rejection
Phase II		
Upon commencement of Phase II		Provide additional draw-down retainer of $xx (per development schedule)
Pursuant to development schedule	Present Deliverable 1 (detailed Specification for modified product)	
Within 5 working days of presentation of Deliverable 1		Provide a written notice of acceptance or rejection
Within 10 working days of presentation of Deliverable 1		Provide 6 units of existing CP product and components called for in specification
Within (per development schedule) working days of acceptance of Deliverable 1	Present Deliverable 2 (2 prototypes of modified CP product)	
Within 10 working days of presentation of Deliverable 2		Provide a written notice of acceptance or rejection
Within 30 working days of the acceptance of Deliverable 2		Make payment of $xx (per development schedule)

EXHIBIT 8.3

Example of a Performance Matrix.

generic performance matrix relating to a contract under which a contracting party (CP) engages a developer (Developer) to one of its existing products (CP Product). Respective columns are provided for event timing, and actions by each of the parties. Each sequential performance event is represented by a successive row. Details with respect to particular performance events can be particularized if not clear from the matrix.

8.5.5 Ownership and Use of Intellectual Property

If performance under the agreement is likely to result in the development of intellectual property, the issue should be dealt with explicitly in the agreement. Ownership of intellectual property resulting from performance under an agreement tends to be highly negotiated. Some considerations include:

- Is the purpose of the contract to develop the IP or is the development ancillary to performance?
- Is the "preexisting" IP of one party being used as the basis for development of the new IP?
- Was the preexisting IP maintained as a trade secret to which the other party would not have access but for the agreement?
- Is the development effort entirely by one party or is it a cooperative effort?
- Is the originator of the new IP being fully compensated for its contribution to the development effort?

In some cases, particularly in development agreements, the developer may assign new technology or IP resulting from the agreement to the other party, but retain ownership to its preexisting technology. If the preexisting technology (or any unassigned technology or IP) is incorporated in the deliverables, the other party is typically granted a license to use it. Again, the agreement should include a grant provision clearly defining the scope of that license.

It should be recalled that in the absence of an express contractual provision, where independent parties are involved, irrespective of which party pays for development, whichever party actually creates the IP will own it. The circumstances of the agreement may give rise to an implied license to use the IP, but it is not good practice to leave the existence and scope of a license to the courts.

Transfers of rights in intellectual property are typically in the form of an assignment or a grant of a license. Such transfers are encountered in, for example, agreements setting up a business venture, agreements between a company and its employees, and development/consulting, manufacturing, licensing, technical assistance, joint venture, distributorship, and marketing agreements.

8.5.6 Consideration

The provisions in an agreement relating to consideration can take a wide variety of forms. In many cases, the consideration is simply a single lump sum payment.

In other cases, payment is tied to a milestone. Where a performance matrix is employed, such payments are simply reflected as performance events. Other agreements may call for periodic payments, the amounts of which vary on the basis of the marketplace performance (e.g., sales of any party). In those cases, provisions particularizing the computation of the payments, as well as maintenance of accounting records, reporting, and audits of the records are typically included in the agreement. License agreements and distribution agreements tend to be of that type.

8.5.7 Representations and Warranties

Representations are, in essence, a statement of facts relied upon by the parties in entering the agreement. Warranties are, in effect, a promise that a product, service, or aspect of performance will meet certain standards or metrics or do certain things. In many agreements, representations and warranties will be combined, ("Developer represents and warrants that ...") so that the statement is both a representation upon which the other party relied and a promise.

Representations and warranties are a vehicle for ensuring that the expectations of the parties are understood and met. If something is important to a party, inclusion of a representation or warranty on that point can ensure that there is no misunderstanding as to the party's expectations. The breach of a representation or warranty is also readily made an event that would trigger a right to terminate. Representations are often provided with respect to:

- Authority to enter into the agreement without any third-party consents
- The expertise of the parties
- Staffing level to be dedicated to the project
- Level of efforts to go into performance (e.g., best efforts)
- Novelty, confidentiality, and/or originality of deliverables
- Absence of pending or threatened claims or proceedings relating to the subject matter of the contract
- Absence of conflicting agreements
- Trade secret status of aspects of the licensed subject matter
- Quality, functionality, and performance characteristics of deliverables
- Completeness and/or accuracy of information provided to the other party

Representations and warranties may be absolute or qualified. When an exhaustive investigation would be required in order to make a representation, a party can qualify the representation by limiting it to their personal knowledge (e.g., "To the best of Developer's knowledge, there are no infringing third-party uses of the Licensed Technology").

Warranties that are explicitly stated in the agreement are referred to as **express warranties**. The terms of express warranties are totally in control of the parties. However, certain other warranties may be implied by law unless explicitly disclaimed in the agreement.[33] Both the UCC and the CISG provide for a number of implied warranties, including a:

- *Warranty of noninfringement*: That the goods sold do not infringe the intellectual property rights of any third party[34]

- *Warranty of merchantability*: That the product is fit for the ordinary purposes for which it is used[35]

- *Warranty of fitness for a particular purpose*: That the product is fit for the particular purpose of the buyer[36]

The scope of these implied warranties is left open to interpretation of the courts. The better practice is to disclaim the implied warranties and provide express warranties that clearly specify the rights and obligations of the parties.

8.5.8 Indemnity

An **indemnity**, also sometimes referred to as a **hold-harmless**, is another mechanism for allocating risk between the parties. The indemnitor is obligated to pay for the liability of the indemnitee for certain specified events. A typical indemnity agreement shifts liability to the party who is likely to be more actively or primarily responsible for the events giving rise to the liability. However, negligence or fault is not necessarily a prerequisite for coverage.

An indemnification provision is often used instead of a representation or warranty when the subject matter of the indemnity is beyond the knowledge of the parties. If an event renders a representation false, there is a breach giving rise to liability, and perhaps excusing further performance by the other party. If, on the other hand, the same event triggered an indemnification, essentially the same liability would accrue, but there would be no breach; the contract simply proceeds according to its terms.

Indemnifications against infringement of third-party intellectual property rights are also commonly encountered. The scope of intellectual property indemnifications can vary greatly depending upon the terms of the provision. Frequent areas of negotiation include:

- Choice and participation of counsel and control of litigation

- Reimbursement for attorneys' fees

- Infringement not solely due to the deliverables from the indemnitor—i.e., by virtue of additions, modifications, or combinations not provided by the indemnitor.

- Remedies in the event of an injunction

8.5.9 Term and Termination

The **term** of the contract refers to the length of time it remains actively applicable to the parties. When performance involves a single discrete transaction, such as the purchase of goods or the assignment of intellectual property, the duties under the agreement are discharged when the transaction is completed. However, many contracts involve a continuing right or duty, or a series of transactions. In those cases, the term of the agreement should be explicitly stated.

TECH TIPS 8.2

The Term of an Agreement

There are a number of different approaches to establishing the term of an agreement. The term of an agreement can be:

- For a stated period of time
- For consecutive periods of stated duration, automatically renewed unless terminated
- For an indefinite period, but terminable

When a contract is of an indefinite period, the termination trigger events are often crafted to establish an initial period during which the agreement can be terminated only in the event of a breach or failure of condition, and thereafter can be terminated by one or both parties upon notice to the other. Termination is typically effected by a written notice, although termination is often not effective until the lapse of a specified time period after the notice.

Certain types of rights and obligations would not be particularly effective if they were terminated or discharged at the same time as the other aspects of an agreement. Confidentiality and indemnification provisions often fall within this category. It is not uncommon for an agreement to expressly state that designated provisions will survive termination of the agreement.

8.5.10 Miscellaneous Provisions

The so-called **boilerplate** provisions typically address various issues of law that are ancillary to actual performance of the agreement. These issues tend to recur in many types of agreements, and over the years, standardized language, the boilerplate, covering the issues has been developed. The issues include:

- *Notices*: The manner in which formal notices under the contract are to be made

- *Waiver*: The failure to enforce a right in one instance will not constitute a waiver preventing enforcement in subsequent instances

- *Applicable law*: Specifying the body of law under which the contract will be interpreted.

- *Dispute resolution processes*: Admission of jurisdiction/exclusive jurisdiction, trial venue, arbitration, mediation.

- *Integration*: Precluding reference to communications not incorporated into the written agreement.

- *Severability/reformation*: Treatment of an unenforceable provisions.

- ***Assignment***: Whether or not assignment and **delegation** is permitted, and any limitations.

- *Force majeure*: Performance excused if made impracticable by forces outside of the control of the party.

- *Relationship of the parties*: Agreement does not create partnership.

8.6 OPERATING AGREEMENTS

When a business entity is formed, an operating agreement between the principals is desirable, if not necessary. This is particularly true when the business entity is formed to implement a joint venture between separate business entities. A partnership agreement, LLC operating agreement, or corporate shareholder's agreement permits the principals to define the internal rules under which the venture will operate, and directly or indirectly assign management responsibilities. We discussed these topics in more detail in Chapter 4.

There are limits to the extent to which management can be separated from stock ownership in a corporation, particularly S-corporations. However, a shareholder's agreement can require individual shareholders to vote their stock for particular board members, thus effectively defining the management structure. Partnership agreements and LLC operating agreements are also vehicles for defining how profits and losses will be allocated among the principals. Other considerations often addressed in these agreements include:

- Any special powers to be given to the management

- Contributions of intellectual property by the principals—e.g., licenses to the venture

- Ownership and use of intellectual property

- Confidentiality

- Services to be rendered on behalf of the venture by individual participants

- Compensation, if any, to principals for services provided to or work done on behalf of the venture

- Transfer restrictions on ownership interests
- Vesting schedules
- Exit procedures, such as repurchase and buy-sell arrangements and processes for dissolution and winding up
- Rights and obligations with respect to further financings and issuance of additional equity interests—e.g., antidilution provisions and capital contribution provisions
- Indemnifications of the individual principals for actions taken on behalf of the venture
- Special tax allocations

8.6.1 Employment Agreements

Agreements with employees are often essential to obtaining and maintaining rights in intellectual property. In general, such agreements typically require each employee to assign to the company all rights to any intellectual property that relates to the company's business created by the employee during the term of employment. In the absence of provisions in a written agreement to the contrary, the allocation of rights to any inventions made by the employee will depend upon whether the employee was "hired to invent" and when and where the invention was made. The agreement typically also defines an obligation of confidentiality.

8.6.2 Noncompete Agreements

In the sale of a business, or where particularly critical confidential information is involved, or an employee is in a crucial position, noncompetition provisions may be appropriate. Noncompetition provisions tend to be very strictly construed, and are typically unenforceable if they are not reasonable in scope as to geography, time, and precluded activities. Exhibit 8.4 is an example of a noncompete agreement.

8.6.3 Confidentiality Agreements

A confidentiality agreement, sometimes also referred to as a **nondisclosure agreement (NDA)**, should be in place before any third party is given access to proprietary know-how outside of the context of a broader agreement already including confidentiality provisions. The NDA precludes disclosure and also limits the use of the information by the recipient to a specific purpose in connection with the business relationship between the parties. Sometimes, technology entrepreneurs use NDAs when they meet with investors. This is usually not a good idea, as discussed in Tech Micro-Case 8.1.

AGREEMENT NOT TO COMPETE

I, _____, make the following agreement with _____ (the "Company") on this _____ day of _____, 19____.

The Company is engaged in the business of _____, and has developed or otherwise obtained and continues to develop or obtain certain confidential information and proprietary technology related to its business.

I acknowledge that I am, or will be employed by the Company, in a position which shall involve the creation of technology relevant to the business of the company, and under conditions in which I will be given access to and personally entrusted with Confidential Information (as that term is defined in the Agreement regarding Confidential Information & Ownership of Technology between me and the Company dated ___) including that pertaining to: (i) proprietary technology of the Company; and (ii) the technical problems and needs, purchasing habits, idiosyncrasies and internal purchasing procedures of the Company's customers; and the Company will suffer irreparable harm if I were to enter direct or indirect competition with the company immediately following termination of my employment.

Accordingly, in consideration of my employment, [_____ dollars ($_____) paid in hand] and other good and valuable consideration, the sufficiency of which is hereby acknowledged, I acknowledge and agree as follows:

1. I shall not, for a period of eighteen months following termination of my employment with the Company for any reason whatsoever, directly or indirectly, own, manage, operate, control or participate in the ownership, management, operation or control of, or be connected as an officer, employee, partner, director, consultant or otherwise with, or render services to, or have, directly or indirectly, any financial interest in (except for ownership of less than 5% of any publicly traded stock), any business entity or other organization which is engaged in or is planning to engage in the business of design, manufacture, or sale of products of the type designed, developed, manufactured or sold by the Company during the term of my employment (hereinafter collectively referred to as "Competitive Activities").

2. Notwithstanding the foregoing, but subject to paragraph 3 of this Agreement, I may accept employment with such an entity, having a diversified business, if I shall have first furnished to the Company clear and convincing written evidence, including assurances from me and from such entity, prior to my acceptance of employment, that I will be employed in a part of the business not involved in Competitive Activities, and that I will not render, directly or indirectly, services related to or in connection with Competitive Activities.

3. Further, recognizing the difficulty in refraining from unconsciously using Confidential Information under some circumstances, I shall not, during such eighteen months period, directly or indirectly engage in any endeavor, or take employment in any position, in which I may reasonably be expected to call upon, refer to, or use any Confidential Information.

4. Recognizing the specialized nature of the business and CONFIDENTIAL INFORMATION of the Company and the geographic scope of its business activities, I also acknowledge that the geographic limitations (universal), duration and scope of restricted activities are entirely reasonable.

5. Miscellaneous
*****

Name:_____   Date:_____
Address:_____

EXHIBIT 8.4

Sample Noncompete Agreement.

> **TECH MICRO-CASE 8.1**
>
> **NDAs Are a Turnoff to Investors**
>
> Asking investors to sign a nondisclosure agreement sets a tone of distrust and anxiety, according to former venture capitalist George Lipper. He suggests that rather than trying to put a lock on investors, technology entrepreneurs instead should put a lock on their technology through patents or trademarks. Alternatively, technology entrepreneurs should excise sensitive information from the documents they share with potential investors. According to Lipper, "The perfect business plan is one that gets mouths watering without giving up the secret sauce."[37]

Some confidentiality agreements impose an obligation of confidentiality for a specified time period. The confidential status of anything disclosed under the agreement is lost at the end of that time period. The time period is typically triggered by disclosure of the confidential information. However, in some NDAs, the obligation runs from the effective date of the agreement irrespective of when the information was disclosed during the course of the agreement. In either case, the time period is arbitrary as it relates to the actual confidentiality of the information. An alternative approach is to require confidentiality to be observed with respect to information until such time as the information falls within one of the exclusions to confidentiality status. This avoids placing an artificial lifetime on confidential status.

8.6.4 Consulting and Development Agreements

In a consulting agreement, a third party is engaged to apply a particular expertise on behalf of the hiring company. The consulting agreement may relate to a particular project or may be more open ended, with the consultant available for any issue involving the consultant's expertise.

A development agreement is sometimes considered a species of consulting agreement that specifically contemplates that the consultant will create IP. In many cases, the developer is not intimately involved with the operation of the hiring company, and may work on a largely independent basis. A development agreement calling for all intellectual property resulting from the agreement to belong to the hiring company is sometimes referred to as a **work for hire** agreement.

If well written, the consulting or development agreement will include provisions that:

- Impose nondisclosure and nonuse obligations

- Define the scope of engagement and all deliverables

- Define any interactive information exchange and preconditions to providing deliverables and the delivery schedule

- Define acceptance criteria and procedures

- Define the respective rights of the parties as to technology developed during the course of the relationship and to preexisting technology incorporated into the deliverables

- Allocate the risk of infringement of third-party intellectual property rights

In the absence of provisions in a written agreement to the contrary, the rights to any inventions made by, and the copyright to any writings or software authored by, the developer during the course of the engagement will belong to the developer.

Depending upon the nature and subject matter of the engagement, the agreement may also include provisions precluding the consultant from providing similar services to competitors during the term of the engagement, and for a reasonable period after the termination of the agreement.

8.6.5 Maintenance and Support Agreements

Maintenance and support agreements cover a wide variety of circumstances, but are most commonly encountered in connection with software products. The agreements often involve the provision of updates and error correction of licensed software. Some of the issues that are commonly addressed in maintenance and support agreements are:

- Ownership of, and the rights of the customer to, new developments

- Availability of updates or error corrections

- Representations by the service persons pertaining to the software

- Criticality of "bugs," and required response times

- The amount and nature of support to be provided

8.6.6 Manufacturing Agreements

Under a manufacturing agreement, a company hires a third party to manufacture an item according to certain specifications. The arrangement often involves communication of confidential information (e.g., the specification). Appropriate confidentiality provisions and, in some cases, restricted licenses under other types of intellectual property (e.g., patents and know-how) are then included in the agreement. Other areas often addressed in manufacturing agreements include:

- Ordering and delivery procedures

- Quality control and approval procedures

- Ownership of molds and tooling

- Ownership of intellectual property arising out of the agreement

- Representations and warranties as to quality

- Whether or not subcontracting is permitted
- Noncompetition provisions that restrict the manufacturer from manufacturing similar items for competitors or itself becoming a competitor

8.6.7 Assignment Agreements

An **assignment** agreement is a document that effects a transfer of title to specified intellectual property rights. In most instances, to be effective with respect to a patent, copyright, or trademark, an assignment must be recorded with the Patent and Trademark or Copyright Office. In general, there should be a recorded assignment with respect to each patent owned by a company, and with respect to each copyright or trademark acquired from outside of the company.

8.7 LICENSE AGREEMENTS

In a license agreement, the licensor grants to the licensee certain rights with respect to intellectual property of the licensor. A license is to be distinguished from a sale or an assignment in which substantially all commercial rights and title to the intellectual property are transferred to the **assignee**. In the case of a license, the licensor retains title to the intellectual property.

Exclusive rights to use intellectual property are the primary sources of competitive advantage in the marketplace (see Chapter 7). Yet, there are a number of reasons why a company might permit others to use its intellectual property. One obvious reason is obtaining a royalty income in consideration for use of the intellectual property. This is particularly so where the licensor does not itself use the technology or the licensor is unable or unwilling to itself meet the demand for products using the intellectual property. For example, the licensor may lack the capital to expand its production facilities or may have other priorities. It may not be practical or economical for a company to export product into a particular geographical area or to establish its own manufacturing facilities in that area. Or the licensor may not have the resources or contacts to develop a necessary distribution system in the area.

Other reasons for licensing intellectual property are not necessarily obvious. For example, having an additional source for a product may increase its market acceptance. In other instances, the licensor itself might use the licensed product as a part in another product, or sell the licensed product as part of an overall line of products, but find it uneconomical or impractical to manufacture the licensed product itself. The fact that the product may also be available to the licensor's competitors is offset by the compensation paid to the licensor by the licensee. In other instances, licensing a local entity in a given geographical area to manufacture one product may create a market in that geographical area, or at least increase market acceptance, for other licensor products.

The consideration for a license is often a license fee, typically referred to as "royalties." The fee can be a single up-front payment, but is often in the form of periodic royalty payments. The periodic royalty payments can be determined upon:

- A percentage of a designated royalty base (e.g., gross sales of products employing licensed intellectual property)
- A fixed per-unit fee
- A percentage of savings through use of the licensed intellectual property

The license may also be made conditional upon meeting specified milestones and/or performance requirements. Milestones could include such things as resources devoted, promotional expenditures, staffing, facilities, and/or product introduction. Performance requirements could relate to such things as sales level, market penetration, and the like. Generally, the conditions are reflected as termination provisions of the agreement.

8.7.1 Patent Licenses

Under a patent license, the licensor agrees not to enforce the patent against the licensee with respect to certain acts that would otherwise be precluded by the licensor's patent, typically making, using, or selling processes or products covered by the patent. It may be no more than a covenant not to enforce the patent, or may be coupled with a transfer of know-how or technical assistance. Essentially all countries that have patent systems in place recognize patent licenses.

8.7.2 Know-How Licenses

Under a know-how license, the licensor permits the licensee to have access to, and use, the licensed know-how. The know-how may be proprietary or nonproprietary. If proprietary know-how is involved, the agreement would include nondisclosure provisions and, typically, provisions restricting the manner in which the know-how can be used. Such provisions are necessary to protect the licensor's proprietary interest in the know-how. Other subjects that tend to be addressed in know-how licenses include:

- The particular mechanism for transferring the know-how
- The nature of deliverables
- A time schedule for the transfer
- The amount and type of technical assistance

8.7.3 Trademark Licenses

Under a trademark license, the trademark owner permits the licensee to employ the licensed trademark in connection with the licensee's goods or services. It permits the licensee to take advantage of the goodwill associated with the trademark. At the same time, presupposing that the agreement includes the appropriate provisions, the licensor expands its goodwill through, and generates revenue from, the licensee's efforts. Trademark licensing is recognized under the laws of most, but not all, countries. In most instances, however, in order for the trademark owner to retain rights in the mark, the trademark owner must maintain careful quality control over the products of the licensee.

8.7.4 Franchise Agreements

A franchise agreement is a species of license agreement that establishes a close, ongoing relationship between the transferor and transferee. Very often, the franchise agreement relates to a turn key business, including trademark know-how and, sometimes, patent licenses and technical assistance. Franchise agreements are closely regulated throughout the United States and in many countries.

8.7.5 Technical Services Agreements

Technical assistance or technical services agreements are, in general, hybrid consulting-know-how agreements relating to provisions of know-how, instruction, and training to the receiving party. Proprietary or nonproprietary know-how, or both, may be involved. For example, an entity with special expertise may be engaged to assist in implementing or installing an equipment plant or process and to instruct and train the contracting party in the operation and management of the plant or process.

8.7.6 Distribution Agreements

In a distribution agreement, the owner of a product (manufacturer, developer) engages a distributor to market the product in substantially unmodified form. In some instances, the distributor may install, service, or customize the product for an end user. Issues often arise with respect to:

- The extent to which, if at all, the distributor is licensed to use the owner's intellectual property internally

- The extent to which, if at all, the distributor is permitted to use the manufacturer/developer's trademarks

- The allocation of risk of liability for infringement of third-party intellectual property rights

In addition, where the product is predominantly software, the distributor agreement typically specifies the manner in which the software is to be marketed by the distributor to the end user. Alternatives include selling a copy of the software to the end user acting as an agent for the developer company, acting as a broker for an end-user license agreement between the developer company and the end user, or entering into a sublicense agreement (with specified terms) with the end user. The agreement also typically specifies the origin of copies of software that will be delivered to the end user (e.g., provided by the proprietor on an as-ordered basis, provided from an inventory maintained by the distributor, or made by the distributor on an as-needed basis from a "master copy" provided by the proprietor).

8.7.7 VAR and OEM Agreements

Under value-added reseller (VAR) and original equipment manufacturer (OEM) agreements, the owner of technology or a product licenses the VAR or OEM to market the technology/product as part of an overall system or product. OEM and VAR agreements are often distinguished by the extent to which the supplied technology/product retains its separate identity. Where the supplied product is a separate module identified to, or identifiable by, the end user, the agreement is typically characterized as a VAR agreement. Conversely, under an OEM agreement, the supplied technology/product tends to lose its separate identity and becomes an integral part of a product marketed under the OEM's name.

Many of the same intellectual property and compliance issues arise with OEM and VAR agreements as with distribution agreements. The agreement, if competently drafted, will also include provisions allocating risks of liability with respect to defects in the products and provisions relating to the infringement of third-party intellectual property rights. Such provisions are particularly significant in VAR and OEM agreements, and even more so where the VAR or OEM is permitted to modify the supplied product. Indemnity provisions often make the owner liable to infringement claims as to unmodified product and run to his or her benefit to the extent that infringement claims relate to modifications of the product by the VAR or OEM.

8.7.8 Purchase Agreements

In general, a purchase agreement is an agreement under which a vendor (supplier) transfers the title to a product to a purchaser (customer). The terms of the purchase agreement are often established by preprinted forms: purchase order and order-acknowledgment or confirmation forms. Issues tend to arise as to the precise terms of the agreement when the terms in the purchase order conflict with those of the confirmation or acknowledgment form. This is the "battle of the forms" discussed earlier.

SUMMARY

This chapter has explored in detail both the contracting process and the many different types of contracts that technology ventures may enter into. We began by highlighting the sources of contract law. It was noted that common law—those multifarious decisions that have been handed down by judges at all levels for decades—serve as the foundation for contract dispute resolution. Because common law can vary from state to state, the Uniform Commercial Code (UCC) has been created to bring order to interstate commerce. The UCC provides judges and others a basis upon which to interpret the intent or "spirit" of a contract. The UCC provides what are known as "gap fillers" to interpret the intent of contracts where that is not explicitly spelled out in the contract itself.

Next, we explored the elements of contract formation, noting that all contracts require the basic ingredients of offer, acceptance, and consideration. The offer was described as the attempt by one of the parties of the contract to initiate a meeting of minds. There are certain conditions required for an offer to be valid, and these were described. Often, an offer results in a counteroffer. When the parties have a meeting of minds, the party to whom the offer was made accepts the offer. Despite the offer and acceptance, the contract is not considered complete until some consideration is offered. Consideration is simply an exchange of something of value—it could be anything—that signifies a transaction has occurred.

Unfortunately, there is no such thing as a perfect contract. As a result, disputes between parties to a contract can and do often arise. This chapter discussed several steps that parties can take to cancel contract enforcement. If conditions change radically after a contract was signed, or if a mistake was made in its drafting, parties to the contract do have some ability to void the agreement. Additionally, there are occasions when one or more parties are unable to perform as specified in the contract. There are a number of ways to seek relief both in the case of not being able to perform and in the case of being harmed owing to another party's nonperformance.

The anatomy of a standard contract was also explored. This should give aspiring technology entrepreneurs a good primer on standard items that parties expect to see in contracts they execute. Novice entrepreneurs sometimes learn the hard way that high-quality clients simply won't work with companies that don't have the paperwork—forms and contracts—with which they are familiar and comfortable. As such, it is important to understand that most contracts have similar anatomical components and vary only in the deliverables, dates, and other details.

Some of the basic contract types that technology entrepreneurs will use to build their ventures were explored. Operating agreements include employment agreements, noncompete agreements, confidentiality agreements, consulting and development agreements, maintenance and support agreements, manufacturing agreements, and assignment agreements. A number of licensing agreements were also highlighted, including patent licenses, know-how licenses, and trademark licenses. Finally, we explored some other common agreements used in technology ventures, including franchise agreements, technical services agreements, distribution agreements, VAR and OEM agreements, and purchase agreements.

STUDY QUESTIONS

1. Discuss the difference sources of contract law.

2. Discuss the differences between the Common Law, UCC, and CISG.

3. Discuss the requisite elements for an enforceable contract.

4. Jean overhears Willaim offering to sell Rashad his computer for $100, and yells out "I accept." Is there an enforceable contract between Jean and William?

5. Discuss the significance of the relative timing of offer, acceptance, and revocation, under the UCC and CISG.

6. DevCo designs a product for BuyCo which conforms to a contract with BuyCo in every aspect except that it is 0.5 ounce over the weight indicated in the specification. Discuss BuyCo's remedies.

7. What types of contracts are required to be in writing? What constitutes a writing sufficient to satisfy the statute of frauds?

8. Discuss the significance of substantial performance under the Common Law, UCC, and CISG.

9. Discuss the results of conflicts in terms of a purchase order form and responding order-acknowledgment form under the Common Law, UCC, and CISG.

10. Discuss the reasons for including representations and warranties in an agreement.

EXERCISES

1. *IN-CLASS EXERCISE*: For this exercise, students will form into groups of eight. Each group of eight will divide itself into separate groups of four. Each of the groups of four constitutes a party to a contract. One group will be the software venture, the other the client. Teams should develop a contract on the basis of the following scenario:

   "The 'client' is a large industrial company that has annual sales in excess of $1 billion. The client has been seeking a software solution to keeping track of its inventory of hand tools that are used everyday by employees around the world. An audit recently revealed that the company loses track of $5 million in tools each year. The software venture has developed an application that it has deployed in several small companies to help them track their hand tool inventory. It wants to sell its software to the client firm. This would be its first contract with a large company."

   Each of the groups should construct a contract, and then compare the contracts they created. How do they differ? Which contract seems to favor the consulting firm the most? Which one favors the large industrial client? What challenges were encountered in coming to a "meeting of minds"?

2. *OUT-OF-CLASS EXERCISE*: Students should search the Internet to come up with examples of the various contract types that were discussed in this chapter. Students should be individually assigned to find contracts examples for:

   - Employment agreements
   - Noncompete agreements
   - Confidentiality agreements
   - Consulting and development agreements

- Maintenance and support agreements
- Manufacturing agreements
- Assignment agreements
- License agreements
- Patent licenses
- Know-how licenses
- Trademark licenses
- Franchise agreements
- Technical services agreements
- Distribution agreements
- VAR and OEM agreements
- Purchase agreements

Simple Web searches should result in examples of these documents. In particular, web sites such as docstoc.com usually have a rich depository of documents of this type.

Students should bring their contract examples to class and be prepared to discuss them.

KEY TERMS

Acceptance: Acceptance of an offer to enter into a contract is a final and unqualified expression of assent to the terms of the offer.

Accord and satisfaction: An agreement (accord) between parties that substitutes a new duty for the original duty, whereby the performance of the new duty discharges the original duty (satisfaction).

Assignment: A voluntary transfer of rights by an assignor to a third party, in which the assignor agrees to extinguish his or her rights to perform.

Assignee: The party to whom contract rights are assigned.

Battle of the forms: The situation under the UCC where a purchase order form elicits a purchase acknowledgment form with different or contradictory terms.

Boilerplate provisions: Standardized language provisions addressing issues of law ancillary to performance under the contract.

Breach: A wrongful failure to perform in accordance with the terms of a contract, which results in the right to damages by the aggrieved party.

Common law: The body of legal decisions that are made by judges in actual court cases, and which builds a library of legal precedents.

Compensatory damages: Contract damages, which are intended to put the aggrieved party in the position it would have been in had the breach not occurred.

Complete performance: Precisely and completely fulfilling all obligations of a contract in accordance with its terms.

Condition: An occurrence or nonoccurrence of a stipulation in an agreement, which affects the contractual duty of performance.

Consideration: The bargained-for exchange of an agreement.

Contract of adhesion: A contract with terms unilaterally dictated by one party, presented on a take-it-or-leave-it basis, typically a preprinted form.

Counteroffer: A response to an offer proposing that the parties move forward on different terms from those proposed in the offer.

Discharge: The termination of a contractual duty.

Express warranty: An explicitly stated warranty in a contract.

Fraud in the inducement: An agreement induced by fraud.

Force majeure clause: An extreme and unreasonable difficulty or expense caused by an unforeseeable supervening event.

Gap fillers: Provisions that are derived from the UCC and that help complete a contract.

Hold-harmless: A mechanism for allocating risk between the parties.

Inadequate performance: Connotes a significant variance on the part of one or more parties to a contract from the terms of the contract.

Incidental damages: Damages arising directly out of a breach of contract as compensation for expenses incurred as a result of the breach.

Indemnity: An obligation to pay for the liability of another.

Mailbox rule: The rule under common law and the UCC that an acceptance is effective if and when it is placed in the mail.

Material breach: A breach that revokes the value of a contract and relieves the aggrieved party from any further duty under the contract.

Mirror image rule: The rule under common law that an acceptance must correspond exactly to the offer to create a contract.

Mitigation of damages: The rule that an aggrieved party may not recover damages that could have been avoided with reasonable effort and without undue risk or burden.

Mutual assent: An agreement between the parties to a contract to terminate their respective duties.

Nondisclosure agreement: Such an agreement should be in place before any third-party recipient is given access to proprietary know-how outside of the context of a broader agreement already including confidentiality provisions.

Novation: A substitution of one obligation by another by mutual agreement of both parties, which releases the existing promisor from liability as a delegator.

Offer: The first part of a valid contract, the other two parts being acceptance and consideration.

Performance: Fulfilling the obligations of a contract.

Promissory estoppel: A consideration substitute where a person reasonably relies upon a promise to his or her detriment.

Quasi-contract: A fictional contract constructed by the courts to prevent unjust enrichment.

Restitution: Compensation that puts the aggrieved party in the position it was in before the contract was made.

Rescission: The cancellation of a contract by mutual agreement of the parties.

Royalty base: The underlying basis upon which royalities are computed, e.g., gross sales or net sales.

Statute of frauds: Rule that in the absence of a writing, certain contracts are voidable.

Substantial performance: Near fulfillment of the obligations agreed to in the contract but with slight variances from the exact terms, unimportant omissions, and/or minor defects that do not defeat the purpose of the contract. The injured party is not discharged from duties under the contract, but is entitled to damages.

Substituted contract: An amended or new contract agreed to by both parties in satisfaction of their duties under the original contract.

Unconscionable: A contract provision so egregiously one-sided that a court refuses to enforce it.

Uniform Commercial Code (UCC): A code that governs commercial transactions, providing gap fillers that help interpret the intent of the parties in case of a dispute.

Unilateral contract: A contract that involves an offer that requires actual performance for acceptance.

Unilateral mistake: A mistake on the part of only one party to a contract.

Voidable contract: A contract entered into by a party that lacks capacity and therefore has the option to disaffirm the contract.

Work for hire agreement: A development agreement that assigns all IP created during performance on the agreement to the hiring party.

WEB RESOURCES

The web sites below are intended as destinations for your further exploration of the concepts and topics discussed in this chapter:

1. http://smallbusiness.findlaw.com/business-forms-contracts/business-forms-contracts-overview/: This web site is a good contract law reference source for anyone in business or thinking about starting a business. Topics covered run the gamut of concerns that may be faced by an entrepreneur.

2. http://www.cisg.law.pace.edu: The CISG is the United Nations Convention on Contracts for the International Sale of Goods, the uniform international sales law of countries that account for three-quarters of all world trade.

3. http://www.expertlaw.com/library/business/contract_law.html: This is an excellent introduction to contract law, with many topics relevant to the launch and growth of technology ventures.

4. http://www.law.cornell.edu/: This site contains information about the Uniform Commercial Code, which governs commercial transactions in the United States.

CASE STUDY

INTEL and NVIDIA Battle over Licensure Rights

On November 19, 2004, NVIDIA announced the signing of both a broad, multiyear patent cross-license and a multiyear, multigeneration chipset agreement (CLA) with Intel. Under these contracts, Intel gained access to NVIDIA's entire technology portfolio in order to further their development of integrated graphics. In return, NVIDIA was granted licensure to create and market chipsets that are compatible with Intel's processors. However, these agreements between the two leading corporations in their respective areas of expertise would not bind Intel and NVIDIA in a peaceful partnership.

Early in 2007, Intel advised NVIDIA of its plans to release a new processor, code-named Nehalem. In autumn of that same year, the Senior Vice-President of Intel, Pat Gelsinger, named NVIDIA as a partner on the impending Nehalem project in a public announcement. However, when NVIDIA informed Intel of its intent to develop, manufacture, and market a graphics chipset for Nehalem, Intel began to dispute NVIDIA's right to generate these chipsets for any Intel product with integrated memory controller functions, like Nehalem. In a letter dated April 30, 2008, Intel stated to NVIDIA that it had the unilateral right to terminate the licensure agreement that it had with NVIDIA since it had integrated the memory controller into the processor and claimed that NVIDIA would be unable to design a chipset for their new system. Intel then demanded that NVIDIA make a public announcement that NVIDIA was not licensed to produce graphics chipsets for Nehalem.

Not only did NVIDIA refuse Intel's demands, but NVIDIA spokespeople continued to publicly claim their right to make chipsets for Nehalem, citing the 2004 CLA. At this point, Intel requested a civil mediation be held between the two companies since this was a provision made under the CLA. However, just prior to the mediation, Intel sent NVIDIA an accusatory letter, stating that NVIDIA had breached the CLA by publicly defending its claim of licensure to create the Nehalem chipsets. This letter also contained a demand for a joint press release that would highlight Intel's belief that NVIDIA did not hold a license for Intel's new processor. NVIDIA claims that it notified Intel of their rejection of Intel's official position and would not participate in a joint press release; Intel denies that it received a response from NVIDIA.

During this year-long dispute, Intel was not idle. Since Intel was determined to block NVIDIA from making compatible chipsets for Nehalem, the company began plans to manufacture its own graphics engine, code-named Larrabee, for the Nehalem platform. Intel spokesman, Dan Snyder, confirmed Intel's intention to produce Larrabee in an April 2008 interview. Owing to the disputes, Larrabee is the only marketed graphics card available at the time of this study.

Intel was not alone in furthering its technological innovations. NVIDIA advanced its Scalable Link Interface (SLI) patent, incorporating new ways to interface multiple graphics cards to increase their processing power. Intel became interested in harnessing SLI's newfound capabilities; however, the 2004 agreements between NVIDIA and Intel did not cover these new innovations. With the hostile climate between the companies growing, NVIDIA made a bold decision to withhold licensing rights from Intel.

NVIDIA's decision left both companies in a losing situation: Intel could not use the SLI graphics in its Core i7 processing units and NVIDIA was without a processor to run its new technology. Despite the long-standing feud between the two companies over Nehalem, NVIDIA and Intel finally entered into contract negotiations regarding SLI. On February 12, 2009, a new licensing agreement was formed enabling Intel to access NVIDIA's SLI multigraphics processing units.

Only a few days after the signing of the SLI agreements, Intel withdrew from mediation discussions concerning Nehalem chipsets and filed an injunction for declaratory relief in Delaware under the contention that it believed that NVIDIA breached the licensing contracts of 2004. Among Intel's complaints, it alleged that NVIDIA violated the contract by making false or misleading statements concerning NVIDIA's license standings for manufacturing and marketing graphics chipsets for Intel's Nehalem processor and failed to comply with multiple requests to correct the allegedly false or misleading statements. Intel also contended that the allegedly misleading statements have brought irreparable damage to its reputation with Intel's clients and confusion within its marketplace. In Intel's opinion, if NVIDIA was not made to discontinue laying claim to licensing rights to Nehalem chipsets, the benefits that Intel enjoyed from the 2004 contract would be nullified. Intel therefore asked the Delaware court to declare that NVIDIA had no rights to any interest in Nehalem chipsets and that NVIDIA breached the CLA.

NVIDIA released a public statement on February 18, 2009, concerning Intel's court filing where it claimed that Intel has alleged that the 2004 contract does not have the scope to cover Intel's newest generation of processors, known as Nehalem. However, NVIDIA did report its confidence in the contractual licensure covering multiple generations of processors and assured the public that the chipsets in current production are not affected by the dispute.

On the same day, Intel offered a press release claiming that if the courts were to decide in its favor, Intel would have an option to renegotiate the contract between itself and NVIDIA and had not, as of the time of the press release, ruled out the possibility. Intel also claimed that it did not want the dispute to interfere with the working relationship it maintained with NVIDIA despite the difference of opinion relating to Nehalem.

These statements would not assuage NVIDIA's outrage. On March 26, 2009, NVIDIA filed a counterclaim in the Delaware court. In its answer to the original complaint, NVIDIA denied most of the allegations set forth by Intel. NVIDIA then alleged that it was Intel that breached the contracts and that it further breached an implied covenant of good faith and fair dealing. NVIDIA has also claimed that the relief that Intel seeks would dissolve only one side of the CLA, leaving Intel with access to NVIDIA's patent portfolios without a compensatory act for NVIDIA's benefit. Therefore, NVIDIA saw fit to ask the court to either prohibit Intel from making any further attempts to sabotage NVIDIA's license or to sanction Intel for the breach of the CLA and to nullify both of the 2004 contracts.

One point that neither company denied in their respective legal actions was the fact that this dispute came at a critical time in the computing market. Major companies, such as Apple, Dell, and Toshiba, were making decisions concerning the Nehalem processor and whether or not to integrate it into their line of products. Apple reported that it had decided to switch its graphics cards from Intel's to NVIDIA's in 2008; however, Apple also has conceded that it may have to reconsider this decision based upon all of the controversy.

At the time of this study, the case between Intel and NVIDIA remains unsettled since it has yet to enter judicial proceedings.

Questions for Discussion

1. Since the agreement in question is not available to the public, use what you have learned in this chapter to deduce what conditions would have to be within the 2004 contract for Intel to prove its claim that NVIDIA does not have a right to produce the chipset for the Nehalem processor and that NVIDIA has breached the contract. Then consider what conditions would have to be in the 2004 contract for NVIDIA to win its countersuit and prove that Intel breached the contract.
2. On the basis of what you know of this case and what you have learned from this chapter, can you make an argument that either party is entitled to compensatory damages? Incidental damages? Mitigation of damages? Nominal damages? Why or why not?

3. A direct quote from the Redacted Public Version of Intel's Complaint for Injunctive and Declaratory Relief under paragraph 5 is, "Resolution of Intel's Complaint for Injunctive and Declaratory Relief will enable both companies to better understand and adhere to their respective obligations." Could this lead you to believe that there may have been implied-in-fact conditions within the 2004 contract that were misunderstood between the two companies? Why or why not?

Source: Abazovic, F. (2008, June). Intel won't let NVIDIA make Nehalem chipsets. Retrieved from http://www.fudzilla.com; Brown, R. (2009, February). Intel takes chipset dispute with NVIDIA to court. Retrieved from http://www.news.cnet.com; Crothers, B. (2008, April). NVIDIA CEO goes on Intel rant. Retrieved from http://news.cnet.com: Hruska, J. (2009, March). NVIDIA countersues Intel—right on schedule. Retrieved from http://www.arstechnica.com; Intel Suit (2009, February). Retrieved from www.nvidia.com/object/ip_1238021549708.html; NVIDIA and Intel sign broad cross-license and chipset license agreements (2004, November). Retrieved from www.nvidia.com/object/IO_17070.html?_templateId=320; NVIDIA Countersuit (2009, March). Retrieved from www.nvidia.com/object/ip_1238021621363.html; Oliver, S. (2009, February). NVIDIA responds boldly to Intel court filing. Retrieved from http://hothardware.com; Maisto, M. (2009, March). NVIDIA countersues Intel over Nehalem rights. Retrieved from www.eweek.com; McLean, P. (2009, March). NVIDIA strikes back against Intel. Retrieved from www.appleinsider.com; Shilov, A. (2009, February). Intel and NVIDIA finally let SLI multi-GPU technology into latest Intel mainboards. Retrieved from www.xbitlabs.com; Smalley, T. (2009, February). Intel won't rule out re-negotiation with NVIDIA. Retrieved from www.bit-tech.net.

ENDNOTES

1 Louisiana is an exception. Its law is based on a civil code, rather than the precedent of judicial decisions.

2 UCC Article 2. The UCC also includes articles covering, among other things, Negotiable Instruments (Article 3), Bank Deposit (Article 4), Letters of Credit (Article 5), Bulk Transfers (Article 6), Documents of Title (Article 7), Investment Securities (Article 8), and Secured Transactions (Article 9).

3 UCC §2-103(1) (k).

4 There are a number of exclusions from the CISG, including "goods bought for personal, family, or household use." CISG Art. 2(a).

5 CISG Article 11. But see Forestal Guarani, S.A. v. Daros International, Inc. 2008 U.S. Dist. LEXIS 79734, (D. NJ, 7 October 2008), Zhejiang Shaoxing Yongli Printing & Dyeing Co. v. Microflock Textile Group Corp., 2008 U.S. Dist. LEXIS 40418, 2008 WL 2098062 (S.D. Fla. May 19, 2008), Chateau des Charmes Wines Ltd. v. Sabaté USA, Sabaté S.A 328 F.3d 528 (9th Cir. 2003).

6 See UCC §204 C. For example, price is not an essential term. UCC §305. See also CISG Art. 14(1).

7 See Restatement (Second) of Contracts §63, UCC §206(1).

8 CISG Art 18(2).

9 Restatement (2d) Contracts §59. CISG Art. 19. CISG Art 19(2) indicates that additional or different terms that do not materially alter the terms of the contract constitutes an acceptance and, unless objected to without undue delay, become part of the agreement. However, CISG Art. 19(3)

denominates most relevant terms as material, including: "terms relating, among other things, to the price, payment, quality and quantity of the goods, place and time of delivery, extent of one party's liability to the other or the settlement of disputes."

10 See Janky v. Batistatos, 86 USPQ2d 1585 (N.D. Ind. 2008).

11 UCC §2-207 See *Architectural Metal Systems, Inc.*, 58 F.3d 1227, 1230 (7th Cir. 1995), Janky v. Batistatos, 86 USPQ2d 1585 (N.D. Ind. 2008), Uniroyal, Inc. v. Chambers Gasket and Mfg. Co., 380 N.E.2d 571, 575 (Ind. Ct. App. 1978).

12 UCC §2-207(A).

13 UCC §2-207(B).

14 UCC §2-207(C) 14a See e.g., UCC §§2-303, 2-305, 2-307 to 2-312, 2-314, 2-315, 2-326, 2-327 and 2-207(C).

15 See Restatement (Second) of Contracts §42, CISG Art. 22.

16 See Restatement (Second) of Contracts §63, UCC §206(1).

17 CISG Art. 18(2).

18 See Restatement (Second) of Contracts §§17, 20.

19 CISG Art. 11.

20 UCC §2-201.

21 UCC §2-201.

22 The CISG defines a breach as "fundamental" if it "results in such detriment to the other party as substantially to deprive him of what he is entitled to expect under the contract, unless the party in breach did not foresee and a reasonable person of the same kind in the same circumstances would not have foreseen such a result." CISG Art. 25, Art. 49.

23 Restatement (Second) of Contracts §352.

24 See Restatement (Second) of Contracts §344, UCC 1-106, CISG Art. 74.

25 See Restatement (Second) of Contracts §350, UCC §2-715(2)(a), CISG Art. 77.

26 UCC §2-715(2)(a).

27 See Restatement (Second) of Contracts §373.

28 See Restatement (Second) of Contracts §374.

29 See Restatement (Second) of Contracts §376.

30 See Restatement (Second) of Contracts §359, UCC §2-716.

31 See CISG Arts. 46, 62. See also CISG Art. 28.

32 See Restatement (Second) of Contracts §184.

33 See UCC §§2-312, 2-314, and 2-315; CISG Art. 35(2). Under the UCC, the disclaimer must be conspicuous. UCC §2-316.

34 UCC §2-312, CISG Arts. 42–44.

35 UCC §2-2-314, CISG Art. 35(2)(a).

36 UCC §2-2-315, CISG Art. 35(2)(b).

37 Herbert M. Confidentially Speaking. *Inc.* April 2005;**27**(4): 50.

Negotiating Fundamentals

After reading this chapter, students should be able to:

- Understand the fundamentals of negotiations
- Identify different negotiating approaches
- Analyze the role of emotions in negotiations
- Enumerate the various elements vital to negotiating contracts
- Recognize how to manage negotiated contracts

TECH VENTURE INSIGHT

Skillful Negotiations Can Make or Break a Business

Michael Dell has been named entrepreneur of the year many times and has a net worth today estimated at $17.3 billion. His story begins in 1984 when he took a $1,000 loan from his grandparents to set up a business in his dorm room at the University of Texas, selling custom-made computers directly to the consumer. Dell adopted this unique approach to avoid channel mark-ups and, more importantly, to pass savings on to customers who were able to buy some of the lowest priced computers in the marketplace.

Dell went on to build on this innovative approach by infusing a customer service concept into his business. PC Limited, as it was initially named, was the first computer company to send a technician out to a customer's home to service his or her personal computer. Dell's company provided excellence in service, thus creating a loyal customer base, which allowed the company to go public in 1988. Thereafter, the company went through numerous name changes as it moved into different product offerings other than computers: from PC Limited to Dell Computer Corporation, and then subsequently Dell, Inc.

The achievements of Dell, Inc., centered not only on its customer service and custom-built personal computers, but also on its ability to negotiate and manage contracts. From

285

the start, Michael Dell understood the importance of negotiating fundamentals, such as understanding worth versus cost, talking less and listening more, as well as understanding one's limits. This steady approach enabled the company at a very early stage to obtain the best prices with suppliers and to establish favorable service agreements with customers.

The company's growth over the years is testimony to its mastery of the basics in writing and managing corporate contracts, including how simple contracts have great potential to lead to more complex ones with larger returns. Today, Dell has an entire department and a plethora of salespeople dedicated to contract management. It provides a fleet of computers to Fortune 500 companies and has negotiated with Wal-Mart to be its exclusive in-store service center, posing a direct competition to Best Buy's Geek Squad.

While Dell has been successful in managing contracts and negotiations, it has also engaged in some significant negotiation blunders. Among the most notable was its failure to negotiate an agreement with Apple Computers to sell iPods via Dell's online stores. The inability to arrive at a mutually amenable agreement with Apple resulted in Dell's losing a significant opportunity to increase its market share. Dell's successes and failures at reaching agreements thus serves as an excellent example of how negotiations can be pivotal to making or breaking deals that increase customer bases.

Source: Adapted from Maich S. Dell: from genius to moron and back. Maclean's 2006;119(22):30; Murphy V. The song remains the same. Forbes 2004;174(4):54–5.

INTRODUCTION

When people think of negotiations, they picture a scene in a movie where the art of negotiation has been made glamorous by Hollywood hostage negotiations or grandiose business deals. Others recall something in the paper about the latest negotiation of a significant merger or acquisition. What few realize is that negotiation plays a part in everyone's daily life. From employees attempting to negotiate higher wages from their employers to corporations trying to negotiate better prices on supplies to simply bartering with a sibling to get out of chores or trying to convince a friend to eat Chinese rather than Italian, negotiation is continually part of the decision-making process.

Growing an enterprise requires the ability to negotiate. Even though it is important, negotiation is usually not a skill well developed by the technology entrepreneur. As one woman entrepreneur stated, "A woman entrepreneur should develop her negotiating skills as quickly as possible. It is so important in starting and particularly in expanding a business. Women tend to be weaker in negotiating skills than men. This may reflect the level of issues typically negotiated at home versus in a business situation. These skills can be learned and then just be practiced."

This chapter will examine the process of negotiation by which parties attempt to resolve a problem by agreement. You will learn that although a resolution is not always possible, the process of negotiation identifies the critical issues in the disagreement and is therefore central to business dealings. We begin by examining the fundamentals of negotiations, both simple and complex.

9.1 NEGOTIATION FUNDAMENTALS

Negotiating is generally defined as "a process in which two or more parties attempt to reach acceptable agreement in a situation characterized by some level of real or potential disagreement." In entrepreneurial ventures, negotiations can and regularly do take place between individuals, between an individual and a group or organization, and between organizations.[1] Entrepreneurs negotiate nearly every day because the environment in which they are working is characterized by **scarcity**. Scarcity means that resources are limited, and they must be acquired, applied, and managed efficiently. Entrepreneurs negotiate with potential employees to join the firm. Armed with limited resources for salary the entrepreneur must negotiate with stock, job titles, and promises of a potentially exciting future. Entrepreneurs negotiate with landlords for office space. With no or little operating history to provide confidence to the landlord, the entrepreneur must negotiate for lease terms that meet the needs of both parties. Entrepreneurs negotiate with angels and venture capitalists to invest in their ventures. Since new ventures have limited capital entrepreneurs must negotiate with individuals and firms that have it to motivate them to invest in their ventures.[2] Entrepreneurs also negotiate with many others, including customers, vendors, suppliers, and investors.

Regardless of the context or parties to a negotiation, all negotiations consist of four basic elements:

1. *Interdependence*: There is always some degree of interdependence among parties; each party is in some way affected by, or depends on, the other. This implies that there is usually a power relationship between the parties.

2. *Conflict*: In most negotiation settings, some degree of conflict exists between the parties. Each wants either something the other has or a part of some resource that is common to both. The conflict may be real or perceived.

3. *Opportunistic interaction*: Negotiation involves attempts to influence and persuade. Each party must believe there is opportunity for their efforts to be successful in order for them to want to enter into negotiations. This perception of the ability to have influence is referred to as opportunistic interaction.

4. *Possibility of agreement*: In addition, for parties to desire to influence each other, they must also believe that it is possible to reach an agreement. In the absence of the possibility of agreement, there is no need to negotiate.

Even the most mundane negotiations generally will have these elements, although they may not be explicitly recognized. For example, in negotiating who will pick up

CHAPTER 9 Negotiating Fundamentals

the tab at lunch there is interdependence in the existence of a mutual debt, there is conflict in that someone's wallet will be a little lighter, there is the opportunity to influence (didn't I pick up the check last time?), and there is the possibility of agreement (no one wants to wash the dishes). These situations are usually handled quickly and people move on to other things. In contrast, negotiations over, say, the distribution of assets in a bankruptcy can be characterized by high levels of conflict among multiple parties, the opportunities for influencing others (the bankruptcy judge), and the potential for agreement (the final ruling by the court).

9.1.1 Process, Behavior, Substance

Negotiation typically involves three elements: process, behavior, and substance. The process concerns how the parties negotiate, the tactics used, and the location of the negotiations. The behavior refers to the relationships among the parties, the stages, and how emotion plays out in the discussions. The substance entails what the parties are actually negotiating from the agenda to the issues, options, and eventually determining the final agreement.

The process for negotiations is relatively easy to understand. It is to determine what a party wants. The behaviors and the substance have several key complex components including preparation, mindset, position versus interest, alternatives to a negotiated agreement, and limitations. A key element in each component is listening. The importance of listening and preparation is indicated in the diagram of the overall negotiation process indicated in Exhibit 9.1.

9.1.2 Preparation

The best negotiators know that the endeavor is 90% preparation and only 10% actual bargaining. It is vital that you know the other side of the issue as well as your own. Do your homework and be able to see yourself in the other person's shoes. If you were on the other side of the table, what would you ask for? What would your expectations or concerns be? By being realistic about the demands of the other party, one can be significantly more prepared upon entering the negotiation. Preparation could also include being aware of any cultural norms that need to be addressed if the business deal is occurring outside of your home country. The goal is to determine the voluntary exchange zone indicated in Exhibit 9.2.

9.1.3 Mindset

Being prepared also allows you to develop the appropriate mindset. It is important to be confident and courteous and to stay calm at all times. One should not be afraid to ask questions and be curious. It is important that you are aware of what you want but also be very clear on what the other party is offering. To ensure you have a proper mindset at all times in the negotiation process, do not hesitate to take breaks.

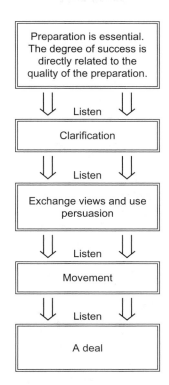

EXHIBIT 9.1

Preparation for Negotiations.

Source: Old Handbook. National Association of Sessional GPs: http://www.nasgp.org.uk/z_old_handbook/7.htm (Retrieved August 3, 2008).

Understanding Your Limitations

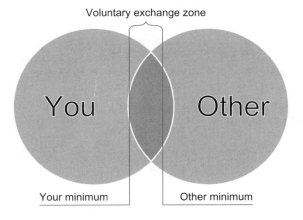

EXHIBIT 9.2

Understanding Your Limitations.

Source: Success Manifesto: http://www.successmanifesto.com/blogs/young-entrepreneur/archives/negotiation.gif (Retrieved November 3, 2008).

9.1.4 Emotions

After adopting an approach or style, it is equally important to be in control of your emotions throughout the negotiation. Though emotions can provide vital feedback on preferences and interests in a negotiation, particularly in trying to understand the opposing individual or team, it can also lead to disastrous results.[3] A sudden burst of anger resulting from anything such as your counterpart showing up an hour late, making an offensive joke, or issuing a careless threat, can threaten the relationship and also affect the ending results of the negotiation process. It is understandable that you feel hurt or angry or upset. However, it is important not to allow these strong emotions to influence how the talks unfold or allow them to distort the balance in the negotiation process. A strategy used by some is to make the counterpart angry and flustered and subsequently take advantage of the situation.

Emotions are something that every negotiator must be aware of. If emotions are running high in a situation, be sure to manage them. Feel free to take some time or ask for a recess or break in the negotiations. Be sure to acknowledge the emotions and feelings and then attempt to address where they are coming from. In doing so, you can re-enter the negotiation in a much calmer and controlled manner to reach a mutually agreeable solution.

If there is a particular element of negotiation that is causing outbursts by either counterpart, be prepared to take the time to assess different or multiple options and come back to the table. Having multiple options available to discuss will result in more of a "brainstorming" session and logical dispassionate thinking rather than heated negotiations.

9.1.5 Position versus Interest

What a person wants is called his or her "position." For example, an employee could be demanding a 20% raise. Though this may seem like an unreasonable request, the important thing is to understand why the person is asking for the raise. Understanding "why" means to understand the individual's interest. With this understanding, a fair and reasonable solution for both parties can be determined.

To determine interest, it is important to ask probing questions. In the employee raise example, things like: Why does he or she want the raise? Why did he or she pick 20%? It could be because the employee knows other employees are being paid more or may need to save more because his or her child is going to college or may feel underpaid for the skills being provided. By obtaining a deeper understanding, the employer has a greater understanding and the ability to offer different solutions, rather than simply a raise (e.g., a college savings plan program where the company matches any funds contributed by the employee). In this way, the needs are met without giving a raise.

The employer should also know the degree of interest in the situation before entering the negotiation. This can come in answers to the following two questions: Is the employee worth the raise or an alternative? Is it in the best interest of the company to let the employee leave? With both party's motivations in mind, one can structure the negotiations around these and achieve better results by ensuring that both sides leave the negotiating table with what they want. Tech Micro-case 9.1 highlights a negotiation between two large utilities that created a win-win outcome.

TECH MICRO-CASE 9.1

Negotiating Creates Win-Win and Helps the Salmon

Southern California Edison Co. and Bonneville Power Administration created a joint partnership through creative problem solving whereby each party met its objectives. In 1991, the two parties conceived of a way to help the Columbia River salmon in the Pacific Northwest, and they managed to achieve that without spending any money.

During the summer months, Bonneville Power increased the flow of water into the Columbia River. This automatically increased the amount of hydroelectric power generated by California Edison. The increased flow of water allowed the young salmon to swim through the channels more easily. It also increased their survival rate because a weaker current made them more prey to becoming lost or more vulnerable to being devoured by predators. Later, in the fall and winter months, California Edison returned the power back to Bonneville Power Administration that it had borrowed during the preceding summer months. As a result, Bonneville had very little need to run its coal-fired and oil plants during the summer months.

This was truly a win-win agreement. The exchange of power, roughly equivalent to about 100,000 households, improved the migration of the salmon, thereby increasing their survival rates and expanding the fish population.[4]

9.1.6 Establishing Your BATNA

A negotiation skill that is commonly recommended by scholars and practitioners is in-depth preparation for the negotiation session.[5] Negotiating is often viewed merely as a process of give and take, and the importance of prenegotiation preparations is ignored.[6] One effective preparation objective is to determine your **best alternative to a negotiated agreement (BATNA)** before entering the negotiation process. If you know your BATNA, which is also known as your **walk-away alternative**, you will know whether the negotiation is worth continuing and you will have established a bargaining floor.[7] Research has shown that conflict is greater in negotiations when one party has a BATNA that is significant to them.[8]

TECH TIPS 9.1

Tips for Using Your BATNA in Negotiations[9]

1. Your power lies in your BATNAs. Make sure that you have real, viable options that don't require an agreement:

- You'll be empowered to support your interests.
- Your confident attitude will compel others to listen to and meet your interests. They'll realize that they have to if they intend to obtain agreement.

2. Do not disclose your walk-away alternatives. When you remind others of the options you have should they not acceptably satisfy your needs, your commitment to negotiation falls into question, and the environment becomes hostile. This draws the attention away from underlying needs, and the climate becomes less conducive to the development of creative options.

3. Figure out the BATNA of the other parties. Knowing what options they have if no agreement is reached will help you construct options that are favorable relative to their specific negotiation. In other words, you'll be able to construct an agreement that improves on their BATNA without giving away too much.

A BATNA is a benchmark for evaluating all offers that need to be determined before entering into a negotiation. It is something that a single party could do without having to negotiate with another party. It is not the best option but one that is realistic and feasible for one party, rather than agreeing to unacceptable negotiating terms. For example, instead of outsourcing the hiring process to an HR consultant who intends to charge high rates, various people within the company can take on some HR responsibilities to achieve the same results in the shortterm.

> *"Having a BATNA is a tremendous source of power in a negotiation," says McLean Parks, a professor of organizational behavior at the John M. Olin School of Business at Washington University in St. Louis, Missouri. You may never need it, but just knowing it's in your back pocket gives you peace of mind. Without one, you can become anxious, appear desperate, and settle for a less than an ideal solution. This is especially true if the other side has more power.[10]*

However, using the BATNA too early or to force a deal is not productive. Remember it is just a back-up plan that can be used as a last resort. Even if one has established a BATNA, it is also important to have limitations and know at which point simply to walk away. When a situation becomes unbalanced or the other party is not being cooperative, it is okay to walk away; particularly if it is taking up a tremendous amount of valuable time and energy. Remember that in every negotiation you should never want anything so much that you cannot walk away.[11]

In determining your limitations and walk-away position, map a range for yourself. What is the best alternative and what is the aspiration for the negotiation? If it is a negotiation on price, what is the lowest you can give a buyer before you have to walk away? Determining the range before entering into the negotiation allows you to be more prepared to walk away. Also, before entering the negotiations, be aware and think about the opposition's range. In doing so, you can be better prepared to see where the two ranges may overlap and thus reach an agreement.

9.2 NEGOTIATION APPROACHES

Besides gaining an understanding of the negotiating fundamentals, it is important to understand that there are also several different personal approaches to the negotiation. A diagram of different negotiation approaches is indicated in Exhibit 9.3. It is important to determine and adopt a style that adequately fits your personality, yet effectively persuades a person or party of your needs and wants. Several negotiation tactics exist, from simply laying out demands to intimidation tactics or cherry picking, which means only presenting one side of the data or argument that legitimizes only one side of the equation.

It is not necessary to choose only one of these negotiating styles. Some people have a predisposition to one style over another, which in certain situations can change. It is important, however, to be aware of your preferred style as well as the other four styles so that another style can be adapted if needed.

- *Accommodating*: Individuals who want to solve problems and get "what they want" but also seek to preserve personal relationships use an accommodating

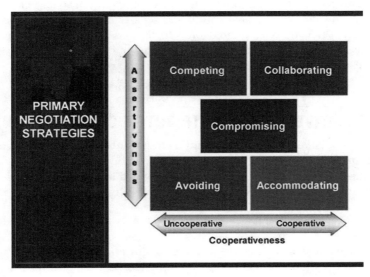

EXHIBIT 9.3

Different Negotiation Approaches.

Source: Global Negotiation Forum: http://www.globalnegotiationforum.com/Portals/0/negotiation-strategies. jpg (Retrieved November 5, 2008).

style. These types of negotiations are sensitive to nonverbal cues like body language and eye contact and even tone of voice. These negotiators are extremely sensitive, and the party using this style can be taken advantage of if the other party places little or no emphasis on the relationship.

- *Avoiding*: People who do not want to negotiate but are forced to, either by a boss or because the situation is such that negotiations are vital to the business, use the avoiding style. They typically avoid eye contact and dodge or sidestep important and difficult issues. They may be viewed as diplomatic and tactful and are typically the most difficult style to negotiate against, as they are predisposed to not making a decision right away and refuse to commit to terms immediately.

- *Collaborating*: Those individuals who enjoy negotiations and love solving difficult questions use the collaborating style. These individuals are "out of the box" thinkers, ask good questions, and really seek to understand the concerns of both parties before making decisions in creative ways. Their downfall is that sometimes they make problems more difficult or complex than necessary.

- *Competing*: The competing style of negotiation assumes that there is a winner and loser in every negotiation. These individuals love to negotiate because they love the thrill of trying to win and are extremely strategic achieving this. Their downfall is that they can dominate the bargaining process and can neglect or ignore the importance of relationships.

- *Compromising*: Individuals who use the compromising style seek equality and fairness throughout the negotiation and seek to close the deal in an equitable manner. When there is limited time to discuss a deal, these individuals are vital as they are able to identify key elements of contention quickly. However, they can also make concessions too quickly and rush the process.

9.3 INTEGRATIVE VERSUS DISTRIBUTIVE BARGAINING

Negotiation often entails one party attempting to get another to do something the first desires. At times, this second party will not act unless they obtain something of interest in return. This motive for a return must be understood in the negotiation process and can best be understood in terms of integrative bargaining and distributive bargaining.

9.3.1 Integrative Bargaining

Integrative bargaining involves **cooperative negotiation** between the negotiating parties. In this situation, the technology entrepreneur is willing to let the other side achieve its desired outcome while maintaining a commitment to his or her own goals.[12] In one sense, integrative bargaining is joint problemsolving based on the concept of rational decisionmaking. A model of this process depicts the negotiation flow moving from establishing objectives to establishing criteria to analyzing the cause-and-effect relationships involved to developing and evaluating alternatives and then, finally, to selecting an alternative and an action plan. To be complete, this process requires that the outcomes be measured once implementation begins. The effectiveness of the decisions made through this process depends on the adequacy of the information and the commitment of the parties involved implementing the decisions reached.

Several steps are involved in implementing this approach. First, the problem must be clearly identified with all parties in the negotiation exchanging views without blaming the other parties or demanding a specific outcome. Each party involved must be genuinely interested in solving the problems and understanding the underlying concerns of the others involved. Meetings need to be scheduled frequently with advance notice of each meeting given to allow all participants to prepare. Each meeting should have an agenda with a majority of the items having a high probability of resolution. Each agenda item should be so stated that it opens discussion and integrative bargaining that lead to mutually agreeable solutions.

The second step in effective integrative bargaining is searching for alternative solutions. Here, the technology entrepreneur and other parties involved should try to uncover areas of agreement so they can realize areas of divergence. This requires that concrete terms be used and appropriate ground rules that encourage trust and respect be established. Through frequent preliminary discussions, tentative solutions and ideas can be brought up and refined before being formally considered. This informal, exploratory discussion of a possible solution reduces the likelihood of immediate rejection. The techniques and frameworks used should differ depending on the specific problem at hand.

Integrative bargaining occurs more quickly when the parties involved sense a greater likelihood of success. This requires that easier items be dealt with first, with any difficult items laid aside temporarily if an impasse is reached. No rigid formula should be used since creative solutions do not come from following a systematic procedure. Finally, as indicated in the model of rational decisionmaking, the best alternative solution should be selected. This requires the establishment of an open, honest relationship that allows each party to indicate when something is wrong and to report preferences and range of latitude in each problem area. This is the area most susceptible to distortion and further problems.

9.3.2 Distributive Bargaining

Even more important for the successful expansion of a business is the technology entrepreneur's effectiveness in distributive bargaining. In contrast to integrative bargaining, **distributive bargaining** does not allow the other party to achieve his or her goals. There is a fixed pie to be divided, which means that the larger the opponent's share, the smaller the entrepreneur's. Since there is no trust between the parties, a solution can only be reached through a series of modified positions of compromise and concession. In this competitive adversarial bargaining arena, each party tries to discover the other's goals, values, and perceptions. The amount of conflict usually revolves around the differing economic objectives and attitudes of the

parties involved. When the technology entrepreneur and the other party have conflicting economic objectives, distributive bargaining can be successful only if each
party compromises his or her objectives to some extent. Directly competing claims
on a fixed, limited economic resource (such as the amount of the 100% equity in the
venture) requires that all parties make concessions in allocating the shares.[13]

The key to successful distributive bargaining is for the technology entrepreneur
to discover the goals and perceptions of the other party through direct and/or indirect methods. Indirect methods include discussing the person with anyone who has
had previous contact, such as your own employees, the party's employees, or outside individuals; carefully reviewing all previous written and verbal correspondence
with the party; and walking around the party's business to get a feel of its vitality
and direction. The technology entrepreneur should employ each of these indirect
methods to the extent possible to get a feel for the other party's goals and drives so
that a strategy can be planned before the actual negotiation commences.

The technology entrepreneur should also use more direct methods to obtain
some insight into the other party. Whenever possible, he or she should meet informally with representatives of the other company, probing them to determine their
levels of preparation. Frequently, insight can be obtained from responses to relaxed,
almost innocent questions.

Several other direct methods employed during the negotiation process can also
reveal significant clues. One of these is for the technology entrepreneur to exagger-
ate his or her level of impatience. This often forces the other party to prematurely
reveal the amount of bargaining room left. Another method employs the principle
of "nonsequiturization." In other words, at a crucial moment in the negotiation, the
technology entrepreneur should take an entirely different posture than what has
been employed throughout. This can be accomplished by making a sudden shift
in manner, argument, or demands. For example, this shift can take the form of an
attack if a laid-back approach has been adopted previously.

Whether direct or indirect methods are used, a technology entrepreneur should
be skillful at negotiation attempts to determine the **settlement range** of the other
party. The settlement range is the area of values in which a mutually agreeable
solution can be reached. This area is between the resistance points of the parties
involved and can be either positive or negative, as indicated in Exhibit 9.2. In the
positive settlement range, the resistance points of the two parties are compatible.
The opposite is true when a negative settlement range exists. In this situation, the
resistance points are not compatible, indicating that it will be difficult to reach an
agreement that would be even minimally acceptable to the two parties.

The key to good negotiations lies in the technology entrepreneur's ability to
manipulate the other party's perception of a likely settlement point so that this point
results in a more favorable position for the entrepreneur. This can be accomplished
through indicating higher initial demands, indicating an overly strong commitment
to a particular position, or focusing on the other party by either minimizing his or
her opportunity to develop a fixed position or by enabling the party to easily revise
a previously committed position.

Perhaps the best strategy for the technology entrepreneur is to bargain in **good faith**—indicating a willingness to bargain and be flexible. This will mean that the technology entrepreneur will make concessions throughout the negotiation process where warranted, particularly when his or her position is further away from a possible solution than the other party's. Any concessions should be linked to the other party's change or at least to an understanding that certain issues are critical in securing an agreement. This strategy often leads to obtaining a settlement point that is acceptable to both parties, completing the negotiation faster than by using other strategies.

9.4 NEGOTIATION OUTCOME TYPES

Negotiation has been the subject of intense scrutiny, debate, and research because it forms a fundamental part of successful venture management. Every entrepreneur is involved in negotiations and often will be immersed in multiple simultaneous negotiations. For example, it's possible that an entrepreneur will be negotiating an offer with a new employee while at the same time negotiating with a contractor to build out additional office space to accommodate growth. Each of these transactions requires negotiation skills to produce desirable outcomes. In general, three types of outcomes can be produced through negotiating: **lose-win**, **lose-lose**, and **win-win**.

9.4.1 Lose-Win Negotiating

In lose-win negotiating, one or more parties views the negotiation process as a **zero-sum game**. This means that they believe there is no "middle ground" that can result in each party being satisfied. Lose-win negotiating is intended to produce a single satisfied party. In zero-sum negotiating, there is assumed to be limited resources and the goal is to obtain control of those resources. This is also called "distributive negotiating." The term refers to the process of distributing scarce resources. Lose-win negotiating is common, since resources are often scarce. The process of purchasing a car is a lose-win negotiation. The lower price you are able to negotiate, the less cash the dealer gets. Cash is the scarce resource that both parties want.

Lose-win negotiations are also common in organizational settings. This type of negotiation is used when ordering material goods, such as supplies for manufacturing or merchandise for retail display. This approach also characterizes much of the negotiating between management and labor involving wages, benefits, working conditions, and other matters.

9.4.2 Lose-Lose Negotiating

Lose-lose negotiations are also zero sum, and involve tactics that often include a third party as a mediator or arbitrator. High levels of conflict and little direct opportunistic interaction among the parties characterize lose-lose negotiations. Disputes that end up in court often result in lose-lose outcomes. Neither party gets what it wants from the process. For example, in cases where companies are placed on trial for product liability, resulting jury awards can be in the millions of dollars, with punitive damages often awarded to the plaintiffs. These awards occasionally make headlines and the appearance is that plaintiffs have made off with a windfall. However, the reality is much different. Instead of making the payment demanded by the court, large companies appeal the rulings, often dragging on the process for years. Plaintiffs in such cases quietly disappear from the headlines and often do not collect the amount awarded to them. Meanwhile, the company continues to spend enormous resources in discovery and appeal. Neither party has "won."

9.4.3 Win-Win Negotiating

Win-win negotiating takes a positive-sum view of the negotiation process. Parties to a win-win negotiation are both satisfied with the outcome and don't have a sense of loss.[14] In contrast to the lose-win negotiation type, a win-win approach is based on the assumption that the parties are engaged in a **positive-sum game**. A positive-sum game is one where the negotiating parties create new substantive and relationship value as a result of negotiating than existed prior to their meeting.[15] For example, two parties negotiating over the price of goods may find that they have mutual business interests that go far beyond their interest in winning the price negotiation. In fact, many new ventures will offer services to large clients at a loss to gain the positive-sum effect of using them as a reference with other clients. These **bellwether customers** can be important in convincing other potential clients of the viability of the new venture. Their thinking is something like: "If Company X (the bellwether) uses them, they must be good."

TECH TIPS 9.2

Rules for Developing Win-Win Outcomes

The actions in each step are guided by five general rules, called the Five Cs:

1. *Concentrate on each side's interests*: Define the nature of the problem and the possible ways to resolve it.
2. *Create an appropriate climate*: This means each side should do whatever is necessary to establish an efficient, effective process, channeling emotion toward solutions, rather than greater conflict.
3. *Communicate clearly*: Make sure each side understands the other's position and eliminate any barriers to that understanding.
4. *Carry out a practical process*: This means concentrating on whichever step the process is in, be it foundation building, the exploration of options, or the quest for solutions.
5. *Consider your effectiveness*: This means evaluating your own actions or words to make sure they are directed at achieving a solution.[16]

Win-win negotiating doesn't necessarily mean that all parties get everything they want, but it does suggest that they are better off as a result of the negotiating process than they were before. Effective win-win negotiation is accomplished in three steps: building a foundation, exploring the possibilities for a solution, and seeking agreement.

9.5 NEGOTIATING GAMBITS

Successful negotiators always have their goals in mind, but they are able to view the negotiation process as a game. Like any game that people play, negotiating has rules, outcomes, and gambits. A negotiating **gambit** is a tactical ploy that a party uses during the negotiation to better their position. Remember, we have advocated using a win-win approach to negotiating. That is the goal or outcome that is desired. At the same time, to achieve that outcome both parties will employ various gambits to ensure that their interests are maintained.

Note that negotiating gambits are not deceptions or unethical behaviors. In most cases, they are ways of ensuring that each party has the opportunity and motivation to state its interests and to walk away from the negotiations feeling as though they made the best deal possible. Negotiating gambits differ depending on the stage of the negotiation. For clarity, we will discuss only three types of gambits: opening, bargaining, and closing.

9.5.1 Opening Gambits

To get a negotiation started one or the other party must state their interests and/or basic position. There are a number of **opening gambits** that negotiators can use to start a negotiation, including:

- *Ask for more than you can get*: Asking for more than you can get ensures that you don't start negotiation from a too-low position. If you start low and the other party immediately agrees, you will likely believe that you could have gotten more and will not be satisfied with the outcome—despite getting what you asked for!

- *Offer less than you are willing to pay*: This is a technique that is known as **bracketing**. For example, if a seller has a car for sale at $20,000 and you are willing to pay $17,000, your opening offer might be $14,000. You bracket your target price between your opening offer and the seller's stated price.

- *Never say "Yes" to the first offer*: Saying yes to the first offer will leave the other party unsatisfied, believing he could have done better or that there is a hidden problem with the object of the negotiation.

- *Flinch at the first offer*: A good way to motivate another party in a negotiation is to openly "flinch" at the first offer. If you are on the telephone (negotiating a salary offer, e.g.) the "flinch" can be an audible gasp or sigh. This lets the other party know that it is going to have to do much better to create a win-win outcome.

By using these opening gambits, you can get the negotiation moving toward a win-win outcome. Another key to opening a negotiation is to avoid conflict. Remember, everything is negotiable and a firm "No" is only an opening position in business negotiations. Of course, once the negotiation is started new gambits are available to each party. We explore some of these bargaining gambits next.

9.5.2 Bargaining Gambits

Once initial negotiating positions have been articulated and understood, the parties enter into **bargaining**. The best negotiators and dealmakers master two tools: **conditions** and **concessions**. Stated simply, in a negotiation a concession is what you give and a condition is what you get. Creative use of conditions and concessions is one hallmark of an effective negotiator. For creative dealmakers each concession constitutes a mini-deal. They give something up in order to get something for which they have a greater need. Five generic strategies based on the conditions and concessions concepts are:

- *Refuse to make any concessions until all the demands are known*: This is an effective use of the condition. It should become a standard part of the entrepreneur's negotiating strategy. If not, the all-too-common result is that just when it seems an agreement is near a new demand will arise.

- *Negotiate how you will negotiate*: The entrepreneur should condition participation in negotiations on a time, place, number of participants, and agenda. Too often, entrepreneurs place themselves at a disadvantage by conceding these "administrative" items to the other party.

- *Condition any agreement on approval of a "higher authority"*: This strategy enables the entrepreneur to separate him or herself from the final terms. Since the entrepreneur is, by definition, the CEO of the venture, the "higher authority" may be a board of directors, management team, or trusted advisor. Note that the "higher authority" doesn't need to exist for this to be an effective strategy.

- *Never make a concession without asking for one in return*: There is no risk for a party to ask for a concession in return for another concession. This tactic has two results: (1) it elevates the value of the concession given; (2) the other party just may agree to a reciprocal concession request.

- *Condition all concessions on one another*: This is also known as a "package deal" and provides the other party with a strong incentive to close since making more demands may cause the whole deal to collapse.[17]

In addition to these bargaining gambits, which are based on the concepts of conditions and concessions, there are several others that entrepreneurs can use to increase the likelihood of creating win-win negotiating outcomes. Several additional bargaining gambits include:

- *Good guy/bad guy*: When the other party makes an offer that is close to acceptable, the entrepreneur can motivate a slightly better offer using the good guy/bad guy approach. Using this approach, the entrepreneur may say "I'd love to do the deal at this price, but my board will kill me. Could you add another 5%? That way we won't have any trouble with my board." Note that the entrepreneur has reserved the "good guy" role for himself, and the "bad guy" role for an unseen and anonymous "board."

- *Maintain a "poker face"*: Poker-faced bargainers who do not care if people like them have an uncanny way of besting those who do care.[18] A poker face refers to the ability to negotiate without revealing internal emotional states.

- *Display patience and persistence*: Since the goal of negotiating is to produce an outcome, the ability to outlast the other party can be an effective negotiating gambit. At least, you should strive to appear more patient than your negotiating counterpart.

- *Appeal to common ground*: Negotiations can often get "off track" as parties engage in various gambits and make offers and counteroffers. If you believe that the negotiation is going badly, a good way to recover is to return to assumptions and goals that are shared by each party. Being the party to initiate this appeal to common ground confers a temporary advantage.

The list of effective bargaining gambits is nearly endless, and frequent negotiators are constantly inventing new approaches. Tech Tips 9.3 highlights some additional gambits you can use in negotiating. Let's turn our attention next to gambits negotiators can use to close a negotiation.

TECH TIPS 9.3

Tips for Effective Negotiating

1. Begin the bargaining with a positive overture—perhaps a small concession—and then reciprocate the opponent's concession.
2. Concentrate on the negotiation issues and the situational factors, not on the opponent or his or her characteristics.
3. Look below the surface of your opponent's bargaining and try to determine his or her strategy.
4. Do not allow accountability to your constituents or surveillance by them to spawn competitive bargaining.
5. If you have power in a negotiation, use it—with specific demands, mild threats, and persuasion—to guide the opponent toward an agreement.
6. Be open to accepting third-party assistance.
7. In a negotiation, attend to the environment and be aware that the opponent's behavior and power are altered by it.

9.5.3 Closing Gambits

As bargaining positions harden and, hopefully, move toward a mutual agreement, the parties have an opportunity to close the negotiation. Although the major conditions and concessions are set during the bargaining stage, the closing stage can be a time when significant concessions are won or lost. The savvy negotiator knows that the close is a sensitive time when both parties, hopeful that the end is near, are vulnerable to last-minute changes. This vulnerability can be used to the entrepreneur's advantage through the following **closing gambits**:

- *Nibbling/whittling*: This gambit is generally used once negotiations are finalized (although it can occur during bargaining). Nibbling/whittling is an effort by one of the parties to win a small concession from the other. It is based on the belief that the other party is motivated to close and will not return to bargaining over the concession request. The party will simply agree to the concession to finalize the agreement. For example, car buyers can use this gambit to acquire accessories, such as floor mats, near the end of the negotiations on the purchase of a car.

- *Re-frame the deal*: This is a tactic used frequently in real estate. Essentially, it involves re-framing the terms of the deal in a manner that makes any remaining differences seem trivial. In the case of a home loan, a realtor can re-frame the deal by pointing out that the difference in price amounts to a mere 35 cents per day. The realtor may say, "You're not going to let 35 cents a day stand in the way of this deal. Are you?" In reality, that 35 cents per day over a 30-year mortgage is more than $7,000. Re-framing the $7,000 figure to 35 cents per day may motivate the homebuyer to close.

- *Appeal to ego*: No one wants to feel like the loser in a negotiation. One of the primary obstacles to a close in selling is the potential for what is referred to as *buyer's remorse*. This is where buyers feel, after a sales transaction is completed, that they did not negotiate hard enough and, as a result, they did not get a good deal. A good way to overcome this barrier is to compliment the buyers on their negotiating savvy and highlight the concessions that they won. This appeal to the other party's ego can help them feel comfortable about negotiated terms and accelerate the close.

As should be apparent from the discussions above, negotiating has been thoroughly studied and analyzed. There are a number of books and articles in the popular media about negotiating effectively, and every entrepreneur should arm himself or herself with the tools to be an effective negotiator. This chapter provides you with a starting point for building and improving negotiating skills for launching and operating an entrepreneurial venture.

9.5.4 Vendor Tactics

While being aware of the company's bottom line, it is also important to be aware of different tactics used by vendors and to have a game plan to deal with them. Rushing the company to make a decision due to "tight deadlines" or proposing end-of-year deals that will only last a few days is a common strategic play. Do not let the vendor push the company into making a choice. It is perfectly acceptable to come to a decision at a later date.

Another common tactic used by vendors is proposing a free trial period. The vendor hopes to capitalize on the busy schedule of the company, and hopes that the company forgets to cancel the product or service. If cancelled after this time, the vendor charges a high cancellation fee. Moreover, claiming that the price in the first bid is the best price the vendor can offer is also a tactic used. Be prepared to have multiple rounds of negotiations to achieve what is best for the company.

9.5.5 Negotiations after the Fact

Once the contract is signed, the work just begins and is not over. Putting systems in place to manage a contract is equally as important as negotiating the contract itself. A team or person should be in charge of monitoring the signed contract. This will ensure compliance with the contract terms and conditions that were so hard to obtain during the negotiation process. An individual needs to review the contract and the service and ensure that no discrepancies exist. Also, as the contract ends, this person will need to prepare for a new negotiation process with ample time and preparation.

In instances where the contract terms and conditions are not respected, it is important to be aware of the company's legal rights or options. If the company is outsourcing some of its work to another country, it will also be important to understand the legal climate in the country where business is being conducted. Just because a company's legal rights will be respected through signing of a contract in the United States, these may not be upheld in a particular country.

9.6 NEGOTIATING CONTRACTS

The technology entrepreneur should be aware of some fundamental elements of negotiation and the negotiation approach in the actual business setting. Some of the most relevant negotiations in business center on negotiating sales and complex project contracts.

9.6.1 Sales Contracts

Sales contracts involve the relationship between a supplier and a buyer. The fundamental negotiating techniques can be employed in strategic ways to achieve best pricing and contractual terms in acquiring anything from office supplies to labor contracts. In performing sales contracts, however, it is even more important to value relationships, as the two parties that ultimately sign the agreement will be working together for the length of the contract. Being respectful and equitable is an important part of the sales contract negotiation process.

The goal of the sales contract is for your company to get the best possible pricing for the best product(s) and service(s) from a vendor or supplier. The essential components to negotiate—initial discounts, annual support costs, future discounts, warranties, and training and consulting—along with their priority and desired outcomes are indicated in Exhibit 9.4. In order to do so, it will be important to determine the components of these services or prices that cannot be negotiated, as well as determining the budget. When these elements are decided, it will be easier to review the returned RFPs (request for proposal). An RFP is a document sent out by a company to request services or products to meet a company's needs, anything from office supplies to temporary labor. The company will also then be equipped to affect such things as the final purchase price, the support costs, and the terms and conditions with the vendor.

The focus should be on getting for the company any economies of scale, the standardization of products, and asking for what is really necessary. The larger the company is, the more its ability to leverage economies of scale and achieve deeper discounts from vendors. A smaller company can achieve similar discounts by doing additional research and discovering the amount of margin that exists in the product being acquired as well as the degree of competition in the industry. In hardware, for example, you can expect to see discounts ranging from 20% for desktop systems depending on volume to 60% for servers. If discipline is used along with one of the fundamentals of negotiations—being prepared—significant discounts can be achieved.

Also, never believe that there is a contractual term that cannot be negotiated. Continually ask questions to determine what the vendor might be prepared to give. This preparation can lead to creating longer, more beneficial contracts.

Component	Level of criticality	Outcome desired
Initial discounts	High	Go for 30–90% off list price
Annual support cost	High	Aim for 15–20% of net cost with 5–8% cap on increases per year
Future discounts	High	Apply to all products available under agreement for 3–5 years
Warranties	Medium	Specify performance measures and remedies
Training and consulting	Medium/Low	Discount off normal training offered by vendor; on site and off site

EXHIBIT 9.4

Essential Components to Negotiate.

Source: Kuo K, Wilson N. The scientific art of contract negotiations. Educause Quart 2002;May 15:32–8.

9.6.2 Complex Project Contracts in Technology

Negotiations are also important in managing the complex project contracts in technology ventures. Though the strategies and fundamentals are the same as in any negotiation, it is important to be prepared to deal with several problems unique to the technology world.

There are two significant issues in project contracts for technology: gaps in technology awareness and technological uncertainty. For gaps in awareness, it is important to note that some parties will have more knowledge of the way complex systems work or how a particular technology works as compared to others. Thus, it will be important to make sure everyone is at the same level of understanding before negotiations begin. Are there terms used in the industry that can be defined prior to the negotiation? Are there new trends in the marketplace that everyone needs to know about? It will be extremely difficult to come to any sort of agreement if one party does not understand the technical requirements or specifications of the systems involved and frustrating if terms or processes need to be explained at length throughout the negotiation process.

Technological uncertainty refers to the ambiguity surrounding whether the hardware, software, or other systems or products being discussed can truly meet expectation, from a time, cost, and performance perspective. If it is a new product that has not been around or does not have a "track record," it will be important to establish how it will work and why and how some parts might need some tweaking. Also, it will be important to make sure that the innovator of the product, whether he/she is directly or indirectly involved, has defined the needs and expectations for the product or service and how it can effectively be rolled out and sold.

SUMMARY

Entrepreneurs find themselves in negotiations with other parties regularly during the early stages of venture growth. Negotiating with bankers, landlords, vendors, and customers is routine as the venture struggles to make its cash flow sufficient to meet its obligations. Developing effective negotiating skills can help the entrepreneur manage cash flow through difficult times and sustain the relationships that will be important to long-term success.

Negotiating is often neglected as it is assumed, wrongly, that everyone has bargaining skills and that, if they know what they want, they are able to negotiate effectively. But as we discussed, negotiating skills can be developed by understanding the elements of successful negotiation. Negotiating effectively includes considering not only the substantive outcome of the bargaining, but also the effects on the relationship between the negotiating parties. Seeking win/win negotiating outcomes is predicated on the assumption that the parties want the relationship to continue on good terms. Achieving that is not possible if either party negotiates from a win/lose perspective, seeking to derive as much substantive advantage as possible.

In this chapter we examined many of the fundamentals associated with effective negotiating. It was pointed out that all negotiations consist of four basic elements: interdependence, conflict, opportunistic interaction, and possibility of agreement. We also highlighted that negotiation in practice consists of process, behavior, and substance. That is, negotiations typically follow formal or informal processes, people affect negotiations by their behavior, and all negotiations are about something of substance. We also examined several issues regarding personal effects of negotiations, such as the need for preparation, getting into the right frame of mind, and the role of emotions. We also taught you to think about the interests of negotiating parties as separate from their positions.

Establishing your best alternative to a negotiated agreement or BATNA is a good way to establish a firm negotiating mindset. Being aware of one's walk-away alternative enables one to better keep emotions out of the negotiating process. Several approaches to negotiating were highlighted, including accommodating, avoiding, collaborating, competing, and compromising. We emphasized that people have different style preferences, and that you should learn the style that you prefer. We also emphasized that you should learn to recognize the style of others and to adapt your style when necessary to further your interests.

Several negotiating outcomes were explored, including lose-win, lose-lose, and win-win. It was pointed out that win-win is the preferred outcome if it is important to retain the relationship after the negotiation is completed. Lose-win is based on the assumption that negotiating is a zero-sum game, and lose-lose is a "scorched earth" approach where one or both parties believe that if they do not win then nobody will.

A variety of negotiating tactics or "gambits" were also explored. We examined opening, bargaining, and closing gambits. It is important to know these gambits to become a more effective negotiator. Gambits can be used to increase one's chances of attaining one's negotiating goals. They also can be recognized and responded to when in use by one's negotiating partners.

Finally, this chapter examined several contract types that likely will be encountered by technology entrepreneurs. These contracts are important components of a venture's commercial success, and need to include terms that are favorable to the venture and that provide it with recourse in case the other parties do not meet their obligations. Sales contracts and complex project contracts must be negotiated with care, and they must be monitored after they are signed to ensure terms are being met.

STUDY QUESTIONS

1. What is meant by "negotiation"?

2. Discuss the five different negotiation approaches and a situation where each one would be appropriate.

3. Compare distributive bargaining with integrative bargaining.

4. Discuss the aspects of negotiating a sales contract versus a more complex technology contract.

5. What are the three outcome types of negotiations? Explain the difference between substantive results and relationship results of negotiations.

6. Why does an entrepreneur need to develop effective negotiating skills? Describe a context in which an entrepreneur may need to negotiate.

7. What are the four components of nearly all negotiations? Explain the meaning of each.

8. What is meant by the term "BATNA"? How is a BATNA established?

9. What is meant by the terms "condition" and "concession"? How can an entrepreneur use these concepts to negotiate effectively?

10. Explain what is meant by the term "environment of scarcity." How does that concept affect an entrepreneurial venture?

EXERCISES

1. *IN-CLASS EXERCISE*: Students should take the negotiating skills assessment below. The assessment scores are provided at the end of the Web Resources section of this chapter. After taking the assessment and learning their scores, students should be asked to volunteer their thoughts on where they think they need to improve to become a better negotiator. They should also provide some thoughts on how they might develop the needed skills over time.

NEGOTIATING SKILLS ASSESSMENT

Instruction: Choose the number for each item 1—10 that most closely reflects your current skill in each item. In this exercise "1" is "Not at All" and "10" is "Highly Capable."

1. I am sensitive to the needs of others.	1	2	3	4	5	6	7	8	9	10
2. I will compromise to solve problems when necessary.	1	2	3	4	5	6	7	8	9	10
3. I am committed to a win-win philosophy.	1	2	3	4	5	6	7	8	9	10
4. I have a high tolerance for conflict.	1	2	3	4	5	6	7	8	9	10
5. I am willing to research and analyze issues fully.	1	2	3	4	5	6	7	8	9	10
6. Patience is one of my strong points.	1	2	3	4	5	6	7	8	9	10
7. My tolerance for stress is high.	1	2	3	4	5	6	7	8	9	10
8. I am a good listener.	1	2	3	4	5	6	7	8	9	10
9. Personal attack and ridicule do not unduly bother me.	1	2	3	4	5	6	7	8	9	10
10. I can identify bottom-line issues quickly.	1	2	3	4	5	6	7	8	9	10

Scoring for In-Class-Exercise: If you scored 80 or more, you have characteristics of a good negotiator. You recognize what negotiating requires and seem willing to apply yourself accordingly. If you scored between 60 and 79, you should do well as a negotiator but have some characteristics that need further development. If your evaluation is less than 60, you should go over the items again carefully. You may have been hard on yourself, or you may have identified some key areas on which to concentrate as you negotiate. Repeat this evaluation again after you have had practice negotiating (adapted from Ref. 19).

2. *OUT-OF-CLASS EXERCISE*: For this exercise, students should work in teams of 3–5. Each team should identify one or more individuals in the community who are involved in technical selling. The objective of this exercise is to meet with these individuals and ask them questions about their negotiating style and any tactics they use to win client contracts. Students should take notes during the interview. After the interview, teams should discuss among themselves the most likely outcome the individual strives to achieve: lose-win, lose-lose, or win-win. Be prepared to discuss and defend in class. The in-class discussion should not name individuals interviewed or their companies, but instead focus on their role and industry.

KEY TERMS

Accommodating negotiation: A negotiation approach used by individuals who want to solve problems and get what they want while preserving personal relationships.

Avoiding negotiation: A negotiation approach used by individuals to sidestep important and difficult issues.

Bargaining: The part of negotiations where the parties use a variety of gambits to achieve their respective aims.

Bellwether customers: Large company customers that often are served at a loss for the positive-sum gain of the reference quality of their brand name.

Bracketing: An opening gambit whereby a party offers less than they are willing to concede and asks for more than they will settle for.

Buyer's remorse: Where a buyer feels, after a sales transaction is completed, that he or she did not negotiate hard enough and, as a result, did not get a good deal.

Closing gambits: Techniques a negotiator can use to bring negotiations to an end (close).

Collaborating negotiation: A negotiation approach used by individuals who like negotiations and really like solving problems.

Competing negotiation: A negotiation approach that ends with a winner and a loser.

Compromising negotiation: A negotiation approach where the parties involved seek quality and fairness throughout the process and have the end results equitable.

Concession: In a negotiation, a concession is what you are willing to give to the other party.

Condition: In a negotiation, a condition is what you are asking for.

Cooperative negotiation: The integrative bargaining approach to negotiating where the entrepreneur is willing to let the other side achieve its desired outcome while maintaining a commitment to his or her own goals.

Distributive bargaining: A process by which at least one of the parties does not achieve the desired goals.

Gambit: A tactical ploy that a party uses during the negotiation to better his or her position.

Good faith bargaining: A strategy that demonstrates that party to a negotiation is willing to be flexible and to bargain.

Integrative bargaining: Cooperative negotiation between the parties involved where each is able to achieve the goals set before the negotiating began.

Lose-lose: A negotiating outcome where neither party gets what it wants.

Lose-win: A negotiating outcome where one party gets all it wants and the other party gets none of what it wants.

Negotiation: A process in which two or more parties attempt to reach acceptable agreement in a situation characterized by some level of real or potential disagreement.

Negotiation process: The method through which parties go to identify and resolve the critical issues so an agreeable solution can be obtained.

Opening gambits: Techniques that can be used to get a negotiation started.

Positive-sum game: A perspective on negotiating that assumes each party can get a portion of or all of what it wants from a negotiation.

Scarcity: Resources are limited, and they must be acquired, applied, and managed efficiently.

Settlement range: The area of values in which a mutually agreeable solution can be reached. This area is between the resistance points of the parties involved and can be either positive or negative, as indicated in Exhibit 9.2.

Win-win: A negotiating outcome where both parties get what they want.

Zero-sum game: A perspective on negotiating where only one party can get what it wants. It assumes no "middle ground" where both parties may be satisfied in achieving some of what each wants.

WEB RESOURCES

The web sites below are intended as destinations for your further exploration of the concepts and topics discussed in this chapter:

1. http://www.queendom.com/tests/access_page/index.htm?idRegTest=1397: This web site provides an online negotiation skills test that is free. It is a good indicator of one's current negotiating skill set and gaps in one's overall negotiating repertoire.

2. http://www.researchchannel.org/prog/displayevent.aspx?rID=11074&fID=345: This web site features a video of Professor William Ury, the Harvard University scholar who wrote the best-selling book "Getting to Yes." In this video, Dr. Ury discusses his new thoughts on "The Power of a Positive No."

3. http://www.globalnegotiationresources.com/: This web site for Global Negotiation Resources provides a wealth of tools, books, and other resources to help you become an effective negotiator in the global economy.

4. http://www.fmcs.gov/internet/: This is the web site of the Federal Mediation and Conciliation Services (FMCS). The Federal Mediation and Conciliation Service, created in 1947, is an independent agency whose mission is to preserve and promote labor-management peace and cooperation. Headquartered in Washington, DC, the agency provides mediation and conflict resolution services to industry, government agencies, and communities.

CASE STUDY

Canon: A Master of Negotiating Contracts

As stated in this chapter, "Negotiating fundamentals and a company's ability to hone them can really make or break a deal or reach more customers." A cooperative negotiation is one way that benefits both parties involved. Cooperative negotiations are part of integrative bargaining, a concept discussed throughout the chapter. Integrative bargaining and a cooperative negotiation are evident in the following situation.

Canon is a world-wide company that is focused on marketing products and digital solutions that enable "businesses and consumers worldwide to capture, store, and distribute information." The company originated in Tokyo, Japan, and was actually a branch of Precision Optical Instruments Laboratory. In 1935, the name "Canon" was the new trademark name for the company. Canon U.S.A., Inc., as its own corporation, was founded in 1966, and focused on camera making. The corporation soon became a global multimedia corporation. In 2007, the corporation ranked third in being patent holders in technology in the United States and earned global revenue of $39.3 billion.

In May 2008, a negotiation was made between Canon U.S.A., Inc., and the state of Virginia. The negotiation resulted in Canon expanding their operations in the United States within the Hampton Roads, Virginia, area. This expansion provided new facilities and 1,000 job opportunities for this area. A $600 million investment was made by Canon U.S.A. to Canon Virginia, Inc., a subsidiary of Canon U.S.A. This was done to build a new research center for the "development of automated and robotic manufacturing technologies for the Americas region … and to expand its efforts in the recycling and reclaiming of toner cartridges and related materials at its Industrial Resource Technologies, Inc. (IRT), operations in Gloucester County, Virginia." Because the negotiation was a success, both Canon U.S.A. and the state of Virginia were benefited.

For Canon U.S.A., Inc., and Canon Virginia, Inc., the negotiation was smart because the cooperation as a whole was making relations throughout the U.S. manufacturing businesses, but also made impacts on the cooperation internationally. Tsuneji Uchida, president and chief operating officer of Canon Inc., said, "The combination of producing, selling, collecting and recycling cartridges locally, and eliminating the need to transport products around the world, will allow us to have a positive environmental impact by helping to reduce CO_2 emissions worldwide."

Canon Virginia established an alliance with workforce development resources throughout Newport News to build a skilled technical workforce for the new expansion. The Canon franchise's success is strongly influenced by its ability to build firm relationships with other businesses and/or their employees and suppliers in the United States, and therefore, having successful negotiations with them. The corporate philosophy of Canon is *kyosei*. A concise definition of this word would be "Living and working together for the common good." Canon's goal is to contribute to global prosperity and the well-being of mankind, which will lead to continuing growth and bring the world closer to achieving *kyosei*.

Among the alliances made with the state of Virginia is the Virginia Community College system. With this expansion for Canon, there was also an expansion of over 1,000 job opportunities for the people of Virginia. These jobs provided availability for people who were "retired from the military, advanced technology students and others seeking a new opportunity with a college curriculum that has been explicitly designed to meet Canon's demanding job specifications."

This negotiation was successful because it resulted in benefits to both companies, a key aspect of integrative bargaining. Not only was there such an outstanding benefit for Canon, but also there was a huge benefit for the state of Virginia as well.

For 23 years, Canon U.S.A. and the Commonwealth of Virginia have been corporate partners. Because of this, Canon has employed about 1,500 Virginians. It's negotiations such as these made with the state of Virginia that make Canon accessible to growth; this gives them more of an ability to influence the consumers of the United States and consumers all over the world, everyday.

Questions for Discussion

1. Before Canon U.S.A., Inc., was able to be successful with this negotiation; they needed to establish an alliance with the state of Virginia. What is a key example that shows an alliance was made between the two parties and that helped earn Virginia's trust?
2. When hoping to make a cooperative negotiation with another party, why is it important to create an alliance beforehand?
3. Taking into account the definition of "preparation", why was it important for Canon U.S.A., Inc., to prepare before the negotiation with the state of Virginia, knowing they had such a strong alliance with them?
4. Because emotions play a huge role in any negotiation, why, even with a strong alliance created between Canon U.S.A., Inc., and the state of Virginia, is it even more important to be aware of the other parties' emotions when hoping to make a successful negotiation?

Source: Adapted from Canon Global. http://www.canon.com/about/history/02.html; Governor Kaine Announces More Than 1,000 Jobs for Virginia. http://www.newportnewsva.com/images/GovernorsCanonsPressReleases.docTail; Canon Inc and Canon U.S.A. to Expand U.S. Operations in Virginia with the Help of Virginia State and Local Officials. http://news.thomasnet.com/companystory/545193; Canon U.S.A., Inc. Profile. http://www.zoominfo.com/Search/CompanyDetail.aspx?CompanyID=250 24186&cs=QE Bfda6Lw&p=indeed.

ENDNOTES

[1] Budjac-Corvette BA. Everybody negotiates, but it doesn't mean negotiation is easy. *Dispute Res J* 2007;**62**(3):88.

[2] Inderst R, Muller HM. The effect of capital market characteristics on the value of start-up firms. *J Financ Econ* 2004;**72**:319–56.

[3] Shu L, Roloff M. Strategic emotion in negotiation: cognition, emotion, and culture. In: Riva G, Anguera MT, Wiederhold BK, Mantovani F, editors. *From communication to presence: cognitions, emotion and culture toward the ultimate communicative experience*. Amsterdam: IOS Press; 2006.

[4] Retrieved from "Creative problem solving in negotiations," The Negotiation Experts. http://www.negotiations.com/case/problem-solving/.

[5] Lepine R. Negotiation basics. *CMA Manage* 2004;**78**:16–17.

[6] Peterson RM, Lucas GH. Expanding the antecedent component of the traditional business negotiation model: pre-negotiation literature review and planning-preparation propositions. *J Mark Theory Pract* 2001;**9**(4):37–49.

[7] Kim PH. Choosing the path to bargaining power: an empirical comparison of BATNAs and contributions in negotiations. *J Appl Psychol* 2005;**90**(2):373–81.

[8] Carpenter J, Rudisill M. Fairness, escalation, deference, and spite: strategies used in labour-management bargaining experiments with outside options. *Labour Econ* 2003;**10**:427–42.

[9] Retrieved from the website http://www.batna.com.

[10] Fee S. (n.d.). **5** Steps to better negotiation. Retrieved November 1, 2008, from Susan Fee: http://www.susanfee.com/articles/business/.

[11] Sebenius JK. Negotiating in three dimensions. *Negotiation* 2004;**February**:3–5.

[12] Johnston JS. The return of bargain: an economic theory of how standard form contracts enable cooperative negotiation between businesses and consumers. *Mich Law Rev* 2006;**104**(5):857–98.

[13] Van den Abbeele A, Roodhooft F, Walrop L. The effect of cost information on buyer–supplier negotiations in different power settings. *Account Org Soc* 2009;**34**(2):245–66.

[14] Lewis BJ. A leader's challenge: converting win-lose decisions into win-wins. *J Manage Eng* 2000;**May/June**:9.

[15] Buchanan JM. Game theory, mathematics, and economics. *J Econ Methodol* 2001;**8**(1):27–32.

[16] Williams DG. Negotiating skills—Part I. *Prof Builder* 2000;**January**:155–8.

[17] Diener M. A tug of war. *Entrepreneur* 2003;**September**:81.

[18] Lichtblau J. Where nice guys do finish last: in negotiations. *BusinessWeek* 2000;**March 18**.

[19] Maddax RB. *Successful negotiation*. Menlo Park, CA: Crisp Publications, Inc.;1988; 19.

Technology Venture Strategy and Operations

Launching the Technology Venture

After reading this chapter, students should be able to:

- Write a business plan
- Conduct the research necessary to support assertions of a business plan
- Network for resources, including capital and human
- Aggregate resources necessary to launch a venture
- Create performance standards to ensure venture goals are met
- Establish an operating foundation for long-term venture success

---

TECH VENTURE INSIGHT

Networking Opens up Opportunities

In computer lore, Stephen Wozniak is well recognized as the co-founder of Apple Computers (now Apple, Inc.). Apple was a huge contributor to the computer revolution of the 1970s, particularly because of the success achieved by both Apple I and Apple II.

An interesting element to this tale is that Wozniak is known to be an introvert, who finds the fame associated with Apple, Inc., somewhat annoying. Nevertheless, his rise to the public arena began when he tried to impress his friends and members of the "Homebrew Computer Club" in Berkeley, California, with the makings of a first computer, having no real aspirations to go commercial with it. Another individual by the name of Steve Jobs, however, saw the potential and was convinced that with a fully assembled "P.C. board," this new device could revolutionize the technology industry.

To make this enterprise a reality, Wozniak and Jobs both needed cash, so they started selling off their possessions, including their HP scientific calculators and a Volkswagen van, which all added up to $1,300. In 1975, after much hard work and sacrifice, they

---

launched their first computer, a more advanced version of Altair 8000, which, at the time, was the best computer seller on the market. Apple I was a fully assembled and functional unit that boasted a $20 microprocessor on a single circuit board with ROM, and outfitted with a RAM, keyboard, and monitor. In April 1976, Apple Computers was born. Wozniak quit his job at HP and became the vice president for research and development at Apple Computers.

In the early years, networking was key to Apple's success. To acquire parts for building the Apple I, Wozniak convinced a small computer shop supplier to give him the parts on credit. Wozniak was a frequent customer, so the owner agreed to buy 50 fully assembled computers at $666.66 (Wozniak liked recurring numbers). Similarly, the launch of Apple II was also a product of networking and forging ties with key individuals. In April 1977, Apple Computer took part in the West Coast Computer Fair. Here, Jobs and Wozniak introduced themselves to a gentleman by the name of Mizushima Satosh. Not long after, Satosh became the first authorized Apple dealer in Japan and, to boot, business advisor to both budding computer entrepreneurs.

The relationship with Satosh was not only instrumental to the final distribution of Apple Computers, but also helped in acquiring crucial funding. Through Satosh's networking group, Wozniak and Jobs met Mike Makkula. Since the banks were reluctant to finance an audacious idea such as "computers for ordinary people," Makkula agreed to put up the cash and co-sign a loan worth $250,000.

The story of Jobs and Wozniak highlights how networking and building on relationships can lead to amazing partnerships and lasting success.

Source: Adapted from MacMillan D. Where tech got its start. BusinessWeek Online 2007;June 8:24; Farivar C. Apple's first 30 years. Macworld 2006;June:14–15; Brand S. We owe it all to the hippies. Time (Special Edition) 1995;Spring:54–6.

INTRODUCTION

Launching a technology venture doesn't happen in an instant, but rather is a process that requires energy, discipline, and determination. Of course, there will be a moment in time when your business is officially "launched," but there are a large number of milestones that must be achieved prior to that auspicious moment.

Throughout this text, we have discussed most of the challenges you will face in launching and operating a venture. In this chapter, we focus only on the launch and prelaunch activities that will be required of most any technology venture.

One of the activities that technology entrepreneurs will need to undertake is the development of a business plan. The business plan serves as the roadmap for the venture over time, and also serves as a reminder to investors and other interested parties of the objectives the business has set for itself.[1] Over time, these objectives will or won't be met. Of course, the business plan will change and be updated periodically as the venture grows and changes, but the prelaunch business plan is a vital document that sets the foundation for what lies ahead.

Another critical launch and prelaunch activity for the technology entrepreneur is networking. Networking will be critical to recruiting the talent that you will need to round out the operating team, to raising the capital that you will need to fuel the venture's growth, and to generating a level of "buzz" about the venture you are launching.

The technology entrepreneur must also be a talented resource aggregator. The resources required to start a venture include money, people, office space, office equipment, and many other things. We will discuss some of the most essential resources and how they can be acquired.

Finally, we discuss essential ingredients that should be present at the start of a new venture that provide it with a foundation for future success. The policies, quality objectives, and organizational culture should be in place at the beginning and provide a long-term competitive advantage. Many successful technology ventures are established by visionary founders whose unique spin on existing business practices not only gives them a competitive edge in products, but also attracts talented people who help the venture sustain its uniqueness. One need only think of Google, with its quirky name and creativity-inspiring work environment to know that organizational culture can play a role in long-term success. Several key elements of laying the foundation for long-term operating success are explored. First, we will discuss the business plan, a tool that many believe to be essential in launching a technology venture.[2]

10.1 THE BUSINESS PLAN

The **business plan** has been called the entrepreneur's "road map" as it lays out the competitive terrain, sources of fuel, and direction for the growing business. Some entrepreneurs get by without a plan, for a while. But they're the lucky ones. Most businesses that lack a detailed plan fail for reasons that often would have been obvious if they had gone to the trouble of writing a business plan. For example, one of the most common problems faced by a young and growing business is cash flow shortages. A thorough business plan requires the entrepreneur to develop pro forma financial statements that project at least 3 years of operations. Detailed cash flow analyses usually reveal potential problems, such as seasonality of revenues, before they happen in the real world. Through planning, entrepreneurs can anticipate different scenarios, good and bad, and prepare in advance for them.[3] Without careful planning, the entrepreneur spends an inordinate amount of time in "reaction" mode.

There are other reasons that an entrepreneur needs to have an up-to-date business plan besides its utility as a road map. If the entrepreneur wants to acquire capital from external parties, whether they are investors, bankers, or friends, a business plan is usually requested. Can you imagine giving your money to someone simply based on his or her idea? You probably can't imagine it. The same thing happens when the entrepreneur seeks funding for his or her business. The investor or lender wants to know if the entrepreneur has thought through the risks confronting the business, and whether strategies and tactics have been developed to manage the risks and make a profit—enough profit at least to pay back the loan or provide a return on the invested capital.

10.1.1 Writing a Business Plan

Creating and building a successful technology venture requires effective planning. As research indicates, poor management and management inexperience are primary causes of new-venture failure.[4] Of the various management functions that entrepreneurs must perform, planning probably contributes the most to new-venture performance. Planning provides a well-thought-out roadmap of action for the critical first months of the new venture. This activity is vital because when resources are slim in the early days of the business, mistakes can be costly or even fatal. Careful planning reduces the chances of major mistakes; it also forces the entrepreneur to examine the business's external environment, competition, potential customers, strengths, and limitations.[5] But despite the importance of planning, many entrepreneurs don't like to plan because they believe planning hinders their flexibility.[6]

Writing a business plan is perhaps the most difficult task for new entrepreneurs, but also one of the most essential. For many entrepreneurs, writing a business plan is similar to having dental problems: it's painful and often requires repeat visits for follow-up work. The major difference is that a trip to the dentist usually ends quietly, knowing that you'll probably be back within a year. Completing the business plan, however, can result in a feeling of exhilaration as your idea comes alive in the words and numbers.

A business plan is a roadmap for starting and running your business. It's also a sales document, since the business plan is used to convince bankers, venture capitalists, family, friends, and even yourself to invest in the venture. Actually, there are seven good reasons that you should write a business plan if you want to be an entrepreneur:

1. *To sell yourself on the business*: This is a "reality check." The most important stakeholders in any business venture are the founders. The founders must be convinced that the business idea is sound, so they develop a passion to make it a reality.

2. *To obtain bank financing*: Since the bank failures of the 1980s and 1990s, it's more necessary than ever for entrepreneurs to have a sound business plan if they're seeking bank financing. Getting bank money may be tougher now than it has been in a long time. A well-written, well-researched business plan can make the difference between getting the money needed to start the business and being rejected.

3. *To obtain equity financing*: The business plan is the price of admission to the equity capital evaluation process. Rare is the private investor who will provide the seed capital a new business requires based merely on an oral presentation

or executive summary. Even if you do get to talk with an equity investor about your deal, you'll be required at least to submit a full financial plan writing.

4. *To arrange strategic alliances*: Many small companies seek alliances with larger companies to get some of their expertise in key areas, or to offer their services. Despite the corporate need for many such services, there usually are more vendors than needed. To help them select who they'll work with, large corporations usually want to see business plans of prospective small business partners.

5. *To obtain large contracts*: When small companies are seeking substantial orders or ongoing service contracts from major corporations, it helps to have a business plan to convince the corporation of the long-terms prospects of the small company. Through a business plan, corporate decision makers can see that the small company expects to be in business 3 years, 5 years, and more into the future. The business plan helps convey a feeling of partnership and commitment.

6. *To attract key employees*: A new business startup will need talented, flexible people who are willing to take the risks associated with a new venture. Even in today's volatile job environment, many key executives are drawn to jobs with large corporations. A written business plan can assure prospective employees that the entrepreneurs have carefully thought through key issues facing the company and have a plan for dealing with them.

7. *To motivate and focus the management team*: As small companies grow and become more complex; a business plan helps the management team stay focused on the same goals. Many companies lose their way when they begin to grow. Management, including the founding entrepreneur, needs to plan growth to ensure the business has cash flow to pay the bills, people to handle the volume of work, and goods to meet the demand they've created.

Research has demonstrated that solid business planning can help entrepreneurs avoid unnecessary mistakes.[7] A business plan should include the items listed in Exhibit 10.1.

- Executive Summary
- Company Information
- Product/Service Description
- Competitive Analysis
- Market Analysis
- Industry Analysis
- Management Team
- Marketing Plan
- Financial Projections

EXHIBIT 10.1

Basic Outline for a Business Plan.

The executive summary is probably the most important section of any business plan. Many of the people to whom you will present your plan will read little more than the executive summary. Since this is the first, and often only, part of your plan that people read, it requires careful attention. The executive summary must present a realistic appraisal of your business, usually in one to two pages.

TECH TIPS 10.1

Tips for Writing the Executive Summary of Your Business Plan

Here are some tips for writing an effective executive summary to your business plan:

- *Write everything else first*: Although the executive summary is the first thing that people will read, it is the last section of your business plan that you will write. Since it is a summary, it must draw from the details that go into the rest of your business plan.
- *Write in plain language*: The executive summary should avoid technical jargon and/or confusing language.
- *Describe the business*: It is of primary importance that you be able to describe your business in a paragraph or less. This means that you will have to think—really think—about what you intend to do and how you intend to do it.
- *Describe the market size*: Most of the people who will read your business plan are potential investors. You should find the most recent hard data on market size and report it in the first or second paragraph of the executive summary.
- *Identify your target market share*: After you have stated the size of the market, you should next identify the market share that your company must attain to be successful. Investors will want to know how much of the existing market you must capture to reach profitability.
- *Identify competitive advantages*: In the executive summary you do not want to analyze the competitors in your industry, but you should note the competitive advantages that your business will develop.
- *Describe the investment opportunity*: The executive summary should conclude with a clear statement about the investment opportunity.

The body of the business plan will expand on each of these elements contained in the executive summary. In the next section, we will explore in detail the information that should be included in each section of the business plan.

10.1.2 Company Information

This section of the business plan includes basic details about the company such as mailing address, phone number, web site, e-mail address, and any other contact information. It also includes details about the company's legal form (e.g., partnership and corporation), and state or country in which the business is incorporated. If stock has been issued, the type of stock issued and the current shareholders can be mentioned in this section. It is often useful to reveal the ownership structure of a business to ward off any potential concerns by parties that may be considering an investment or loan. Accurate and up-to-date information is needed so that anyone who reads the plan and has an interest in contacting the venture can quickly do so. The company information should also include a named individual who will be the point of contact for anyone interested in learning more about the company.

This section should also include the company's mission statement and a brief overview of any operating history the company may have. If the venture is a start-up with no operating history, this should also be mentioned. A brief statement about how the venture was conceived can be useful, especially if there is an interesting story or compelling business need to convey.

10.1.3 Product/Service Description

This section of the plan describes the product/service in detail sufficient to enable a working understanding of what the company sells. This may include photographs, schematic drawings, and descriptive scenarios about how the product/service is used by customers. The business plan should specify the salient features of the product/service that distinguish it from competitors. These features will form the basis of the business strategy. As such, this section should also describe how the features are protected from competitors. If there are patents, copyrights, or other forms of intellectual property protection in place, this section should highlight the status of those measures.

This discussion should also focus on the benefits of each of the salient product/service features. "Features" refer to what the product is, and "benefits" refer to what the product does. What benefits will customers gain as a result of the particular feature set of the new venture's products/services? The benefits they gain should be those that are of high value to the new venture's market. Later, the business plan section on market analysis should emphasize why the benefits of the new venture's products/services are of high value to customers. Tech Micro-case 10.1 describes a very successful business plan that was not recognized as such when it first was crafted.

TECH MICRO-CASE 10.1

The FedEx Business Plan Got a "C" Grade in School

FedEx is arguably one of the world's most successful businesses. Its 2008 sales topped $35 billion and its brand is recognized around the world. Fred Smith, the FedEx CEO, founded the company in 1971 after a tour of duty as a U.S. Marine in Vietnam. Prior to his military duty, Smith was an undergraduate student at Yale, where he wrote a business plan that was to become the idea behind FedEx. As Smith recalls it, he was given the assignment by an economics professor in 1965 to solve a fundamental problem that was becoming a concern in the early days of workplace computerization. In his plan, Smith pointed out that growth in the computer industry was dependent upon the development of a new type of logistic system, one that could speed computer parts to remote corners of the country to keep businesses up and running.

Smith's ideas were not well received by the professor, and legend has it that it earned no more than a C grade. When he was in Vietnam, Smith recognized that the same, outdated logistics system used to transport goods in the United States was being used by the military. He described the system as "supply-push." After his duty, at the young age of 27, Smith raised money from friends and family and set up a 25-city package delivery system. He leased some airplanes and delivery trucks and, on April 17, 1973, delivered his first package.[8]

10.1.4 Competitive Analysis

This section includes a review of the firm's top competitors and their relative market share. It should discuss the strategies employed by each competitor and should also compare their products/services to one another, including discussing the relative advantages/disadvantages of each offering.

Each competitor also has a business strategy that it follows to promote its products/services in various markets. This section should examine the markets the competitors serve and the strategies they employ in those markets. It's likely that the new venture will encounter various **barriers to entry** as it attempts to distribute its products/services in markets already served by existing competitors. These barriers should be described and strategies for overcoming them should be articulated in detail. In particular, the competitive analysis should describe the expected response from competitors when the new venture enters the existing markets. Will they be passive? Will they fight aggressively to maintain their market share? While it's not possible to be certain about the competitive response, it is useful to make reasonable judgments about what might occur and counter-strategies that the venture can use to stay on course toward its business objectives.

One effective technique for the competitive analysis is to design a features and benefits matrix that highlights the new venture's product/service in the context of competitor offerings. An example of a features and benefits matrix is given in Exhibit 10.2.

10.1.5 Market Analysis

The market analysis section of the business plan is designed to provide insights into the target market for the products or services offered by the venture. A target market is nothing more than the group of people the entrepreneur believes to be most likely to purchase the venture's products or services. For example, if the entrepreneur was launching a new mobile phone service, the target market would be the people in the region to be served who potentially would want and could afford the service.

Of course, identifying a target market is slightly more complicated than that. In the case of the mobile phone service, the target market is not simply "all people likely to purchase the service." That would use too broad a brush to identify a true target market. The entrepreneur behind the mobile phone services needs to identify his target market in more precise terms. For example, the phone service may be specifically set up for business users. In that case, the users would want the variety of features

	Sample Features/BenefitsTable for a Competitive Analysis					
Features	**Benefits**	**Company**	**Competitor A**	**Competitor B**	**Competitor C**
Feature 1	(describe benefits to customers)				
Feature 2					
Feature 3					
	Possible Features Include:				
	Price				
	Quality				
	Delivery				
	Credit terms				
	Reliability				
	Return policies				
	Technical support				

EXHIBIT 10.2

Sample Features/Benefits Table for a Competitive Analysis.

necessary for daily business communications, including voice, data, and other features. On the other hand, the service might target young people. In that case, popular features might include such things as text messaging, online games, and shopping deals.

Entrepreneurs need to think very hard about their target market in order to ensure the best use of limited marketing resources. There are several common ways to think about the market for a new venture, including:

- Demographics
- Psychographics
- Geography
- Sociographics

"Demographics" is the term used to refer to readily identifiable characteristics of an individual or a group. These characteristics include many of the variables that are covered in U.S. Census data. Some examples of demographic characteristics include:

- Age
- Income
- Sex
- Race
- Ethnic background
- Marital status
- Home ownership
- Educational background

Many marketing groups closely monitor data on these characteristics. Products as diverse as movies and soft drinks are marketed to specific groups, based on detailed demographic information.

Psychographics is used to refer to group or individual characteristics that are less readily observable. Terms such as "attitude," "taste," or "values" are used to refer to these variables; they are terms that refer to psychological states or tendencies associated with individuals and/or groups. For example, a psychographic profile of college males might include their "preference" for sports, beer, and loud music. While this does not describe all college males, someone who had a product that would appeal to a group with that psychographic profile might try the college male market.

Markets that are confined to a particular location can be described using terms from geography. A restaurant, for example, describes its target market using a radius drawn around a central point—the location of the venture. Other venture types that are sensitive to geography include retailing, sports teams, and agriculture or natural-resource-related businesses such as wine or coal mining.

Finally, sociographics refers to the tendency for people to identify with a particular cause or social group.[9] Many firms have developed distinctive competencies in recent years by marketing directly to individuals based on their social values. Tom's of Maine is a good example of a company that uses this approach. Tom's of Maine makes products that it sells to people who prefer to buy organic or so-called "natural" products. Tom's promises its customers that its products are developed using "pure and simple ingredients from nature." This message carries enormous appeal to people within a specific sociographic community, and gives Tom's of Maine substantial pricing power.

10.1.6 Industry Analysis

The industry analysis section of a business plan focuses on identifying the current health, maturity, and future prospects of the industry in which the venture will participate. "Industry health" refers to whether the industry is marked by innovation, growth, and dynamism, or whether it is in decline. The maturity of an industry in part is a reflection of its health. An emerging industry is often characterized by rapid growth and frequent and dramatic changes. On the other hand, an industry that has peaked and is on the decline will have less opportunity for growth. The future prospects of an industry are linked to customer trends and buying patterns. For example, future prospects of the cola industry are strong, but it is a mature industry and likely inhospitable to new ventures.

Industries follow a life cycle that can be readily observed and analyzed. This life cycle is usually illustrated and described as shown in Exhibit 10.3. As the exhibit shows, an emerging industry often has a long period where it struggles to establish itself in the minds of its primary customers. A business that is attempting to sell a brand-new product or service, for example, may have to spend a lot of time and money to develop a market. Entering an emerging industry that doesn't have an established market often means that the company will need substantial capital to sustain itself during the market-building period.

Many entrepreneurs in emerging industries overestimate the customer response to their product or service, and underestimate the amount of cash they will need to

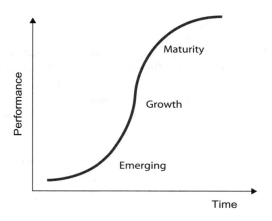

EXHIBIT 10.3

The Standard Industry Life Cycle.

develop a market that is large enough to sustain the business. For example, many of the early Internet companies wildly overestimated the market potential because they believed their online presence created a worldwide market. In reality, many of these dot-com ventures failed because they did not have the cash to develop a loyal customer base. Today, as the e-commerce industry matures and people become more comfortable with online buying, many business models that were abandoned just a few years ago are now being used by successful ventures.

A growth industry is one that has achieved critical mass in terms of market size, and now has enough active customers to sustain a growing number of competitors. A growth industry presents a new set of challenges for the entrepreneur. During the emerging industry stage, the major issue confronting the entrepreneur was whether enough cash could be generated to survive day by day. During the growth stage, the challenge is whether the company can scale itself quickly enough to capture a significant market position. Despite the growth in the industry, many new ventures fail during this phase of the industry life cycle, because they attempt to grow faster than their systems and skills allow. When that happens, a company may wind up losing customers due to lack of available product, poor customer service, or poor product/ service quality.

Steady or only slowly growing industry-wide revenues characterize the mature stage of an industry. This is the stage in which a few well-established brands dominate the industry and control much of the market share. Despite the lack of growth in the industry, the mature stage can be attractive to a new venture because of its stability. Entering a growth phase industry means that innovation could make a large difference between winners and losers. During the mature phase, innovation is less important than outright execution. Customers in a mature industry have well-established buying patterns and preferences. It's usually very difficult to get customers to switch their buying habits based on a product/service innovation. What will matter more to customers in this phase is execution—delivery of the product or service in a manner that meets the customers' needs.

An industry that is in decline is characterized by falling industry-wide revenues, consolidation of competitors, and falling prices and profit margins. Declining industries tend to eliminate competitors who are least efficient in their delivery of products and services. The falling prices and smaller profit margins mean that loose operations are likely to experience cash flow problems. Two consequences of that are employee layoffs and declining salaries and bonuses. A declining industry is not all bad, however. Many entrepreneurs are very proficient at building businesses in declining industries. Some seek out distressed companies and purchase their assets at a fraction of their former value. Entrepreneurs skilled at "turnarounds" are able to reverse the course of the distressed company and return it to a profitable enterprise. If the entrepreneur is successful, the revitalized company can then be sold to an established competitor for a profit.

The main point of understanding the stage of the industry life cycle is that no phase is "better" than any of the others from the perspective of whether profits can be made. The point is that the entrepreneur needs to understand the industry life cycle phase in order to develop an appropriate business strategy.[10]

Another variable to examine in the industry analysis section of the business plan is the relative size of the industry. Industry size is characterized by the total revenues generated by all of the competitors in the industry. You can determine the size of your industry in a number of ways. Many industries have what are called "trade associations." Trade associations are organizations—usually nonprofit organizations—designed to promote and conduct research on an industry.

Another industry variable that a business plan should address is the future prospects of the industry. It's one thing to know the historical background of an industry, as told by the revenue numbers from the past. It's another thing altogether to forecast where an industry is going and how large (or small) it's going to get. Forecasting industry trends can be done in several ways. A direct means of forecasting is to plot industry size over the past 5 to 10 years on a graph, and then extrapolate the graph forward. Another, perhaps more reliable method, is to examine what industry analysts are saying about an industry's growth potential.

10.1.7 The Management Team

The management team running a new technology venture is vital to its success. Many investors have commented that they scrutinize the management team as much as or more than they do the products or services offered by the venture.[11]

The discussion of the management team in the business plan should emphasize the relevant expertise that each team member brings to the enterprise. The section should not include each person's resume, but rather a brief (one or two paragraphs) biographical description of each management team member's experience, education, and duties. It's important to highlight the strengths each team member brings to the venture. For example, if someone has been designated as the "Chief Marketing

Officer," the biographical sketch for that person should include relevant information about his or her experience in developing and managing marketing campaigns—preferably, those that are similar to the marketing effort required for the venture.

In addition to the corporate officers, the business plan should list and highlight the strengths of other stakeholders who will contribute management or consulting expertise. For example, many firms establish a board of directors to watch over the company. Having an experienced and stable board can be a powerful advantage for a new venture, and can send a message to the investment community that the venture will be well managed. Board members should be listed and brief personal bios should be prepared for each member. In addition, it is useful to provide details about what the various board members contribute to the organization. Some will provide essential expertise while others will have key industry contacts or access to growth capital.

Some entrepreneurs elect not to establish a board of directors for a variety of reasons. The entrepreneur may not have enough ongoing business to entice people to join a board. Or the entrepreneur may simply not want to have to answer to a board of directors. Still, there can be many benefits to surrounding yourself with experienced and well-connected people willing to provide advice.[12] An alternative to a board of directors is an advisory board. An advisory board does not have legal responsibility for the financial health of the firm the way a board of directors does. Instead, an entrepreneur may elect to convene an advisory board purely for advisory purposes. This board would meet on an irregular basis, and would provide advice to the entrepreneur. Legal responsibility for the consequences of that advice lies solely with the entrepreneur.

If the firm has appointed legal counsel or works closely with any other type of professional services firm, these also can be listed in the "Management Team" section of the business plan. If the new venture has an accounting firm, management consulting firm, or other professional services that provide it with value, these should be listed. The various roles of the professional service providers should be briefly described.

10.1.8 The Marketing Plan

The marketing plan should describe how the firm intends to enter its chosen markets and how it intends to acquire customers. Market entry strategy describes the approach the firm will take to introduce its products. There is a wide range of market entry strategies that firms can use, including onslaught, guerrilla, and feint. These and other market entry strategies are covered in more detail in Chapter 11.

In addition to discussing the market entry strategy, the marketing plan usually also specifies the techniques that will be used to market the product to target customers, and the budget that will enable those activities. The techniques used to market the product/service to customers may range over such things as advertising, public relations, and promotions. Advertising includes the use of print, radio, television, billboards, flyers, and other media for alerting customers to the product/service, its features, and how/where they can purchase it. The marketing plan will not detail every advertising strategy the firm will use, but it should make clear that a strategy for the firm's emerging stage has been developed and is based on reliable marketing tactics.

Promotions and public relations supplement advertising in spreading the word about a new venture and its offerings. Promotions may include such activities as a grand opening, special sales events, or support of a favorite charity. Public relations involves such activities as speaking at industry events, writing articles for a local business journal, or sitting on nonprofit boards.

The marketing plan budget should specify the likely costs of the early stage advertising and promotions campaigns. A good marketing plan for a start-up venture will specify how the firm intends to leverage as much "free" advertising as possible. During the early stages of a venture, when money is likely to be tight, investors like to see the entrepreneur leveraging relationships, contacts, and other low-cost avenues to spread the word about the venture.

The marketing plan should be no more than three to four pages long. If the venture has developed some advertising, brochures, or other marketing collateral, it may be helpful to include some of the material in an appendix to the business plan.

10.1.9 Financial Projections

The financial projections included in a business plan usually include a sales forecast, income statement (also known as a profit and loss statement), cash flow, balance sheet, and capital expenditures budget. These various statements provide a forward-looking picture of the enterprise and how it will earn or lose money over time.

The sales forecast is a straightforward document that lists the various products/services that the company intends to sell, the price they will sell at, and the number of units of each item that will be sold each month.

The income statement is the document that records the venture's gross revenues per month and subtracts all monthly expenses, both variable and fixed. Variable expenses are usually those associated with the COGS. As we discussed in Chapter 3, subtracting cost of goods from revenue produces a figure that is referred to as gross profit (or gross margin, when expressed as a ratio of gross profit divided by revenue). Fixed costs are subtracted from gross profit to produce earnings before interest, taxes, depreciation, and amortization (EBITDA). Each fixed cost item (or line item) is listed on the income statement and its monthly cost to the enterprise is listed in the appropriate column.

The cash flow statement tracks the movement of cash into and out of the venture. This is an important statement for the business owner because it helps them track the venture's liquidity. Cash is an important resource to track because it is the firm's lifeblood. Cash is used to pay bills, purchase raw materials, invest in capital improvements, and pay employees. A firm can acquire cash in a number of ways, including income from operations, bank loans, equity investments, and income from investing activities. Each of these sources is tracked in the cash flow statement.

Finally, the balance sheet provides a snapshot of how the venture is using the money it earns through operations and money that has been invested or loaned. The balance sheet is a point-in-time view of the assets and liabilities of the organization. It is called a "balance sheet" because the firm's assets should be perfectly balanced with the firm's liabilities and owner's equity. That is, the balance sheet equation is:

$$\text{Assets} = \text{Liabilities} + \text{Owner's Equity}$$

The balance sheet enables a quick review of the debt to equity position of the company, and other so-called "key ratios." Using ratios enables entrepreneur to compare his venture to others in the industry. In addition to debt/equity, other key ratios that are regularly tracked by entrepreneurs from the balance sheet are liquidity, which reflect the firm's ability to meet current obligations as they become due.[13]

These various pro forma financial statements provide investors with a perspective of how the firm will evolve and grow over time. They are discussed at length in Chapter 13. The entrepreneur is wise to develop conservative financial statements that present numbers that can be achieved. Of course, if the entrepreneur develops numbers that are too conservative, he or she might risk losing potential investors because the deal is not "rich" enough. The entrepreneur must be cautious to develop financial estimates that are reasonable, yet compelling.

10.2 NETWORKING

Most entrepreneurs would agree that networking is an important part of their daily activities. **Networking** is defined as the art and practice of attending social events and connecting with individuals who may be able to assist with the entrepreneurial venture. Entrepreneurs use their networking skills for a variety of reasons: acquiring new business; finding new employees and other human resources; learning new ways of doing things that may be better than the old ways; and obtaining resources such as money, office space, second-hand telephone systems, and many other things.[14]

Scholars of entrepreneurship have been paying increasing attention to the social environment in which entrepreneurs build their ventures. In particular, they have begun to assess the importance of the entrepreneur's **social network** as a means to gain support, knowledge, and access to distribution channels. Research has shown that an entrepreneur's social network varies from time to time depending on the different phases of the venture's life. One study examined how entrepreneurs build and use social networks throughout various phases of new-venture development. The venture development phases identified by the investigators are:

- *The motivation phase*: This is the phase where entrepreneurs discuss the initial idea and develop their business concept.

- *The planning phase*: This is the phase where entrepreneurs seek the knowledge and resources needed to launch the venture.

- *The establishment phase*: This is the phase where entrepreneurs actually establish and run the venture.

Entrepreneurs need capital, skills, knowledge, and labor to start new ventures. The entrepreneur provides some of these resources, and some are gathered from the entrepreneur's social network. The webs of contacts that help bring about success are the **social capital**. Over time, entrepreneurs accumulate social capital, which is essential for starting and growing new ventures.[15] An entrepreneur's social network has several characteristics. One of these characteristics is the size of the network. Entrepreneurs can enlarge their social network to gain access to crucial resources. Another characteristic of an entrepreneur's social network is **positioning**. Entrepreneurs can position themselves within the social network so as to shorten the pathway to crucial resources.

Networking is usually a deliberate act on the part of the entrepreneur to increase his or her social capital and to improve positioning within the social network. As mentioned, the research is fairly clear that social networking is an important part of an entrepreneur's business success. Fortunately, there are some straightforward techniques that the entrepreneur can use to be effective in networking.[16] Below is a list of eight such techniques.

TECH TIPS 10.2

Effective Networking

1. *Make a great first impression*: When you meet someone for the first time, whether it is a male or female think "S.H.E. is the key." The acronym "S.H.E." stands for "Smile, Handshake, Eye Contact."
2. *Remember names*: A good way to remember names is to repeat the name aloud followed by a social comment such as "It's nice to meet you."
3. *Ask open-ended questions*: Open-ended questions tell people that you are interested in them, and they allow people to talk about themselves. Some examples include:
   - Tell me about your business.
   - What services do you provide?
   - How is your industry doing?
   - What are your biggest challenges?
4. *Be an active listener*: Active listening means paying attention to the person you are talking to and really understanding what he or she is saying.
5. *Establish common ground*: From the base of common interests, a deeper relationship can be developed that may include business and friendship. To find common ground, use techniques 3 and 4 above.
6. *Seek to help others first*: The most successful networkers think of the many ways they can help others before they help themselves.
7. *Be able to describe how your venture helps others*: When you tell others about your business, tell them how it helps others.
8. *Follow up*: Develop a system for recording information, including contacts and relevant notes, following a networking event. The notes you record could include such items as the common ground established with a contact, key dates and phone numbers, and names of people referred by contacts.[17]

---

1. Pin your nametag on your right shoulder so that when you shake someone's hand, her line of sight is aimed at your nametag.
2. Write your name in large letters on your badge.
3. Eat a snack before you go so you can better focus on working the crowd instead of juggling a plate of food.
4. If you smoke, opt for a nicotine patch. In today's business environment, smoking can make a negative first impression and it causes bad breath.
5. When attending meetings, seminars, or socials with a friend, plan to split up and meet new people. When you're by yourself you're more likely to approach others.
6. For an easy conversation starter, stand in the bar or food line where you can strike up a conversation with the person in front of and/or behind you.
7. Be friendly and initiate conversations with strangers. That's why you're there.
8. Wait for a pause in conversation before you begin speaking; interrupting someone is bad way to start a new relationship. Also, always be positive and upbeat. People are attracted to success.
9. When the other person's eyes begin to wander or you get the feeling she would rather be somewhere else, exit gracefully by thanking her and moving on.
10. Follow up on every promise or statement you make--the very next day. If you told someone you would send him a report, get it in the mail along with a handwritten note.

EXHIBIT 10.4

Tips for Effective Networking.

Source: DeFatta MC. Working the room. Office Pro 2001:June/July:16–18.

The results of networking can occur in a few months or over several years—nothing is guaranteed, but the potential is unlimited. Once someone knows who the entrepreneur is, evaluates his or her work ethic, and determines what the individual can do, the entrepreneur becomes a contact for that person—and vice versa. Exhibit 10.4 provides some additional tips for effective networking.

10.2.1 Serendipity

Entrepreneurs usually network with a purpose in mind, but not always. Sometimes it's important to network at specific types of events merely because the right people are attending. **Serendipity** is the term that refers to finding something that you need, but weren't necessarily seeking. Entrepreneurs often attend social functions and events that are linked to their industry based on the possibility of a serendipitous encounter.[18] They may meet someone who is looking for the products/services offered by the venture. They may meet someone who knows how to solve a difficult problem that the venture has been facing. Or they may meet someone who has a key contact with a customer they've been trying to land. The seasoned entrepreneur is always networking because a serendipitous encounter may happen anywhere, including on an airplane, at the symphony, during lunch, or standing on line at a baseball game.

10.2.2 Using the Internet to Network

In addition to the social networks that entrepreneurs develop in the physical world, there are a number of resources they can now use to build their networks in the cyber-world. For example, the following web services provide entrepreneurs the opportunity to connect with a wide range of other entrepreneurs, professional service providers, and consultants via the Internet:

- *LinkedIn*: LinkedIn is a provider of private online networking tools used by over 800,000 people in all industries and across the globe. By managing their existing network on LinkedIn, people can discover inside connections they never knew they had and reach the people they need to meet through referrals from others they already know and trust.

- *Biznik*: This site enables entrepreneurs to join their local business communities to connect and collaborate with nearby entrepreneurs. It also lists local seminars and events for entrepreneurs.

- *Econnect.entrepreneur*: This site, run by the magazine *Entrepreneur*, enables entrepreneurs to share content, join groups, and promote their respective businesses in an online environment.

- *GoBigNetwork*: This site focuses on fast growth ventures and enables them to network online to find capital, talent, and other resources.

Each of these services, and others like them, enable entrepreneurs to connect to resources that may not be needed in the present but may be useful in the future.

10.2.3 The Primary Objective of Networking

Remember, a large part of networking is simply "being present" so that the contacts that are being made can be leveraged if and when they are needed. As part of networking, the entrepreneur should also readily be willing to offer support, advice, and contact information to others who are also seeking to build their network. In fact, some networking consultants advise that the primary objective of networking should be to gain an understanding of others' concerns and problems. Then the entrepreneur can make an assessment regarding whether the contact would have any interest in the solutions the venture offers. Most people waste the few moments they have with new and existing contacts by focusing on themselves. Better to spend most of that time asking questions and collecting information about them.[19]

Establishing key contacts is the goal of networking. Once those contacts have been established, the entrepreneur must develop the relationship for business purposes. Each party will have business interests that will occasionally overlap and present the opportunity for working together.

10.3 RESOURCE AGGREGATION

The final prelaunch activity that we will discuss in this chapter is the acquisition of resources that are essential to the success of the venture.

10.3.1 Capital Resources

Technology entrepreneurs are nearly always preoccupied with the acquisition of capital resources. For most, there is never enough cash or long-term capital. The strategies involved in acquisition of capital to launch and operate a technology venture were discussed at length in Chapter 5. Here, we intend only to reiterate the importance of fund raising to the technology venture and the many sources of capital.

In the early stages of most ventures, the only source of capital is the technology entrepreneur's own personal resources, and perhaps those of family and friends. Later, as the venture begins to establish an intellectual property portfolio and, hopefully, revenue, other funding sources begin to open up. Angel capital is available in most major cities, and most have clear objectives of the types of ventures they want to invest in and how much capital they are interested in putting at risk. The ACA is a nationwide organization in the United States that lists most of the angel groups and their interests.

Venture capital becomes available to the technology venture after it has achieved certain milestones and is poised for rapid growth. This is often the stage where the founding entrepreneur is also in need of additional expertise in managing and operating a high growth venture. VC firms normally invest large amounts of capital, and usually are also heavily involved in managing the company thereafter.

The technology entrepreneur can also take advantage of debt financing, and should seek to develop banking relationships early on. Debt capital typically looks to the venture's cash flow when deciding on whether to lend, but that is not always the case. Developing a banking relationship before capital is required can pay off in the future. Some banks provide SBA loans that are backed by federal guarantees.

Other sources of new-venture financing include grants, such as the SBIR and STTR programs of the federal government. These can be sources of substantial capital, but also come with strings attached.

The technology entrepreneur should actively cultivate multiple sources of capital. Each source has strengths and drawbacks, and will need to be balanced with the goals of the venture, and the investing goals of those who already have capital committed to the venture.

10.3.2 Human Resources

The human resources we refer to here are top-level executives and board members—the talented people who add significant value to the venture and that often are referred to as **human capital**. During the early stages of a venture, individuals recruited to serve as executive officers will often request to see the business plan before making a decision about joining the firm. A venture that lacks such a plan will not be of much interest to seasoned, high-demand executives. They are interested in associating themselves with a winner, and can't afford to waste time with ill-conceived and/or poorly managed ventures.

In addition to executive officers, the venture may want to establish a board of directors and/or an advisory board. The difference between the two boards is primarily centered on fiduciary responsibility. The board of directors has it; the advisory board doesn't. Recruiting well-qualified candidates to a venture's board of directors can be an important factor in success.[20] For most board candidates, a business plan will be required for them to evaluate their willingness and ability to assist the entrepreneur. Although most firms protect board members from liability through appropriate insurance, well-qualified board candidates will take their board duties seriously and will not want a poorly planned venture to affect their reputation. Thus, the entrepreneur will likely be required to share his or her business plan with potential board of directors member candidates.

Although the advisory board does not have fiduciary responsibility, most individuals who are qualified to provide advisory assistance to a new venture will also want to review the venture's business plan. In fact, as a rule to thumb, if the entrepreneur asks a person to serve on the venture's advisory board and that person does not ask to see the business plan, the entrepreneur should think twice about including that person on the advisory board.

10.3.3 Organizational Resources

Organizational resources are those items that are required for people to do their jobs. This includes the normal tools of office work, such as desks, chairs, copy machines, and other things that people normally expect. Imagine showing up for work on day one of the launching of a new venture and there were no desks or other office equipment or furniture. Maximum productivity requires the normal tools for getting work done. Technology ventures often operate on very tight budgets in the early stages, and these items can be expensive.

Notice, however, that we did not say these standard office items need to be top of the line. Often, technology ventures are launched in the homes and/or garages of founders. Still, makeshift workstations should be set up, and opportunities for focused work should be created. For most office furniture and equipment, good deals can be found where there have been bankruptcies or liquidations of other ventures. In addition, most office furniture and equipment can be rented or leased. These can be less expensive short-term financing options, but might cost more in the long run.

Another organizational resource is office space. Office space can be a major hurdle for young technology ventures as they often require long-term commitments. Good office buildings in major cities will normally require the entrepreneur to sign a 3 year lease. In addition, as early stage ventures often have difficulty with cash flow, they will on occasion have trouble paying regular monthly bills. Landlords know this, and they attempt to ward off problems with lease payments through several well-known tactics. One tactic that is commonly used in early stage ventures is to require the lease applicant to submit a business plan as part of the application process. The landlord knows that a business without a plan is more likely to fail. Another tactic landlords often use to lessen the risk of leasing office space to early stage ventures is to require that the founders provide a personal guarantee on the lease. That means the founders will be required to continue making lease payments even if the business fails.

10.3.4 Technology Resources

The modern technology worker is accustomed to using a wide range of technology gadgets and tools to get work done. Technology ventures must provide these tools to compete with others in their industry, and to attract the talented workers who expect these tools to be part of their work environment. In most technology ventures, it is requisite that each employee be outfitted with his or her own personal computer, which may also include a desktop workstation and a separate laptop computer. In addition, today's mobile workforce also expects to be outfitted with the tools required for the virtual office, including BlackBerry devices, cellular telephones, personal digital assistants, and other things.

The technology entrepreneur will need to determine which tools are required for the unique products and services offered by the venture, and the unique work processes that are required to produce them. As with the organizational resources discussed above, the requisite technology resources can also be obtained via lease, lease to own, or financing options. Companies such as Dell Computer have divisions that cater to the emerging venture, offering each of these options as a means of limiting cash outflows in the short run. Of course, in the long run such options cost more, but to the small venture the preservation of precious cash in the near term may outweigh the long-term costs.

The technology required to operate a new venture likely will include various software applications as well. If the venture is ready for sales, it may want to investigate various sales force automation and customer relationship management packages. These are often available online and have a wide variety of purchasing options. From a straight-up commercial off-the-shelf software option, to individual and enterprise licenses, these software tools can help the emerging venture appear bigger than it is, and help it become more efficient.

10.4 NEW-VENTURE OPERATIONS

It is important for the technology entrepreneur to lay the foundations for operating success early in the venture's life. The foundations are put in place at launch; if they are appropriately tuned to the type of venture desired, the type of people required, and the economic environment in which the launch is taking place, explosive growth can occur. Some of the most well-known technology giants of the modern era, including Oracle, Dell, Apple, Microsoft, eBay, and many others are still led by the founders. The founders of these technology ventures established high-performance standards, unique organizational cultures that foster creative work, relentless focus on customer satisfaction, and attention to quality that separated them from the pack of competitors.

Launching a venture without performance standards and other foundational elements of long-term success runs the risk of squandering resources. Even worse, it is not difficult for the technology entrepreneur to lay effective foundations, prelaunch, that will pay dividends over the long run. Let's begin this discussion with an overview of how to set performance standards.

10.4.1 Performance Standards

A venture's **performance standards** are derived from its objectives and have many of the same characteristics. Like objectives, standards are targets; to be effective, they must be stated clearly and they must be measurable. Standards are the criteria that enable entrepreneurs to evaluate future, current, or past actions. They are measured in a variety of ways, including physical, monetary, quantitative, and qualitative terms.

Performance standards should be aligned not only with overall venture goals but also with the firm's reward and compensation plans. Research has determined that employees are more willing to accept and live up to performance standards if they are reasonable given the compensation and other rewards they receive. Standards that are overly rigorous compared with the compensation are likely to be ignored or violated. In setting performance standards, entrepreneurs must be aware both of organizational goals and of how they intend to reward people for achieving them.[21]

10.4.2 Information and Measurement

Information that reports actual performance and permits appraisal of that performance against standards is necessary. This usually involves some type of **objective metrics** but also could include subjective or **qualitative metrics**. Objective metrics are most easily acquired for activities that produce specific results. For example, production and sales activities have easily identifiable end products for which information is readily obtainable. The performance of a marketing department, research and development unit, or human resource department is more difficult to evaluate, however, because the outputs of these units are difficult to measure.

Qualitative metrics are those where information is gathered and can be analyzed, but not with precision. For example, many service businesses such as restaurants or hotels use customer surveys to gauge satisfaction. These surveys often are not very precise in what they measure, but they can tell the manager whether customers are generally satisfied or not. The "Research Reveals" element details five common flaws with this type of feedback.

While objective measurement of performance is necessary to manage a venture, too much measurement of performance can actually be detrimental to performance. Some organizations compile information in too much detail, needlessly driving up reporting expenses. Managers must learn to balance their need to know with the operating results of the company.[22]

Corrective action depends on the discovery of deviations from performance standards and the ability to intervene when necessary to improve performance. Individuals responsible for taking corrective action must know that (1) they are indeed responsible for the business process and (2) they have the assigned authority to take corrective actions.

Corrective action involves implementing methods that will provide answers to three basic questions:

1. What are the planned and expected results?

2. By what means can the actual results be compared with the planned results?

3. What corrective action is appropriate from which authorized person?

Often, the corrective action that is required involves disciplining an employee or employees. If someone is not performing according to expectations, entrepreneurs are obligated to step in and take corrective action. It's important for entrepreneurs with little managerial experience to remember that performance-based corrective action should not be construed as punishment.[23] Rather, corrective action should be delivered and received in a constructive manner. Yet the entrepreneur should leave no doubt about the performance expectations of a job. Consequences of nonperformance should be clear but not bellowed or broadcast to employees. Entrepreneurs should help employees focus on the rewards of high performance (the carrot) more than the consequences of nonperformance (the stick). Employees generally respond more favorably to promises of reward than threats of punishment.

10.4.3 Quality Standards

Despite the many perspectives on quality, the customer is the key perceiver of quality because his or her purchase decision determines the success of the venture's products or services, and often the fate of the venture itself. A consumer's perception of a product/service's "excellence" is generally based on the degree to which the product or service meets his or her specifications and requirements. Specifically, a consumer perceives "excellence" by evaluating one or more dimensions of quality, which are summarized in Exhibit 10.5.

In judging the quality of a television set, for example, a prospective buyer may examine how well the television set performs its primary function. (Is the picture sharp, the color vivid, the sound clear?) The set's extra features such as picture-in-picture may also be evaluated. Reliability may be a factor as well as serviceability— the convenience and quality of repair should a breakdown occur. The manufacturer's reputation for product quality may also influence the consumer's evaluation of the set's overall quality (perceived quality).[24]

Two points are noteworthy concerning a consumer's "perception of excellence" or quality. First, consumers emphasize different dimensions of quality when judging a product or service. Because of these differences in consumer preferences, a venture may choose to emphasize one or a few dimensions of quality rather than compete on all eight dimensions (Exhibit 10.6).

Second, perceiving "excellence" can be highly subjective. Some dimensions such as reliability or durability can be quantified by simply reviewing the product's

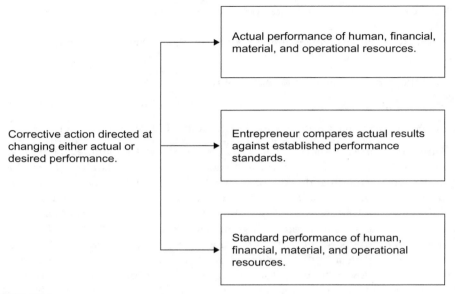

EXHIBIT 10.5

Basic Elements of Venture Control.

Dimension	Example
Performance:	The product/service's primary operating characteristics
Features:	Secondary, "extra" characteristics
Reliability:	Consistent performance within a specific period
Conformance:	Degree to which design and characteristics meet specific standards
Durability:	The length of a product/service's useful life
Serviceability:	The speed, courtesy, competence, and ease of repair
Aesthetics:	The looks, taste, feel, sound, smell of a product/service
Perceived quality:	Quality conveyed via marketing, brand name, reputation

EXHIBIT 10.6

Dimensions of Quality.

Source: Garvin DA. Competing on the eight dimensions of quality. Harvard Business Review 1987;November–December:101–9.

records. However, other dimensions such as aesthetics depend on personal likes and dislikes, which are highly subjective. Differences in preferences and the subjectivity of perceptions underscore the need for ventures to obtain accurate market information concerning consumer perceptions and preferences.[24]

10.4.4 Managing for Quality

Quality has been defined as a function of customer expectations. The greater the perceived value of a product or service, the higher is customer expectations for quality. A trip to Disney World inspires a higher perception of perceived value than a trip to a local amusement park. This expectation translates to pricing, management of the customer experience, and control methods and measurements used. Effective management of quality depends on a number of factors:

- Policy
- Information
- Engineering and design
- Materials
- Equipment
- People
- Field support

An integrated quality control system must focus on each of these factors. Below we explore each factor in the context of an entrepreneurial venture. As Tech Micro-case 10.2 illustrates, entrepreneurs can distinguish themselves in an already crowded market by focusing on quality.

TECH MICRO-CASE 10.2

Belle Baby Carriers Has Relentless Focus on Quality

Seth Murray, CEO of Belle Baby Carriers, is an avid rock climber. He approaches his business in much the same way he confronts a rock face: Grab only what you can see, touch, and test. When he set up his supply chain for his brand of modern but comfortable baby carriers, he used only local suppliers that he could visit and test. "We have three sewing companies and an order-fulfillment house all within 30 miles," Murray said. "That made it easier for us to develop both the product and the quality control program."

While most consumer goods companies are outsourcing production to China, Belle Baby Carriers spends the majority of its production dollars at home in Boulder, Colorado. On a per-piece basis, it may be more expensive, Murray says, but he has found a financial formula that makes it all worthwhile. While competitors are ordering by the container load from Asia, Murray is taking delivery of very small batches locally, which minimizes his cash needs and keeps inventory low. The 30-year old CEO said his 2008 sales were more than $300,000.[25]

Policy

The entrepreneur must establish policies concerning product and service quality. These policies specify the standards or levels of quality to be achieved and consistently delivered to the customer. Research has determined that quality policies are most effective if they are in place from the launch of the new venture.[26]

Entrepreneurs generally consider three factors in determining policy for quality: the product or service's market, its competition, and its image. An evaluation of the market provides an indication of customer expectations of quality and the price they are willing to pay for it. Quality levels provided by the competition also affect policy because the company's products or services must be competitive qualitywise to succeed in the marketplace.

Besides considering the market and competition, entrepreneurs must also consider the venture's image. Making a product low quality that is inconsistent with a firm's image may damage long-term interests.

Information

Information plays a vital role in setting policy and ensuring that quality standards are achieved. Concerning policy, accurate information must be obtained about customer preferences and expectations and about competitor quality standards and costs. Competitive **benchmarking** is an effective approach to obtaining valuable information about a competitor's quality standards and costs.[27] Also, new computer technology is enabling organizations to quickly obtain and evaluate information about the quality of products while they are being produced.

Engineering and design

Once management has formulated a policy concerning quality, it is the engineer or designer who must translate the policy into an actual product or service. The engineer/designer must create a product that will appeal to customers and that can be produced at a reasonable cost and provide competitive quality. This need to integrate cost, quality, and customer satisfaction has led to a process known as "concurrent engineering." This process replaces the traditional, sequential product design process with simultaneous design by integrated, multifunctional teams operating in cooperation with both customers and suppliers. The objective is to achieve "right the first time" production of new products in less time, with better quality and at lower cost.[28]

Materials

A growing number of organizations are realizing that a finished product is only as good as materials used to produce it. In this regard, many manufacturing companies are implementing a new precontrol strategy with material suppliers. They are reducing their number of suppliers, for example, weeding out the lower-quality vendors and focusing on developing effective, long-term relationships with the better ones.

People

While materials, design, and equipment are important ingredients to quality, the employee is the vital contributor to quality. Working individually, or as teams, employees take the ingredients and process them into a final product or service of quality. Entrepreneurs must train employees not only in the specialized knowledge of producing a quality product or service, but also in maintaining an attitude of quality.

Field support

Often, the field support provided by a venture determines a product's quality image (perceived quality). IBM, Enterprise Rent-a-Car, and Progressive Insurance have reputations for providing strong field support for their products. This is not to say that the products of these firms are necessarily the best in their industries. Many customers select IBM computers, Enterprise cars, and Progressive policies because the field support of these firms is considered excellent.

SUMMARY

This chapter provided the student of technology entrepreneurship with an overview of the elements that will be required successfully to launch a venture. We began with an introduction to the business plan. The business plan was described as a document that provides a roadmap to the entrepreneur, and to potential investors and other stakeholders. The technology entrepreneur should write a business plan that is exciting and inspirational, while at the same time being realistic. The goals, objectives, milestones, and financial outcomes depicted in the business plan can be used as measures of the entrepreneur's and the venture's future success. Setting goals too high means risking coming up short. Setting goals too low runs the risk of having the venture appear to be unattractive or unexciting.

We discussed each of the major components of a standard business plan, and some of the ways in which the entrepreneur should complete these sections. Although each business plan is unique—or should at least strive to be unique—there are some elements that are expected to be part of nearly any plan. The plan's executive summary should be written last, and should be no more than one or two pages in length. The executive summary certainly should highlight the size of the market being pursued, the unique nature of the products and/or services offered by the new venture, and the potential returns for investors and/or employees.

Next, the importance of networking to the success of the entrepreneur and the venture was discussed. Networking is the art and science of meeting other people and finding those who can provide assistance in achieving venture goals. We noted that networking requires the entrepreneur to get the word out about the venture's products/services and its aspirations. Of course, the entrepreneur should not reveal those elements that provide sustainable advantage. Still, in order to identify and qualify individuals who can assist the venture in its growth goals, the entrepreneur must be ready, willing, and able to discuss the venture with others. Tips for successful networking were provided.

Resource aggregation was the third component of prelaunch and launch activity that we discussed. The technology entrepreneur will need to aggregate capital resources, human resources, organizational resources, and technological resources. Each of these various areas of concern was discussed. One of the hallmarks of a successful entrepreneur is resourcefulness. The ability to work with limited resources, and to scrounge up the needed resources when they are needed, is essential to new-venture launch. The entrepreneur who lacks this capability must find others to join the team who are talented in resource acquisition and deployment.

Finally, this chapter examined the components that are necessary to develop a solid foundation upon which to build an operating organization. Many entrepreneurs relish the flexibility and excitement of the venture in its early days when everything is fluid and changing. However, in order for an organization to scale from small start-up to multi-million-dollar enterprise, a foundation must be laid. This foundation includes such things as policies, performance standards, quality control techniques, and organizational culture. Many of today's large technology ventures are still led by the founders whose unique vision has been sustained through time. Oracle, Microsoft (until recently), Apple, Cisco, Google, eBay, and many others are still led by their founders. Their unique visions and the foundations they laid for long-term organizational success were as vital at the beginning as they are today in the more mature organizations they have become.

STUDY QUESTIONS

1. Explain what is meant by the statement: "The business plan is the entrepreneur's 'road map.'"

2. What are the various sections of a business plan? What guidelines should an entrepreneur follow when writing the Executive Summary to the business plan?

3. What should be included in the Company Information section of the business plan? Why should the entrepreneur discuss the ownership structure of the firm?

4. What is the difference between a product's features and its benefits?

5. Explain some of the key elements of effective networking. Why does the technology entrepreneur want to learn how to network effectively? What are some potential outcomes of effective networking?

6. Explain some of the pricing options that likely will be available to the entrepreneur when purchasing organizational resources. When purchasing technology resources, what is the main concern for the emerging venture when examining these various options?

7. What are the technology resources that would be required for a 10-person computer gaming venture? How can the venture acquire and deploy these resources efficiently?

8. Why does a new venture need performance standards? Where do they come from?

9. What level of the organization is concerned about "customer satisfaction"? Explain your answer.

10. How can an entrepreneur use qualitative information to control the performance of a software venture? Of a manufacturing venture?

EXERCISES

1. *IN-CLASS EXERCISE*: One of the ways in which the modern generation of college students measures the importance of a person, topic, or issue is the number of "hits" it generates via a standard Web search. This exercise is meant to drive home the importance of quality management for entrepreneurs and small business owners. As a homework assignment, students should use their favorite search engine to find web sites that pertain to the implementation and utilization of quality management techniques in new ventures and/or small businesses. Each student should select two web sites that are informative about:

   - How entrepreneurs can benefit from using quality management techniques
   - How entrepreneurs should implement quality management in their ventures
   - Case studies of how quality management has worked for new ventures
   - Common do's and don't's for implementing TQM

   In class following the homework assignment, students should discuss the web sites they discovered (this is enhanced if the classroom has a live Internet connection). Discussion should center on the following:

   1. Which of the web sites seems to have the most comprehensive information for entrepreneurs about how to use TQM in a new venture?

   2. What are the primary messages concerning utilizing quality management in new ventures or small businesses? Does inserting a quality management approach seem to be worth the effort?

   3. What role does human capital play in an effective quality management philosophy in a new venture? What does that say about the importance of preliminary control of human resources?

2. *OUT-OF-CLASS EXERCISE*: For this exercise, each student or students working in teams should go to a nearby small business support center to review business plans that they may have on hand. Most large cities have federally supported SBA "one-stop shops," where aspiring entrepreneurs can go to get help to launch their ventures. Often, these support centers will have sample business plans on hand for individuals to review and learn from. Other places to go to review a business plan include federally sponsored Small Business Development Centers (SBDC), which are located throughout the United States. It is also possible to review business plans online. Many web sites supporting small business owners and entrepreneurs have sample business plans to review. For example, the U.S. SBA has a number of sample business plans on its web site at www.bplan.com/samples/sba.cfm. The commercial site www.bplans.com sells business-planning software and also has sample business plans to review.

   Students should review the plans and be prepared to address the following issues at the next class session:

   1. What products/services will the business sell? Were these clearly specified in the Executive Summary? Did the Executive Summary have a compelling proposition?

2. What industry does the firm compete in? How large is the industry? Is the industry growing or declining?

3. What customers does the company serve? What are the characteristics of those customers? How does the company intend to reach its primary customers?

4. What are the financial goals of the company? How many years did the company project its financial future? Do you think the company can reach its financial goals?

KEY TERMS

Barriers to entry: The barriers that exist within an industry that make it difficult for new firms to get started and to become competitive.

Benchmarking: The practice of reviewing the business practices of some best-in-class companies, and then copying those practices inside one's own venture.

Business plan: Explains in detail the business to a variety of stakeholders. It is often referred to as the new venture's roadmap.

Corrective action: Action taken to correct deviations from standard performance.

Human capital: Refers to the value added to the venture through the intelligence, ingenuity, and creativity of its people.

Networking: The practice of meeting new people with the intent of discovering ways to help one another.

Objective metrics: Metrics that have some quantitative value and standard means of measuring.

Performance standards: The standards established by the venture's leadership that the venture strives to achieve on a daily basis.

Positioning: Refers to how the entrepreneur is regarded by others within a network of individuals.

Qualitative metrics: Metrics that are used to evaluate the performance of the venture, but that do not normally have easily measured variables—such as "customer delight."

Serendipity: Refers to the eventuality of finding something useful that one was not necessarily intentionally seeking.

Social capital: The network of people an entrepreneur builds up and upon whom the entrepreneur can rely for advice and other forms of support.

Social network: The network of all individuals associated with an entrepreneur and his or her venture.

WEB RESOURCES

The web sites below are intended as destinations for your further exploration of the concepts and topics discussed in this chapter:

1. www.bplans.com: This site provides a wide sample of business plans in a number of technology and other industries.

2. http://www.sba.gov/smallbusinessplanner/index.html: This site has been set up by the U.S. SBA. It offers a number of references and other useful services to small and entrepreneurial business ventures.

3. www.businessplans.org: Similar to the bplans web site, this site has a number of sample business plans and useful templates for creating your own business plan.

4. http://tech.seas.harvard.edu/home: This is the home page for the Technology and Entrepreneurship center at Harvard University. It is a good starting point for understanding the relationship between high technology and entrepreneurial success in the global economy.

CASE STUDY

A European Concept Comes to America: The Launch of Zipcar

Antje Danielson's trip to Germany during the summer of 1999 turned out to be more than a tourist experience. While in Germany, Danielson took note of the new car-sharing concept that was taking off in many places around Europe. Car-sharing companies provide members with short-term, on-demand use of a fleet of private cars parked in convenient locations around a metropolitan area. Danielson, a PhD biochemist, saw the environmental implications of the car-sharing service, and believed it could also work in the United States.

Not having significant business experience, Danielson recruited her friend, Robin Chase, who has an MBA from the Sloan School of Management at MIT to help her start up a company. Chase agreed that the car-sharing concept could work, and she was confident they could build the technology infrastructure necessary. The women were encouraged to proceed quickly by Chase's former mentor and Dean of the Sloan School, Dr. Glenn Urban. He said, "This idea is much bigger than you are imagining. You have to do this at twice the speed and think twice as big."

During the fall of 1999, Chase refined the business concept, conducted market research, and wrote the business plan. During the spring of 2000, she designed and wrote the company's web site, and began working on the online reservation system with the help of contract engineers.

Chase discerned that the car-sharing concept was best suited to urban locations with dense bases of potential customers, where parking is expensive, and the need to own a car is limited. Her market research indicated that, among urban dwellers, college-educated individuals were most likely to use the system. She noted that the car-sharing concept is best suited for those who log fewer than 6,000 miles per year driving. She also noted that market penetration for car-sharing companies in Europe was low, comprising about 0.01% of all drivers. There were approximately 200 car-sharing companies in Europe, operating in 450 cities. Chase's research also indicated that the U.S. market was potentially large and virtually pristine.

The business plan for the first car-sharing service in North America was completed in December 1999. The service Chase envisioned would deliver convenience, ease of use, travel freedom, and hassle-free transportation for urban dwellers. Chase spent months modeling different pricing assumptions and cost structures. Her first business plan assumed:

- Potential users would be required to become members and pay a $25 nonrefundable fee.
- Members would be required to pay a $300 refundable security deposit.
- Members would also be required to pay a $300 annual subscription fee.
- Members would be charged for driving time at the rate of $1.50/hour and $0.40 per mile.

Users of the company's vehicles were expected to handle light maintenance by themselves. Drivers would be required to put fuel into the vehicles and get reimbursed from the company. They were also required to keep the vehicles clean, and to handle personally any parking or traffic tickets. The car had to be returned to its original location before the reservation time expired. There was a $20 fine for late returns.

With the business plan complete, Chase tested it by asking trusted advisors for feedback. She also used the opportunity to begin the fundraising process. Her list of potential investors included professionals, classmates, family, and friends. Chase called a contact at a VC firm whom she had met socially. The contact agreed to meet with her, but did not fund the venture. Nonetheless, Chase learned a great deal from talking with her VC acquaintance, and modified her business model following her visit.

During the initial fundraising effort, Chase was not taking any salary, and she and Danielson were financing initial expenses out of their own pockets. Eventually, Chase was successful in securing a $50,000 investment from a former MIT classmate in the form of a convertible note. The terms of the investment stipulated that the principal amount would accrete at 1% per month, and the investor would be able to convert the cumulative principal into common stock at the valuation established in the contemplated Series A investment.

With the $50K, Chase and Danielson began to build the technology platform. The vision was to enable users to book a vehicle online, access the parked car, and drive off. In addition, the system was to capture information about when the car was returned, and the number of miles the user had driven. The solution became a wireless system that would transmit data between the car and the central server that would authorize users and log odometer readings, mileage, and time. They found a software engineer who was perfect for the task and developed a prototype system.

Meanwhile, the fundraising continued. Throughout 2000, Chase and Danielson presented their idea to a number of potential investors, including angel groups. However, no one wanted to risk investing in the start-up. Chase recalled this period as one where they were "often behind and occasionally quite desperate." The two founders continued their personal investing, and they stretched payables as long as they could. During this period, Chase was also building the organization, bringing on a management team. She also signed contracts with large institutions for parking their vehicles.

Naming the company was also a difficult task. Chase and Danielson knew they needed to build a brand, and they needed a name that would communicate the value and concept clearly and simply. Eventually, they chose the name "Zipcar." They liked the name because it conveys friendliness, convenience, ease of use, and affordability. They also wanted to appeal to users who considered themselves smart and forward thinking.

By June 2000, Chase changed the Zipcar pricing model. Discussions with potential users revealed the $300 annual fee to be too high. Consequently, the fee was lowered to $75 per year, and a tiered pricing structure was created. Per hour pricing would range from $1.50 to $7.00, depending on the cost of parking in an area. With these changes, Chase assumed the company would generate about $5.50 per customer per usage hour.

By September 2000, Zipcar had raised an additional $300,000 under the same convertible note terms as the original investment. The company officially "launched" in June 2000, although its technology platform had not been completed.

Postscript: In 2007, Zipcar merged with rival Flexcar, retaining the Zipcar name. Since its inception in 1999, Zipcar has signed nearly 200,000 consumer and business members/drivers to share 5,000 vehicles. On average, more than 7,500 new members join Zipcar's service each month. Zipcar also serves more than 600 businesses that benefit from having access to cars dispersed throughout busy city centers. The firm's car-sharing programs are also found on more than 25 university and college campuses.

Questions for Discussion

1. How did the founders come up with the idea for Zipcar? Do you think there are any other valuable ideas waiting to be discovered in this fashion? Explain.
2. What is meant by the term "convertible note"? Why do you think the founders used this approach to the seed stage financing? Do you think they could have made a better choice?
3. Define the value proposition for Zipcar. Try to develop an elevator pitch that could have been used back in 1999 that may have helped in the early fundraising. Share your pitch with the class.

Source: Adapted from Hart M Roberts MJ, Stevens JD. Zipcar: refining the business model. Harvard Business School Case, 2005;May 9; Levine M. Share my ride. New York Times 2009; March 5: MM36.

ENDNOTES

[1] Sahlman WA. *How to write a great business plan*. Cambridge, MA: Harvard University Press; 2008.

[2] Henricks M. Do you really need a business plan? *Entrepreneur* 2008;**36**(12):92-5.

[3] Mainprize B. The benefit: a well-written business plan is to an entrepreneur what a midwife is to an expecting mother. *J Private Equity* 2007;**11**(1):40-52.

[4] LeBrasseur R, Zanibbi L, Zinger TJ. Growth momentum in the early stages of small business start-ups. *Int Small Bus J* 2003;**21**(3):315-28.

[5] Larson E. The best-laid plans. *Inc.*, February 1987, pp. 60-4 Posner BG. Real entrepreneurs don't plan. *Inc.*, November 1985, pp. 129-35.

[6] Osborne RL. Planning: the entrepreneurial ego at work. *Bus Horiz* 1987;**30**(1):20-4.

[7] Horzoni AM, Sutton GS, McMinn RD, Lucio W. Business plans for new or small businesses: paving the path to success. *Manage Decis* 2002;**40**(8):755-64.

[8] Smith F. How I delivered the goods. *Fortune Small Bus* 2002;**November**:2.

[9] Morton LP. Segmenting publics: an introduction. *Public Relations Quart* 1998;**Fall**:33-4.

[10] Ganco M, Agarwal R. Performance differentials between diversifying entrants and entrepreneurial start-ups: a complexity approach. *Acad Manage Rev* 2009;**34**(2):228-52.

[11] Beckman CM, Burton MD. Founding the future: path dependence in the evolution of top management teams from founding to IPO. *Organ Sci* 2008;**19**(1):3-24.

[12] Terry P, Rao J, Ashford SJ, Socolof SJ. Who can help the CEO? *Harv Bus Rev* 2009;**87**(4):33-40.

[13] Rushinek A, Rushinek SF. Using financial ratios to predict insolvency. *J Bus Res* 1987;**15**:74-7.

[14] Bottles K. Focus, exchange, and trust mark successful entrepreneurs. *Physician Exec* 2002;**28**(2):71-3.

[15] Davidsson P, Honig B. The role of social and human capital among nascent entrepreneurs. *J Bus Venturing* 2003;**18**:301-31.

[16] Baron RA, Markman GD. Beyond social capital: how social skills can enhance entrepreneurs' success. *Acad Manage Exec* 2000;**14**(1):106-16.

[17] Takash J. Networking success: discover the tools you need to get to the top. *Bus Credit* 2004;**106**:24-5.

[18] Cooper AC, et al. Entrepreneurship. *Acad Manage Proc* 1992:74-89.

[19] Cook C. 7 Secrets of effective networking. Retrieved from: http://www.charliecook.net/networking-secrets.html.

[20] Shultz SF. Developing strategic boards of directors. *Strateg Finance* 2003;**85**:1-4.

[21] Kane JS, Freeman KA. A theory of equitable performance standards. *J Manage* 1997;**23**(1):37-58.

[22] Kofman C. How measurement can undercut performance. *Bank Strategies* 1997;**July/August**:44.

[23] Mills B. Getting comfortable with disciplining employees. *J Bus* 2001;**January 25**:B10-11.

[24] Garvin DA. Competing on the eight dimensions of quality. *Harv Bus Rev* 1987;**November/December**:101-9.

[25] Worrell D. Paying for quality over quantity. *Entrepreneur* 2008;**36**(8):55.

[26] Duffy G. Quality from scratch: a model for small business. *Qual Prog* 2004;**37**(7):27-35.

[27] Ann Klaus L. Benchmarking is still a useful quality tool. *Qual Prog* 1997;**November**:13.

[28] Singh KJ, Lewis JW. Concurrent engineering: institution, infrastructure, and implementation. *Int J Technol Manage* 1997;**14**(6-8):727-38.

Going to Market and the Marketing Plan

After reading this chapter, students should be able to:

- Appreciate the need for developing a strong go-to-market strategy
- Identify the best market segment to enter
- Recognize different aspects of a marketing plan
- Understand all aspects of the marketing mix—product, price, distribution (channel and physical), and promotion

---

TECH VENTURE INSIGHT

Great Products Also Need Great Marketing to Be Successful

Born in New York City, but raised by his aunt and uncle in Chicago, Larry Ellison first encountered computer programming at age 20 and loved it. He decided to pursue his passion and enrolled at the University of Chicago. After only one term, though, he found the college environment much too rigid and stifling. When Ellison heard of programming opportunities in Silicon Valley, he decided to drop out of school and move to California.

In California, Ellison joined the Ampex Corporation, where one of his major projects was working on a database for the CIA, which he named "Oracle." While building his knowledge about computer programming at Ampex, Ellison was inspired by a paper titled, "A Relations Model of Data for Large Shared Data Banks" written by Edgar F. Codd. Ellison had discovered that many companies lacked services that could store, process, and extract insights from large databases.

Ellison eventually quit Ampex, and with $2,000 in hand and two of his Ampex colleagues on board, he founded Software Development Laboratories (SDL). He took the CIA project with

---

him, aiming to create a new type of relational database management system (RDBMS). SDL finished the CIA project a year ahead of schedule and used the remaining time and money to work on the first RDBMS for a commercial market. Ellison had heard about the IBM System R database, which was also based on Codd's theories. His idea was to make his system compatible with IBM's. IBM, however, had decided to keep its error codes secret, thus forcing Ellison to create an independent system.

In 1979, SDL was renamed Relational Software Inc. (RSI), whose main focus was the fast-growing minicomputer market. Ellison's vision was to create portable software that would be compatible with IBM's structured query language (SQL), but aimed at minicomputers. Oracle 2 (there was no release of Oracle 1, suggesting that bugs and kinks had been worked out) was the company's first release. It dominated the mid-range computer and microcomputer arena because of its flexibility and usability factor for businesses that used the product for departmental and decision-support activities on their VAX minicomputers.

With a great product, however, Oracle lacked knowledge of marketing techniques. For instance, it tried to convince potential customers to buy large amounts of software all at once. It would also book the value of the sale in the current quarter to increase its quarterly bonus, even if it had not yet been completed. If the sale didn't happen, the books were not accurate, and one year almost all the revenue recorded had not been realized. Ellison has admitted that the sales issue was an "incredible business mistake," highlighting an important truth that affects many technology companies that fail to develop a solid marketing plan.

Source: Adapted from How he made his pile: Larry Ellison. Management Today December 2004:3; Cortese A. My jet is bigger than your jet. Business Week August 18, 1997:126; The house that Larry built. Economist February 19, 1994:73.

INTRODUCTION

Marketing is a central part of every society, influencing everyone to some extent.[1] Marketing has an increasing impact on the decisions being made not only by buyers and sellers but also by a diverse group, including physicians, lawyers, politicians, and even the clergy. For example, a politician may be trying to decide how to allocate his or her promotional budget during the last 2 weeks of an election campaign. Should an expensive television advertisement be used? Should there be more radio spots? Or would additional advertisements in various newspapers be a better use of the available money? Other decisions need to be made on where he or she should spend valuable time and resources.

   In this chapter, you will learn about how buyers of technical goods are confronted with related yet different aspects of marketing. Should sources of supply be switched because of the superior quality and lower price of a new supplier?

Will a new firm be consistently able to supply the quantity needed at this price? Will the new supplier deliver the product on time, in order to avoid any costly plant shutdowns that would result from an out-of-stock condition? An industrial buyer must carefully weigh many aspects of marketing in selecting the correct product and supplier.

11.1 MARKETING

What exactly is this activity of marketing that has such an impact on our lives? The answer to this question depends on the perspective of the individual. Law, finance, economics, operations, and consumers all view marketing from a different vantage point.[2]

A broad definition of marketing is more applicable in today's rapidly changing hypercompetitive environment. Marketing actually begins with an idea for a product or service and ends only after the consumer has had sufficient time to evaluate the product thoroughly, which might be months or years later. The American Marketing Association defines the term "marketing" as: "an organizational function and a set of processes for creating, communicating, and delivering value to customers and for managing customer relationships in ways that benefit the organization and its stakeholders."[3]

In standard marketing classes, students are taught the so-called **Four P's: product, price, placement** (commonly referred to as **distribution**), **and promotion**. The Four P's are comprehensive of the considerations that a venture must address when attempting to market a product. As indicated in Exhibit 11.1, each element has its own variables where decisions are required to achieve customer satisfaction. Each of these will be discussed individually later in the chapter.

The product area includes all aspects that make up the physical product or service being offered for sale. Decisions need to be made on quality, assortment, the breadth and depth of line, warranty, guarantee, service, and the package. All these aspects make the final product or service more (or less) appealing to the target market.

Closely related to the product and its mix is the price. While probably the least understood of the elements, the price of the product greatly influences the image of the product as well as whether or not it will be purchased. The price that is established needs to take into consideration the 3Cs—cost, competition, and the consumer.

The third basic element of the marketing mix—distribution—covers two different areas. The first area—channels of distribution—deals with the institutions such as wholesalers and retailers that handle the product between the firm and the consumer. Physical distribution, the second area, deals with the aspects of physically moving the product from the firm to the consumer. This includes such things as warehousing, inventory, and transportation.

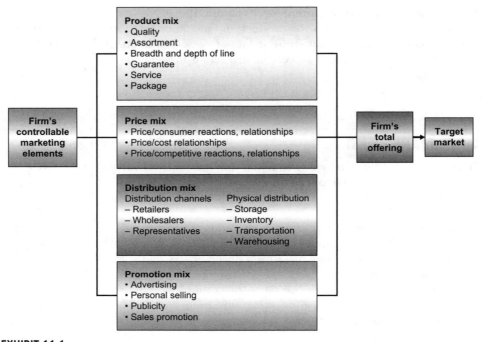

EXHIBIT 11.1

Elements of the Marketing Mix.

The final element—promotion—involves policies and procedures related to four areas:

1. *Personal selling*: Emphasis on personal selling and the methods to be employed in the manufacturer's organization and in the trade.

2. *Advertising*: Policies and procedures relating to budget, message, and media.

3. *Sales promotions*: Policies and procedures relating to budget and types of consumer and trade promotions and displays.

4. *Publicity*: Policies and procedures relating to a comprehensive program for good media coverage and a strong company image.

Most technology entrepreneurs are familiar with various promotional techniques, but unsure about how they work or how to create a favorable mix of options.

In this chapter, we will discuss elements of a marketing plan. You will learn how to set a budget for marketing expenses, how to set a basic marketing strategy, and how to develop a marketing message that is consistent with your chosen strategy. Let's turn our attention first to what is called the "marketing concept."

11.1.1 **The Marketing Concept**

More than any other functional business area, marketing helps develop the goals and direction of the organization. Overall, the basic philosophy motivating the firm's management should focus on the satisfaction of the target customer—the heart of the marketing concept. Under the marketing concept, the customer becomes the dominant focal point of the firm, with all the resources and activities in the firm directed at generating customer satisfaction.

The adoption of a **marketing concept** by many firms in the United States as well as the rest of the world was precipitated by several factors.[4] The marketing concept holds that firms should first consider needs of customers before producing products and services. This is in contrast to other perspectives, such as the production concept which first asks whether a product can be produced, or the sales concept which first asks whether a product can be sold.

The marketing concept has taken root over the past few decades for a number of reasons. First, there has been an increase in the intensity of competition in both national and international markets, forcing organizations to place greater emphasis on customers and their satisfaction. A second factor that has caused the adoption of the marketing concept is the high level of customer knowledge and sophistication. Customers are more aware of product alternatives and price than ever before. Just count the number of shoppers carrying calculators and coupons in your local supermarket.

A high level of customer awareness has evolved in the United States and other developed countries due, in part, to good communication systems that inform them of faulty products, price increases, and fraudulent advertising. The resulting increased awareness has enabled customers to be much more discriminating in their purchase decisions.

An increase in production capabilities in conjunction with the development of worldwide markets has also led to the adoption of the marketing concept. Increased production capacity has led to **economies of scale** in production—a decrease in per-unit production cost as the total number of units is increased. The increased units produced can only be sold through successfully selling to mass markets by focusing on customer satisfaction.

11.1.2 **Market Segmentation**

The process of **market segmentation** is extremely important.[5] Technology entrepreneurs use market segmentation to identify a target market. The difference between the market and a venture's **target market** is profound. The **total available market** (TAM) for a product or service is everyone in the world who would purchase the product or service if they could afford it. The target market, on the other hand, consists of the subset of potential customers that can be made aware of the venture's products and services and that can be persuaded to purchase those offerings. The target market is defined as some fraction or portion of the overall market.

Once the target market has been correctly defined, it is much easier to develop a marketing strategy and the appropriate combination of product, distribution, promotion, and price to reach that market effectively. The target market that a venture sells into is known as the **served available market (SAM)**. Even though a market is rarely oversegmented, care should be taken to delineate a market as necessary.

Segmentation techniques for all three types of markets (consumer, industrial, and government) are indicated in Exhibit 11.2. The basic segmentation criteria—demographic, geographic, psychological, volume of use, and controllable marketing elements—can be effectively used to define a target market, whether the overall market be consumer, industrial, or government. It is particularly important to recognize and implement the capability of segmentation in the industrial and government markets. Firms in the consumer markets are generally much better at segmenting their markets.

The first step in determining the venture's target market is to segment the overall market. This is a process of identifying various subsets of the overall market, and

Segmentation criteria	Basis for type of market		
	Consumer	Industrial	Government
Demographic	Age, family size, education level, family life cycle, income, nationality, occupation, race, religion, residence, sex, social class	Number of employees, size of sales, size of profit, type of product lines	Type of agency, size of budget, amount of autonomy
Geographic	Region of country, city size, market density, climate	Region of country	Federal, state, local
Psychological	Personality traits, motives, lifestyle	Degree of industrial leadership	Degree of forward thinking
Benefits	Durability, dependability, economy, esteem enhancement, status from ownership, handiness	Dependability, reliability of seller and support service, efficiency in operation or use, enhancement of firm's earning, durability	Dependability, reliability of seller and support services
Controllable marketing elements	Sales promotion, price, advertising, guarantee, warranty, retail store purchased service, product attributes, reputation of seller	Price, service, warranty, reputation of seller	Price, reputation of seller

EXHIBIT 11.2

Market Segmentation Techniques by Type of Market.

then determining which of these is going to provide the greatest opportunity for the venture's products and services. The venture's market can be segmented using a variety of criteria. One of these is **demographics**. The term "demographics" refers to the characteristics of the target market, usually defined in quantitative/statistical or qualitative categories. In business to business marketing, demographic data can be collected on firms in categories such as:

- Age of the firm
- Market size served
- Total sales
- Operating income
- Number of employees

There are many companies that specialize in providing demographic statistics compiled from government agencies and private studies by firms such as Gartner Group, Forrester Researcher, or E-Marketer.com. Demographic data can be just as important to ventures that market products and services to industrial customers as it is often to ventures that market directly to consumers.[6]

Other criteria that entrepreneurs use to segment markets and identify a target market are **geographic location** of the market and **psychological profile** of the market. Geographic location, as might be guessed, concerns where the customers are physically located. A market's psychological profile refers to the activities, interests, opinions, attitudes, and values of the people that comprise a market segment. Understanding these characteristics of a market segment is an important part of developing a volume strategy whether or not the target market is chosen for psychographic reasons.[7]

In the case of segmentation based on benefits, markets are identified by the differential benefits created by the same product. For example, a new line of laptop computers might be marketed to young people by virtue of the processing power for computer games that is built into each unit. On the other hand, that benefit would not necessarily appeal to seniors. In that case, the power of the processor to render pictures of grandchildren and enable long-distance communications may be emphasized instead.

Segmenting based on controllable marketing elements refers to the potential to leverage communication and distribution channels. For example, some products are easily marketed through television shopping networks. This type of marketing appeals to a segment of the overall market, which is predictable to a high degree. The people who manage home shopping can usually determine with a high degree of accuracy whether a product is a fit for that type of marketing, and how well it likely will do in terms of net sales.

Based on this segmenting process, the entrepreneur can determine which segment should be the target market. In addition, the process of segmentation provides details about important target market characteristics. With this information, the entrepreneur can determine the most effective way to bring the products and services to market, and the most effective means for positioning the products and services in the context of any competitors that may already be pursuing the same segment.

11.2 PRODUCT

The concept of product varies from one venture to another as well as from one customer to another. In order to innovate and manage products successfully, the technology entrepreneur must define what an acceptable product is from the perspective of the customer. It is important to understand the nature of the product—the total offering of the firm that satisfies the needs of the customer. This concept of product is broad and focuses on the most important aspect, the need to satisfy a customer. This satisfaction can be derived from such things as the benefits of product use, pride of ownership, or dependable, reliable performance. As the term **product** is defined here, it can mean a physical product or a service. In addition, the product encompasses all those ancillary items that make it what it is, such as the package, the brand, the service, the quality, the options, the breadth of the line, the guarantee, and the warranty. It is the basic unit of sale for a venture.

11.2.1 Product Planning and Development

One of the most useful tools for growing a business is the product planning and development process (see Exhibit 11.3). Although the actual process and the time and sales involved in each step vary greatly, not only from industry to industry but within a given industry, the overall process still provides a framework for evaluating and developing new products as well as formulating basic marketing strategies. In order to implement the product planning and development process, it is essential that an effective method for obtaining new product ideas be established.

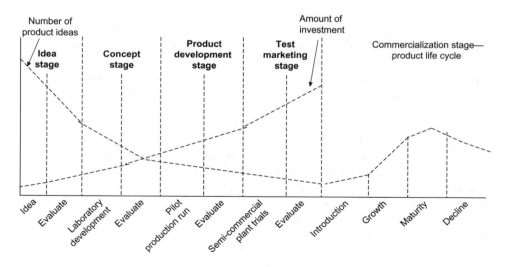

EXHIBIT 11.3

The Product Planning and Development Process and Product Life Cycle.

Ideas for new products can emerge from a wide variety of sources—customers, competition, distribution channels, federal government, research and development, and company employees.[8] Many firms ask their customers to submit ideas either informally or through an established process. (These ideas can even be in the form of complaints.) New ideas often emerge from the company's attempt at solving the problem. For example, General Foods developed a new cereal package after determining from customer complaints that their cereal box was too tall for the standard shelf and could tip over. Tech Tips 11.1 provides some additional ideas for generating new product concepts.

TECH TIPS 11.1

Tips for Generating New Product Ideas

1. Solve an existing problem for people. There are thousands of problems in the world. Create a product that can provide a solution to one of those problems.
2. Find out what the current hot trend is. You can find out what the new trends are by watching television, reading magazines, and surfing the Net. Create a product that's related to the current hot trend.
3. Improve a product that is already on the market. You see products at home, in ads, at stores, etc. Take a product that's already out there and improve it.
4. Create a new niche for a current product. You can set yourself apart from your competition by creating a niche. Your product could be faster, bigger, smaller, or quicker than your competitor's product.
5. Add on to an existing product. You could package your current product with other related products. For example, you could package a football with a team jersey and football cards.
6. Reincarnate an older product. Maybe you have a book that's out of print and is no longer being sold. You could change the title, design a new front cover, and bring some of the old content up-to-date.
7. Ask your current customers. You could contact some of your existing customers by phone or e-mail and ask them what kinds of new products they would like to see on the market.

Once a new product idea has evolved, it is very important for the idea to receive a careful and thorough evaluation. Entrepreneurs should screen out as many new product ideas as possible in the early stages, allowing more time and money to be spent on products with a greater probability for market success.

The new product ideas that successfully make it to the concept stage should be further evaluated, analyzed, and where appropriate, refined for market success. This often takes the form of determining consumer acceptance by presenting drawings or explanations of the new product to various consumers and members of the distribution channel. Even when there is no actual product prototype, this procedure will obtain valuable information from consumers.

Once the new product concept has received a positive evaluation, it is necessary to transform the concept into a physical product or prototype. Whenever possible prior to mass production, the technical and product feasibility of the item should be determined through the development of a prototype that has customer appeal.

Test marketing is not used for all products prior to commercialization. It is a costly and time-consuming process that gives the competition time to react to your initiatives. Test marketing is frequently used for industrial products, regional products, and specialty goods where the information gained is worth the cost of the test. In some highly technical, capital-intensive industrial products, a test market is not even physically possible.

Only a few new products actually reach the market. The capital expenditures for launching the product into the market are extensive and should only take place for those new products that have a high probability of success. In addition to the costs of new equipment and facilities for production, the company has the marketing expenses of training salespeople, extensive introductory advertising, and sales promotion in the trade.

11.2.2 Product Life Cycle

There is probably no marketing concept more widely known and used than the product life cycle. This is the process that a product enters upon commercialization or introduction into the market. The life cycle has five stages. Each stage—introductory (or market development) stage, growth stage, competitive turbulence stage, maturity (or saturation) stage, and decline stage—requires a specific marketing strategy so that sales and profits can be maximized.

The technology entrepreneur should plan a marketing strategy that will be implemented for the product at each stage before the product is ever introduced. In addition, most new technical products need some significant recycling after a 1½- to 3-year period if market leadership is to be maintained. Additional flavors, added packs of items, new packaging, new features, and new advertising have all been used to positively affect this needed recycling. Exhibit 11.4 is a standard depiction of the product life cycle.

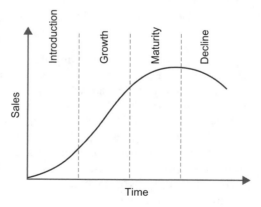

EXHIBIT 11.4

Product Life Cycle Curve.

11.3 **PRICE**

Even though price may be considered the most important of all the elements in the marketing mix, it is usually something that technology entrepreneurs know the least about. In fact, in many technology companies, few executives even know how the price of the product is established. In order to begin to understand pricing, it is important first to have an overall view of the economic dimension underlying price and price theory.

11.3.1 **Economic Dimension of Pricing**

In a free market system, such as the one operating in the United States, price is an important factor in controlling the supply and demand of goods. Although a totally pure economic system of supply and demand rarely exists in the short run, it is important to understand its basis as this will provide the foundation for understanding the economic dimension of pricing.

To have an economic free market system in an industry, several factors need to exist: a large number of manufacturers (sellers), a larger number of buyers, no dominant or very large manufacturer that can control price, manufacturers' ease in entering or leaving the industry, and large differentiating factors between product or producers. While rarely do all these factors exist in a specific industry situation, when they do, a rising price will entice more volume from present manufacturers and attract more manufacturers into the industry. Similarly, a decreasing price will cause sales volume to rise in the industry as the goods will have increasing demand due in part to the lower price. More goods will be sold at the lower price than at a higher price. If the supply of an item is very large in relation to the demand, the price tends to decrease. When market conditions allow the total revenue to increase due to a greater number of units being sold as the price goes down, the price is elastic.

11.3.2 **Pricing Objectives**

The economic dimension of pricing requires taking into consideration the pricing objectives and the 3Cs of pricing—cost, competition, and the customer.

There are many different pricing objectives that can be used on a single product, a product line, or on all product lines of a company. While some companies use different objectives in different parts of the company's operations, other companies use the same general objectives across all products and operations. The similarity of the pricing objectives used depends on the product and market conditions. Specific pricing objectives can be classified in terms of market penetration, market skimming, early cash recovery, return on investment, and other nonfinancial objectives (product line promotion, market share, meeting competition, and image).

Cost

Cost, the first C of pricing, is the floor that is used in determining the price of the product. In the long run, the price of the product cannot be below its cost if the firm expects to achieve a profit. Every product need not provide an established return, but it is rare that a product would be priced below the amount needed to recover its costs. Cost considerations should reflect the historical costs of the product as well as anticipated costs given volume production.

One method for evaluating the cost and its impact on price is **cost-plus-pricing**, which can be done in several ways depending on the cost basis used. The cost base most frequently used, particularly by manufacturers, is total cost. In cost-plus-pricing, a percent markup is applied to the total cost of the product to determine its selling price. For example, if the percent markup is 100% of the cost to produce the product, which is $25, it would be priced at [$25+100% ($25)]. The manufacturer is using a 100% markup on the $25 cost, yielding a $25 profit margin and a $50 selling price to either the end user or someone in the distribution system.

Many industrial firms establish a price to achieve a target return on investment, such as 20% before taxes. Then the marketing strategy is developed to achieve this rate of return using the price that will achieve the target rate of return specified.

Firms must also understand that customers might be willing to pay some of the costs involved with products and services. For example, Tech Micro-case 11.1 discusses a situation in which Merck was able to transition its customers from paying a price that included shipping to the same price with shipping as an extra fee.

TECH MICRO-CASE 11.1

Merck Switches Shipping from Free to Fee

Merck is one of the world's largest pharmaceutical companies. Despite its size, it is constantly on the lookout for ways to improve its profit margins. In one of the company's product categories, its French subsidiary had a long-standing tradition of including delivery in its product price for customers. Because specialty chemicals are high in value but low in volume, Merck had never questioned its responsibility to assume transportation and insurance costs, which represent a tiny fraction of the amount invoiced. And because no shipping costs were itemized, customers were unaware of the value Merck provided. A few years ago the company put this tradition to a test: Managers randomly selected 100 customers and changed the terms of delivery from "shipping and insurance paid" to "ex works," though the bottom line barely changed. Ninety percent of those customers readily paid the additional charges, seemingly without noticing. Of the 10% that recognized the change, only half insisted on returning to the prior terms of payment. Merck re-established the original terms for those customers—but it had succeeded in managing the transition from "free to fee" for the other 95%. Once the new billing terms had been rolled out to the entire customer base in France, Merck's profitability in this product category improved significantly, even though the cost to customers was minor. Switching services from free to fee clarifies the value of the assets involved for both firms and customers.[9]

Competition

In addition to cost, it is important that the technology firm evaluate the competition, the second C of pricing. The prices of competitive products establish the benchmarks for a pricing decision. You have to decide whether to meet, undersell, or oversell the market prices in existence, knowing that the consumer will compare your price to the prices of those other products, depending to some extent on the amount of product differentiation.

The price of competitive products is also important in evaluating the competitive reaction to a possible price change being considered. You need to determine the most probably competitive reaction and its consequences when your price is changed. Will competition follow your price change? What will be the impact if they do or do not? The small firm should constantly monitor competitive prices so that appropriate timely action can be taken.

Customer

The final C—the customer and the customer's reaction to the particular price—is the most important factor in establishing the price of the product or service. The customer (whether in the consumer, industrial, or government market) really determines whether the price established is a good one. You should carefully analyze the possible customer reaction to the product and the price, taking into account the utility of the product and its return to the purchaser. A product having a high utility to the customer can usually have a higher price than one that does not.

The relative importance of price with respect to the rest of the marketing mix depends in part on the position of the product on the continuum from **commodity** to unique product. As a product approaches becoming a commodity, price becomes an increasingly important factor in proportion to the other elements in the marketing mix as there is little or no product differentiation. Since there is nothing that differentiates one product from another, the price is often the only factor that leads to the sale of the product. For example, for many people a basic home computer is something of a commodity today. Prices for home PCs, relative to processing power, have dropped dramatically over the past decade.[10]

On the other hand, if the product or some other element of the marketing mix is very unique, then this element becomes more important than the price in affecting product sales. These elements can be in the form of the physical product itself, the package, advertising, sales force, distribution, or location in the store.

11.4 DISTRIBUTION

Physical distribution—the movement or flow of the product from the firm to the customer—primarily involves the transportation and storage of the product. The area of physical distribution is very important today reflecting the high cost of energy. In addition, the main focus here will be on the two components of physical distribution—transportation and storage.

11.4.1 **Functions of Physical Distribution**

Getting the right goods to buyers at the right time and at the lowest possible cost is an important aspect of every good marketing program. The functions of physical distribution can be classified in four major areas:

1. Location of distribution centers; these may be company-owned centers, public warehouses, or centralized distribution centers where products are stored for longer periods.

2. Development and maintenance of an inventory control system.

3. Development and maintenance of an order-processing system and a customer service department.

4. Determination of the best transportation method.

Physical distribution involves the cost of moving goods to consumers from a variety of facilities such as factories, subassembly plants, company-owned warehouses, public warehouses, or trucker-owned warehouses, as well as the additional costs involved in storing and handling the inventory in each of these locations.

Physical distribution is directly related to customer service, the lack of which can result in several costs. First, when the warehouse is out of a product, in-store shelves will not receive the needed stock, causing initial lost sales that can be permanent if the customer switches to a competitive product. Second, there may be a lack of inventory to handle periodic demand generated by such things as customer-initiated promotion. This out-of-stock will result in customer irritation. Third, goods damaged in transit or while stored in the warehouse can result in nonsalable merchandise or damaged merchandise that can also cause negative customer reaction.

Physical distribution is important due to several factors. Warehouse costs have increased significantly due to increases in labor and material costs. Transportation in a world of variable energy costs has become more expensive. Also, money for financing inventory is more costly and often difficult to obtain. A company "inventory" sheet covering most variables of physical distribution is indicated in Exhibit 11.5.

11.4.2 **Modes of Transportation**

Five basic modes of transportation are available to the technology entrepreneur to move the goods forward to the consumer: trucks, railroads, airlines, oceangoing vessels and vessels on inland waterways, and pipelines.

Trucks are considered the most flexible carrier and the most suitable for moving small quantities relatively short distances. Since trucks eliminate much of the in-transit unloading and loading, they provide a fast service without the handling costs. While truck rates are usually reasonable for the service rendered, for some higher-valued commodities, truck rates are sometimes equal to or lower than rail rates.

Railroads are generally used to move products long distances. Their rates, where water is not accessible, are almost always lower than truck rates.

Transportation costs

Total costs by company, operating division, or major product group
- Boat
- Plane
- Rail
- Truck—common carrier
- Truck—company-operated
- Comparison of above costs to sales this year versus last year

Total plant to warehouse costs
- To individual warehouses
- Each as percent of sales, this year versus last

Total warehouse to customer costs
- By areas
- Each as percent of sales, this year versus last

Inbound freight costs
- Major material categories
- Cost per dollar of material

Penalty costs
- For partial shipments
- For other than lowest method of transportation

Warehousing costs

Public warehouse costs
- Per square foot
- Per dollar of product handled

Company-operated warehouse costs
- Total operating cost
- Per square foot
- Per dollar of product handled

Inventory performance

Actual versus planned inventory
- By major category
- By location

Inventory versus sales
- By major item
- As a group for low-volume items

Cost of carrying inventories

Quality of service costs

On-time order performance
- Percent of orders shipped on time

Severity of late orders
- Customer problems resulting

Comparison with competitive services offered
- Reliability
- Speed
- Timely delivery

EXHIBIT 11.5

Company Inventory Sheet on Various Aspects of Physical Distribution.

The airlines are, of course, the fastest, but also the most expensive means of transportation. Air freight requires little capital, inventory, and warehouses. In some cases, transporting products by air is comparable in cost due to other factors in physical distribution, such as the need to develop and manage a warehouse or maintain a significant amount of inventory to ensure no out of stocks occur.

Ships are usually the least expensive carriers. Their rates are often one-third of those of rail, but they are slow and tie up money in "floating" inventory. In addition, the amount of product necessary to fill a ship is considerably more than the amount necessary to fill a rail car. Oceangoing vessels are of course used extensively in international marketing.

Pipelines are used primarily to transport crude oil or oil products and natural gas and to move coal slurry (a liquid coal product) from mines to points of processing.

It is important to evaluate each possible type of transportation in terms of its cost, speed, absolute and percent delivery time variability, and the amount of loss or damaged products in order to make sure the product is always available at the lowest possible cost.

11.4.3 Types of Transport Services

There are a variety of transport services available that the technology entrepreneur should know about. The two most important are freight forwarders and warehouses.

Freight forwarders are brokers of air, ship, rail, and truck transportation who make their living on the differences between full car or truckload, which they obtain from the less than carload (L.C.L.) or less than truckload (L.T.L.) rate paid by each shipper. Occasionally they charge an additional small fee. Although paying the higher L.C.L. or L.T.L. rate, freight forwarders have the advantage for faster service and less handling than what occurs in full car and truckload lots. For a small firm, the freight forwarder actually becomes the firm's traffic manager, determining the best way to ship the product. For small manufacturers or retailers who are shipping numerous small quantities to many different places, freight forwarders are extremely valuable. Even in large companies, their services are often used in overseas shipments.

In most large cities, there is usually more than one trucker-owned warehouse. As the name implies, each of these warehouses is owned by a trucking company that usually specializes in local delivery within the metropolitan area. For many companies, these can be used to develop the most economical method of delivery—a combined trucking and warehousing service. The company can transport in either carload or truckload lots directly to the trucking company's warehouse, where the trucker's fleet of trucks distribute to local customers. This provides the company with cheaper rates for local delivery. Often, these warehouses can furnish full services such as billing and collection as well.

Public warehouses are warehouses providing almost any kind of space (refrigerated or nonrefrigerated) or service needed by the manufacturer. Warehouse receipts can be obtained from public warehouses and used for financing inventories. Many companies prefer to own their own warehouses. These warehouses are operated by the firm's personnel.

11.4.4 Managerial Practices in Physical Distribution

There are several practices in physical distribution that should be understood in order to develop the best physical distribution system for a firm. In terms of customer service, the major factors that are affected by physical distribution are:

1. The length of time from the placement of an order to the delivery date

2. The percent of "out-of-stock" orders

3. The quantities of merchandise stocked to cover special promotions or emergency needs of the customer

4. The availability of parts and/or installation services of the manufacturer

5. The condition and care with which merchandise is delivered to the customer

6. The manufacturer's willingness and promptness in replacing defective merchandise

Each of these factors should be evaluated to determine the cost of imperfect customer service. This figure can then be used to evaluate the many issues involved in speed and size of inventories.

While accurate forecasting is the key to managing a business efficiently and controlling all the marketing factors, it is also certainly the key to maintaining a minimum level of inventory and the established target level of customer service. There is probably no other factor more important for minimizing costs in physical distribution than accurate forecasting. Accurate forecasting can avoid excess rental of warehouse space as well as unnecessary inventory. A decrease in both factors helps increase money turnover.

While too much inventory can result in significantly higher holding costs, too little inventory can result in lost sales. Some factors that need to be considered in developing an inventory model include: the repetitiveness of the inventory decision, whether the source of supply is inside or outside the company, the amount of knowledge available about future demand, knowledge of the amount of lead time needed, and the type of inventory system in use.

11.5 PROMOTION

Promotion and the promotion mix are used to communicate the positive facts about a product to prospective consumers. While promotion must be carefully coordinated with all the other elements of the marketing mix, promotion itself has its own mix consisting of advertising, personal selling, sales promotion, and publicity, each of which includes various elements.

Advertising is a paid nonpersonal presentation and perhaps the most visible component of the promotion mix. Many firms rely heavily on advertising as the primary element in their promotion strategy. The primary components of advertising are media and message. Such media as radio, television, magazine, newspaper, direct mail, outdoor, and transit are selected in light of the message to be communicated, the budget, and the target audience.

Unlike advertising, which is nonpersonal in nature, personal selling involves a personal presentation on an individual or group basis. While personal selling is frequently done face-to-face, it also occurs via telephone. Personal selling involves such activities as selection, training, compensation, territories, and control.

Sales promotion is perhaps the most diverse area in the promotion mix, consisting of all those nonrecurring activities that are not considered advertising or personal selling. Its mix consists of such things as coupons, samples, trade contests, sweepstakes, price-off-packs, point-of-purchase material, and retail promotion allowances. The use of these activities as part of the promotion strategy has significantly increased in the past several years. As the cost of advertising personal selling continues to escalate and more creative attention is devoted to this area, sales promotion will continue to have an increasing share of the promotion budget.

Publicity is a nonpersonal form of promotion that is not paid for by the sponsoring organization. It generally involves the favorable presentation of the sponsoring organization in a print or broadcast medium. Specifically, this is generated through news releases, product releases, or company events covered in feature stores. Speeches by employees of the organization are also an effective form of publicity.

11.5.1 The Marketing Mix

Most entrepreneurs elect to use a variety of tools to maximize exposure to potential customers. The variety of tools selected to communicate with customers is referred to as the **marketing mix**. Marketing strategy involves the careful selection of marketing tools to ensure maximum exposure of the venture's message to the target market.[11]

Of course, as was the case with determining market size, a major constraint on the new venture's marketing mix will be the scarcity of resources—mainly, money. The new venture's limited funds make it imperative that its investments in marketing tools be focused and effective. The entrepreneur cannot afford to waste money on experiments—and should certainly avoid volume tools that have not been proven to be effective. There are many techniques that entrepreneurs can use to communicate with potential customers, including:

- Direct salespeople
- Employees
- Television
- Radio
- Print
- The Internet
- Direct mail
- Marketing collateral
- Billboards
- Signage
- Location
- Trade shows

Direct salespeople

Direct salespeople are probably the most effective of the marketing tools, but it is also the most expensive. Direct salespeople are effective because they have personal contact with customers. Because of this direct salespeople can build relationships with customers and "sell" themselves as well as the venture's products and services. Salespeople can build trust with the customer and know when the customer is dissatisfied with the product or service.[12] A good salesperson will strive to maintain regular contact with the customer and correct any breakdown in customer satisfaction.

Direct salespeople are expensive because of their salary, benefits, and other cost items such as travel and entertainment. These high costs, coupled with the limited number of sales calls that can be accomplished within a given period of time, make the cost per direct contact expensive. For most new ventures, the use of a direct sales force is not feasible from a cost perspective unless the venture's offering is a big-ticket item or there is a large repeat purchase series from the initial sale.

Employees

One often neglected, yet very effective, volume tool is the employees of the venture—especially those on the front line who have **customer facing** jobs. Customer facing jobs are those that have direct contact with customers. It's important for the entrepreneur to recognize that employees in these roles often represent the venture's only direct contact with customers. If the people filling these roles are unhappy, unhelpful, or just plain surly, what message do you think that conveys to customers? Worse, what action is the customer likely to take as a result of a negative encounter with a disgruntled employee at one of these "low-level" jobs? Customer facing employees are the venture's ambassadors. It is wise for entrepreneurs to make sure the people in these key roles are well versed in the company's products and are motivated to treat customers right.[13]

Television advertising

Television advertising is another effective marketing tool for many new ventures. Most television stations are local and broadcast over a limited geographic area. Ventures that have defined their target market using geography—such as a computer repair service or HVAC venture—can take advantage of this. In fact, many cable television operators enable precise targeting of advertisements tied to zip codes.[14] This ensures that television ad buyers spend money only on ads that will reach their target markets.

For ventures that don't have a target market defined by geography, a mass-market approach may be effective. In large cities, the coverage area of a local television station will include tens of thousands of viewers—some of whom are also potential customers.

The rates for advertising on television vary according to the number of viewers and the popularity of the station in comparison with competing stations in the area and its affiliation with nationwide networks such as CBS or NBC. In general, advertising on cable-only channels will be less expensive than advertising on programming offered by local broadcast stations.[15] The cost of advertising on television varies as well during the time of day or night the commercial is aired. The **prime time** of day or evening will result in the highest advertising charges. In addition to this media time charge, a television advertisement has production costs. Therefore, a venture must look carefully at its target market, its offerings, and the likelihood that a series of ads will be cost-effective. The company must look carefully at the particular audience of the television station being considered and at the target audience it is attempting to sell before deciding to employ this tool of volume.

Radio advertising

Radio advertising has many of the same characteristics of television advertising, but is generally less expensive than television.[16] Because the cost of radio advertising is less than television advertising, it is possible for a venture to use a "saturation" technique when a product or service is new to the marketplace. This technique uses repetition of the radio ad—perhaps on multiple stations—as a means of getting the potential customer's attention.

Radio advertising is sold in much the same way as television ads are sold. Most stations have sales representatives who will be able to provide the entrepreneur with a **rate sheet.** The rate sheet will outline the charges for ads of varying lengths, and for varying times of day or for the different radio shows that air on the station. For example, many radio stations syndicate the shows of nationally known radio celebrities, such as Howard Stern or Laura Ingraham. Ad space on such shows—which presumably have very well-defined listener groups—will generally cost more than local programming.

Print advertising

Newspapers and magazines present the entrepreneur with a more targeted marketing tool than radio or television advertising. Publishers of print media study their readership carefully and make the data available to advertisers. If the entrepreneur has done his or her demographic homework, the choice of which print publication will be most effective can be straightforward. Based on the data provided by the publisher, the entrepreneur can match his or her demographics with the market profile of the print publication.

Newspapers are generally local, but there are some that are nationwide—such as *USA Today* and *The Wall Street Journal.* Still, even these nationally distributed newspapers have opportunities for businesses to advertise only in select regions. They generally set aside a proportion of ad space for local distribution only. This technique enables them to reduce the costs for local distribution, while still providing lower-budget ventures the prestige of advertising next to nationally known brands.

Magazines are usually national in distribution, but many regional magazines have been established in recent years. With reduced production costs due to digital technologies, it is not difficult to produce a monthly or weekly magazine with a local focus. Some are distributed for free at local merchants. Entrepreneurs have many choices in print publications, and must be careful to monitor the return on each investment. Placing ads in freely distributed magazines may ensure wider ad distribution, but it may also be a less targeted market. People will pick up a free magazine regardless of its content, but will not purchase a magazine unless they have a specific interest in its editorial focus.

Other less expensive print media can also be effective volume tools. For example, neighborhood newspapers and trade publication are usually specifically targeted toward certain audiences. When the target market of a company matches the target market of one of these publications, the resulting volume is cost-effective. Neighborhood newspapers are most effective when the venture's target market is localized and its product or service is appealing to the readers. **Trade publications** have the advantage of being targeted to the special interests of the group to which they are distributed. Some examples of trade magazines include:

- *U.S. Banker*
- *Call Center Magazine*
- *American Printer*
- *Beverage World*
- *Restaurant Business*
- *Packaging World*

Each of these publications circulates within a narrow interest group. If that interest group comprises the venture's target market—or at least a subset of the target market—advertisements in trade publications can be cost-effective. Most professionals who read trade publications do so to help them solve problems inherent to their profession and industry. If the venture provides a solution to one of those problems, it has a high likelihood of being noticed and of motivating action.

The Internet

The Internet as a marketing tool is most effective when dealing with a product or service that does not require personal inspection or handling prior to purchase. The steps to creating an effective presence on the Internet are straightforward and similar to traditional advertising. As with any effective marketing technique, it takes understanding how the market evaluates and purchases a venture's products to create the most effective Web presence.

One of the biggest differences between Internet marketing and traditional marketing is that the time and space considerations are different. With a web site, the entrepreneur has more space to work with than a newspaper ad and can include more details, graphics, and examples to support a description of the venture's offerings. The entrepreneur also has more time to communicate a message through a web site than, say, via a radio or television ad. Visitors to a venture's web site can spend as much time interacting with the information presented as they need. The key to an effective Web presence is an intuitive navigation strategy. Users of a firm's web site should be able to find the information they need without having to spend a lot of time searching.[17]

An excellent way to take advantage of the space available on a web site is by providing materials that educate as well as sell. For example, a company that sells power supplies could provide a white paper discussing the most common types of power outages and line surges. A company that sells accounting software could provide a fact sheet outlining the differences between cash-based and accrual accounting. The best type of educational material is the kind that complements the more sales-oriented Web content, building a picture of a company that can offer knowledge and solutions as well as merely provide product.

Direct mail

Direct mail is an especially effective tool if it can be targeted to a reasonably specific target market. It has become a much more expensive method as both postal charges and printing costs have increased. The largest drawback for this approach to volume is the large number of users has made the customer more insensitive to receipt of the communication.

Marketing collateral

The term **marketing collateral** refers to the material a company develops to communicate with the target market that can be physically distributed and left behind. This includes such items as brochures, handbills, business cards, and other items. Hand bills are a very cost-effective substitute for direct mail when the target market is local. Handbills are usually delivered door to door, placed on vehicles in shopping centers, or handed out to pedestrians or shoppers. In the case of a student-started business for pet care, Club Pet built its initial volume by distributing handbills on cars parked in the shopping center where it was located. The company initially spent $5,000 advertising in a neighborhood newspaper and local magazine that resulted in three customers. In a meeting at the store, the owner was reporting these results to his mentor when he said in a fit of despair, "If just 5% of the cars parked in that parking lot would bring in their dogs, I'd be a success." The next day, the young entrepreneur prepared a handbill that offered shoppers a 10% discount on pet grooming if they would drop off their pets while shopping in the neighboring stores. The marketing tool worked, and the cost of the hand bills was less than $300.[18]

Billboards

Billboards are an effective tool if they are located in a high-volume traffic route. The great strength of billboard advertising is the constant repetition to the potential customers who drive past the billboard regularly. The limitation on the use of billboards is the commitment of resources. Billboard contracts usually span several months and several locations. The better locations are usually assigned to the largest users. It is frequently necessary for a new venture to accept some marginal locations in order to get prime locations later. As with other volume tools discussed in this chapter, billboard ads are evolving and new technologies are offering greater opportunities to communicate more information.[19] Entrepreneurs who have a geographically defined target market can use billboards to great effect.

Location

When a customer must visit the business location in order to make a purchase, the most important thing to remember is that the customer must feel comfortable.

For example, retailers go to great lengths to present merchandise in a manner that makes shopping easy and intuitive for customers. If customers are confused or frustrated because they can't park near the business or they can't locate products easily within the store, they may use that as an excuse not to purchase the venture's products or services. Location can impact customer behavior.

Entrepreneurs can reduce the impact of location by enabling customers to order without having to visit the business. The Internet has been the solution for many companies seeking to reach customers who are unable to or simply don't want to visit their physical location. Everything from home electronics to automobiles can now be purchased online. The entrepreneur should recognize that some portion of the venture's target market will prefer to visit the location prior to any purchase, and some will be satisfied with an online purchase. The wise entrepreneur presents customers with multiple ways to interact with the company and purchase its products and services.

Signage

Signage at the business location is vital when the location of a business is important to the businesses success. The customer's impression of the business is affected in part by the appearance of the signage. The size, color, and design of the company's signs communicate a first impression of the business and its product. The decision to stop and buy or pass the location can depend on the appearance of the signage.

Point-of-sale (POS) displays are used by businesses whose products are sold through retail outlets. No matter the size of the retail outlet, if the product is to compete with other products like it, shelf space and POS displays are vital. The POS display must attract the attention of the shopper and it must explain the product and its benefits. Marketers have gone high tech with POS displays, using special sensors to determine when a shopper is looking at a particular product and offering to provide advice—using a prerecorded voice—to help the shopper decide.

Trade shows

Trade shows are an excellent volume tool for new products. The audience is targeted to the specific type of product or service. It is an expensive tool, but because of its audience characteristics, it is effective for building volume in a new company. Trade shows provide an outstanding communication vehicle because they allow direct salespeople to talk one on one with interested potential customers, and they allow a large number of direct sales presentations in a single day. Trade shows are particularly effective in introducing new products or in establishing new dealerships or distributorships.

11.6 THE MARKETING PLAN

The **marketing plan** created by the technology entrepreneur designates how the venture will communicate its basic value proposition to customers and potential customers. The plan should highlight how the venture intends to address the "Four P's"—product, price, placement, and promotion. We have discussed these elements in the sections above. The plan should also specify the capital resources that will be invested in marketing—the marketing budget. Below, we discuss various common techniques used by entrepreneurs to establish a marketing budget. Finally, a marketing message must be crafted—one that distinguishes the venture's products and services from the others in its environment. This is a critical decision, and will play a large role in determining how much money is spent on marketing, and where that money is allocated. For example, a venture that chooses to be the premiere product in its industry will likely spend marketing money on high-end print advertising before it purchases billboards. Let's begin our discussion of the marketing plan by examining techniques for building a marketing budget.

11.6.1 The Marketing Budget

One of the most important and yet difficult questions for a technology entrepreneur is determining how much to spend on marketing. Total marketing spending varies greatly from 40% of total sales in the cosmetic industry to about 15% in the industrial machinery industry. Several key factors need to be considered in determining the size of a firm's marketing budget. They include the nature of the market, the nature of the product and its stage in the life cycle, the channels of distribution, and the objectives.

The characteristics of the market determine the intensity of the marketing effort. When an organization is marketing its product or service in a concentrated geographic area to a very homogeneous customer group or in an established market with existing favorable consumer attitudes, less effort is needed. More marketing effort is needed when an organization is trying to obtain a market position in an area dominated by a strong competitor. Such was the case when Procter & Gamble expended significant promotion dollars to introduce Folger's Coffee in the New England area which was then dominated by General Foods' Maxwell House Coffee.

The nature of the product and its stage in the life cycle affect the size of the marketing budget. Different kinds of products require different kinds and amounts of marketing. For example, convenience goods purchased frequently at the most convenient location must be supported by a great deal of advertising; other products such as, say, wooden pallets used in transporting goods require very little.

Consumers need to be aware that the convenience goods are available and will fulfill a need, whereas pallets are ordered on a regular basis usually from the same two or three suppliers. A given product's promotion needs also vary with the stage in its life cycle. During the introduction and growth stages of the product, more marketing effort is needed to inform the target market about the existence and benefits of the product. Significant promotional expenditures are often needed in the start-up phase as well as in the maturity stage when competition becomes intense. In the decline stage, of course, promotion becomes less effective and should be used sparingly.

The final factor affecting the size of the marketing budget—the marketing objectives—is usually the most important. Objectives not only help determine the size of the budget but also provide a basis for evaluating the results of the promotion effort. For example, if a college wants to attract 10% more students with higher SAT scores, then a promotion budget needs to be established to achieve this objective. An evaluation of the applicants will indicate whether or not the promotion campaign achieved the results.

11.6.2 Determining a Marketing Budget

Given these factors, how can you determine the marketing budget for your firm? There are five methods for determining a promotion budget: arbitrary determination method, competitive parity method, objective and task method, percent-of-sales method, and various quantitative methods.

In the **arbitrary determination method**, the marketing budget is determined in a seemingly arbitrary manner. When using this method, company executives rely on their intuition and past to establish the budget. This is a very unsophisticated method, but more firms use this method than any other. A common benchmark used is 10% of last year's sales.

In the **competitive parity method**, a company uses the marketing expenditures of competitors to establish its own budget. This information on the competitor's budget is also useful in evaluating the overall competitive environment, and can be obtained through carefully analyzing a company's financial statements or through a service that monitors all the marketing efforts in a given product/market area for a fee. The competitive parity method has some drawbacks as it employs historical data of competitive expenditures and assumes that all have very similar marketing situations.

Probably the best method for establishing a marketing budget is the third method—the **objective and task method**. When using this method, you establish definitive marketing objectives and then determine the amount and cost of the marketing required to reach these objectives by costing each element of the marketing mix needed. Not only does this method allow for each element in the marketing mix to be effectively used, but it also makes sure that the amount of money to be spent on marketing is commensurate with the task at hand. A clearer basis for evaluation of the marketing expenditures also results.

The fourth method—the **percent-of-sales method**—is also widely used mainly because it is easy to use. This method involves applying a fixed percentage to either past or future sales figures. If last year's sales were $200 million and the percentage used is 3%, the promotion budget for this coming year would be $6 million. The sales figure used can be the sales for the last year or an average of sales achieved over the past several years. By using an average sales figure, the impact of any erratic sales fluctuations on marketing expenditures is eliminated. You can also use a percentage of future sales, basing the marketing budget on future conditions, not on past events. When you use the percent-of-sales method for determining the marketing budget, you should compare it to the percents of leading competitors as well as industry norms.

The final method for establishing the size of the marketing budget—the **quantitative method**—involves implementing one or several mathematical models or more sophisticated quantitative techniques. While these add a more vigorous dimension to the budgeting process, their use requires a significant amount of data and expertise. As the costs of marketing continue to increase, quantitative methods will continue to have increasing use, usually in conjunction with one of the other four methods.

Regardless of the size of the marketing budget, the expenditures need to be carefully evaluated to determine if the target audience is being effectively reached, the objectives accomplished, and the desired sales achieved.

11.6.3 The Marketing Message

To be effective, the marketing message must be responsive to the characteristics of the customers in the target market. Persuasive messages should help differentiate the entrepreneurial venture from existing competitors, while at the same time accurately communicating the offering. This is a creative process that may take several iterations to refine the persuasive message. The entrepreneur should be prepared to alter his or her persuasive message based on customer feedback. If the results expected are not being achieved, the reason may be that the message is not motivating potential customers as anticipated. In such cases, it may be wise to alter the message slightly and gauge the effects of the new approach. The goal is find the message that produces the best results.

Although there are multiple ways to describe a product, in reality marketers have very few choices about how to differentiate their offering from competitors. Research into product differentiation has defined three strategies for differentiating a product:

1. *Cost/price leadership*: This is a strategy where the venture claims to be the low-price leader in the target market for particular products and services.

2. *Quality leadership*: Under this strategy, the venture makes no claims about price but rather caters to a market that prefers upscale, luxury, or prestigious offerings and brands.

3. *Niche strategy*: Many ventures identify a niche market such as a particular ethnic group and claims that its offerings uniquely meet that market's needs.

Each strategy has advantages and disadvantages. Claiming to be the price leader will win immediate interest from the market, but requires that the venture keep its operating costs low so that it can make a profit. Quality leadership (often referred to as the "differentiation" strategy) means the venture has pricing power, but it will have a limited market size and will compete within that limited market with other upscale brands. Niche marketing (also referred to as the "focus" strategy) can be effective, but could limit the venture when it wants to grow beyond the defined niche.[20]

Many entrepreneurs hire marketing consultants, advertising agencies, or public relations firms to assist in the development of a persuasive message. Using this approach, the entrepreneur must instruct the service provider about his or her understanding of the nature of the target market and the benefits provided by the venture's offerings. One benefit of using a vendor to help craft the persuasive message is that the entrepreneur will also begin to understand his or her business better. The marketing vendor has been trained to ask penetrating questions that force the entrepreneur to think differently and in a more focused manner about the true **value proposition** of the venture's offerings. The value proposition comprises the features and benefits—in the context of price—which the customer can expect to obtain from the venture's offerings. Marketing scholarship has identified three basic value propositions:

1. Operational excellence
2. Product leadership
3. Customer intimacy

Operational excellence refers to the ability to provide products or services that are produced more efficiently and offer time or cost savings to customers. Product leadership refers to the ability to lead in the quality of a product or service. Customer intimacy means that the venture is touting its ability to serve customers as the primary reason for doing business with it. The idea behind the concept of a "value proposition" is to choose one of these three as the primary focus of the venture, and perform at industry standard levels on the other two. Thus, if a firm chooses customer intimacy as its primary value proposition, it will still operate efficiently and provide quality products and/or services. However, its level of customer service and the depth of its relationship with the customer will lead the industry.[21]

The persuasive message should attempt to appeal to as many human senses as possible. Effective messages use visual images, words, and sounds to influence, stimulate, and ultimately persuade the customer. Words are used to convey technical details, but should also tell a story about how the products will benefit the customer. A good persuasive message will use words, sounds, and images that invoke a sense of urgency, persuading the customer to take action to gain the benefits of the venture's offerings.

Appealing to the senses is not the only important aspect of a good marketing message. The customer must also receive information necessary to make a decision about the offering. To that end, the persuasive message must include information about how to order the product or service. Far too often a venture develops a catchy and persuasive message but fails to make it easy for the customer to purchase the advertised product. Think about how long a marketing message stays in your mind after you hear it or read it in a print publication. Not long. For a persuasive message to be effective, the ordering instructions should be simple and clear. Tech Tips 11.2 provides some ideas about how to create a compelling value proposition.

TECH TIPS 11.2

How to Create a Compelling Value Proposition

Strong value propositions describe tangible results such as:

- Increased revenues
- Faster time to market
- Decreased costs
- Improved operational efficiency
- Increased market share
- Decreased employee turnover
- Improved customer retention levels

Finally, it's important that the message be repeated. Many customers do not respond to a message the first time they are exposed to it. An axiom of the advertising business is that any message that is not repeated at least three times is wasted.

SUMMARY

This chapter deals with many of the most important aspects of going to market and the marketing plan. The definition of marketing and its primary aspects of exchange and customer satisfaction while accomplishing the objectives of the organization were discussed. Particular emphasis was placed on such important issues as the marketing concept and market segmentation.

The four aspects of the marketing mix—product/service, price, distribution (physical and channel), and promotion (advertising, sales promotion, publicity, and personal selling) were described in depth along with examples of their use. Technology entrepreneurs often learn about the various promotional opportunities by trial and error. While this can be an effective way to learn, in the long run, it can also be expensive in the short run. Spending money on marketing and promotions that don't pay for themselves by generating new sales is unsustainable. Technology entrepreneurs must carefully monitor their marketing efforts to ensure they are achieving the objectives that have been set.

Segmenting markets using the techniques discussed in this chapter is one way to ensure that marketing dollars are being spent wisely. Ventures that don't segment their markets carefully may be using promotional techniques that are a waste of time and money, or they may be using a marketing message that is unproductive. Once a market is segmented, the technology entrepreneur can select a marketing mix and message that is designed to achieve the venture's sales objectives.

The marketing plan designates how much capital a venture intends to invest in marketing and what its overall marketing message will be. This chapter discussed various techniques for setting up a marketing budget. The marketing budget constrains the available channels and frequency of their use, but it is essential to establishing a disciplined approach to marketing.

Finally, the marketing message is a fundamental part of the venture's overall go-to-market strategy. New ventures can generally choose one of three approaches: high quality, niche, or low cost. Each of these approaches demands a different overall marketing budget and marketing mix. New ventures must choose one or the other of these alternatives to ensure a consistent message to its target market. The coherence of a venture's message as well as the consistency of its delivery to the target market are essential elements of a successful marketing plan.

STUDY QUESTIONS

1. Why is marketing important to the technology venture?

2. Identify the "Four P's" of marketing. How should the technology entrepreneur use the Four P's as part of an overall marketing effort?

3. What are the 3 C's of pricing? What is their purpose?

4. Define the different ways of establishing a marketing budget.

5. What are the different methods available to the technology entrepreneur for communicating with potential customers? Describe how a new software venture might use the various tools.

6. What is a value proposition? What are some key elements of a compelling value proposition?

7. Explain how a technology entrepreneur can decide among different modes of transportation for shipping products. What are some of the key concerns?

8. What is meant by the term "marketing concept"? How does it contrast with "production concept" and "sales concept?"

9. How does a technology entrepreneur use marketing collateral as part of an overall marketing mix? If you were an entrepreneur, what types of marketing collateral would you consider to be essential?

10. What are the three primary marketing messages the technology entrepreneur can choose among? Which message does Dell Computers utilize? Explain.

EXERCISES

1. *IN-CLASS EXERCISE:* This exercise is designed to be carried out in class by students organized into groups of five to seven. The goal of this exercise is to pick a

business or nonprofit organization, identify a target market for it, and then craft a persuasive message that is designed to motivate the target market. The business or organization selected for the exercise should be a local brand. It could be the college or university, it could be a local car dealer or service organization, or it could be a popular local nonprofit organization such as a festival or charity.

Students should be organized in groups of five to seven, and within 30 minutes, they should be able to:

1. Identify a target market for the organization. The target should be defined either using demographic data, geographic location, or psychographic profile.
2. Determine the needs of the target market and develop a value proposition that is targeted to those needs.
3. Create a persuasive marketing message that communicates the value proposition to the target market.
4. Develop a 30-second radio ad to communicate the persuasive message.

For the last 15–20 minutes of class, student groups should share their findings regarding the target market and read their radio ads aloud. Each group leader should be able to explain how the group arrived at its conclusions regarding the target market and how they identified the needs of that segment.

To make the exercise a little more fun and competitive, students should vote at the end on which group created the most compelling radio ad.

2. *OUT-OF-CLASS EXERCISE*: For this exercise, students should discuss radio advertising as a means of reaching a target market. They should identify the three or four most notable radio advertisements in their local market and answer the following questions:

1. What makes these ads distinctive?
2. To whom do the ads appeal?
3. What is the frequency of the ads on the local airwaves?
4. How many people hear the ads each day?

Students should contact the local radio station on which the ads are being run and request a rate sheet. The instructor should help the students estimate how much money the entrepreneur has spent to run the ads. In light of this calculation, discuss the following questions:

1. How many people need to respond to the ad and make a purchase each week?
2. Is it reasonable to suspect that the entrepreneur's investment in radio advertising is paying off?
3. What other advertising does the venture use?
4. How important is radio in the overall marketing mix?

KEY TERMS

Arbitrary determination method: A technique for establishing a marketing budget that is largely arbitrary.

Commodity: Products that are mostly similar—in the eyes of customers—and are purchased primarily based on their price.

Competitive parity method: This is a method for establishing a marketing budget based on analysis of competitor expenditures.

Cost-plus-pricing: A price setting technique whereby the price of a product or service is determined by adding up all the costs associated with creating it and adding a profit margin.

Cost/price leadership: A differentiation strategy that positions a venture as the low-cost alternative in a market.

Customer facing: Employees of a venture who have jobs that are in direct contact with customers and can be leveraged for additional sales.

Demographics: Refers to statistical characteristics of markets, such as average age, that can be used as a segmentation technique.

Economies of scale: Occurs when a venture achieves production volumes that result in a per-unit decrease in costs.

Four P's: The basic elements of marketing, which are product, price, placement (distribution), and promotion.

Geographic location: A market segmentation technique that determines a market by virtue of its geographic location.

Marketing collateral: The materials a company creates to communicate with customers and which can be left behind—such as brochures.

Marketing mix: The variety of tools a venture uses to communicate with customers.

Marketing plan: This is the overall plan that a venture creates to how much it will spend on marketing, where those dollars will be spent, and the primary message that will be communicated to customers.

Market segmentation: Dividing the total market into groups based on some desired characteristic or attribute.

Marketing concept: Focusing all the marketing activities of the firm on the satisfaction of the customer.

Niche strategy: A strategy that positions a venture as the leader in targeting a specific market niche.

Objective and task method: A method for setting a marketing budget that relies on setting clear objectives for marketing and determining how much it will cost to achieve them.

Percent-of-sales method: A method for setting a marketing budget that uses a percentage of overall sales revenue to set the marketing budget figure.

Placement: An element of the marketing mix that refers to how a venture distributes its products and/or services to its target market.

Price: An element of the marketing mix that refers to the price that a venture charges for its basic unit(s) of sale.

Prime time: Key times of day when the most viewers or listeners are watching television or listening to radio and during which advertising rates are the most expensive.

Product: The venture's offering, which is either a product or a service or a combination of both—the product is the basic unit of sale.

Promotion: An element of the marketing mix that refers to the methods a venture uses to communicate with its target market about the features and benefits of its offerings.

Psychological profile: A market segmentation technique that identifies different markets by virtue of their different psychological characteristics.

Quality leadership: A differentiation strategy that positions a venture as the highest quality alternative in a market.

Quantitative method: A technique for setting a venture's marketing budgets that uses sophisticated mathematical formulae and statistics.

Rate sheet: A document provided by media to potential advertisers that communicate advertising rates.

Served available market (SAM): The proportion of the TAM that a venture actually serves.

Total available market (TAM): The total market that might be willing and able to purchase a venture's products and services.

Trade publications: Specialized publications for niche markets that can be effective promotional channels for ventures with the same target market.

Value proposition: Comprises the features and benefits of a venture's offerings that meet the needs/wants/desires of the target market.

WEB RESOURCES

The web sites below are intended as destinations for your further exploration of the concepts and topics discussed in this chapter:

1. http://www.marketingpower.com/Pages/default.aspx: This is the web site for the American Marketing Association, the premiere marketing trade and academic

community. There are a number of useful resources on this site, including definitions of all the key marketing terms.

2. http://www.marketresearch.com/: This web site has free and fee-based market reports on a wide range of industries.

3. http://www.gartner.com/: The Gartner Group is recognized around the world as a leader in the analysis of market and economic data to detect and describe market trends.

4. http://www.emarketer.com/: This is a for-profit market research site that is a leading destination for technology ventures to understand market trends.

5. http://www.census.gov/: This site provides the latest data from the U.S. Census on demographic trends in the United States and around the world.

CASE STUDY

Massive Goes to Market with In-Game Advertising Network

In October 2004, Massive, Inc., completed the beta launch of its online video-game advertising network. The market reaction to the venture's value proposition was generally positive, but the path to that acceptance was exceedingly complex. Katherine Hays, Massive's cofounder and chief operating officer, pointed out that to gain market acceptance the company had to win over three distinct audiences: game publishers, advertisers, and gamers.

Game publishers were experiencing margin pressures due to the rapid introduction of new games and the proliferation of competitors. Massive needed the game publishers to be willing to adapt their games in subtle ways to accommodate the advertising network it was building. Advertisers and media buyers normally adapt quickly to new approaches to distributing ads. What would it take to get the Massive business model accepted by this audience? Finally, gamers were a key constituency. Market research that Massive had conducted discovered that, if the gamers did not embrace the concept, it was doomed from the beginning.

During the 2004 beta rollout, Massive knew that it needed to continue to experiment, ask questions, listen to its various audiences, and learn what each constituency needed. When she had joined the company in July 2003, she anticipated serving only a 5-month term to work with the CEO, Mitch Davis, to help him select among a handful of products the company was developing. At the time, Massive's products were primarily "middleware" designed to assist game developers. The ad network concept was among Massive's potential products, but it was not a favorite of the venture's development team.

Hays conducted multiple interviews with game publishers and developers, and also with gamers and advertisers. She wanted to develop a deep understanding of the issues each of these potential audiences faced. As a result of her conversations with game developers she realized that they liked the idea of middleware that could accelerate the game development process, but they were reluctant to give up any control over the core components of their code. She reasoned that, even if the company were to win over the game developers with its middleware products, the market opportunity was not large enough to interest venture capital investors.

Davis was at first reluctant to accept Hays' analysis that the middleware products the company was developing should be scrapped. However, several subsequent investor presentations helped him to realize that focusing on a single product—the advertising network—was a more compelling story. Several investor groups that Davis presented to were unsupportive of the "multiple products" business concept. They preferred to hear a more focused pitch. Hays insisted that the company build itself around the advertising network concept.

In September 2003, Davis allowed Hays to pitch her idea to the Massive board of directors. She recommended that the company confine itself to two products. One product was a hardware/software testing tool called "Testbox" that developers use to replicate bugs in the quality control process. The second product—the online game advertising network—would take more time to develop, but provided the venture with a potential "home run." She noted that the ad network solved a real problem identified by game publishers and advertisers. Publishing partners would get a new source of revenue, and advertisers would get a new channel into the coveted 18–34-year-old male demographic. She recommended to the board that Massive focus the bulk of its energy on this product. She also recommended a shift in market positioning, from a technology company to a media company, to coincide with this shift in product focus. The board agreed, and Hays decided to stay on with the company permanently.

Massive's only competitor in 2004 was a company called Transplay. However, Transplay's business model was significantly different from Massive's. Transplay did not build in-house capability for advertising sales, back end campaign management, or measurement and reporting. Instead, Transplay licensed its technology to publishers as a traditional software company for imbedding into the publishers' games. By mid-2004, Transplay had gone out of business.

Having no competition in the online game advertising market was nice, but it also posed a problem. As Hays noted, not having others competing in the same market space meant a slower sales cycle as Massive was solely responsible for educating customers on the concept. By late 2004, however, two new competitors had entered the market: Israeli-based DoubleFusion and Canada-based BiDamic. And why not? Massive estimated the market for in-game advertising to be as much as $2.5 billion.

Massive's re-focusing as a media company began with a product definition process. This was informed in large part by the interviewing that Hays had done during the summer of 2004. It was clear that there was demand for the ad network product by both the game publishers and advertisers.

But Massive also knew that it did not want to use the model that Transplay had used. Hays' research revealed that game publishers didn't want anything to do with selling ads or measuring their effectiveness. They wanted to stick to their own core competency—publishing hit online games. Massive designed its product to deliver ads in batches to a game via the Internet. Advertising would be delivered to "game zones" such that, when a player achieved a new zone, the ads would be sent.

The advertisers required a system that would reach a large, aggregated audience on a given night, day, or week. It also needed to reach the target demographic not through a single online game, but through multiple games simultaneously. Advertisers also had to be able to measure the effectiveness of their advertising campaigns.

Massive designed the product to appeal to both of these users. It developed an end-to-end solution for the game publishers, while providing a television-like medium for advertisers that enabled them to broadcast to multiple games and genres. Massive addressed the advertisers' need for data by establishing a system based on ratings points, reach, and frequency.

Even with all the market research behind its initial product design, Massive still had difficulty convincing the game publishers to use its product. To help the publishers get over their resistance, Massive set up a carefully orchestrated product introduction plan that it used with each potential new client. As a new game was being developed, a Massive engineer would be sent in to work with the game developers to insert the code that enabled the ad network. In addition, Massive would conduct focus group interviews with gamers on every game prior to going live to ensure that ad placement was optimized and nonintrusive to the gaming experience. Finally, Massive did quality assurance testing to make sure all code worked properly. All games, once live, were supported by Massive's ad operations unit, overseeing campaign delivery and traffic to optimize revenue, as well as to manage billing and reporting.

By virtue of spending so much time and attention on the needs of its customers, Massive was able to achieve significant success during its market launch. The case demonstrates the value of knowing the market in depth and designing solutions and messaging that solve real problems for customers. In 2006, Microsoft recognized the value of what Massive had created and acquired the firm for an amount reported to be as high as $400 million.

Questions for Discussion

1. Describe some of the things Massive did to learn more about its market. Do you think these efforts were essential to its success? Explain.
2. In their go-to-market strategy, Massive placed an engineer onsite with game publishers. What was the purpose of that? What potential problems do you think might arise from such a strategy?
3. Massive decided to focus on only two products when it went to market, with one product—the ad network—being their primary focus. Why do you think this approach makes sense? Do you think this is a general lesson for all technology ventures? Explain.

Source: Adapted from Lassiter III JP, Gilbert C, Winston VW. Massive incorporated. Harvard Business School Case Study *February 5, 2007; Cole V. Microsoft munches Massive, Inc.* JoyStiq *April 26, 2006; Microsoft acquires Massive, Inc. Stanford University, Case Wiki, May 4, 2006.*

ENDNOTES

1 Shultz II CJ. Examining the interactions among marketing, markets, and society. *J Macromarketing* 2009;**29**(1):3–4.

2 Schwartz M, Tsadik R, Maddox K. AMA's definition of marketing stirs debate. *B to B* 2008;**93**(2):41.

3 Retrieved from the American Marketing Association online dictionary at www.marketingpower.com.

4 Lewandowska A. Is the marketing concept always necessary? The effectiveness of customer, competitor, and societal strategies in business environment types. *Eur J Mark* 2008;**42**(1/2):222–37.

5 Foedermayr EK. Market segmentation in practice: review of empirical studies, methodological assessment, and agenda for future research. *J Strateg Mark* 2008;**16**(3):223–65.

6 Powers TL, Sterling JU. Segmenting business to business markets: a micro–macro linking methodology. *J Bus Ind Mark* 2008;**23**(3):170–7.

7 Lam MD. Psychographic demonstration. *Pharm Exec* 2004;**1**:78–82.

8 Gofman A, Moskowitz H. Steps towards a consumer-driven innovation machine for "ordinary" product categories in their later lifecycle stages. *International Journal of Technology Management*, 46(1/2): 2009; 349–363.

9 Reinartz W, Ulaga W. How to sell services more profitably. *Harv Bus Rev* 2008;**86**(5):90–6.

10 For more on this, see: http://www.intel.com/technology/mooreslaw/.

11 Garber Jr LL, Dotson MJ. A method for the selection of appropriate business to business integrated marketing communication mixes. *J Mark Commun* 2002;**3**:1–16.

12 Young L, Albaum G. Measurement of trust in salesperson–customer relationships in direct selling. *J Personal Selling Sales Manage* 2003;**Summer**:253–69.

13 Caulfield B. How to win customer loyalty. *Business 2.0* 2004;**March**:77–8.

14 Larson M. Comcast recasts ad sales. *MediaWeek* 2004;**2**:8–9.

15 Ourand JP. Broadcast ad gap, no problem, cable nets say. *Cable World* 2004;**February 16**:18–19.

16 Kurts R. Mad as hell about a rates. *Inc.* May 2004:22.

17 Dailey L. Navigational web atmospherics: explaining the influence of restrictive navigation cues. *J Bus Res* 2004;**July**:795–803.

18 Personal experience.

19 Sager I, Lowry T. Coming: way more info on billboards. *Bus Week* 2004;**February 23**:16.

20 Porter M. *Competitive strategy*. New York: Free Press; 1998.

21 Treacy M, Wiersma F. *The discipline of market leaders: choose your customers, narrow your focus, dominate your market*. New York: Perseus Publishing; 1997.

Financial Management and Control

12

After studying this chapter, students should be able to:

- Perform in the role of financial manager for a start-up technology venture
- Develop a start-up venture financial plan
- Analyze financial statements using ratio analysis techniques
- Project cash flows into and out of a venture
- Develop credit and collections policies to ensure adequate cash flows
- Manage a venture's working capital and establish a capital budget

---

TECH VENTURE INSIGHT

Balancing Growth with Profitability

To achieve their objectives, technology entrepreneurs have to manage their venture's financing. Zorik Gordon, 36, cofounder and CEO of an online ad agency called ReachLocal, Inc., was not thinking about profits when he sought to raise venture capital for his rapidly growing company.

Founded in 2004 with $750,000 from a single angel investor, ReachLocal quickly proved that online advertising could be used successfully to promote small local merchants. The company helps businesses with many aspects of their online presence, from selecting keywords for search advertising to building landing pages and creating and analyzing the results of the campaigns.

ReachLocal's biggest competitor is the nearly ubiquitous Yellow Pages. To compete against this long-time market incumbent, Gordon knew he would need a significant equity investment. According to Gordon, profit was something that he would worry about later—*after* he received equity financing. In the meantime, he became busy building the venture's top line and expanding rapidly.

While ReachLocal's focus on sales growth made it an ideal target for equity investors, it was a poor candidate for bank debt. It did not generate a steady cash flow to take on a significant bank loan, according to Jamie Westmoreland, commercial relationship manager at Bank of York. "Ideally, the bank wants to see historical cash flow at levels that adequately service the proposed debt. Revenue growth, profits and EBITDA are great, but historical cash flow demonstrates the ability to collect receivables and properly manage inventory and payables," added Westmoreland.

Gordon was successful in raising significant equity capital with the largest round of financing adding up to more than $55 million that closed in October 2007. But, in order to raise that capital, Gordon had to explain how ReachLocal could eventually become a more traditional, "bankable" company. "These later-stage investors are looking for a business that is going to grow really fast and throw off a lot of cash."

These days, ReachLocal is well on its way to finding the balance between growth and cash. In 4 years, the company grew revenue from zero to nine figures. Now Gordon is turning his attention to profitability. "At a certain point," he says, "you start getting held to your projections."

ReachLocal's story proves the model that winning early sales takes precedence over profits. Later, when significant revenue is needed, profitability takes on more importance.

Source: Adapted from Worrell D. Playing favorites: striking a balance between sales, profits, and cash flow may mean giving one the upper hand. Entrepreneur 2008:September; Kaplan D. ReachLocal snags $55.2 million for local search marketing. Venture Beat 2007:October 8.

INTRODUCTION

Every venture needs capital (money) to start, operate, maintain itself, and grow. Money, preferably in the form of cash, fuels the venture. To be successful, a venture must take in more cash than it needs to operate. That cash either must be generated via sales or it must come from external investors and/or lenders. Businesses need money to purchase or lease equipment; build up inventory; and pay the utilities, employees' wages, taxes, and rent. Without cash, a business cannot survive. A well-funded venture can be *unprofitable* for a while until it builds a customer base and moves toward profitability. However, without the *cash* needed to pay its creditors, employees, and others, it cannot last very long at all.[1]

No venture, no matter how large or how small, is isolated from the need for sound financial management. Often, when firms get too large and lose track of important details of their day-to-day financial condition, they encounter problems that may take months or even years to fix—and some cannot be fixed at all. The bankruptcy of large financial firms such as Bear Stearns, Lehman Brothers, and AIG during the credit crisis of 2008 has placed a new emphasis on financial control and reporting. These firms had long histories, great brands, and were, prior to their bankruptcies, worth billions of dollars. The fact that they went bankrupt should highlight the need to pay attention to the financial health of the venture at all times.

In this chapter, we define accounting and finance and describe the role of the accountant and the financial manager through launching, operating, and growing a technology venture. Then we look at the basic financial statements that are used both to estimate how the venture will perform in the future and to analyze how it actually does perform. Next, we discuss how to analyze financial statements to determine whether a venture is achieving its goals and to provide insight into what is working and what is not working. Finally, we review some techniques for managing a venture's cash and finances, including accounts receivable, inventory, credit and collections policies, and several other items. We begin by examining the function of accounting and how it influences venture performance.

12.1 ACCOUNTING—DEFINITION AND PRACTICES

Busy entrepreneurs and aggressive business-people sometimes deride accounting as being little more than "bean counting." However, most successful entrepreneurs and people in the business community will attest that a basic understanding of accounting principles is an essential entrepreneurial tool. In a word, accounting is the language of business. Anyone who aspires to launch and operate a technology venture must, to some degree, speak that language. Businesses run on their numbers, and accounting plays an important role in recording and validating the numbers that drive business decisions. The American Accounting Association (AAA) defines **accounting** as "the process of identifying, measuring, and communicating economic information to permit informed judgments and decisions by users of information."[2]

Over the years, accounting has become increasingly standardized, but there are always new and unexpected transaction types that must be understood in accounting terms. It's important for business-people to realize that accounting, although based on rules, is an evolving discipline. In general, accounting practitioners adhere to what is known as **generally accepted accounting principles (GAAP)**. The Financial Accounting Standards Board (FASB) develops the principles in GAAP. FASB consists of seven board members who issue statements and guidelines regarding accounting practices. As new financial reporting and measurement systems arise, FASB board members consider how business transactions should be recorded.

Accounting can be divided into two categories: financial accounting and managerial accounting. These types of accounting differ primarily by virtue of the people they are designed to serve. **Financial accounting** is intended primarily for use by external decision makers such as investors, creditors, and the Internal Revenue Service. **Managerial accounting** is mainly used by internal decision makers such as company managers. In this chapter, we will be concerned primarily with financial accounting.

Accounting information is useful to a venture's managers, investors, creditors, advisors, and others. Current investors use it to review the performance of the company in which they have interest and to determine whether to maintain, increase, or liquidate their investment. Potential investors use accounting information to help them make investment decisions. Creditors use financial information to evaluate credit applications and to make decisions about candidates for loans. In most cases, when making a loan decision, lenders will look at a firm's historical financial records (usually, at least 3 years). Most lenders focus on **cash flow** when making a loan decision to ensure that the firm has sufficient **liquidity** to make principal and interest payments.[3]

The venture's managers are the most frequent users of its accounting information. They must have reliable information to make decisions about allocating resources, cutting expenses, and investing in growth. Entrepreneurial managers also use accounting information to assess the consequences of alternative business strategies. In addition, accounting information can be used to compare actual financial results with expectations.

12.1.1 The Accounting Cycle

A primary purpose of accounting is to communicate the results of business transactions. The **accounting cycle** is a sequence of six steps used to keep track of what has happened in the business and to report the financial effect of those events. The steps in the accounting cycle are depicted in Exhibit 12.1.

The accountant first analyzes business transactions to determine which should be recorded and at what amount. Typically, accountants record only transactions that can be measured and verified with some degree of precision. For example, the purchase of a truck for making deliveries to customers can be accurately measured and easily verified. However, an event like the resignation of a key employee, also a business transaction, would not be recorded. Although such an event may represent an economic loss to a company, it is difficult to determine accurately what the amount of the loss would be.

Accounting transactions are recorded chronologically in a **journal**. A journal may be either a book (in a manual accounting system) or an electronic file (in a computerized accounting system). Each entry contains the date of the transaction, its description, and debit and credit columns. Transactions are recorded in the firm's general journal or in specialized journals. A general journal is a book or file in which transactions are recorded in the order they occur. As businesses expand, they may adopt specialized journals to record particular types of business transactions (e.g., credit sales or capital expenditures).

Businesses need to know the balances of various financial statement elements (assets, liabilities, owners' equity, revenues, and expenses) at any point in time. **Accounts** are used to summarize all transactions that affect a particular financial statement component. A business maintains separate accounts for each of its assets, liabilities, equities, revenues, and expenses. All transactions affecting an account are

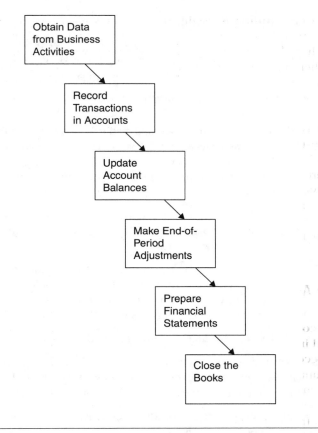

EXHIBIT 12.1

The Accounting Cycle.

posted (recorded) from the general journal or specialized journals to the account. All the accounts of a business are summarized in the **general ledger**.

The final steps of the accounting process involve preparing financial statements. **Financial statements** present a company's financial position, results of operations, and flow of cash during a particular period of time. The financial statements that are prepared include the income statement, balance sheet, and statement of cash flows. They can be prepared for any time interval (e.g., daily, weekly, monthly), but they are always prepared annually according to the venture's fiscal year. Investors, creditors, company managers, and others interested in a firm's financial position all rely heavily on financial statements. In preparing these statements, accountants check the balance sheet to determine whether their entries balance. If not, adjustments are made to how the numbers have been entered until a balance is achieved and the books can be closed.

12.1.2 The Accounting Equation

The **accounting equation** indicates a company's financial position at any point in time. On its framework rests the entire accounting process. According to the accounting equation, a company's assets equal its liabilities plus owners' equity. Thus:

$$\text{Assets} = \text{Liabilities} + \text{Owners' Equity}$$

Any recorded business transaction can be analyzed in terms of its effect on the accounting equation. Also, business transactions must be recorded to maintain the equality of this equation. This equality is reflected in the balance sheet, one of the financial statements a firm is required to prepare. These are some terms and concepts that businesses use to keep records:

- **Assets** are anything of value owned by the business and used in conducting its operations. Examples include cash, investments, inventory, accounts receivable, and furniture and fixtures.

- **Liabilities** are debts owed by the business to its creditors, including obligations to perform services in the future. Liabilities include accounts payable and notes payable (e.g., when a firm uses credit to purchase machinery), wages payable to employees, and taxes payable.

- **Owners' equity** represents the claims of the owners against the firm's assets.

- **Revenues** are inflows of assets resulting from the ongoing operation of a business. Businesses generate revenues by sales of goods and services, interest earned on investments, rents, royalties, and dividends.

- **Expenses** are costs incurred to produce revenues. Expenses include the costs of goods sold (the goods or services that the firm used to generate revenues), salaries, utilities, taxes, marketing, and interest payments. Revenues and expenses are components of owners' equity. Revenues result in an increase, while expenses result in a decrease in owners' equity. Net income, or the bottom line, is the excess of revenues over expenses. Investors, creditors, and company managers closely watch net income, a chief barometer of business performance.

12.1.3 Working with Accountants

Most entrepreneurs do not spend their time actually doing the venture's accounting. Rather, they generally hire an experienced certified public accountant (CPA) to manage the venture's accounting. Most large cities have multiple CPA firms that work with small- to medium-sized companies and who will tailor their pricing and billing to match the budgets of these smaller ventures.

Technology entrepreneurs may want to select a CPA or CPA firm with experience in its industry. The rules that govern accounting are constantly changing, and they are especially rapidly changing in technology-based industries. A good CPA will know the rules that apply within a particular industry and will use them aggressively, yet ethically, to the best advantage of the venture. A good way to find an accountant qualified to work with a particular company in a particular industry is to ask others for referrals. The best accountants usually acquire most of their new clients via referrals, which is an endorsement of their capabilities and service quality.

Accountants serve to record and report the financial transactions of the venture. They are responsible for assuring that the venture conforms to the accounting rules of its industry. The finance function, by way of contrast, uses accounting statements to determine whether the venture has adequate capital, whether the capital it has is being used efficiently, and whether additional capital needs to be obtained. This is the role of the financial manager, which we turn to next.

Alternatively, for those entrepreneurs who do not want to outsource their accounting to an external company, there are sophisticated software tools that help them manage their finances. These tools can, at least in the early stages of the venture, provide substantial support in establishing accounts and the venture's journals and general ledger.[4]

Regardless of the method used to set up and manage a venture's accounts, the research is very clear that entrepreneurs who develop financial controls into the venture early on are more likely to succeed than those who do not.[5]

12.2 FINANCE—DEFINITION AND PRACTICES

Because a venture must have a sufficient supply of cash to operate, the area of cash management requires special attention. Entrepreneurs must determine the venture's cash needs for both the short and the long term, and then find sources to provide the necessary cash. Cash management within the firm is the entrepreneur's responsibility.

As a discipline within a business school, **finance** is the study of how to manage money—either from a personal perspective or from a business perspective. At the personal level, finance is concerned with household wealth management, including things like savings accounts, mortgages, mutual fund investments, and others. Within a business, finance is the functional area charged with locating and acquiring capital for the business, managing capital in a disciplined manner, and determining the best uses of capital given the strategies and goals of the venture. There are vast differences between personal finance and business finance. And there are substantial differences between corporate finance and entrepreneurial finance. The latter, of course, will be our primary concern in this chapter.

12.2.1 The Financial Manager

The **financial manager** is the individual responsible for the venture's finance function. It is common for large firms to use the title of chief financial officer (CFO) to refer to the person in charge of a firm's financial affairs. Many entrepreneurial ventures do not formally appoint a financial manager, but the financial management tasks must still be performed. In entrepreneurial ventures, one of the managers or owners often will handle these tasks. The tasks assigned to the person responsible for financial management include:

- Business plan development
- Risk management and insurance
- Benefits and compensation
- Equipment leasing and financing
- Facility and real estate needs analysis
- Budget development
- Tax planning
- Expansion planning
- Cash flow management
- Owner compensation planning
- Commercial real estate financing and planning
- Financial system design and implementation
- Business sale preparation

Another option available to early stage ventures that cannot afford a full-time CFO is to work with a firm that provides "virtual CFO" services.[6] These firms work with early stage ventures in a manner similar to accounting firms. A virtual CFO firm will assign one of its financial experts to work with the client venture. In turn, the client is billed for services normally at an hourly or monthly rate. Often, the venture will list the individual performing virtual CFO duties as the "acting CFO" in its business plan and investor documents, and/or it may list the retained firm in those documents.

12.2.2 The Financial Plan

Entrepreneurial finance differs from corporate finance in several important ways. The first and most obvious difference is that of scale. Corporate finance is concerned with the management of large amounts of capital. It includes such activities as investing the firm's capital to ensure that the business is earning optimum returns and that its capital is not sitting idly in a noninterest-bearing savings account. The second major difference is the relative time spent on activities associated with financial management. The corporate financial manager spends a considerable time thinking about how to invest the company's money, and about where and how to acquire capital for large projects and investments. The entrepreneurial financial manager spends far more time acquiring capital and relatively little time investing.[7]

To be effective, the financial manager in the entrepreneurial venture should develop and follow a financial plan.[8] A sound financial plan will minimally require the following tasks to be performed:

- Estimate month-by-month flow of funds into the venture.
- Estimate month-by-month flow of funds out of the venture.

- Compare monthly inflows to monthly outflows.

- If excess funds exist, plan how to use these funds for growth or investing.

- If funds are short, adjust inflows or outflows and/or look for other sources of funding.

- If other funding sources are needed, analyze alternatives to find the most efficient source.

- Establish a system routinely to monitor and evaluate the results of this process.

Over time, a technology venture's financial plan can become very complex, with investors, creditors, suppliers, and customers all adding or subtracting cash. These inflows and outflows of cash must be tracked diligently. The monthly expenses that a business incurs to operate, regardless of where the funds come from, are commonly referred to as the venture's **burn rate**. Investors and lenders will all examine a venture's burn rate and its cash on hand to determine how long it can operate. For example, if a venture's burn rate is $30,000/month and it has $300,000 in the bank, it can operate for another 10 months before more cash will be required.

In addition to the monthly burn of cash, the future growth strategies and cash requirements of the venture have to be considered. The technology venture's financial manager must work with the financial statements produced by the firm's accountant to monitor the burn rate. The financial manager must also work with the executive team to understand the strategic objectives of the venture and the amount of capital that will be required to achieve those objectives in the future.

Technology entrepreneurs face challenges similar to any other entrepreneur, but raising capital can be considerably more challenging for some types of technology ventures. For example, bio-pharmaceutical and medical device companies may require large injections of capital before they reach cash flow breakeven.[9] This unique challenge requires diligence and persistence on the part of the entrepreneur.[10] It also requires developing relationships with investors and/or lenders before capital is required. Smart technology entrepreneurs cultivate potential investors in advance of their actual capital requirements. That is, the entrepreneur should develop key relationships with investors and lenders in advance so that they can develop an understanding of the business and a sense of trust for the management team.[11] This will provide a slight edge in the fundraising process over first-impression meetings when capital is critical to carry on operations. Recall that the process of fundraising was covered at length in Chapter 5.

12.3 FINANCIAL STATEMENTS

A venture's cash needs will change over time, and having sufficient capital for continued operations can become an issue even for long-established ventures. For example, in 2008, the U.S. financial system was sorely challenged by a sudden loss of confidence in many mortgage-backed securities. The so-called "subprime crisis" of 2008 was caused by a frenzy of innovation in mortgage originations to homebuyers.[12] Many people who, under normal circumstances, would not qualify for a mortgage loan were being approved.

The factors underlying the subprime crisis are exceedingly complex, and will take years for investigators and regulators to sort out. The outcome of the crisis, however, was very real for many formerly highly liquid financial firms on Wall Street. One very visible victim of the crisis was the venerable financial services firm, Bear Stearns. Bear management realized during the spring of 2008 that the firm's reliance on mortgage payments from subprime borrowers had placed it at risk. Many of these borrowers were unable to pay their mortgages, and the flow of cash was drying up. The crisis for Bear reached a fever pitch in March 2008 when senior managers realized the investment banking house was about to fail. On the weekend of March 15–16, Bear management, the U.S. Federal Reserve, and others met to decide the company's fate. On the morning of March 17, it was announced that Bear would be sold to J.P. Morgan at the rock-bottom price of $2 per share—less than one-tenth of the value of the firm's shares at the close of business the previous Friday.[13]

To avoid such problems, a venture must plan its cash flow. A detailed financial plan helps ensure the long-run success of the venture. Entrepreneurs are usually required to estimate how the venture will perform financially in the future. This is done through four basic financial statements:

1. The sales forecast
2. The income statement
3. The cash flow statement
4. The balance sheet

These basic financial statements (often referred to as "pro forma financial statements") are normally required as part of a complete business plan.[14] Investors and lenders will almost always require that entrepreneurs provide financial projections for 3 years or more into the future.[15] Note that the process of estimating future financial performance is different from what accountants do. Accountants record actual performance and allocate the cash flow from transactions to the various financial statements. The process of estimating future cash flows is usually not a job done by the venture's accounting firm. Rather, it is the job of the financial manager to estimate sales, expenses, cash needs for growth, and future performance. Later, we'll examine how financial managers use the actual financial statements prepared by accountants to analyze past performance. Here, we examine the four statements in the context of estimating how the venture will perform in the future. We begin with the sales forecast.

12.3.1 Sales Forecast

The finance function in a business is responsible for acquiring the funds the business needs. Sound financial management involves determining how much money is needed for various time periods and the most appropriate sources of funds. The most obvious source of funds for a venture is the revenue generated by the business. However, as stated earlier, sometimes a venture cannot generate adequate cash inflows from sales to operate and grow. This is especially true of start-up ventures, and may also occur during the venture's high-growth phase. To determine the amount of external financing needed, financial managers estimate the revenues the venture can expect on a month-to-month basis, and subtract all of the expected costs.

The primary tool for estimating a venture's cash inflows from selling activities is the **sales forecast**. The sales forecast is the foundation of the various financial statements that entrepreneurs use to project their venture's financial performance into the future. The sales forecast is foundational for the pro forma financial projections because the numbers reflected here flow automatically to the other financial statements. A sales forecast consists of the following basic components:

- Units of sale
- Pricing for each of the various units
- Volume of sales for each unit in each accounting period
- Variance in sales volume between accounting periods due to exigencies such as:

  - Market penetration rates (referred to as **ramp rate**)
  - Seasonality (variations in sales patterns based on seasons)
  - Marketing investments (variations in sales due to marketing investments)

- Total (or "gross") sales (this line will be carried to the top line of the income statement (revenue) without change

Exhibit 12.2 provides a basic template for a sales forecast. Many entrepreneurs use Excel to generate these statements from scratch, but there are commercially available software packages as well as free shareware and Excel templates available on the Internet. Notice the template is divided into monthly accounting periods, which is standard. Notice also that the template contains the basic components listed above. Each cell of the template is formula driven, other than those that establish price and ramp rate. These must be based on reasonable assumptions.

All assumptions used to develop the pro forma financial statements should be listed on a separate document. The formula-driven nature of the sales forecast enables the entrepreneur to run through various "what-if?" scenarios to see the effects on gross sales of changes in assumptions about price, ramp rate, or both.

The entrepreneur must determine the venture's basic units of sale. The price of each of these various units is recorded as an assumption in the sales forecast. The ramp rate of sales per item is also recorded as an assumption for each unit of sale and for each accounting period. That is, the ramp rate may change from period to period as market penetration succeeds. Entrepreneurs must think hard about how slowly or quickly their offerings will be adopted by the market. The ramp rate should be justifiable—i.e., it should be based on research rather than guesswork.[16] Anyone interested in looking at the venture's financial projections will want to know why the assumed ramp rate was used.[17]

Year 1

Item	Month 1	Month 2	Month 3	Month 4	Month 5	Month 6	Month 7	Month 8	Month 9	Month 10	Month 11	Month 12	
Product A Unit Sales	50	50	101	201	402	804	1,206	1,809	2,714	4,070	6,105	9,158	26,670
Product A Revenue	$500	$503	$1,005	$2,010	$4,020	$8,040	$12,060	$18,090	$27,135	$40,703	$61,054	$91,581	$266,699
Product B Unit Sales	0	0	0	100	175	245	294	323	356	391	430	473	2,788
Product B Revenue	$0	$0	$0	$2,995	$5,241	$7,338	$8,805	$9,686	$10,654	$11,720	$12,892	$14,181	$83,512
Product C Unit Sales	0	0	0	0	0	0	0	1000	1000	1000	1000	1000	5,000
Product C Revenue	$0	$0	$0	$0	$0	$0	$0	$12,000	$12,000	$12,000	$12,000	$12,000	$60,000
Product A Ramp Rate	0.000%	0.500%	100.000%	100.000%	100.000%	100.000%	50.000%	50.000%	50.000%	50.000%	50.000%	50.000%	
Product B Ramp Rate	0.000%	0.000%	0.000%	0.000%	75.000%	50.000%	10.000%	10.000%	10.000%	10.000%	10.000%	10.000%	
Product C Ramp Rate	0.000%	0.000%	0.000%	0.000%	75.000%	40.000%	20.000%	10.000%	10.000%	10.000%	10.000%	10.000%	
Product A Price	$10.00	$10.00	$10.00	$10.00	$10.00	$10.00	$10.00	$10.00	$10.00	$10.00	$10.00	$10.00	
Product B Price	$29.95	$29.95	$29.95	$29.95	$29.95	$29.95	$29.95	$29.95	$29.95	$29.95	$29.95	$29.95	
Product C Price	$12.00	$12.00	$12.00	$12.00	$12.00	$12.00	$12.00	$12.00	$12.00	$12.00	$12.00	$12.00	
GROSS SALES	$500	$503	$1,005	$5,005	$9,261	$15,378	$20,865	$39,776	$49,789	$64,422	$85,946	$117,762	$410,212

EXHIBIT 12.2

Sales Forecast Template.

In addition, to pondering how many units of each item the venture will sell, the entrepreneur must think hard about how long it will take to sell each new unit. The length of time required to obtain a new sale is referred to as the **sales cycle**. The sales cycle will vary from industry to industry. For example, getting physicians to adopt a new medical instrument or pharmaceutical will likely take longer than getting consumers to purchase a new computer game. Experienced entrepreneurs know that it is far easier and less expensive to sell more products to existing customers than it is to acquire a new one. Selling more to existing customers creates **recurring revenue**. Recurring revenue is even better if the customer is set up to purchase continuously and automatically. For example, consider your mobile phone service provider. Essentially, the provider is selling you service every month. And, if you are like many customers, they automatically debit your credit card to get their payment.

Often entrepreneurs will develop multiple sales forecasts, reflecting worst-case, base-case, and best-case scenarios. The projections developed in the sales forecast flow into the other financial statements. The gross sales figures on the bottom line of the sales forecast flow through to the income statement as the top line or gross revenue. Tech Tips 12.1 has some additional suggestions for crafting a sales forecast.

TECH TIPS 12.1

Tips for Developing a Sales Forecast

- *Develop a unit sales projection*: Where you can, start by forecasting unit sales per month. Not all businesses sell by units, but most do, and it's easier to forecast by breaking things down into their component parts.
- *Use past data if you have it*: Whenever you have past sales data, your best forecasting aid is the most recent past.
- *Use factors for a new product*: Having a new product is no excuse for not having a sales forecast. Nobody knows the future—you simply make educated guesses. So break it down by finding important decision factors or components of sales. If you have a completely new product with no history, find an existing product to use as a guide. For example, if you have the next great computer game, base your forecast on sales of a similar computer game.
- *Break the purchase down into factors*: For example, you can forecast sales in a restaurant by looking at a reasonable number of tables occupied at different hours of the day and then multiplying the percent of tables occupied by the average estimated revenue per table.
- *Be sure to project prices*: The next step is prices. You've projected unit sales monthly for 12 months and then annually, so you must also project your prices.[18]

12.3.2 Income Statement

Operating costs (usually referred to as "expenses") for a venture are projected or recorded on what is referred to as the **income statement**. The income statement reports the gross sales of the venture as revenue on its top line. All of the expenses associated with operating the venture are recorded and subtracted from revenue in each accounting period (monthly and annually) to generate bottom line profits or losses.

Costs associated with operating a venture can be divided into categories of variable and fixed. **Variable costs** are those that vary directly with production and output. For example, a venture that makes electric automobiles will need to increase production as demand for its products increases. Increasing production will require that the firm purchase additional raw materials, energy to run the plant, and other items whose costs will vary directly with the increased production. In the language of accounting, variable costs are listed on the venture's income statement as cost of goods sold, or "COGS."

Technology ventures will have widely divergent variable costs, depending on the industry. For example, it's clear that manufacturing electric cars will have very high variable costs, due to the relatively high cost of raw materials for manufacturing cars. On the other hand, software companies usually have relatively low variable costs as producing another copy of a software program can cost as little as a blank CD ROM disk.

Fixed costs, by definition, do not vary directly with production. For example, the car manufacturer will use the same factory and equipment to produce more or fewer cars over a period of time. It will also use the same labor and other items that are referred to in the venture's income statement as "General and Administrative" (G&A) expenses. Sometimes, these costs are also referred to as "overhead." Note that it's possible that the venture will need to hire more people and/or build new plants to handle its increased output. These costs are still "fixed" in that they don't vary directly with the increased output. Once the labor is hired or the factory built, its cost remains fixed on the venture's income statement across accounting periods.

A template for a basic income statement (also often referred to as the "profit and loss statement" or, simply, "P&L") is provided in Exhibit 12.3.

The result of subtracting G&A expenses from gross profit is a line item typically entered as "EBIT" or "EBITDA." These abbreviations stand for "earnings before interest and taxes" and "earnings before interest, taxes, depreciation, and amortization." Depreciation is a noncash expense that accounts for wear and tear on the venture's capital assets (e.g., machinery, real estate). Amortization is the interest paid on loans. Depreciation and amortization are considered business expenses and can be subtracted from a venture's revenue.

The amount of money that is left after all expenses are subtracted from revenue is the venture's net income. Net income represents the venture's taxable income. If the venture is structured as a pass-through entity (i.e., it is not a C-corporation), then this amount is passed through to owners and taxed at the individual income tax level. If the venture is a C-corporation, then this amount is taxed at the prevailing corporate income tax rate.

The amount that remains after taxes is the venture's profit. This is what the venture's officers can now decide to reinvest into the venture as retained earnings or distribute to shareholders as dividends.

YEAR 1

	Month 1	Month 2	Month 3	Month 4	Month 5	Month 6	Month 7	Month 8	Month 9	Month 10	Month 11	Month 12	
Revenue													
Total Revenue	$500	$503	$1,005	$5,005	$9,261	$15,378	$20,865	$39,776	$49,789	$64,422	$85,946	$117,762	$410,212
COGS													
Raw Materials	$200	$201	$402	$2,002	$3,705	$6,151	$8,346	$15,910	$19,916	$25,769	$34,378	$47,105	$164,085
Shipping	$75	$10	$20	$100	$185	$308	$417	$796	$996	$1,288	$1,719	$2,355	$8,269
Total COGS	$275	$211	$422	$2,102	$3,890	$6,459	$8,763	$16,706	$20,912	$27,057	$36,097	$49,460	$172,354
GROSS PROFIT	**$225**	**$291**	**$583**	**$2,903**	**$5,372**	**$8,919**	**$12,102**	**$23,070**	**$28,878**	**$37,365**	**$49,848**	**$68,302**	**$237,858**
Percent	**45.00**	**58.00**	**58.00**	**58.00**	**58.00**	**58.00**	**58.00**	**58.00**	**58.00**	**58.00**	**58.00**	**58.00**	
G&A													
CEO	$3,500	$3,500	$3,500	$3,500	$3,500	$3,500	$3,500	$3,500	$3,500	$3,500	$3,500	$3,500	$42,000
Sales Salary	$2,000	$2,000	$2,000	$2,000	$2,000	$2,000	$2,000	$2,000	$2,000	$2,000	$2,000	$2,000	$24,000
Burden	$1,045	$1,045	$1,045	$1,045	$1,045	$1,045	$1,045	$1,045	$1,045	$1,045	$1,045	$1,045	$12,540
G&A Allocated Rent	$2,500	$2,500	$2,500	$2,500	$2,500	$2,500	$2,500	$2,500	$2,500	$2,500	$2,500	$2,500	$30,000
Travel	$2,500	$2,500	$2,500	$2,500	$2,500	$2,500	$2,500	$2,500	$2,500	$2,500	$2,500	$2,500	$30,000
Insurance	$400	$400	$400	$400	$400	$400	$400	$400	$400	$400	$400	$400	$4,800
Professional Services	$300	$300	$300	$300	$300	$300	$300	$300	$300	$300	$300	$300	$3,600
Telephones	$500	$500	$500	$500	$500	$500	$500	$500	$500	$500	$500	$500	$6,000
Office Supplies	$500	$500	$500	$500	$500	$500	$500	$500	$500	$500	$500	$500	$6,000
Office Equipment	$300	$300	$300	$300	$300	$300	$300	$300	$300	$300	$300	$300	$3,600
Postage/Courier	$50	$50	$50	$50	$50	$50	$50	$50	$50	$50	$50	$50	$600
Total	$13,595	$13,595	$13,595	$13,595	$13,595	$13,595	$13,595	$13,595	$13,595	$13,595	$13,595	$13,595	$163,140
EBITDA	($13,370)	($13,304)	($13,012)	($10,692)	($8,223)	($4,676)	($1,493)	$9,475	$15,283	$23,770	$36,253	$54,707	$74,718
Depreciation	$0	$0	$0	$0	$0	$0	$0	$0	$0	$0	$0	$0	$0
Interest	$0	$0	$0	$0	$0	$0	$0	$0	$0	$0	$0	$0	$0
Taxes	$0	$0	$0	$0	$0	$0	$0	$2,842	$4,585	$7,131	$10,876	$16,412	$41,846
NET INCOME	**($13,370)**	**($13,304)**	**($13,012)**	**($10,692)**	**($8,223)**	**($4,676)**	**($1,493)**	**$6,632**	**$10,698**	**$16,639**	**$25,377**	**$38,295**	**$32,871**

EXHIBIT 12.3

Income Statement Template.

12.3.3 Cash Flow Statement

A very important financial statement to entrepreneurs is the **cash flow statement**. The cash flow statement shows the flow of cash into and out of a business during a period of time. You may have heard the expression that, to the entrepreneur, "cash is king." What that means is cash is required to pay bills and to invest in venture growth. This statement also highlights the difference between "cash" and "profits."

Profits can be shown on an income statement without the venture having received any cash at all. For example, according to the rules of accounting, a venture can record that a sale has occurred and book revenue on the income statement even though no cash has been transferred. This would be common for credit transactions, where a sale is made to a customer who promises to pay at some future date. Until that future date arrives, the venture's income statement will indicate a profit on the transaction, even though no cash has been received. That is why it is vital for the entrepreneur to monitor the actual cash position of the venture. Only cash can be used to pay bills.

The cash flow statement monitors the inflow and outflow of cash from the venture's operating, investing, and financing activities. Cash flow from **operating activities** measures the cash results of the firm's primary revenue-generating activities. *Investing activities* include buying fixed assets, buying stock in other companies, and selling stock held as an investment in another company. **Financing activities** include issuing new stock, paying dividends to shareholders, borrowing money from banks, and repaying amounts borrowed. The statement of cash flows also shows the net change in cash for the period. Exhibit 12.4 is a template for a basic cash flow statement.

As a summary of the effects on cash of all the firm's operating, financing, and investing activities, the statement of cash flows allows entrepreneurs to see the results of past decisions. The statement may, for example, indicate a great enough cash flow to allow the firm to finance projected needs itself rather than borrow funds from a bank. Or management can examine the statement to determine why the firm has a cash shortage, if that is the case. Investors and creditors can use the statement of cash flows to assess the firm's abilities to generate future cash flows, to pay dividends, and to pay its debts when due, as well as its potential need to borrow funds.

12.3.4 Balance Sheet

The **balance sheet** lists everything a company owns and everything it owes at a specific moment in time. It shows where all of the venture's money has come from (sources of capital) and what it has been used for. In the end, a balance sheet must balance because all of the sources of money to the venture must equal the uses of that money. The previous financial statements discussed earlier all represent how the venture performs over time. The balance sheet, by way of contrast, is a "snapshot" of the financial health of the venture at a particular moment. Balance sheets normally are prepared at the end of each month and at the end of each fiscal year.

YEAR 1

	Month 1	Month 2	Month 3	Month 4	Month 5	Month 6	Month 7	Month 8	Month 9	Month 10	Month 11	Month 12
Cash Flow from Operations												
Income	($13,370)	($13,304)	($13,012)	($10,692)	($8,223)	($4,676)	($1,493)	$6,632	$10,698	$16,639	$25,377	$38,295
Add Depreciation	$0	$0	$0	$0	$0	$0	$0	$0	$0	$0	$0	$0
Net Cash from Operations	($13,370)	($13,304)	($13,012)	($10,692)	($8,223)	($4,676)	($1,493)	$6,632	$10,698	$16,639	$25,377	$38,295
Cash Flow from Investing												
(Increase) Decrease	$0	$0	$0	$0	$0	$0	$0	$0	$0	$0	$0	$0
Net Cash from Investing	$0	$0	$0	$0	$0	$0	$0	$0	$0	$0	$0	$0
Cash Flow from Financing												
Short-Term Debt	$0	$0	$0	$0	$0	$0	$0	$0	$0	$0	$0	$0
Long-Term Debt	$0	$0	$0	$0	$0	$0	$0	$0	$0	$0	$0	$0
Common Stock	$100,000	$0	$0	$0	$0	$0	$0	$0	$0	$0	$0	$0
Net Cash from Financing	$100,000	$0	$0	$0	$0	$0	$0	$0	$0	$0	$0	$0
Net Increase (Decrease) in Cash	$86,630	($13,304)	($13,012)	($10,692)	($8,223)	($4,676)	($1,493)	$6,632	$10,698	$16,639	$25,377	$38,295
Beginning Cash	$0	$86,630	$73,326	$60,314	$49,622	$41,399	$36,723	$35,230	$41,862	$52,560	$69,199	$94,577
Ending Cash	$86,630	$73,326	$60,314	$49,622	$41,399	$36,723	$35,230	$41,862	$52,560	$69,199	$94,577	$132,871

EXHIBIT 12.4

Cash Flow Statement Template.

The sources of money for a venture include owners and creditors. Creditors have loaned money to the venture and have thus created liabilities for it. That is, the venture is liable for paying back the money it has borrowed, plus interest. Owners have invested their own money into the venture or they have allowed the venture's earnings to be retained rather than drawing them out in the form of dividend payments. As equity is understood, the venture "owes" its owners their pro rata share of the accumulated equity (invested capital plus retained earnings).

A template for a balance sheet is provided in Exhibit 12.5. As depicted, the balance sheet has two sides: one lists all of the venture's assets (what it owns); the other lists its liabilities and equity (what it owes). The basic equation that represents the balance sheet is:

$$\text{Assets} = \text{Liabilities} + \text{Equity}$$

The assets side of the balance sheet represents how the venture has used its capital. The other side represents where the capital came from. If everything has been accounted for properly, the two sides will be equal and are said to "balance."

In Exhibit 12.5, both assets and liabilities are listed in decreasing order of liquidity. **Liquidity** refers to the readiness with which an asset can be converted to cash or to the urgency with which a liability requires cash. On the asset side, the most liquid asset, of course, is the cash the venture holds in its bank accounts. Accounts receivable is listed next. This is the amount of money that is due to the venture from sales to customers—sales that have been made on credit. Additional items on the asset side include inventory and fixed assets. These are usually the least liquid of a venture's various assets.

ASSETS		LIABILITIES	
Cash	$1,000	Accounts Payable	$1,500
Accounts Receivable	$3,000	Accruals	$500
Inventory	$2,000	Current Liabilities	$2,000
Current Assets	$6,000		
		Long-Term Debt	$5,000
Fixed Assets		Equity	$2,000
Gross	$4,000	Total Capital	$7,000
Accumulated Depreciation	–$1,000		
Net	$3,000		
		TOTAL LIABILITIES	
TOTAL ASSETS	$9,000	AND EQUITY	$9,000

EXHIBIT 12.5

Balance Sheet Template.

Liabilities usually begin with accounts payable. These are listed first because they represent the vendors that provide the venture with supplies and materials needed to operate and produce. If these bills go unpaid, the venture may be unable to meet customer demand. Second on the list of liabilities is what is referred to as "accruals." Accruals are simply expenses associated with transactions that are not entirely complete at the close of an accounting period. For example, suppose a monthly accounting period ends on a Wednesday, but a venture pays its employees on Friday. The accrued salaries for those employees must appear on the end-of-month balance sheet, which is developed on Wednesday. If those accrued wages were not recorded in the end-of-month balance sheet, the actual liabilities of the venture would be misstated. Current liabilities are defined as those requiring payment within 1 year, and long-term debt consists of the venture's obligations to lenders.

The equity in the venture consists of two items: that which has been directly purchased by investors as common stock and that which represents the retained earnings. Money retained by the venture is considered to be a contribution by the owners. That is, the owners decide whether to take earnings out of the venture in dividend payments or they leave it in the venture as retained earnings to fuel future growth.

12.4 FINANCIAL STATEMENT ANALYSIS

Financial statement analysis compares (or finds relationships in) accounting information to make the data more useful or practical. For example, knowing that a company's net income was $50,000 is somewhat useful. Knowing also that the net income for the previous year was $100,000 is more useful. Knowing the amounts of the company's assets and sales is better yet. To this end, several types of financial statement analyses have been developed. Here, we discuss two of the primary techniques entrepreneurial financial managers use to analyze the venture's financial statements: breakeven analysis and ratio analysis.

12.4.1 Breakeven Analysis

Breakeven analysis is normally done for two different variables: profit and cash. Profit breakeven occurs when the venture sells enough of its goods and services to cover its operating costs. Cash breakeven occurs when the venture is able to collect sufficient cash from sales of its goods and service to cover all of its outgoing cash payments. Each is calculated differently. The profit breakeven point is calculated using the income statement, and the cash breakeven point is calculated using the cash flow statement.

In Exhibit 12.6, the profit breakeven point can be seen to occur around the time the venture is selling 25,000 units per year. It is the point where the venture estimates that its sales will bring in greater revenue than its variable and fixed costs combined.

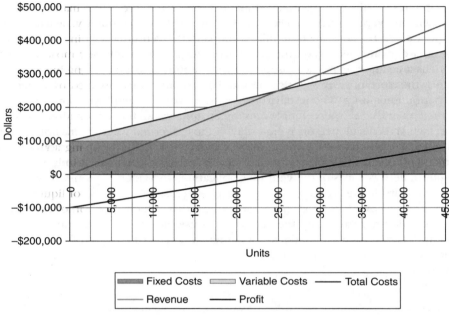

EXHIBIT 12.6

Profit Breakeven.

The cash flow breakeven point for a venture generally occurs after the profit breakeven point—sometimes many months after. That is because the venture may make at least a portion of its sales using credit terms. The lag between profit breakeven and cash breakeven is the time it takes to collect the accounts receivable. There may also be some portion of payments that never all collected, increasing the time lag between profit breakeven and cash breakeven.

12.4.2 Ratio Analysis

Ratio analysis examines the logical relationships between various financial statement items. Items may come from the same financial statement or from different statements. The only requirement is that a logical relationship exists between the items. Ratios are very important to investors and other stakeholders in evaluating the financial health of the venture. Comparing company ratios to industry standards can indicate areas in which the company is successful—and those in which the company is below standard. Using ratios enables a venture to compare itself to others in its industry, regardless of the size of the comparable firms.[19] For example, a small medical devices venture may compare itself to firms such as Medtronic or Smith-Kline.

Credit managers, bankers, and financial and investment analysts use several hundreds of different ratios to analyze the health of a venture. Only a handful of these ratios are useful in interpreting a venture's financial standing, and many more may give misleading information. The ratios presented here—liquidity, activity, profitability, and debt—are some of the so-called **key ratios** that are in common use. Industry standards on certain key financial ratios can be found for most industries through trade associations, investment analyst reports, or from leading business publications. Trends in a company's performance are also easily spotted by comparing ratios from the current period with ratios from earlier periods.

It is important to note that different industries have distinct standards for ratio adequacy. What is an adequate ratio for a company in one industry may not be adequate for a different company in another industry. Beware of judging a ratio solely on the basis of a universal standard; ratios should be viewed against industry norms. For instance, the banking industry and manufacturing industry would differ significantly in terms of current assets. Banks carry an enormous amount of liquid current assets, whereas manufacturing firms have a much more even division of current and long-term assets. So a comparison of a bank and a manufacturing firm in terms of any ratio that contains current assets would not be realistic.

12.4.3 Types of Ratio Analyses

There are four basic types of ratio analyses commonly used to determine the financial health of a venture. They are given in the following sections.

Liquidity ratios

A firm's ability to pay its short-term debts as they come due is measured by **liquidity ratios**. Liquidity is a measure of how quickly an asset can be converted to cash. Highly liquid firms can more easily convert assets to cash when needed to repay loans. Less liquid firms may have trouble meeting their obligations or obtaining loans at low cost. Two common liquidity ratios are the current ratio and the quick (acid-test) ratio:

- The **current ratio**, which is current assets divided by current liabilities (two balance sheet items), indicates a firm's ability to pay its current liabilities from its current assets. Thus, the ratio shows the strength of a company's working capital. Current ratio is expressed using the following equation:

$$\text{Current Ratio} = \frac{\text{Current Assets}}{\text{Current Liabilities}}$$

- The **quick ratio**, also called the acid-test ratio, divides quick assets by current liabilities. It measures more immediate liquidity by comparing two balance sheet items. Quick assets do not include inventory, because to convert inventory to cash, merchandise must be sold and a receivable collected. **Quick assets** are cash or those assets the firm expects to convert to cash in the near future. The quick ratio is expressed in the following equation:

$$\text{Quick Ratio} = \frac{\text{Quick Assets}}{\text{Current Liabilities}}$$

Activity ratio

How efficiently the firm uses its assets to generate revenues is measured by **activity ratios**. These ratios indicate how efficiently a firm uses its resources. A common activity ratio is accounts receivable turnover. **Accounts receivable turnover** is the number of times per year that the average accounts receivable is turned over (collected). An income statement item (net sales) is compared with a balance sheet item (accounts receivable). We calculate this ratio by dividing net sales by average net accounts receivable as shown below:

$$\text{Accounts Receivable Turnover} = \frac{\text{Net Sales}}{\text{Average Net Accounts Receivable}}$$

The allowance for doubtful accounts is first deducted to arrive at net accounts receivable. Average net accounts receivable is computed by adding the accounts receivable amounts at the beginning and end of the year and then dividing by two.

Profitability ratios

A company's overall operating success—its financial performance—is measured by **profitability ratios**. These ratios measure a firm's success in terms of earnings compared with sales or investments. Over time, these ratios can indicate how successfully or unsuccessfully management operates the business. Two common profitability ratios are return on sales and return on equity.

Return on sales measures a firm's profitability by comparing net income and net sales, both income statement items. It is computed by dividing net income by net sales:

$$\text{Return on Sales} = \frac{\text{Net Income}}{\text{Net Sales}}$$

Return on equity (ROE) measures the return the company earns on every dollar of shareholders' (and owners') investment. Investors are very interested in this ratio; it indicates how well their investment is doing. We compute return on equity by dividing net income (an income statement item) by equity (a balance sheet item).

$$\text{Return on Equity} = \frac{\text{Net Income}}{\text{Equity}}$$

Debt ratios

Companies measure their ability to pay long-term debts by **debt ratios**. These ratios try to answer questions such as (1) Is the company financed mainly by debt or equity? and (2) Does the company make enough to pay the interest on its loans when due? Potential investors and lenders are very interested in the answers to such questions. Two ways to answer them are the debt-to-equity ratio and the times-interest-earned ratio.

The **debt-to-equity ratio** measures a firm's leverage as the ratio of funds provided by the owners to funds provided by creditors, both balance sheet items. Total debt includes current liabilities and long-term liabilities. The debt-to-equity ratio demonstrates the risk incurred by the owners of the firm. The higher the debt-to-equity ratio is, the greater the chance that the firm will be unable to meet its obligations.

$$\text{Debt to Equity} = \frac{\text{Total Debt}}{\text{Equity}}$$

Creditors need to know whether a borrower can meet interest payments when they come due. The **times-interest-earned ratio** compares cash received from operations with cash paid for interest payments. It measures how many times the firm earns the amount of interest it must pay during the year.

$$\text{Times Interest Earned} = \frac{\text{Income Before Interest and Taxes}}{\text{Interest Expense}}$$

12.5 FINANCIAL MANAGEMENT

The entrepreneurial venture's financial manager is responsible for maintaining the proper flow of funds into and out of the venture. They manage the use of funds and find the appropriate sources of funds as required. They also invest excess funds to earn additional income for the company. In performing these duties, the financial manager has to estimate the venture's start-up costs, manage its working capital, develop capital budgets, and develop appropriate financial controls.

12.5.1 Start-up Costs

A common mistake made by new ventures is failing to plan for sufficient start-up capital (the money needed to start a business), assuming that revenue will provide the cash needed to operate and grow. Insufficient capital is a problem not only for new ventures, but it can also be a problem for long-established firms.

Start-up costs can be projected and managed by creating a start-up budget. When seeking initial capital to launch the venture, the entrepreneur should ensure that sufficient capital is raised to cover these start-up costs, and the costs of operating until breakeven or until additional capital is raised. A template for the start-up budget is provided in Exhibit 12.7. Note that the categories of costs listed are somewhat generic, and any given technology venture may have quite different associated start-up costs.

12.5.2 Working Capital

If a firm's current liabilities (obligations that must be paid within a year) are subtracted from its current assets, the result is the value of working capital. **Working capital** represents the amount of capital available for the day-to-day running of the

Start-up Expenses	
Legal	$50,000
Stationery, etc.	$2,000
Brochures	$0
Consultants	$3,000
Insurance	$3,000
Rent	$9,300
Research and development	$0
Expensed equipment	$0
Custom cad software	$5,700
Logo design	$1,000
Management salaries	$56,000
Other	$0
Other	$0
Other	$0
Total Start-up Expenses	**$130,000**

EXHIBIT 12.7

Start-up Budget Template.

firm. Sufficient working capital is obviously important to the effective management of a firm's operations.

In managing current assets, the financial manager needs to concentrate on three assets: cash, accounts receivable, and inventory. The primary concern with cash is that it should never be left idle; it should always be working. Funds not immediately needed should be invested and be earning interest. At the very least, an interest-earning checking account should be used.

12.5.3 Accounts Receivable

Accounts receivable are really promises of cash from customers of the firm. Until this cash is in hand, the firm has only the promise. One task of the financial manager is to speed up the collection of accounts receivable as much as possible. This, of course, must be done without offending customers and with the understanding that, in many cases, providing credit is necessary to generate sales.

In managing accounts receivable, the financial manager needs to date accounts receivable so that overdue accounts are flagged immediately and appropriate action is taken. The financial manager also wants to speed up the conversion of received payments into cash in the company's account. When received at the office, they must be processed and then sent to the bank. This means that the cash may not be credited to the company's account for two more days. To speed this up, many companies use electronic transactions to enable their customers to deposit money directly into a bank account.

Many ventures also elect to enable their customers to use credit cards to make payments. Credit cards speed their time to the receipt of cash, but at a price. Companies that use credit cards must pay a small part of each transaction to the credit card company—usually between 3% and 5%. Banks also require a special account, called a **merchant account**, to process credit card transactions. Most banks charge a monthly and annual fee to maintain a merchant account. For many firms, these fees are worth their expense because they facilitate the timely receipt of cash.

12.5.4 **Credit and Collections**

The venture must develop credit and collections policies, and it must execute those policies consistently to maximize revenue. In fact, a venture's credit and collections policies and procedures can have dramatic effect on revenue. For example, a credit policy that is overly lenient allows customers to take ownership of a product or service without cash being transferred to the business. Credit is based on the assumption that the customers will pay later, usually with an interest charge. An overly lenient credit policy may attract a lot of customers. The problem is that some of these customers are credit risks—they will not pay what they owe, or they will pay only at their own pace. Such a lenient credit policy can result in a lot of "sales," but it can be very damaging to the cash flow stream of the growing venture.

The decision about how to use credit as a component of a venture's sales strategy is an important one, especially for companies that sell big-ticket items. Credit must be extended to customers for goods such as housing and automobiles. Without credit most customers would not be able to afford such purchases. The challenge for big-ticket businesses is not whether to extend credit to customers, but on what terms credit will be offered.

Granting credit without an established collection policy has ruined many businesses. **Collection policy** refers to the system for collecting from customers who do not pay on time. **Accounts receivable** is the term used to refer to payments due from customers. Accounts receivable (also known as "receivables") arise as a result of selling inventory or services on credit terms that allow delivery prior to the collection of cash. Inventory is sold and shipped, an invoice is sent to the customer, and cash is collected at a later date. The receivable exists for the time period between the sale and the receipt of cash. Tech Tips 12.2 provides some pointers to an effective credit regime in an entrepreneurial venture.

TECH TIPS 12.2

Suggestions for an Effective Credit Regime

Have policies: Even if you have a full-time experienced credit manager on staff, you need to set policies for extending credit and making collections.

Keep it legal: You need to be sure that anyone making credit decisions on your company's behalf understands the legal issues involved. For example, when determining credit eligibility, you need to be sure that factors such as race, sex, or ethnic origin are not considered.

Use the phone: When monies owed you are overdue and the amount is in excess of a few dollars, your best bet for collection activity is via the phone. Phone calls are simply much more effective than letters.

Keep it positive: Always keep a positive, upbeat attitude when you are pursuing collections, whether through written correspondence or over the phone. Maintaining a positive, but firm, rapport with people who owe you money is the most effective means toward possible payment.

Size is no guarantee: Just because a customer is large doesn't mean it can pay its bills. So, especially if you are considering extending credit to a large company, run a credit check first.

Tier credit decisions: For small credit requests, set up criteria so that a lower-level clerk can quickly make a decision. For large credit requests, have a credit manager or accounting manager make the decision. For very large credit requests, have your top financial person make the decision. And for the largest requests, you, the owner of the business, should become involved.[20]

Keep salespeople away from credit decisions: Salespeople are invariably too loose with credit. Pay very little attention to their credit recommendations, no matter how much seniority or experience they have.

Receivables exist because most industries, with the exception of retail, offer their customers payment terms other than cash on delivery. A company that refuses to offer credit terms will lose some customers because they will purchase their goods from competitors who do offer such terms. Credit terms are quoted in a variety of ways, such as:

- Net 10
- Net 30
- 2% in 10 days, net 30

The first term requires payment in 10 days from the invoice date. The second term requires payment within 30 days from the shipment date. And the third set of terms offers a bonus for early payment. It offers 2% discount from the invoice amount if it is paid within 10 days of the invoice date. Beyond the 10 days, up to 30 days, the customer pays 100% of the invoice. The discount is an incentive to the customer to pay early.

The existence of receivables indicates that the company, instead of collecting cash, invested cash into receivables that, in effect, are loans to customers. If a company gives 30-day terms, it should collect its receivables in 30 days. One method of measuring the quality of receivables is to compare the actual collection period to the stated terms. The average actual collection period is known as **Days Receivable** and can be calculated as follows:

- *Days Receivable = Actual Accounts Receivable/Sales Per Day*, where *Actual Accounts Receivable* = Average levels of receivables on the balance sheet during the period being evaluated

- *Sales Per Day* = Annual sales/360

"Aging" of accounts receivable identifies problem customers, and also allows the firm to manage its credit policies based on industry standards. If the venture's accounts receivable are abnormally long, it must work harder at collecting. If, on the other hand, accounts receivable are abnormally short, it may be able to increase sales by easing its credit policies.

Once a business understands the concept of Aging and Days Receivable, it is easier to understand other aspects of collection. There is one universal maxim that every entrepreneur should understand: The longer an account goes unpaid, the more difficult it becomes to collect. No account, with the exception of government or certain medical claims, should be uncollected for more than 90 days. After the original billing goes out, a notice should be sent once the due date has passed. Notices should be sent out at 30, 60, and 90 days. During this time, phone calls should be made to reach the party. If no arrangements are made and there has been no response, then there are a few options. They include:

- *Precollect notice*: A precollect notice is a notice sent by a collection agency to the "debtor" (if they don't pay, they cease being called "client" or "customer") stating, "We are monitoring this account, please make arrangements to pay or we'll take the account over." The debtor then has the opportunity to pay the business directly or to deal with the collection agency.

- *Collection agencies*: Generally, collection agencies focus on collecting past due accounts for businesses. They are considered by law to be "third-party collectors." This means they must follow state and federal laws in the collection of a debt. One of the laws they must follow is the Fair Debt Collection Practices Act, or FDCPA. Depending on the state, there may also be certain state laws that apply.

- *Court*: According to the FDCPA, a legal debt collector may sue consumers for the purpose of obtaining court judgments for debts but, only in the judicial district where the consumer resides or signed the contract, except that an action to enforce a security interest in real property that secures the obligation must be brought where the property is located.

Despite the occasional need to take action to motivate a customer to pay, retaining the customer as a client must always be a consideration. If the customer doesn't have a regular history of late paying, the venture may want to be tolerant of overdue payments. On the other hand, it may be wise for the venture to "fire" those customers who are routinely late in paying and who must be contacted repeatedly to honor their obligations.

12.5.5 Inventory Management

Inventory is an investment in future sales. Until sold, however, inventory represents a cash use for the technology venture. The financial manager needs to continuously review inventory levels to pinpoint any excess inventory and work with production and marketing to alleviate the condition. Of course, understanding inventory's importance to sales, the financial manager also works with production and marketing to make sure sufficient inventory is available to satisfy customer needs.

Many inventory models use computer programs and company information to determine the best level of inventory for different levels of sales. These models also help determine the best time to order additional inventory and the amount of inventory to order. The auto industry has begun working more closely with its suppliers to reduce the lead time needed for deliveries. The goal is to achieve just-in-time deliveries, which are materials that arrive at the plant just when they are actually needed for production.

Inventory needs are further complicated by the fluctuations in demand that occur in various businesses. A retail business such as Wal-Mart will have to build up its inventory for the Christmas season when sales are brisk, and then hold down inventory levels in January and February when sales are slower. A farmer must purchase his inventory of seed and fertilizer all at one time but can expect no income from his crops until they are harvested. While a vegetable canner will have to purchase the vegetables for canning when the vegetables are harvested, i.e., all at one time of the year, sales of that inventory will be stretched over the entire year. A detergent manufacturer, on the other hand, will experience a relatively stable product demand and material supply and therefore will experience much less fluctuation in inventory needs than the other firms mentioned.

12.5.6 Purchase of Capital Assets

Although capital assets such as land, buildings, and equipment often have to be purchased by a firm at start-up, these same assets must be periodically replaced and upgraded. And as business increases, additional assets may be required. A firm may need to open a second plant, purchase another delivery truck, or buy additional machinery. The company may also want to plan for the future by purchasing land for expansion when the real estate market is most favorable. All of these represent major expenditures and therefore major uses of company funds. If these expenditures are not planned for, the company may have to borrow unnecessarily or at high interest rates, or the firm may have to forgo the purchase of new equipment, expansion of the plant, or purchase of delivery equipment.

12.5.7 Payment of Debt

Most firms need to borrow money at some time or another. They may borrow to make major purchases or to get over a particularly tight cash period during the year; however, the financial manager has to consider the payment of interest and principal on any outstanding debt as a use of funds and needs to add this debt service into the calculation for funds usage.

12.5.8 Payment of Dividends

Dividends are payments made to the firm's shareholders as a form of earnings on their stocks. Although stock usually does not require the payment of dividends, most firms pay dividends to keep their stock attractive to potential investors and to show that the firm is financially sound. As a use of funds, the payment of dividends must be planned for.

12.6 CAPITAL BUDGETING

Capital budgets represent the funds allocated for future investments of the firm's cash. These may be plant expansion, equipment improvement, acquisitions, or other major expenditures. The process of **capital budgeting** involves comparing and evaluating alternative investments.

Capital investments are generally long-term investments and therefore involve long-term sources of funds. When evaluating different capital projects, the financial manager looks not only at the amount of money required to do the project but also at the incremental cash flow the project will produce. These cash flows are looked at to determine when the project will have paid for itself (generated sufficient cash to pay for the initial investment) and what the long-term rate of return will be.

Determining the long-term rate of return can be difficult, because it depends on factors such as customer response, competitive reactions, the state of the economy, and other environmental factors. Therefore, benefits are difficult to gauge in advance. Managers generally look at the most likely circumstances and try to estimate returns based on these. However, this approach does not always work.

SUMMARY

This chapter has examined the basic elements of financial management for a start-up technology venture. As we have stated, all businesses must be concerned with effective financial management, whether or not they have a dedicated financial manager. Effective planning, capital acquisition, cash management, and financial control are the key elements of an effective financial structure.

Entrepreneurs must develop a financial plan for their venture. The financial plan will include tracking and recording the day-to-day operations of the business. The recording of the actual transactions of the venture is done using the rules and standards of accounting and is generally managed by an accountant. Entrepreneurs must also develop estimates or projections of how their venture will perform over time, and its expected growth and cash needs into the future (usually required to be

3 years). Estimating the future financial performance of the venture uses the same financial statements that the accountant produces using the actual transactions of the venture. However, the future-looking statements are normally not generated by an accountant, but by the venture's financial manager. Often, the financial manager and the founding entrepreneur are one and the same person—at least in the early days. Many entrepreneurs either hire a financial expert to assist in the financial management of the venture or hire an external consultant such as a virtual CFO firm.

As the venture matures and grows, financial management is critical to success. Technology ventures often require large amounts of capital to make significant progress. Thus, technology entrepreneurs spend more time analyzing cash needs and making presentations to investors and bankers than other types of venture entrepreneurs. Technology entrepreneurs should gain the trust of investors and bankers before they actually need their capital. Projecting financial needs into the future is one component of the fundraising and confidence-building process. Managing ongoing financial affairs is the other. Using techniques such as breakeven analysis and ratio analysis, a venture can develop a clear picture of how it is performing over time, compared to its peers and plan.

Financial management also requires establishing effective policies for inventory control, credit and collections, and capital budgets. Technology entrepreneurs often learn these various techniques of financial management by doing, not by reading a textbook. That is possible, and many entrepreneurs subscribe to the "ready, fire, aim" approach to building a venture. However, financial management should not be taken lightly, and the health of a venture's finances is one of the key indicators of the health of the venture as a whole. If an entrepreneur wants to sell or otherwise exit the venture, having so-called "clean" financial statements and records is vital. As one technology entrepreneur told us: "You should run your venture every day as if you were going to sell it tomorrow." That includes focusing on the details of accounting and financial management.

STUDY QUESTIONS

1. Explain the role of accounting in new venture operations. What is the accounting equation? Can you explain the components? Explain the accounting cycle.

2. How does accounting differ from finance? Who performs the role of "financial manager" in the start-up technology venture?

3. Define the four fundamental financial statements that are used to estimate the future performance of the venture. Explain how each statement is used to manage the venture over time.

4. What is the role of the new venture's financial plan? What data should the financial plan track? What type of useful information can be extracted from the plan?

5. How is it possible that a venture could reach its profit breakeven point sooner than its cash flow breakeven point?

6. Explain the role of ratio analysis in managing the new venture's financial plan. Define the different ratio types and the role they play in managing the venture.

7. How should the technology entrepreneur manage the venture's accounts receivable? Provide some specific details.

8. Explain the difference between variable and fixed costs. How can the technology entrepreneur minimize variable costs during times of rapid growth?

9. What are the basic elements of a sales forecast? Explain each and how they interact with one another.

10. Define the terms: "assets," "liabilities," "equity," "revenue," and "profit." How does profit differ from cash?

EXERCISES

1. *IN-CLASS EXERCISE*: For this exercise the class should be divided into teams of four to six individuals. The task will be to develop a 1 year sales forecast for a simple business model as defined below. The exercise is designed to focus on the underlying assumptions that each group uses to develop its sales forecast.

The Business Model

The business model for this exercise is selling hamburgers at a stand located near the college campus. The units of sale and prices should be as follows:

1. Hamburger: $2.00
2. Hot dog: $1.75
3. Soda: $1.25
4. Chips: $1.25

The Assignment

Each team should specify the underlying assumptions that form the basis of their sales forecast. The assumptions that each team should specify include but are not limited to:

1. Location
2. Foot traffic at the location
3. Average number of customers/day
4. Average sale/customer
5. Hours of operation

The exercise can be completed in class or as a homework assignment. Each team should develop a spreadsheet and list of assumptions. Each team should present its spreadsheet to the rest of the class and be prepared to discuss and defend its assumptions.

2. *OUT-OF-CLASS EXERCISE*: This exercise is based on the assumption that there is no better way to convince students of the value of financial planning than to hear it from a practicing entrepreneur. For this exercise, students should visit with an entrepreneur to address the following questions:

1. What are the most challenging cash flow issues that confront your industry?
2. What credit terms are commonly offered to customers?
3. How does your company manage its cash?
4. How often do you review your company's financial statements?
5. Who else in the company reviews the financial statements?
6. Who else outside the company reviews the financial statements?
7. What role do financial statements play in setting strategy?
8. Students should visit with entrepreneurs individually or in groups outside of class. They should report back on their findings on the above questions.

KEY TERMS

Accounting: The process of identifying, measuring, and communicating economic information to permit informed judgments and decisions by users of information.

Accounting cycle: The series of steps taken, from a business transaction, through entering the transaction in the general ledger for the business.

Accounting equation: Equates a business's assets to its liabilities and equity combined, commonly expressed as:

$$Assets = Liabilities + Owners' Equity$$

Accounts: Used to summarize all transactions that affect a particular financial statement component.

Accounts receivable: Refers to promises by customers to pay for sales that have occurred at some point in the past.

Accounts receivable turnover: The number of times per year that the average accounts receivable is collected.

Activity ratios: How efficiently the firm uses its assets to generate revenues.

Balance sheet: Lists everything a company owns and everything it owes at a specific moment in time.

Burn rate: The amount of cash that a venture uses each month to pay for operations.

Capital budgeting: The process of comparing and evaluating alternative investments.

Cash flow: The flow of actual cash into and out of a venture.

Cash flow statement: Shows the flow of cash into and out of a business during a period of time.

Current ratio: Indicates a firm's ability to pay its current liabilities from its current assets.

Days receivable: The average collection period for a company's accounts receivable.

Debt ratios: The way companies measure their ability to pay long-term debts

Debt-to-equity ratio: Measures a firm's leverage as the ratio of funds provided by the owners to funds provided by creditors, both balance sheet items.

Expenses: Costs incurred to produce revenues.

Finance: The study of how to manage money—either from a personal perspective or from a business perspective.

Financial accounting: Accounting that is intended for use by a venture's external decision makers.

Financial manager: The individual responsible for the venture's finance function.

Financial statements: Standard statements that are used to record revenue, expenses, and profits and losses of a venture. They include an income statement, cash flow statement, and balance sheet.

Financing activities: Revenue generated by a business through investing excess cash in other businesses as well as share sales or dividend payments to shareholders.

Fixed costs: Those costs associated with operating a venture that do not vary directly with output. Also referred to as "overhead."

General ledger: The repository of closed transactions that have occurred during an accounting period (e.g., monthly, annually).

Generally Accepted Accounting Principles (GAAP): The accounting principles established by the FASB and to which all companies in the United States must comply.

Income statement: The financial statement that records both revenue and expenses for the venture across accounting periods, and is sometimes called the "profit and loss statement."

Investing activities: The income and expenses associated with a venture generated by investing its excess cash into assets that are not associated with its primary business (e.g., an interest-bearing savings account).

Journal: Used to record day-to-day transactions of a venture.

Key ratios: Particular financial ratios that companies in an industry generally track to gauge their relative health compared to their peers.

Liabilities: Debts owed by the business to its creditors, including obligations to perform services in the future.

Liquidity: The ability of a venture to meet its near-term financial obligations.

Liquidity ratios: The measure of a firm's ability to pay its short-term debts as they come due.

Managerial accounting: Accounting that is used primarily by a venture's internal decision makers.

Merchant account: A type of bank account that accommodates credit card transactions.

Operating activities: The income and expenses generated by a venture during the course of its normal business activities.

Owner's equity: Owner's equity represents the claims of the owners against the firm's assets.

Profitability ratios: These ratios measure a firm's success in terms of earnings compared with sales or investments.

Quick assets: Cash or those assets the firm expects to convert to cash in the near future.

Quick ratio: Also called the "acid-test ratio," divides quick assets by current liabilities. It measures more immediate liquidity by comparing two balance sheet items.

Sales forecast: The foundation of the various financial statements that entrepreneurs use to project their venture's financial performance into the future.

Ramp rate: The rate of market penetration, i.e., an assumption built into the sales forecast.

Ratio analysis: Examines the logical relationships between various financial statement items. Items may come from the same financial statement or from different statements.

Recurring revenue: Revenue that is generated by selling products/services to existing customers on a regular basis.

Return on sales: Measures a firm's profitability by comparing net income and net sales, both income statement items.

Return on equity: Measures the return the company earns on every dollar of shareholders' (and owners') investment.

Revenues: Inflows of assets resulting from the ongoing operation of a business.

Sales cycle: The amount of time required to sell products or services to a new customer.

Times-interest-earned ratio: Compares cash received from operations with cash paid for interest payments.

Variable costs: Costs associated with a venture that vary directly with increased output.

Working capital: Represents the amount of capital available for the day-to-day running of the firm.

WEB RESOURCES

The web sites below are intended as destinations for your further exploration of the concepts and topics discussed in this chapter:

1. http://www.inc.com/resources/finance/: This web site is provided by the entrepreneurial magazine *Inc*. The site has a number of resources and articles that focus on the financial challenges that entrepreneurs face.

2. http://www.entrepreneur.com/money/index.html: This web site is provided by the entrepreneurial magazine *Entrepreneur*. It has a variety of resources that can be explored to gain a better understanding of entrepreneurial finance.

3. http://www.aoef.org/: The Academy of Entrepreneurial Finance is an academic organization that offers publications, conferences, and workshops on topics central to financing start-up and high-growth ventures.

4. www.investopedia.com: This web site is similar to the popular web site "Wikipedia" except it focuses exclusively on finance and investing concepts. It is very useful to the entrepreneur whenever he or she encounters a term in finance, accounting, or investing that is new.

5. http://aaahq.org/: For those with an interest in developments in the field of accounting, this link is to the AAA. The AAA is the leading accounting professional society in the United States.

6. http://www.fasb.org: This is the web site for the FASB. The FASB sets rules and develops accounting standards that affect businesses of all sizes.

7. www.morebusiness.com: This web site has a wealth of resources for the technology entrepreneur, including sample business plans and financial statement templates.

CASE STUDY

Nimble Entrepreneurs Adjust When the Bank Says "No More"

Tara Olson and Sherrie Aycock are the co-owners of market research firm AllPoints Research, Inc., in Winston-Salem, NC. Over the years, the women had grown their company to over $2.5 million in revenues. Despite these impressive revenues, however, the firm struggled from time to time with cash flows. Many of the venture's large clients have payment terms that make it difficult for the small market research company to manage cash. For example, Olson said, "We have situations where we'll be working on a project that is worth maybe a couple hundred thousand dollars, but that client, because of their payment policy, is going to pay in two installments and it's going to take them 60 days to process the first invoice."

Of course, during the time period when the client is processing the invoice, AllPoints Research is working on the project and incurring expenses. To offset the cash flow lag between expenses and client payments, Olson and Aycock had a line of credit with a local bank in the amount of $250,000. To their surprise, however, the bank with whom they had been doing business for a number of years suddenly informed them that it was not renewing AllPoints' credit line. Instead, the bank offered the women a short-term loan plan that had significantly higher interest payments than the former credit line.

Rather than accepting the new short-term loan program, Olson and Aycock decided instead to explore their financing options. They reached out to several banks and asked them to make bids to win their business. The banks they approached included both local or regional commercial banks and several major national banks. In the end, the women chose to accept the offer they received from Wachovia—a bank that was not even part of their original search effort. A Wachovia business development representative had just happened to call on the venture during its search for a new bank.

Olson said, "We had this mindset that we had to be with a small bank. But Wachovia just blew us away." Wachovia offered the women a credit line that was twice as big as the one they had had before, and at a significantly lower interest rate. Olson said, "That experience really educated us on what we should expect from our bank."

AllPoints learned that business does not have to be bad for a bank to decide to change its lending terms. Banks are businesses too, and they have to respond to their competitive environment in a manner that enables them to continue to make profits. During the economic downturn of 2009, for example, credit for small and big businesses alike practically dried up.

In reality, there are a number of possible reasons that banks must act to call in a loan or decide to nonrenew a loan. For example, sometimes loans are called or nonrenewed because the venture was unable to achieve specific covenant terms outlined in the loan agreement. On the other hand, banks will occasionally drop a client simply because they are not making enough money to service the account. For example, a small business that has a line of credit but rarely uses it or pays it back quickly when it does use the credit line does not generate much interest income for the bank. A credit line that doesn't generate interest income may not make sense for a bank.

Another reason that a bank may change its lending interest in a particular venture is that it may have merged with another financial entity and is altering its industry focus. In cases like the latter, the venture may be performing well, using its credit line sufficiently and reliably, but the bank simply does not want to serve the industry any longer.

One thing that Olson and Aycock learned as a result of their nonrenewal experience is that it pays to sit down regularly with their banker to evaluate how things are going. Asking a simple question, such as, "What do we need to do to maintain our credit line?" might have prevented the bank from acting as it did. On the other hand, the business owners also learned that better terms are available in the marketplace if one searches for them.

Olson and Aycock also learned that it is important to avoid burning bridges with the old bank when being courted by a new one. If a bank decides to sever its ties, it is valuable to keep the lines of communication open. As was the case with AllPoints, the bank was willing to provide short-term credit while Olson and Aycock sought a new bank. In general, banks will provide a business with 30 days' notice prior to cutting off funds.

Although they did secure a new line of credit, it wasn't without challenges. Olson and Aycock had to explain the cash flow peculiarities of their business time and again. Olson said, "One of the things we had to communicate to these different institutions was the dynamic nature of our sales." As they pointed out to bank after bank, collections were never an issue with the Fortune 500 clients AllPoints provided services to. The problem was the variable payment terms these firms use.

Questions for Discussion

1. What types of things will a banker look for in a technology venture before providing it with a line of credit? How much of a credit line should a venture request?
2. Do you think it is better for a technology venture always to be seeking the best banking terms, or developing a long-term banking relationship even though maybe not at the best terms? Explain.
3. How can a venture manage the payment terms offered by clients? What are some techniques to ensure such terms are, at least in part, favorable to the venture?

Source: Adapted from Detamore-Rodman C. Bounce back. Entrepreneur 2007;June;56–60; *Detamore-Rodman C. Just your size.* Entrepreneur 2005;April:59–61.

ENDNOTES

[1] Brush CG. Pioneering strategies for entrepreneurial success. *Bus Horiz* 2008;**51**(1):21-7.

[2] Carmichael DR, Whittington R, Graham L. *Accountant's handbook: financial accounting and general topics*. Hoboken, NJ: John Wiley and Sons; 2007.

[3] Carter S, Shaw E, Lam W, Wilson F. Gender, entrepreneurship, and bank lending: the criteria and processes used by bank loan officers in assessing applications. *Entrepreneurship: Theor Pract* 2007;**31**(3):427-44.

[4] Bressler LA. How entrepreneurs choose and use accounting information systems. *Strat Finance* 2006;**87**(12):56-60.

[5] Davila A, Foster G. Management control systems in early-stage startup companies. *Account Rev* 2007;**82**(4):907-37.

[6] Campian M. Part time CFOs: a concept worth considering. *Small Bus Assoc Michigan (SBAM Focus)* 2008;**May**:10-12.

[7] Berman K, Knight J. *Financial intelligence for entrepreneurs: what you really need to know about the numbers*. Cambridge, MA: Harvard Business School Press; 2008.

[8] Shane S. *The illusions of entrepreneurship: the costly myths that entrepreneurs, investors, and policy makers live by*. New Haven, CT: Yale University Press; 2008.

[9] Baeyens K, Vanacker T, Manigart S. Venture capitalists' selection process: the case of biotechnology proposals. *Int J Technol Manage* 2006;**34**(1/2):28-46.

[10] Schrager JE. Strategies and techniques for venture investing. *J Private Equity* 2004;**7**(2):1-2.

[11] Timmons JA, Sander DA. Everything you (don't) want to know about raising capital. *Harv Bus Rev* 1989;**67**(6):70-3.

[12] Crouhy MG, Jarrow RA, Turnbull SM. The subprime credit crisis of 2007. *J Derivatives* 2008;**16**(1):81-110.

[13] Sorkin AR. JP Morgan pays $2 a share for Bear Stearns. *The New York Times* 2008;**March 17**:2.

[14] Sahlman WA. How to write a great business plan. *Harv Bus Rev* 1998;**75**(4):98-108.

[15] Mason C, Stark M. What do investors look for in a business plan? A comparison of the investment criteria of bankers, venture capitalists, and business angels. *Int Small Bus J* 2004;**22**(3):227-48.

[16] Reynolds PL, Lancaster G. Predictive strategic marketing management decisions in small firms: a possible Bayesian solution. *Manage Dec* 2007;**45**(6):1038-57.

[17] New study shows six critical business plan mistakes. *Business Horizons* 2003;**46**(4):83-4.

[18] Berry T. Creating a sales forecast. *Entrepreneur.com* 2005:May 1.

[19] Patrone FL, Dubois D. Financial ratio analysis for the small business. *J Small Bus Manage* 1981;**19**(1):35-40.

[20] *Streetwise tips on credit and collections. BusinessTown.com* (retrieved from: http://www.businesstown.com/accounting/credit-advice.asp).

Venture Management and Leadership

13

After studying this chapter, students should be able to:

- Formulate skills to become an effective entrepreneurial manager
- Perform in the roles required of entrepreneurial managers
- Appreciate and act on the difference between leadership and management
- Understand and develop ethical principles of entrepreneurial leadership
- Use effectuation theory as a means of understanding entrepreneurial expertise
- Recognize various entrepreneurial strategies and apply them as appropriate

TECH VENTURE INSIGHT

New Ventures Often a Product of Experience and Initiative

Todd Basche wasn't always an entrepreneur. In fact, he spent many years working in big companies before he decided to strike out on his own. As Apple Computer's vice president of application software, Basche's responsibilities were always entrepreneurial in nature. In addition to his creative days at Apple, Basche had also worked in product development at Hewlett-Packard. Despite the creativity of his day job, however, Basche dreamed of starting his own venture one day.

One day, just after he turned 50, Basche had a breakthrough idea as he was trying to come up with an intuitive number for the combination lock to his backyard pool. He suddenly realized that words and *not* numbers were more conducive to memory and recall. People were more likely to remember letter-based codes that had personal significance rather than a random string of numbers.

"The numerical left-right-left lock was invented in 1862 and has remained virtually unchanged," Basche stated, "and there are not a lot of other places in your life where you're using 1862 technology." This set into motion Basche's vision to capture a $1 billion consumer market with revolutionary potential.

While his new product did not exactly fit into the high-tech market, Basche used his experience in the computer industry and the prevalence of user IDs and passwords, most of which are letter-based codes and easy to recall, to launch "Wordlock." Basche and his wife, Rahn, developed and licensed their initial product to Staples for a trial run. Quickly realizing the enormous growth potential of this venture, they quit their jobs in January 2007 to focus solely on their new enterprise.

During its first year, the Santa Clara, California–based company was self-funded and operated out of Bausche's home. By 2008, it had experienced "astronomical" growth. "We moved out of the house, and now we're up to 10 full-time employees," says Basche, adding that in the first quarter of 2008, Wordlock was in 900 retail locations. By the end of the second quarter, it was in 12,000 stores. Basche predicts to be in 50,000 next year.

Basche sees potential in other markets, as well. "There are other consumer product areas where the market giants have been sleepy," he says. "We now have the expertise, and we can grow this into a line of innovative products that brings ease of use to the consumer."

The learning curves and challenges for an entrepreneur are considerable, but according to Bashche, "when the product is your own, versus a corporation's, it's an amazing feeling to watch it go from your head to the store shelf."

Source: Adapted from Boorstin J. Staples lets customers do the designing. Fortune April 18, 2005; Wang J. A step down that's a step up. Entrepreneur November 14, 2008.

INTRODUCTION

A common problem that arises for many technology entrepreneurs is the lack of preparation as the venture grows, and the founder must make the transition from lone entrepreneur to manager of a growing enterprise.[1] Many technology entrepreneurs have not had formal training in managing/leading others, and some are happiest when working in isolation on challenging technical problems. That is not to say that they are incapable of developing and even mastering the skills of effective management.[2] Simply stated, most technology entrepreneurs spend their careers learning and working within highly specific technical niches, but do not focus on developing managerial talents and skills.

The good news is that it does not require an MBA to be an effective manager/leader of a technology start-up. Many technology entrepreneurs simply take the "on-the-job-training" approach. However, this can be hazardous too because start-up ventures often have very little room for failure and must execute effectively from the beginning to succeed. Not only can the venture suffer if the entrepreneur is learning management for the first time "on the job," but investors and other stakeholders may have little patience for the learning process.[3]

Although it is imperative that entrepreneurs learn at least some of the art and craft of managing and leading from experience, a few lessons from books and advisors can also be helpful. In this chapter, we provide some fundamental ideas about managing and leading a technology venture. The basic concepts and managerial techniques discussed here are intended to provide a framework for understanding the variety of issues that inevitably arise in start-up ventures. These ideas should be applied to leading and managing, and the results of applying them should be monitored and assessed. Most likely, each of the skills, roles, and techniques described in this chapter will need to be adapted and modified for unique situations.

We begin our discussion of management and leadership by reviewing several of the higher-order skills that most entrepreneur managers will need to organize and control the venture work environment.

13.1 ENTREPRENEUR MANAGERS

Performance in a start-up or growing technology venture does not just happen. Committed and skillful entrepreneurs carrying out specific roles as **managers** make it happen. The entrepreneur as a manager of a growing venture influences performance by defining objectives, recognizing and minimizing obstacles to the achievement of those objectives, and effectively planning, organizing, and controlling resources to attain high levels of venture performance. This section focuses on the management skills that must be applied to everyday situations experienced in start-up technology ventures.

Learning to manage effectively as the venture grows can be a difficult challenge for many technology entrepreneurs. Most first-time technology entrepreneurs have never received formal training in management. Their only point of reference may be the individuals who managed them at some point in their careers. And *those* individuals may also never have been exposed to formal management training.

It's not necessary to have had formal management education or training to be an effective manager, but the concepts and tools provided via formal training can lead to more effective management performance. Entrepreneurs who do not adapt to managerial roles often have to cede control of their venture to more experienced managers. This can work, and many technology entrepreneurs have done so. But for those entrepreneurs who want to stay engaged with their venture over the long haul, developing personal management and leadership skills will be essential.[4]

Most investors and experienced entrepreneurs will readily attest to the importance of **execution** in start-up ventures.[5] The term "execution" simply means the ability to make plans, organize resources to achieve those plans, and to lead others in the quest to achieve specific goals. Most of the key skills and abilities necessary to execute are learnable, and the technology entrepreneur who is willing to learn has the best chance to succeed.[6] For most entrepreneurs, learning to manage and lead people should occur both via reading and understanding, and through on-the-job training.

When it comes to judging the success of a start-up technology venture, actual performance is all that matters. Good intentions, promises, and wishes will not matter to the entrepreneur or to investors if solid execution and performance are lacking.

To **perform** means "to do, to accomplish." In entrepreneurship, this means the creation of value for a particular market or set of markets, and a steady increase over time in overall **enterprise value**. Entrepreneurs as managers are responsible for effectively utilizing the resources of their venture to create and deliver value to the customer. The entrepreneur manager must design workflow; purchase and use raw materials; and produce, market, and sell the organization's products or services. How well the entrepreneur utilizes the resources of the organization (performance) will have an effect on how much value can be created from those resources (productivity). **Productivity** is a term that applies to service businesses as much as it does to manufacturing businesses. Ventures of all types must be concerned about the productivity of the assets used to serve customers—including people, cash, machinery, and other assets.

In addition to creating customer value, the entrepreneur must be focused on enhancing enterprise value over time. That means setting up repeatable systems, acquiring high-value customers, establishing recurring revenue streams, defining and protecting competitive advantage, etc. Investors and shareholders are primarily interested in a steady increase in enterprise value over time.[7]

13.2 BASIC MANAGEMENT SKILLS

Regardless of the venture type, entrepreneur managers must possess and seek further to develop many critical skills. A **skill** is an ability or proficiency in performing a particular task. Management skills are learned and developed. In general, all technology entrepreneurs as *managers* should seek to develop skills in the following areas:

- Analytical
- Decision-making
- Communication
- Conceptual
- Team building

13.2.1 Analytical Skills

Analytical skills involve using repeatable approaches or techniques to solve management problems. In essence, **analytical skills** are concerned with the ability to identify key factors affecting venture performance, to understand how they interrelate, and how they can be managed to achieve venture goals. Analytical skills include the ability to diagnose and evaluate the issues that face the venture on a daily basis. They're needed to understand problems and to develop action plans for their resolution.

Analytical skills also include the ability to discern and understand how multiple complex variables interact and to conceive of ways to make them interact in a desirable manner. These skills include the ability to analyze one's own talent, as well as the talent of others associated with the venture. The entrepreneur who is able to analyze and accept his or her own strengths and weaknesses will be in a better position to achieve performance goals. The entrepreneur who has accurately analyzed his or her own capabilities will hire those who complement strengths and compensate weaknesses. Tech Tips 13.1 has a few suggestions for building what is referred to as "You, Inc."

TECH TIPS 13.1

Tips for Building "You, Inc."

Take stock of yourself: You have to buy into yourself if you want others to value your brand.

Build from the inside out: Seek to be authentic in everything you say and do.

Find your true purpose: Most happy people have a purpose in life. Once you know what your purpose is, you are well on the way to discovering it.

Accept your strengths: Your personal brand will be more authentic when you leverage your strengths.

What do you stand for? People buy valuables from people they value.

Carve out your personal niche: Another integral part of brand building is the ability to specialize.

Think extraordinary, not ordinary: Aspire to push your brand out to as many people as possible.

Be on a mission: It is one thing to have a personal mission statement, but it is another thing to be *on* a mission.

Value your unique gifts: It is up to you to value your unique gifts and offer them to the rest of humanity to create value.[8]

Most technology entrepreneurs already possess strong analytical skills, but they are mostly focused on technology issues rather than business issues. Running a venture requires shifting focus from analyzing the venture's key technology as a technology to analyzing it as a business. Technically, a venture's offerings may be the best in the world. From a purely business perspective, in contrast, the entrepreneur must assess and analyze the potential market, the costs associated with bringing the product to market, the scalability of the enterprise over time, the financial resources that will be required to build and grow the venture, and many other things. We have discussed many of these analytical tools throughout this book. Successful technology entrepreneurs learn to be adaptable in their thinking, developing an ability to shift their analytical focus between technical and business issues as required.[9]

13.2.2 Decision-Making Skills

All entrepreneurs must make decisions, and the quality of these decisions determines their degree of effectiveness. An entrepreneur's **decision-making skills** in selecting a course of action are greatly influenced by his or her ability to deal with ambiguity. One of the hallmarks of start-up ventures is their ambiguity.[10] They are often ambiguous with respect to their market, their value proposition, their competition, and even their ability to persist. Entrepreneurs must learn to resist pressures for a quick fix when problems and issues arise. They must learn to live with uncertainty and ambiguity, and to recognize subtleties in what works and what does not.[11]

Decision-making in start-up ventures is almost always done in the face of irresolvable ambiguity. This is something that the technology entrepreneur must learn to accept. Yet, the ambiguous nature of many of the issues the entrepreneur faces must not lead to inaction or paralysis.[12] The entrepreneur must have a predilection toward action and must be able to make decisions in the face of incomplete information. The term that is often used to describe this situation is **satisficing**. This means that the entrepreneur as manager must choose the best solution to a problem despite incomplete information about both the problem and the likely outcome of the decision that is taken.[13] Choosing the best alternative is a vastly different decision from choosing the correct alternative. Making a decision among alternatives in running a start-up venture is different from selecting among alternatives on a multiple-choice exam or in crafting a solution to a problem posed in a textbook. In the world of the start-up, there usually is no correct course of action, but there often is a best one—that is, the one the technology entrepreneur must learn how to identify, select, and act upon.

The analysis of decision alternatives likely will alternate between technology and business issues. Technology entrepreneurs must make choices about the technology development road map and the business development road map. This will require flexible analytical skills as we discussed above. But being flexible should not lead to an inability to decide firmly and move ahead. Most technology entrepreneurs are very good at analysis. To be entrepreneurial, however, requires an equal ability to decide and take action. Tech Tips 13.2 highlights some decision-making traps to avoid.

TECH TIPS 13.2

Some Decision-Making Pitfalls to Avoid

Paralysis by analysis: Emphasizing discussion at the expense of action.

The knowing-doing gap: Failing to execute the decision.

The anchoring trap: Giving disproportionate weight to the first piece of information received.

The status quo trap: Maintaining things as they are rather than venturing outside of one's comfort zone.

The sunk cost trap: Also known as "escalation of commitment," where the decision maker perpetuates a bad decision merely because money, time, and effort have already been expended.

The confirming evidence trap: Seeking information that confirms an original decision and discounting information that is disconfirming.

The overconfidence trap: Overestimating the accuracy of forecasts.

The recent event trap: Giving undue weight to a recent event.

The prudence trap: Being overcautious and risk averse.[14]

13.2.3 Communication Skills

Since entrepreneurs must accomplish much of their work through other people, their ability to work with, communicate with, and understand others is important.[15] Effective communication, written and verbal, is vital for venture performance. The skill is critical to success in every venture, but it is crucial to entrepreneurs who must achieve results through the efforts of others. **Communication skills** involve the ability to communicate in ways that other people understand and to seek and use feedback from others to ensure that one is understood.

The entrepreneur's communication skills also will be tested among investors, shareholders, and other stakeholders. One of the primary ongoing tasks for the technology entrepreneur is fundraising. Fundraising requires that the entrepreneur be able to tell a succinct story about the venture's offerings and market, its value proposition, and its business model. This short story about the venture is sometimes referred to as its **elevator pitch**. The elevator pitch is so named because it is a short and clear overview of the venture that might be told to a potential investor during a chance meeting on an elevator. Generally, an elevator pitch should take no more than one minute and should be clear and even a bit exciting. Tech Tips 13.3 highlights some elements of crafting an effective elevator pitch.

In addition to the elevator pitch, the entrepreneur must be able to communicate the intent and direction of the venture via a written business plan. The business plan is a required document when speaking to potential investors, lenders, or other key stakeholders. Business plans should be supplemented by a one- to two-page executive summary.[16] Substantial writing skills are required to boil an entire business plan down to an executive summary. There are firms that provide entrepreneurs with fee-based support in writing and editing business plans and executive summaries, but research has indicated that it is better for the entrepreneur to write these documents.[17] Not only is there potential value to the entrepreneur in the research and learning that goes into business plan writing, there is also a need to constantly update and/or adjust the plan.

Finally, the entrepreneur must be skilled in presenting the plan to investors, lenders, and others. The verbal ability to articulate the intent and goals of the business, as written in the plan, requires the ability to understand and adapt communication style to the audience. For example, many technology venture entrepreneurs must present their business plans to others who are not as technically adept as they are. Speaking in technical jargon or using complex graphics to convey the business idea will likely

---

TECH TIPS 13.3

Crafting an Effective Elevator Pitch

Figure out what is unique about what you do: The whole idea behind a great elevator pitch is to intrigue someone. It's an icebreaker and a marketing pitch—all rolled into one.

Make it exciting: A superior elevator pitch speaks to who you really are and what excites you about your business. What is it about your business that really motivates you? Incorporate that.

Keep it simple: A good elevator pitch doesn't try to be all things to all people. Rather, it conveys a clear idea in a short amount of time. Keep it under 30 seconds.

Write it down: Write down your pitch, say it out loud, rewrite it, and then rewrite it again.

Practice, and practice some more: The first few times you try out your elevator pitch may be a bit uncomfortable, but it gets easier. After a while, it will become second nature to you, and when it does, you will be glad you practiced.

---

not produce the desired results in such cases. The technology entrepreneur must adapt communications to match the ability of the other party.

13.2.4 Conceptual Skills

Conceptual skills consist of the ability to see the big picture, the complexities of the overall organization, and how the various parts fit together. Entrepreneurs use their conceptual skills to develop long-range plans for their companies. Conceptual skills enable the entrepreneur to look forward and project how prospective actions may affect a company 1, 3, or even 5 years in the future. The single most important foundation of entrepreneurial success today is leadership—especially visionary leadership. It is critical that every venture develop a vision and mission to guide the many choices that it must make in the present and into the future.

Visionary companies possess a core ideology from which their values spring that is unchanging and that transcends immediate customer demands and market conditions. The unifying ideology of visionary companies guides and inspires people. Coupled with an intense, "cultlike" culture, a unifying ideology creates enormous solidarity and *esprit de corps*. Last, visionary companies subscribe to what some have called "big, hairy, audacious goals" that galvanize people to come together, team, create, and stretch themselves and their companies to achieve greatness over the long haul.[18]

Technology ventures that have achieved the highest levels of success have often been launched by visionary founders. Steve Jobs of Apple, Next, and Pixar is perhaps the most visible of these visionaries. His overarching vision for Apple is to create "insanely great products." This simple vision has elicited incredible loyalty and effort from employees, and has provided the world with a stream of novel and path-breaking products for more than two decades.

Other visionary leaders have used their focused visions to revolutionize industries as diverse as package delivery, grocery retailing, and airlines. Fred Smith founded FedEx on the vision of delivering packages overnight across the entire United States. His vision was that people would call FedEx for packages that "absolutely, positively had to get there overnight." Whole Foods founder John Mackey created a new concept in grocery retailing on the basis of his vision of "whole foods, whole people, whole planet." This vision permeates the culture, operations, and merchandise of Whole Foods. Tech Micro-case 13.1 discusses how JetBlue founder David Neeleman's vision has created and sustained one of the leading passenger airline companies.

13.2.5 Team Building Skills

Rare is the lone-wolf entrepreneur. Most technology ventures are too complicated for a single person to operate without help from others. Successful entrepreneurs are usually talented team builders. They are able to attract other people to

---

TECH MICRO-CASE 13.1

Neeleman's Vision Drives JetBlue to Success

The power of a vision in developing and guiding an entrepreneurial venture has been proven again and again. David Neeleman founded JetBlue airlines at the age of 39 on the simple vision to "bring humanity back to air travel." Neeleman had been involved in the travel and airline industries for most of his adult life. After dropping out of the University of Utah in his senior year, he ran his own travel agency, was vice president of a regional airline, and ran a software company that improved the electronic ticketing process for passenger air carriers.

When he founded JetBlue, Neeleman envisioned that his new airline would combine the low fares of a discount air carrier with the comfort of a cozy den in people's homes. His goal was to make JetBlue the best "customer service" company in the world, not just the "best airline." To drive customer service, JetBlue was built from day one on what Neeleman calls the "3Ps":

> At JetBlue we live by our belief that great **people** drive solid **performance** which generates **prosperity** for all.

JetBlue was launched on February 11, 2000, with a ceremonial flight between Buffalo, NY, and New York City. Later that day, it cut the ribbon on its first commercial route between New York City and Fort Lauderdale, FL. Today, JetBlue is among the leading innovators in commercial air travel, and one of the few that are profitable. It still offers free television in passenger seat backs, free snacks, and more legroom than on most other airlines. Neeleman's commitment to his original vision continues to create satisfied customers and employees.[19]

their vision, and then build them into a coherent team, all focused on pursuing the same goals. **Team building** is based on identifying gaps in talent and skills that are required for the venture to succeed and then finding people with these necessary traits. Successful entrepreneurs are able to attract employees, advisors, and investors who provide them with missing talent that is required to achieve venture goals.

Building successful teams requires each of the skills mentioned above, and also a large helping of humility. Successful technology entrepreneurs attest to the importance of hiring and motivating people more talented than themselves. In fact, the strongest performing entrepreneur managers are not afraid to hire people who are more talented than they are. This statement may seem obvious, but it is not uncommon for technology entrepreneurs to feel intimidated by people who are more talented. After all, throughout their formal technical education, they competed against other technically talented people. That competitiveness doesn't just disappear; one has to reorient one's thinking to overcome the tendency to compete. Technology entrepreneurs must replace their feelings of competitiveness among their peers with competitiveness as a business manager. That means swallowing one's ego and hiring talented people who will help the venture compete in its market to achieve its goals. Analytical skills in evaluating, selecting, and motivating talented people are immensely important in venture performance.

The foundation of successful teams is simple: clarity of goals and responsibilities. Highly talented people are usually self-motivated to a high degree. They want to do a good job and enjoy working on goals that are clearly defined and are measurable. Technology entrepreneurs have learned that they can expect a high degree of commitment from talented people without a lot of managerial intervention. That is, talented people usually perform at their peak when their managers provide them with their goals and then get out of their way. Teams organized around specific projects and goals will often self-organize. Google, for example, allows its engineers to choose the projects that they work on, with very little day-to-day oversight on their work. This relaxed approach to managing can only work in an environment where the vision and goals are clear and understood by all and where rewards and incentives are directly linked to performance that helps the venture achieve its objectives.

13.3 ENTREPRENEURIAL LEADERSHIP

Leadership is an important and necessary skill for achieving individual, group, and venture performance. The entrepreneur influences the attitudes and expectations that encourage or discourage performance, secure or alienate employee commitment, reward or penalize achievement. Entrepreneurial leaders must be able to influence others to work as hard as they do in order to create a valuable company. Some scholars have noted that the skills that make entrepreneurs successful in the early stages of a company actually hinder them as the company grows.[20]

Research has determined that many entrepreneurs are effective in creating and running a company on their own. However, when growth requires additional employees, the entrepreneur doesn't have the necessary skills to motivate and inspire followers. Entrepreneurs who lack leadership skills hang on to their independent status for as long as they can. Their growing companies are often characterized by high employee turnover and general employee dissatisfaction. Usually, the entrepreneur who lacks leadership skills either eventually leaves the company or finds someone else who can lead employees day to day while the entrepreneur takes on a different role within the company—perhaps as chairman or as an externally oriented CEO.

In this chapter, we define leadership as "the ability to influence through communication the activities of others, individually or as a group, toward the accomplishment of worthwhile, meaningful, and challenging goals." First, this definition indicates that one cannot be a leader unless there are people (e.g., coworkers, followers) to be led. Second, leadership involves the application of influence skills. The use of these skills has a purpose, to accomplish goals. Finally, an objective of leadership is to bring about influence so that important goals are achieved. This influence is brought about not only directly through authority or motivation, but also indirectly through role modeling. Research has shown that employees generally have higher expectations of leaders as models or exemplars of the organization.[21]

Leadership is a general concept that applies in many different social contexts: sports, politics, organizations, and entrepreneurial ventures. You probably will agree that the traits and behaviors that make a person an effective leader in politics may be quite different from those that make a person an effective leader of a sports team. The same is true of leaders in business. The traits and behaviors that are necessary for leaders of large organizations are different from those that are required for leaders of entrepreneurial ventures.

13.3.1 Influence

The exercise of **influence** is the essence of leadership behavior. Leaders use influence as their primary tool to move the venture toward its goals. Of course, the entrepreneur must also use such tools as compensation, employee feedback and evaluation, and organizational structure to move a venture forward.

Entrepreneurial leaders need to learn a variety of influence strategies. As their business grows, they cannot rely solely on exercising the power they possess by virtue of their position as founder and owner. Research indicates that employees demonstrate higher levels of motivation if they're allowed to influence the way the organization works. Scholar Noel M. Tichy said, "The ultimate test for a leader is not whether he or she makes smart decisions and takes decisive action, but whether he or she teaches others to be leaders and builds an organization that can sustain success even when he or she is not around."[22]

Entrepreneurial leaders must be able to exercise their influence without developing arrogance or an air of superiority. This is difficult for novice entrepreneurs who may be unfamiliar with power and authority. Suffice it to say that tyrants rarely inspire great effort among their people.[23] In fact, people respond to and are

---

TECH TIPS 13.4

Seven Influence Strategies for Entrepreneurial Leaders

Reason: Using facts and data to develop a logically sound argument

Friendliness: Using supportiveness, praise, and the creation of goodwill

Coalition: Mobilizing others in the organization

Bargaining: Negotiating through the use of benefits or favors

Assertiveness: Using a direct and forceful approach

Higher authority: Gaining the support of higher levels in the hierarchy to add weight to the request

Sanctions: Using rewards and punishment[25]

---

influenced by leaders who can demonstrate concern for their personal growth. Of course, in business, a leader cannot be concerned about personal growth as it pertains to nonbusiness areas, such as a person's romantic life or personal financial acumen. Rather, people respond to leaders who are able to provide them with a continuing stream of challenging projects that are appropriate to their current skills, abilities, and temperament.[24] Tech Tips 13.4 provides seven influence strategies that entrepreneurial leaders can use.

13.3.2 Leadership versus Management

The distinction between leadership and management is controversial. No clear line separates the two. It goes without saying that managers must often be leaders and that leaders must often be managers. Nonetheless, there does seem to be some truth to the idea that management is distinguished from leadership.

Harvard scholar Abraham Zaleznick supports this notion. He thinks that the distinction between leadership and management lies in the primary focus of each role. In their role as managers, entrepreneurs must focus on day-to-day venture performance. In their role as leaders, they must focus on the long-term goals of the venture. In Zaleznick's terminology, managers are primarily concerned with process and leaders are primarily concerned with substance.[26] Tech Micro-case 13.2 highlights this distinction.

TECH MICRO-CASE 13.2

Michael Dell Brings in Executive Talent

Michael Dell is the CEO of Dell Computers. He started the company from his dorm room at the University of Texas, selling made-to-order computers over his telephone. Realizing that the demand for his computers far outstripped his ability to supply them, he quit school and launched his company. As his company grew, Dell became immersed in endless details about workflow, product quality, and supply chain management. It wasn't long before he realized that he was in trouble. Dell Computer experienced difficult times due to production problems, questionable product quality, and uncertain market focus. Dell realized that he was better as a leader than as a manager. He had the vision that created the company, but he wasn't very good at managing the day-to-day details involved in operating the growing company.[27]

This is a common problem many entrepreneurs like Dell face. They are good visionaries and they are very influential, but they often aren't very good at leading the growing venture. Fortunately, for Dell, he realized his limitations before the company got into too much trouble. He went out and hired the best operations, financial, and other executive officers he could find to handle the day-to-day management of his company. Today, Dell Computers is a major international competitor and one of the most respected companies in America.[28] Michael Dell serves as the company's chairman. He handpicked Kevin Rollins, 12 years his senior, to serve as president and CEO. Dell was able to set his ego aside for the sake of his company, realizing he didn't have the management skills to operate his rapidly growing venture.

Typically, good entrepreneurial leaders sacrifice micromanagement of the bottom line in favor of a macroscopic understanding of the enterprise, its associates, and its strategic direction. Although it has never been conclusively proven that leaders produce lower profits than managers, they do tend to create more inspired, more empowered associates—willing to go the "extra mile"—and leaders are significantly less likely to be deemed a "workaholic," "ogre," or "taskmaster" by their associates or colleagues.[29]

13.4 EFFECTUATION AND ENTREPRENEURIAL EXPERTISE

Effectuation is a relatively new term in entrepreneurship research that refers to a pattern of behavior and decision making common to what is referred to as "expert entrepreneurs." Expert entrepreneurs are individuals who have been successful in launching and operating ventures, usually more than once. The pattern of behavior that has been observed and documented to predominate among these experts differs from the patterns observed in nascent entrepreneurs. Expert entrepreneurs have learned that the road to venture success is long, unpredictable, and often loaded with ambiguity. The goals that originally may have been intended may take longer to achieve than expected, or they may need to be altered altogether.[30] The expert realizes that the chief question to ask when launching the venture is not "what goals should I attempt to achieve?" but rather "what resources do I currently control and how can I put them to use to create value?"

Part of the resource base available to the entrepreneur is personal talents, skills, knowledge, and insights. These must be evaluated realistically to determine what type of venture, and in what industry, the individual may be most likely to succeed. As the originators of the concept of effectuation put it, the chief question that the entrepreneur should ask is: "Given who you are, what you know, and whom you know, what types of economic and/or social artifacts can you, would you want to, and should you create?"[31]

Effectuation also involves a new type of logic that differs from the causal logic that is taught in business schools. Causal logic suggests that a business should establish action plans and then set in motion the forces that will bring about the desired ends. Causal logic places a high degree of emphasis on predictability.

Effectuation logic, by way of contrast, emphasizes control over predictability. The various types of logic are illustrated in Exhibit 13.1.

In addition to effectuation logic, this way of thinking about ventures and the people who launch and operate them includes five principles:

1. *The Bird-in-Hand Principle*: This principle states that the expert entrepreneur starts with who they are, what they know, and who they know rather than with the opportunity.

2. *The Affordable Loss Principle*: This principle states that expert entrepreneurs only risk what they can afford to lose when launching their new ventures, rather than focusing on the expected return.

3. *The Crazy Quilt Principle*: This principle states that the expert entrepreneur actively engages in social network building to increase the resources available to the new venture, as opposed to doing in-depth competitive analysis and attempting to differentiate.

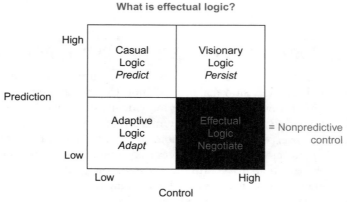

EXHIBIT 13.1

Effectuation Logic.

4. *The Lemonade Principle*: This principle states that the expert entrepreneur has learned to make lemonade out of lemons, that ambiguity and uncertainty are embraced rather than avoided.

5. *The Pilot-in-the-Plane Principle*: This principle states that the future is dependent upon and created by what people do; it is not inevitable.[32]

Not all entrepreneurship scholars accept the concept of effectuation as central to understanding entrepreneurship, but it has drawn an increasingly large following.[33] Expert entrepreneurs, defined as those with 15 years of experience, were examined to determine whether they were more likely to think in terms of effectual or causal logic. The findings revealed the following:

- Expert entrepreneurs use effectual logic significantly more than both novices and corporate managers.

- Organic growth leaders in large corporations, unlike corporate managers in general, appear to be closer to expert entrepreneurs in their use of effectual logic.

- While venture capitalists in general are not likely to use effectual logic, the more experienced they are, the closer they come to expert entrepreneurs in their use of effectual logic.

- Angel investors, who use specific effectual logic more widely, experience a reduction in the number of negative exits they experience, without any reduction in their rate of positive exits.

Expert entrepreneurs do not necessarily begin with an opportunity or with market research. Instead, they start with who they are, what they know, and whom they know. Expert entrepreneurs also may not begin with a clear vision for a new venture. Instead, their strategies are predominantly means-driven and result in new ends that are usually characterized by specific courses of action rather than outcome variables. Expert entrepreneurs are as likely to see entrepreneurship as a way to achieve noneconomic goals (such as attaining a preferred lifestyle or solving societal problems) as they are to see it as a career or as a way to make money.

13.5 ENTREPRENEURIAL ETHICS

Ethics is often a difficult topic to address in a business-oriented textbook. For the most part, people who excel in business are not also deeply familiar with the terminology and concepts that comprise a standard course in ethics. Many businesspeople also regard the topic as "fuzzy" and difficult to comprehend because there are so many differing perspectives on the topic. In fact, many businesspeople live daily by the adage "if you can't measure it, it doesn't exist."

Ethics can be simple or hard. It is simple when individuals decide on a few basic principles that will guide them in their lives and stick with these principles, come what may. It is hard when people believe that ethics can be nuanced, and that situations must be judged independently to know what the right course of action may be. This textbook does not have enough space to take the nuanced approach to understanding entrepreneurial ethics. As such, it takes a more straightforward, principles-based approach to understanding this important topic.

Fortunately, it does not require that you have a master's degree in philosophy to understand some basic principles of entrepreneurial ethics. One very successful media entrepreneur, Karl Eller, summed up entrepreneurial ethics in the title of a book he wrote about his own successful ventures: *Integrity Is All You've Got*.[34] Integrity is defined, basically, as doing what you say and saying what you do.

Integrity is a good starting point for a set of ethical principles. It would be difficult to argue the opposing perspective that integrity is not important or that a lack of integrity is important to business success. In fact, most successful business people will attest to the role that integrity plays in their ability to make things happen. Most business is transacted within a framework of trust. Business people trust that those with whom they are associated will follow through on commitments and contracts to which they are counterparties. Although contracts and commitments are normally also governed by legal rules, business people don't want to have to resort to lawyers, lawsuits, and the court system every time they want to achieve a business goal. Rather, they want to work with people who will deliver on promises. A single failure to fulfill a contractual obligation, or to honor one's word, or to deliver on a promise can ruin a reputation.

Another principle that should be closely aligned to integrity is honesty. In business, the virtue of honesty is also sometimes referred to as "transparency." This simply means that the venture is operating each day in a manner that would pass a formal financial audit, and it is operating generally according to what is referred to as "good faith." This term means that counterparties to a contract or business relationship are using their best efforts to deliver their end of the contract. Clearly, not all business transactions or arrangements live up to their expectations. However, parties that act in good faith and in a transparent manner generally will not suffer any negative legal consequences as a result of a failure. The free-market system is often described as a "profit and loss system." That is, business transactions occasionally fulfill and even exceed profit expectations. On the other hand, oftentimes they do not, and parties to the transaction suffer financial loss. This is normal, and parties can recover to work together again in the future if they believe each has operated honestly and in good faith, and will likely act similarly in the future.

A final ethical principle that we will discuss in this chapter and that seems essential for business success is humility. Humility has often been described as a character trait, but it can also be expressed as an ethical principle. An individual that expresses humility is one that recognizes that many of the good and bad things that happen in life and in business are often a function of chance events. Consider the founders of Google, Sergey Brin and Larry Page. Certainly, these are two shining stars among the technology entrepreneurs of our time. They not only have created a singularly impressive company, but they have achieved incredible levels of personal wealth. Their success, no doubt, can be attributed to their respective talents in computer programming and web site design. These talents are clearly relevant to the technology age in which we live. What if Page and Brin had come of age in 1890 rather than 1990? Would their unique talents have been as applicable in that long-ago time period? Clearly, they would not have founded Google. Not only was there no Internet in 1890, there were no computers, databases, or even electric power. The point is, despite the great success of Google, Page and Brin have to admit they were lucky to have been born during an age when their unique talents are highly prized and rewarded. That would not have been the case had they come into the world 100 years earlier.

Most success is attributable to a complex mix of personal talent, fortunate circumstances, and chance events. Entrepreneurs who recognize that will be able to maintain an authentic humility during times of success, and will also be more balanced personally during difficult times. In fact, several authors have coined the term "egonomics" to refer to the importance of keeping one's ego in check and developing humility.[35] These authors list four warning signs that indicate when an individual's ego has taken over:

1. *Being comparative*: "When we're comparative, we tend to either pit our strengths against another's weaknesses, which may lead us to an exaggerated sense of confidence, or we compare our weaknesses to their strengths, which can cause negative self-pressure."

2. *Being defensive*: "When we can't 'lose,' we defend our positions as if we're defending who we are, and the debate shifts from a we-centered battle of ideas to a me-centered war of wills."

3. *Showcasing brilliance*: "The more we want or expect people to recognize, appreciate or be dazzled by how smart we are, the less they listen, even if we do have better ideas."

4. *Seeking acceptance*: "When we equate acceptance or rejection of our ideas with acceptance or rejection of who we are, we 'play it safe.' We tend to swim with the current and find a slightly different way of saying what's already been said as long as acceptance is the outcome. That not only makes us a bland follower, but an uninspiring leader."

It is important to recognize how humility fits within the spectrum of possible character orientations, from a completely empty ego to egotism.[35] Exhibit 13.2 provides a useful illustration that humility is in the middle of these two ends of the spectrum. According to this illustration, humility represents a healthy and intelligent understanding of one's unique talent and skills, but avoids the destructive potential of overconfidence and egotism.

EXHIBIT 13.2

An Illustration of Humility.

Source: Marcum D, Smith S. Egonomics: what makes ego our greatest asset (or most expensive liability). New York: Fireside Publishing; 2007.

13.6 ENTREPRENEURIAL STRATEGY

Entrepreneurial strategy is the process of determining how best to deploy the assets of the venture to create value. Early on, when the company is very small and may consist of the founding entrepreneur and one or two other founding partners, the only strategy may be to survive until tomorrow. As the business grows, however, and employees and additional investors are brought into the venture, strategy setting is far more complex and challenging. Responsibility for setting and executing strategy falls squarely on the shoulders of the entrepreneur, who is increasingly recognized as the venture's chief executive officer (CEO).

To reduce some of the pressure and uncertainty surrounding strategy setting, the entrepreneur should rely on trusted advisors that he or she has been able to attract to the venture for input and advice. The complexities associated with entering markets, setting prices, defining customer value, and spending scarce cash can be daunting, especially for novice entrepreneurs. Establishing a board of advisors or leveraging the venture's board of directors for strategic advice and consultation can help reduce the unknowns in the venture's environment.

Entrepreneurial strategy has been studied by scholars and invented by practicing entrepreneurs. While the final word will never be written about entrepreneurial strategy, scholars have been able to identify several recurring themes that entrepreneurs tend to use and return to as they invent venture after venture.

In this final section of this chapter, we will explore and discuss several approaches to entrepreneurial strategy. However, it must be noted that all discussions of entrepreneurial strategy necessarily will be incomplete and pertinent to a particular time and set of circumstances. In this light, the discussion below focuses on early twenty-first century technology ventures and the challenges they face in an increasingly global and fast-paced economy. There are several main lines of thinking regarding entrepreneurial strategy that we will explore:

- Real options reasoning
- Resource-based theory
- Porter's strategic options

13.6.1 **Real Options Approach**

The **real options** approach to entrepreneurial strategy is derived from the world of financial management. Financial managers must have some rationale for the choices they make in allocating the scarce financial resources of a corporate venture. To manage that, they often make reasonable guesses about the probability of certain outcomes and the financial implications they would have. Simple multiplication of the financial outcomes by the probability of a particular event enables decision makers to quantify and justify the decisions they make. A real option is defined as the right, but not the obligation, to undertake a certain business decision. In the world of corporate finance, this is normally the decision to make or abandon a capital investment.

The real options approach to entrepreneurial strategy assumes that strategic decision making occurs amidst a great deal of uncertainty. Entrepreneurs must make choices daily about which markets to pursue, which capital expenditures to make, who to hire, and what features and benefits to include and exclude from an offering. The real options approach also assumes that entrepreneurs are action oriented, opportunistic, and capable of coping with ambiguity. The progenitors of this approach to entrepreneurial strategy are Wharton School of Business professors Rita Gunther McGrath and Ian MacMillan.[36]

The foundation of the real options approach to strategy setting is what McGrath and MacMillan call the "entrepreneurial mind-set." There are several characteristics they highlight as essential to this mind-set, including:

1. A passion for seeking new opportunities
2. Opportunities are pursued with enormous discipline
3. Only the very best opportunities are pursued
4. An intense focus on execution—especially what is referred to as "adaptive execution"
5. An ability to engage the energies of others in pursuit of opportunity

Real options reasoning advocates that aspiring entrepreneurs and practicing entrepreneurs alike maintain what is referred to as an **opportunity register**. The opportunity register is a record of business opportunity insights that may or may not become the basis of new revenue opportunities. Under the real options approach, there are five categories of economic opportunity:

- Redesign products and services
- Redifferentiate products or services
- Resegment the market
- Completely reconfigure the market
- Develop breakthrough competencies or areas of competitive strength that create new competitive advantages

Through a process referred to as **directed discovery**, the entrepreneur constantly examines the environment for these types of opportunities. Directed discovery is a process of plotting a direction into an uncertain future and then making adjustments as reality unfolds. The requisite elements of directed discovery are speed, capacity for rapid response, and insight. Speed is necessary as many economic opportunities are fleeting. The capacity for rapid response suggests the entrepreneur has organized the resources necessary to seize an opportunity for economic gain. Finally, insight is required to create a value proposition for the opportunity that differentiates it from existing offerings.

13.6.2 Resource-Based Theory

Another approach to understanding entrepreneurial strategy is the so-called **resource-based theory** of sustained competitive advantage.[37] It has been applied to entrepreneurial ventures, but it has its origins in understanding large corporations. In the latter context, the theory holds that large organizations form to create and sustain some type of competitive advantage over rivals on the basis of the resources they control. The resources that provide advantage are those that have the following four characteristics:

- They are rare.
- They are valuable.
- They are difficult to copy.
- They have no ready substitutes.

When applied to entrepreneurial ventures, these various attributes of resources are exploited to develop a start-up with a sustainable competitive advantage from the beginning.[38] The entrepreneur uses insight into these various characteristics of resources to gather those necessary to exploit an opportunity.

Resources are rare if they are not widely available to potential competitors. Examples of rare resources for technology ventures include things like patents on key technologies, access to inexpensive technical labor pools, and knowledge possessed by only a few individuals.

Valuable resources help the venture implement its strategy effectively and efficiently. In other words, valuable resources help a venture exploit opportunities and minimize threats. Examples of valuable resources to a start-up technology venture may include real property, capital equipment, key people with unique talents and skills, and cash.

Ventures that possess rare and valuable resources will have an advantage over other ventures if those resources are difficult to replicate or copy. Several factors may conspire to make it difficult for ventures to copy each other's key resources, including unique historical conditions, complex social relationships (such as exclusive contracts), and ambiguous cause and effect so that the key resources are difficult to identify amidst a variety of factors.

Finally, nonsubstitutable resources are strategic resources that cannot be replaced by commonplace resources. For example, an expert system may replace a manager's knowledge about running an efficient operation. However, it may be far more difficult for a venture to find a commonplace substitute for the charismatic leadership that may be provided by a founding entrepreneur. There are many instances of the latter. Perhaps no better example exists than computer companies that have attempted to use commonplace resources to substitute for the charismatic leadership of a Steve Jobs, Bill Gates, or Michael Dell. For the most part, these iconic leaders of the computer age are impossible to replace with substitute leaders.

13.7 COMPETITIVE STRATEGY MODEL

Michael Porter's **competitive strategy model** offers several alternative operating strategies.[39] According to the competitive strategy model, ventures can develop **distinctive competence** in three ways: differentiation, cost leadership, and niche (see Exhibit 13.3). The goal of developing distinctive competence is to establish a unique position in the market, preferably a position that provides a **sustainable competitive advantage** over others already in the market or those who potentially may enter. A sustainable competitive advantage is one that is based on unique knowledge, insight, or property.

13.7.1 Differentiation

In an effort to distinguish its products, a firm using the **differentiation strategy** offers a higher-priced product equipped with more product-enhancing features than its competitors' products. Using this strategy, firms seek a premium price for their products and attempt to maintain high levels of customer loyalty. The firm markets

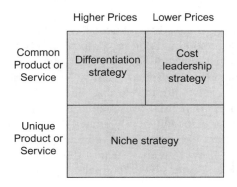

EXHIBIT 13.3

Michael Porter's Competitive Strategy Model.

Source: Porter M. Competitive strategy: techniques for analyzing industries and competitors. New York: Free Press; 1998.

and sells the product to a relatively small group of customers who are willing to pay a higher price for the premium features. This strategy (sometimes called a "premium strategy") leads to relatively high-cost, low-volume production, with a high gross profit margin per item. Often, advertising or marketing adds a perception of luxury that creates demand for the product owing to the psychological value of buying and using it. Mercedes Benz automobiles, Ben & Jerry's ice cream, and Godiva chocolates are marketed under a differentiation strategy.

13.7.2 Cost Leadership

In contrast to the differentiation strategy, the **cost leadership strategy** means low costs, low prices, high volume, and low profit margins on each item. With this strategy, a cost leader attempts to attract a large number of customers with low prices, generating a large overall profit by the sheer volume of units sold. Examples of cost leaders are Dell Computer and Hyundai. It's difficult, though not impossible, to be both lower cost and differentiated relative to competitors. One company that has managed both is Saturn Corporation. Differentiation is a key competitive strategy that has been built in from Saturn's beginnings in 1985. First, Saturn sold the company instead of the car. Second, it developed a relationship with customers. In its first year of operations, Saturn's advertisements focused on employee commitment and low price. Since then, Saturn has focused on the experience of its customers, emphasizing economy and quality at the same time.

13.7.3 Niche Strategy

The **niche strategy** involves offering a unique product or service in a restricted market (usually a geographical region). For example, Dallas-based Southwest Airlines has targeted the point-to-point, low-fare traveler since its beginnings. Many airlines have attempted to compete with Southwest in its niche, but usually meet with failure. Continental Airlines, for example, tried to compete with Southwest by introducing a new point-to-point service it called "Continental Lite." However, Continental learned a painful and expensive lesson by trying to compete in this niche at the same time as it maintained its traditional hub-and-spoke system. Continental Lite failed because the company didn't have the infrastructure to compete head to head in Southwest's niche.

These various approaches to developing distinctive competence have been debated and discussed by strategists, consultants, and scholars. Some have suggested that these three choices are not sufficiently refined—that there are other choices. For example, strategy scholars have suggested refining Porter's three strategic choices into five.[40] Exhibit 13.4 shows how they suggest this should be achieved.

This model expands on Porter's by differentiating the market opportunity between one that is narrow focused and one that is broad. The five competitive strategies in this model are defined as:

1. *Low-cost provider strategy*: Ventures using this approach strive to achieve overall lower costs than rivals, and appeal to a broad spectrum of customers,

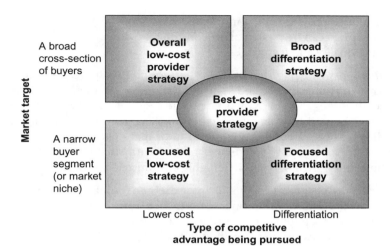

EXHIBIT 13.4

Five Possible Venture Strategies.

usually by underpricing rivals. This was Dell's strategy in its early days, and today it is the strategy deployed by flat screen television makers, such as Vizio.

2. *Broad differentiation strategy*: This strategy is based on differentiating the venture's offerings from rivals in ways that will appeal to a broad spectrum of buyers. One example of a company that uses this approach is Intel. Intel appeals to a broad spectrum of buyer's through its "Intel Inside" advertising campaign. Intel builds products for average users, power users, and business users. Each of these markets had learned to look for the "Intel Inside" logo on computer purchases.

3. *Best-cost provider strategy*: This strategy focuses on giving customers more value for their money by incorporating good-to-excellent product features at a lower price than rivals; the target is to have the lowest prices compared to rivals offering products with comparable attributes. This was the approach adopted by Hyundai on entering the U.S. market. Its cars were very low priced compared to domestic rivals. Hyundai's advertising emphasized that its cars had all the same features and the longest product warranty in the business. After spending years building its brand in the United States with very low-cost vehicles, in 2008, Hyundai pursued upper-scale buyers by introducing a new model, Genesis, that was marketed against Lexus, Mercedes, and BMW models.

4. *A focused low-cost strategy*: Ventures pursuing this strategy concentrate on a narrow customer segment and compete by having lower costs than rivals in this segment. For example, ventures that locate operations near their market

can reduce costs involved in transporting goods to that local market. Those cost savings can be passed on to consumers as lower prices, without compromising quality and feature sets.

5. *A focused strategy based on differentiation:* This strategy concentrates on a narrow buyer segment, and ventures that use it compete by offering customized attributes that meet their tastes and requirements better than rivals. This strategy is used by companies like TutorVista, an online tutoring service that utilizes experts around the world to tutor high-school students on their studies. TutorVista is able to connect people with specific expertise to students who can use their services in a highly customized and personalized manner.[41]

The strategies discussed in this chapter are generic, and offer a starting point for those who are interested in building a technology venture. However, it should also be noted that technology entrepreneurs are notorious for innovation—not only in products and services, but also in strategies. The various strategic frameworks discussed above will no doubt be eclipsed in the coming years with variations that are only dimly recognized today. Savvy technology entrepreneurs will be aware that strategy is always changing, and they will be attuned to understanding existing strategies and adept at inventing new ones.

SUMMARY

This chapter has explored entrepreneurial management, leadership, ethics, and strategy. As we mentioned throughout this chapter, these areas of expertise are built over a lifetime of practice. There are no absolutes in any of these topical areas. Each entrepreneurial manager/leader must find the practices that work best for him or her according to the circumstances encountered.

Despite the lack of absolutes, however, we have explored principles that have stood the test of time and are likely to be highly applicable to whatever technology venture you may be considering or already are operating. For example, the ethical principles of integrity, honesty, and humility likely will apply across any type of venture based on the fundamental rules of good business practice.

Additionally, principles of entrepreneurial management and leadership that were discussed in this chapter should form a basis for continued learning. As we stated, many technology entrepreneurs have not been exposed to formal training in the management or leadership sciences. As such, many learn the art and craft of managing and leading via "on-the-job training." While this is laudable and acceptable, in many cases the venture environment is too competitive to allow this relatively slow and inefficient learning process. While it is not necessary for entrepreneurs to have had formal training in management, it does help to have a conceptual framework available to interpret the events and situations that are experienced when operating a venture. Conceptual frameworks provide some structure to the chaos of managing and leading. The principles discussed in this chapter are meant only as a starting point to further reading and exploration of the concepts and principles that define effective managing and leading.

Effectuation and entrepreneurial expertise were also addressed in this chapter. Effectuation is a relatively new approach to understanding entrepreneurship, but it has drawn a substantial following among scholars and entrepreneurs alike. The notion that entrepreneurs more commonly create opportunities rather than identify and pursue them has intuitive appeal and resonates with what has been discovered about expert entrepreneurs. They are normally very comfortable with ambiguity, and use whatever resources they control to create value and serve markets.

Finally, this chapter examined some approaches to strategy and strategy setting for technology ventures. Entrepreneurial strategy is difficult to address because strategies in start-up ventures often must be fluid and rapidly change to meet unexpected ambiguities and uncertainties of the market and competition.[42] Technology entrepreneurs often must alter their strategies to seize unexpected opportunities or to adjust to competitive responses. Despite the need to be flexible on strategy, ultimately most successful ventures focus on a single strategy. As a venture grows and it is able to offer multiple brands and products, it becomes possible to consider multiple strategic approaches. Consider how most automobile companies are structured. GM offers Chevrolet to low-end buyers and Cadillac to high-end buyers. GM has the scale to maintain its core brand, while offering various products utilizing various strategies to different markets.

The major takeaway for this chapter is to continue to study and learn about management, leadership, strategy, and business ethics. No single book or management guru knows everything about each of these important topics. It is imperative that you not be discouraged by the fact that there are no single, correct answers to management and leadership questions. Rather, this knowledge should lead to a constant pursuit of excellence in managing and leading the technology venture. This constant pursuit will lessen the chance that mistakes will be repeated, and enhance the likelihood of success.

STUDY QUESTIONS

1. What are the five types of fundamental skills that technology entrepreneurs will need to succeed with their venture?

2. What is the difference between technology analysis and business analysis? Which of these is required of the technology entrepreneur? (Hint: This is a tricky question.)

3. Explain what is meant by the term "elevator pitch." Give an example of an elevator pitch for a familiar company such as Microsoft or Nokia.

4. The foundation of successful teams in technology ventures requires clarity of two things for each team member. What are these? Explain why these are essential to effective teams.

5. What behavioral characteristic was described in this chapter as the essence of leadership behavior?

6. Explain what is meant by the term "effectuation." What are its five basic principles?

7. Why do businesspeople sometimes find it difficult to think about ethics?

8. Identify and describe the three basic ethical principles described in this chapter.

9. Explain the concept of real options reasoning. How can the technology entrepreneur use this approach to establish strategy?

10. What are the four elements of the resource-based theory of the entrepreneurial venture? How can these be used to establish strategy?

EXERCISES

1. *IN-CLASS EXERCISE*: This is a short test of a person's entrepreneurial personality. As mentioned in this chapter, self-knowledge is an important part of entrepreneurial success. While this short quiz will not definitively determine whether one can be or is an entrepreneur, it is the beginning of self-knowledge.

	Yes	No	Maybe
I am persistent			
When I'm interested in a project, I need less sleep			
When there's something I want, I keep my goal clearly in mind			
I examine mistakes and learn from them			
I keep New Year's resolutions			
I have a strong personal need to succeed			
I have new and different ideas			
I am adaptable			
I am curious			
I am intuitive			
If something can't be done, I find a way			
I see problems as challenges			
I take chances			
I'll gamble on a good idea even if it isn't a sure thing			
To learn something new, I explore unfamiliar subjects			

I can recover from emotional setbacks			
I feel sure of myself			
I'm a positive person			
I experiment with new ways to do things			
I'm willing to undergo sacrifices to gain possible long-term rewards			
I usually do things my own way			
I tend to rebel against authority			
I often enjoy being alone			
I like to be in control			
I have a reputation for being stubborn			

Source: Adapted from the U.S. Small Business Administration, http://www.sba.gov/starting_business/startup/entrepreneurialtest.html.

Scoring

For each "Yes" response, give yourself 3 points. For each "Maybe" response, give yourself 2 points. For each "No" response, give yourself 0 points.

If you scored between 60 and 75, you can start that business plan. You have the earmarks of an entrepreneur.

If you scored between 48 and 59, you have potential but need to push yourself. You may want to improve your skills in your weaker areas. This can be accomplished by either improving yourself in these areas or by hiring someone with these skills.

If you scored between 37 and 47, you may not want to start a business alone. Look for a business partner who can complement you in the areas where you are weak.

If you scored below 37, self-employment may not be for you. You will probably be happier and more successful working for someone else. However, only you can make that decision.

2. *IN-CLASS EXERCISE*: This exercise involves identifying at least five people who can be referred to as leaders. The exercise is designed to determine if there are any personality traits or physical traits common to each identified leader.

The instructor is to list the various leaders on the blackboard or on an overhead slide in a single column. Beginning with the first name on the list, ask the class to name three personality or physical traits that are important to the person's leadership abilities. Do this for each one on the list, placing the traits in the rows next to the corresponding name.

Once you have collected three traits for each individual, go back over the list of traits and circle the ones that appear more than once. For these traits, ask the following questions for discussion:

1. Which of these traits appears to be necessary for leadership?
2. Which of these traits appears to be important for leadership?
3. What is the difference between a necessary trait and an important one?
4. Do you know any nonleaders who display the traits that have been highlighted?
5. What traits other than the ones listed are important to effective leadership?

KEY TERMS

Analytical skills: Technology entrepreneurs must develop their business analytical skills in addition to their technical analytical skills.

Competitive strategy model: This is the strategy model developed by Michael Porter and includes three basic strategies: cost leadership, differentiation, and niche.

Conceptual skills: Technology entrepreneurs must develop conceptual skills that take a "big picture" approach to understanding the venture and that lead to a captivating vision and mission.

Communication skills: Technology entrepreneurs need communication skills to influence and motivate others toward performance and execution.

Cost leadership strategy: This is the competitive strategy whereby a venture attempts to be the low-price option in a particular industry.

Decision-making skills: Technology entrepreneurs must develop their ability to make clear decisions amid ambiguity and incomplete information.

Differentiation strategy: This is the competitive strategy whereby a venture attempts to provide higher levels of value to a market, without being concerned about being the lowest price option.

Distinctive competence: This refers to a venture's capabilities that are unique and that are difficult for others to imitate.

Directed discovery: Under the real options reasoning approach to strategy, "directed discovery" is a process of deliberately experimenting with a range of approaches to serving a market.

Effectuation: Effectuation is a new understanding of entrepreneurship whereby the entrepreneur is seen as a person who creates opportunity rather than identifies and exploits it.

Elevator pitch: This term refers to a short story that an entrepreneur might tell a potential investor about the venture in the time it takes to ride an elevator.

Enterprise value: Enterprise value is the value of an enterprise to a potential acquirer or to investors.

Entrepreneurial strategy: This term refers to actions, resource allocations, decisions, and goals that are aligned to create value for a particular market.

Execution: This term refers to the venture's ability to take actions that realize its stated goals and objectives.

Influence: This term refers to the ability to achieve goals and objectives through other people.

Leadership: Leadership differs from management in that leadership is concerned with the big-picture issues and strategies, and management is concerned with execution.

Managers: As a manager, the entrepreneur is concerned primarily with meeting goals.

Niche strategy: This is the strategic option whereby a venture develops offerings tailored to a specific subset of a larger market.

Opportunity register: Within the real options reasoning approach, the entrepreneur keeps an opportunity register or notebook of product and service opportunities that the venture might be able to pursue.

Perform: This term refers to the ability of a venture to achieve stated goals.

Productivity: This term refers to how efficiently a venture is able to achieve stated goals.

Real options approach: This approach to strategy is derived from financial management, and focuses on the probability of different outcomes for various strategic options a venture could pursue.

Resource-based theory: This approach to strategy focuses on a venture's resources, especially those that are rare, difficult to imitate, nonsubstitutable, and valuable.

Satisficing: This is a decision-making strategy whereby the decision maker chooses the best among available options, without being concerned about whether the option chosen is the correct one.

Sustainable competitive advantage: A sustainable competitive advantage is a strategic advantage a venture holds over competitors and potential competitors that can be sustained over a period of time.

WEB RESOURCES

The web sites below are intended as destinations for your further exploration of the concepts and topics discussed in this chapter:

1. http://www.entrepreneur.com/management/index.html: This web site is hosted by *Entrepreneur* magazine. This portion of the site provides a lot of information about management and leadership challenges that most entrepreneurs face. There are plenty of articles, tips, and case studies to learn from at this site.

2. http://www.inc.com/resources/leadership/: This site is hosted by *Inc.* magazine. This portion of the site focuses on leadership issues that entrepreneurs face.

3. http://www.ccl.org/leadership/index.aspx: This is the web site for the Center for Creative Leadership. There are many resources available at this site, some of which are clearly useful for novice and experienced entrepreneurs alike.

4. http://www.effectuation.org: This web site addresses the concept of effectuation, which was discussed in this chapter. There are teaching tools, research abstracts, and complete papers.

5. http://www.1000ventures.com/business_guide/crosscuttings/leadership_entrepreneurial.html: This site covers the topic of entrepreneurial leadership exclusively. While not the most visibly appealing site ever posted to the web, it does have a lot of resources, ideas, and leadership tips.

6. http://www.business.illinois.edu/ael/: This site is more serious and academic than the one above, with a lot of information about entrepreneur workshops, conferences, and other events.

ENDNOTES

[1] Hamermesh RG, Heskett JL, Roberts MJ. Note on managing the growing venture. *Harv Bus Rev* 2005;**January**:1–8.

[2] Rubenson GC, Gupta AK. Replacing the founder: the myth of the entrepreneur's disease. *Bus Horiz* 1992;**35**(6):53–7.

[3] Forbes DP. Managerial determinants of decision speed in new ventures. *Strateg Manage J* 2005;**26**(4):355–66.

[4] Todorovic WK, Schlosser FK. An entrepreneur and a leader! a framework conceptualizing the influence of a firm's leadership style on a firm's entrepreneurial orientation–performance relationship. *J Small Bus Entrepreneurship* 2007;**20**(3):289–307.

[5] Kawasaki G. The art of execution. *Entrepreneur* 2008;**36**(4):48.

[6] Singh S. Practical intelligence of high potential entrepreneurs: antecedents and links to new venture growth. *Academy of Management Proceedings* 2008;1–6.

[7] Reid C. Sizing up early stage tech firms to help gauge value. *Mergers Acquisit: Dealmaker's J* 2007;**42**(6):28–31.

[8] Reitz V, Fried RM. Building a brand called "You, Inc.". *Machine Design* 2005;77(13):78.

[9] Dutta DK, Thornhill S. The evolution of growth intentions: toward a cognition-based model. *J Bus Venturing* 2008;**23**(3):307-32.

[10] Companys YE, McMullen JS. Strategic entrepreneurs at work: the nature, discovery, and exploitation of entrepreneurial opportunities. *Small Bus Economics* 2007;**28**(4):301-22.

[11] Ucbasaran D. The fine "science" of entrepreneurial decision making. *J Manage Studies* 2008;**45**(1):221-37.

[12] McKelvie A, Gustafsson V, Haynie JM. Entrepreneurial action: exploitation decisions under conditions of uncertainty. *Academy of Management Proceedings* 2008: 1-6.

[13] Winter SG. The satisficing principle in capability learning. *Strategic Manage J* 2000; **21**(10/11):981-96.

[14] Kourdi J. The deciding factor. *Director* 2006;**60**(2):33.

[15] Darling JR, Beebe SA. Enhancing entrepreneurial leadership: a focus on key communication priorities. *J Small Bus Entrepreneurship* 2007;**20**(2):151-67.

[16] Torres NL. Sounds like a plan. *Entrepreneur* 2005;**33**(3):102-4.

[17] Lange JE, Mollov A, Peralmutter M, Singh S, Bygrave WD. Pre-start-up formal business plans and post-start-up performance: a study of 116 new ventures. *Venture Capital* 2007;**9**(4):237-56.

[18] Collins J, Porras J. Building your company's vision. *Harv Bus Rev* 1996;74(5):65-77.

[19] Adapted from Janet Rovenpor, "JetBlue Airways: can it survive in a turbulent industry?" In: Arthur A, Thompson Jr., AJ, Strickland III, JE. Gamble, *Crafting and executing strategy: the quest for competitive advantage*. New York: McGraw-Hill, 2007.

[20] Hamm J. Why entrepreneur's don't scale. *Harv Bus Rev* 2002;**December**:110-15.

[21] Williams JC. Self-control. *Baylor Bus Rev* 1997;**Fall**:9, 32.

[22] Cited in Vivian Pospisil. Nurturing leaders. *Industry Week* 1997;**November** 17:35.

[23] Sutton RI. *The no asshole rule: building a civilized workplace and surviving one that isn't*. New York: Business Plus; 2007.

[24] Csikszentmihalyi M. *Flow: the psychology of optimal experience*. New York: Harper Publishing; 1991.

[25] Kipnis D, Schmidt SM, Swaffin-Smith C, Wilkinson I. Patterns of managerial influence: shotgun managers, tacticians, and bystanders. *Organ Dyn* 1984;**Winter**:58-67.

[26] Zaleznick A. Leaders and managers: are they different? *Harvard Business Review* 1977:31-42.

[27] Radosevich L, et al. Leaders of the information age. *CIO* 1997;**September 15**:137-81.

[28] Marchetti M. Dell computer. *Sales Marketing Manage* 1997;**October**:50-3.

[29] Zimmerman EL. What's under the hood? the mechanics of leadership versus management. *Supervision* 2001;**August**:10-12.

[30] Sarasvathy SD, Dew N, Read S, Wiltbank R. Designing organizations that design environments: lessons from entrepreneurial expertise. *Organ Stud* 2008;**29**(3):331-50.

[31] Read S, Sarasvathy SD. Knowing what to do and doing what you know: effectuation as a form of entrepreneurial expertise. *J Private Equity* 2005;**Winter**:45-62.

[32] Sarasvathy SD. *Effectuation: elements of entrepreneurial expertise, Vol. March 30*: [Nd city] Edward Elgar Publishing; 2008.

33 Sarasvathy SD. Empirical investigations of effectual logic: implications for strategic entrepreneurship. In: Smith RH, editor. Presented at the Entrepreneurship Theory & Practice Conference on Strategic Entrepreneurship, School of Business, University of Maryland. April 10–11, 2007.

34 Eller K. *Integrity is all you've got*. New York: McGraw-Hill; 2004.

35 Marcum D, Smith S. *Egonomics: what makes ego our greatest asset (or most expensive liability)*. [Nd city] Fireside Publishing; 2007.

36 McGrath RG, MacMillan I. *The entrepreneurial mindset: strategies for continuously creating opportunity in an age of uncertainty*. Boston: Harvard Business School Press; 2000.

37 Barney J. Firm resources and sustained competitive advantage. *J Manage* 1991;**17**:99–120.

38 Dollinger MJ. *Entrepreneurship: strategies and resources*. Upper Saddle River, NJ: Prentice-Hall Publishing; 1999.

39 Porter M. *Competitive strategy: techniques for analyzing industries and competitors*. New York: Free Press; 1998.

40 Thompson Jr AA, Strickland AJ, Gamble JE. *Crafting & executing strategy: the quest for competitive advantage*. 15th ed. Burr Ridge, IL: McGraw-Hill; 2007.

41 Prahalad CK, Krishnan MS. *The new age of innovation: driving co-created value through global networks*. New York: McGraw-Hill; 2008.

42 Wiltbank R, Dew N, Read S, Sarasvathy SD. What to do next? The case for non-predictive strategy. *Strategic Manage J* 2006;**27**:981–98.

Managing Risk and Career Development

Venture Risk Management

14

After studying this chapter, students should be able to:

- Identify various components of venture risk management
- Apply a variety of approaches to manage venture risk
- Recognize the potential liabilities associated with a technology venture, and develop plans for minimizing associated risks
- Distinguish between a board of directors and a board of advisors
- Develop a mission statement for their venture
- Manage human resources in compliance with federal laws
- Terminate employees without raising lawsuit risks
- Understand the role of the Uniform Commercial Code in governing business transactions
- Develop an ability to deal with failure risk

TECH VENTURE INSIGHT

"Puzzle Interviews" Help Screen Job Applicants, but Can Entail Unforeseen Risks

Using intelligence tests to determine whether a prospective employee is a good hire is a fairly common practice. In fact, many professions require some type of evidence of proficiency via testing before one is allowed to practice in that profession. The legal profession, for example, requires that aspiring practitioners pass the bar exam to determine their fitness for practice.

In the high-tech industry, many firms have begun to use an entirely different type of test to screen prospective employees, namely, the puzzle interview. Microsoft is one large technology employer famous for using this type of test. Questions such as:

- How would someone weigh a jet without using scales?
- If a person tosses a suitcase overboard, will the water level rise or fall?

- Why are manhole covers round rather than square?
- How many gas stations are there in the United States?

Companies use puzzle tests to discover the smartest and most creative job applicants. Such tests are designed to find out whether an applicant can think under stress and in a creative manner. Some companies resort to more direct testing of stress management skills, such as by requesting that an applicant open a window that secretly has been glued shut.

Todd Eberhardt, CEO and founder of Minneapolis-based Comm-Works, uses puzzle tests to identify top talent. Eberhardt especially uses such testing in later-round interviews. He likes to tell applicants a story, and have them repeat it to him to test their ability to recall information. He also uses puzzle questions like those listed above. Sales candidates might be asked to explain their thought processes behind a recent major purchase.

Eberhardt says his biggest challenge is evaluating the responses he gets from a group of highly talented applicants. "You may get five different answers that are all equally viable," he stated. In such cases, Eberhardt distinguishes among the candidates by examining the logic they use to derive their answers. "The best employees are already three to ten steps ahead of everyone else…. You really get some insights into the person's thought patterns," says Eberhardt.

The puzzle interview accounts for about 20% of the interview process at Comm-Works. "There's a place for them," says Eberhardt, but he adds, "I would look at it as a part of the overall solution when you're evaluating talent."

Employers need to be careful they don't use puzzle tests inadvertently to screen out a protected class under Title VII of the Civil Rights Act of 1964. For example, employers who wish to use puzzle testing must be prepared to assist disabled applicants in filling out this part of the interview process. Employers should also administer puzzle testing consistently, and they should document everything as preparation in case the hiring decision will need to be justified.

Source: Adapted from Pentilla C. This is a test. Enterpreneur August 2004:72–3; Honer J, Wright CW, Sablynski CJ. Puzzle interviews: what are they and what do they measure? Applied HRM Res 2007;11(2):79–96.

INTRODUCTION

A common belief often expressed in the popular literature is that entrepreneurs are notorious risk takers. Perhaps, this perception of entrepreneurial risk propensity has become commonplace because of the many exemplary entrepreneurs who, in fact, exceed the average tolerance of risk. The risk-taking propensity among these exemplars is then identified as *essential* to their relative success. Of course, some entrepreneurs are risk takers to an extraordinary degree—but research indicates that the frequency of risk taking among entrepreneurs is similar to that of the general population.[1] Instead, expert entrepreneurs have become adept risk *minimizers*. They are able to deal with situations that include elements of risk, but they have learned how to bring those risks down to levels that are tolerable.[2] By way of contrast, someone who has not developed this risk minimization capacity will avoid the situation and its associated risks.

Entrepreneurship scholars have in the past several years begun to examine not only how entrepreneurs recognize economic opportunity, but also how they evaluate the risks associated with that opportunity.[3] Managing risk involves internal and external components. Internally, the successful entrepreneur has learned to live with risk and to adapt to the ambiguity that it usually entails. Externally, the expert entrepreneur has learned to minimize risk through a multitude of actions. Raising capital from external investors, aggregating required resources, honing in on essential and advantage-providing knowledge, etc. are techniques the seasoned entrepreneur routinely employs.

In this chapter, we will explore techniques and strategies for minimizing venture risk, legal risk, and failure risk. Each of these will draw in part on what has been discussed in earlier chapters of this book. In keeping with the general approach in this book, we will keep our discussion focused on practical tools and applications pertinent to launching and operating a technology venture.

14.1 VENTURE RISK MANAGEMENT

We use the term "venture risk management" (VRM) to indicate that our focus in this section is on start-up and early stage technology ventures. In practice, VRM is derived from the well-developed tools and principles associated with what is referred to as "enterprise risk management" (ERM). ERM is a discipline by which the organization "assesses, controls, exploits, finances, and monitors risks from all sources for the purposes of increasing the organization's short- and long-term value to its stakeholders."[4] ERM is practiced at the level of large corporate organizations, and involves a wide range of tools and professional knowledge. ERM consists of monitoring four types of potential risks to the venture:

1. *Hazard risk*: This includes potential problems such as fire and theft, business interruption, disease and disability of key personnel, and liability claims.

2. *Financial risk*: This includes potential problems such as price competition, credit availability, inflation, and liquidity.

3. *Operational risk*: This includes potential problems associated with business operations, leadership and management challenges and concerns, information technology, and business reporting.

4. *Strategic risk*: This includes problems associated with the venture's reputation and consumer perceptions, competitor positioning, and regulatory and political trends.

Managing venture risk is a continuous process that should be part of every venture's management and leadership tactics. In the early stages of the venture, financial risk likely overwhelms the day-to-day thoughts and activities of the leadership team. However, as the venture matures and finances become less of a daily concern, the other risks begin to predominate. For example, as the venture scales and begins to serve customers and generate revenue, operational issues become more prominent. No venture wants to work hard on generating a sale and then lose the deal because operational problems prevented delivery on the order. Hard-won customers are difficult to win back in the event of such a failure to deliver.

We have already examined financial risk management in Chapter 5 and strategic risk management in Chapter 13. Here, we will consider tactics that can be used to manage hazard risks and operational risks. We begin with hazard risks.

14.2 HAZARD RISKS

Every business runs the risk of exposure to liability simply by offering products and/ or services that may not live up to expectations or may even fail. Medical devices, electronics, engineering design, and every other type of product or service could fail to meet customer expectations, or fail as a product—sometimes dramatically.

On August 1, 2007, the I-35W bridge that spans the Mississippi River in Minneapolis collapsed during rush hour. Tragically, 13 people were killed and more than 140 injured. After an exhaustive review of the accident, the National Transportation Safety Board identified a fundamental design flaw as the likely cause of the collapse. Lawsuits filed by victims' families have identified an engineering consulting firm as one of the defendants.[5] The firm had inspected the bridge prior to its collapse and deemed it to be in need of repairs and upgrades, but otherwise stated that it was structurally sound. Although the consulting firm was not involved in the construction of the bridge or its original design, it nevertheless provides a service that makes it possible to be the defendant in a lawsuit.

Liability risk also includes other potential disruptions to business, including theft, natural disasters, and impairment of key personnel. Fortunately, different types of insurance can be purchased to cover most, if not all, of these potential risks. Most businesses should insure against basic theft and liability. This is simply common sense and ensures that the business will not be disrupted in the case of an accident or loss of property. It is important to work with an insurance company that understands the needs of a growing business and has scalable plans to meet the particular needs of the venture.

14.2.1 Product Liability

Product liability involves the responsibility of business firms for negligence in design, manufacture, sale, and operation of their products. For example, workers sued keyboard makers for selling products that cause wrist injuries and for not warning users of the potential risks.[6] Public interest groups constantly seek to broaden the liability of business and manufacturers of such products as breast implants, automobiles, cigarettes, and guns.[7]

In certain instances, product liability laws have been expanded to include cases in which the producer or marketer of the product is not proved negligent. Under **strict product liability**, the business is responsible for any damages that may result regardless of the care it observes to guard against such damages. Strict liability is commonly applied to "ultrahazardous" business activities such as crop dusting, pile driving, and storing flammable liquids. No amount of care can prevent companies in ultrahazardous industries from liability if their businesses should cause harm to a third party.

14.2.2 Liability Insurance

Liability insurance covers accidents that may occur on the venture's premises, as well as accidents that may occur through use of the venture's products and services. Business liability insurance protects a venture in the event of a lawsuit for personal injury or property damages. It will usually cover the damages from a lawsuit along with the legal costs. Depending on business needs, liability insurance can be purchased in three forms:

- *General liability insurance*: This form of business liability insurance is the main coverage to protect a venture from injury claims, property damages, and advertising claims. General liability includes D&O liability and employer liability. **D&O liability** stands for "directors and officers" liability and is intended to cover the acts or omissions of those in the director or officer position. **Employer liability** is also known as "worker's comp," and it is a mandatory form of liability insurance coverage that all businesses must carry.

- *Professional liability insurance*: Ventures providing services may need professional liability insurance. This coverage protects a business against malpractice, errors, negligence, and omissions. For some professions, it is a legal requirement to carry such a policy. Doctors require coverage to practice in certain states. Technology consultants often need coverage in independent contractor work arrangements.

- *Product liability insurance*: Technology ventures selling or manufacturing products should be protected in the event of a person becoming injured as a result of using the product. The amount of coverage and the level of risk depend on the nature of the products and their potential for misuse.

The most likely person to ask about your insurance needs as a start-up venture is your accountant. Accountants are not only trained in helping the ventures manage their financial tracking and reporting, but they also generally are connected to other financial experts, such as insurance providers. As with the venture's accountant, it is advisable to work with an insurance provider who has experience in the venture's industry. It is also advisable to work with a firm that has scalable policies and premium schedules that are a match with the needs of the venture.

14.3 OPERATIONAL RISKS

Although the operational risk to conducting business is substantial, there are direct and specific actions ventures can take to minimize those risks. Here, we discuss four specific approaches that will help a start-up technology venture limit its exposure to legal risks:

1. Venture governance
2. Investor relations
3. Human resources management
4. Founding and ongoing legal counsel

14.3.1 Venture Governance

In the twenty-first century, all firms, large and small, must pay attention to sound corporate governance. In 2008, the world was rocked by rapidly declining stock markets, and a number of high-profile business scandals. In the United States, former NASDAQ chairman Bernard Madoff admitted to running a $50 billion Ponzischeme investment house.[8] The failure of investors and government agencies to uncover his scheme during its more than 20 years of fraud will surely lead to new regulations about governance. In 2000, for example the collapse of Enron led to the development of the **Sarbanes-Oxley** regulations for corporate governance.[9] These regulations have placed significant new governance and disclosure burdens on public companies. The intent of the regulations is to ensure that investors are aware of and can make informed investment decisions based in part on how a company governs itself.

The need for effective governance also is a global issue. In early 2009, India's financial markets were stunned to learn that one of its star performers, Satyam Computer Services Ltd., was based on fraudulent profits.[10] Satyam's founder and CEO B. Ramalinga Raju confessed that he had been "cooking the books" for several years. As a result of this revelation, India is moving quickly to boost its oversight of corporate governance and accounting practices. The nation's leaders are concerned that global investors and business interests will lose confidence in conducting business in India.[11] Sound corporate governance and assertive oversight, including the ability to uncover and punish wrongdoers, is critical to creating that confidence.

Although effective governance begins with compliance with legal regulations, in many cases, it goes far beyond that. For example, organizations around the world are being evaluated by their commitment to "sustainability" and environmental issues. These pressures don't necessarily come from regulatory agencies; they are often local or national political concerns. Still, start-up ventures must be alert to issues that are important to market participants. Running afoul of these issues—such as being perceived as unconcerned about the environment—could be bad for future business. On the other hand, it may be smart to take advantage of prevailing market opinions. For example, currently it is possible to build goodwill for the venture by indicating a commitment to "green" concerns and issues.[12] Of course, this goodwill could immediately be withdrawn if a venture didn't follow through with its stated commitments.

Venture governance practice and philosophy should include a number of tactics and approaches. In the early stages, most technology entrepreneurs have a number of pressing concerns on their minds. Often, the establishment of sound governance procedures is overlooked as other urgent issues capture attention. Yet, experienced entrepreneurs attest to the importance of laying a solid foundation for effective governance during the early stages of the venture's life.[13] This often begins with the formulation of a compelling mission statement and vision for the venture.

Mission and vision

Sound corporate governance begins with a **mission statement**. A venture's mission statement can have significant meaning to stakeholders. On the one hand, simply having such a statement indicates a level of maturity and sophistication that is common to larger, more well-established ventures. The venture's mission statement should be achievable and should be clearly and succinctly written. A mission statement should avoid trite and unachievable assertions, such as "global leader" or "best in the world." These are meaningless and, in practice, would be impossible to measure and evaluate. The best mission statements are focused on customers, and how the venture is dedicated to excelling in serving them. Exhibit 14.1 is the mission statement for biotech firm Genentech.

> "Our mission is to be the leading biotechnology company, using human genetic information to discover, develop, manufacture and commercialize biotherapeutics that address significant unmet medical needs. We commit ourselves to high standards of integrity in contributing to the best interests of patients, the medical profession, our employees and our communities, and to seeking significant returns to our stockholders, based on the continual pursuit of scientific and operational excellence."[27]

EXHIBIT 14.1

The Genentech Mission Statement.

A basic governance philosophy and code of ethics or statement of company values might also be formulated and used to guide the venture's decision making. Exhibit 14.2 provides the governance principles behind Dell Computers.

"The Board of Directors, as representatives of the stockholders, is committed to the achievement of business success and the enhancement of long-term stockholder value with the highest standards of integrity and ethics. In that regard, the Board has adopted these principles to provide an effective corporate governance framework for Dell, intending to reflect a set of core values that provide the foundation for Dell's governance and management systems and its interactions with others."

Ethics and Values

"The Board and management are jointly responsible for managing and operating Dell's business with the highest standards of responsibility, ethics and integrity. In that regard, the Board expects each director, as well as each member of senior management, to lead by example in a culture that emphasizes trust, integrity, honesty, judgment, respect, managerial courage and responsibility."

"Furthermore, the Board also expects each director and each member of senior management to act ethically at all times and to adhere to the policies, as well as the spirit, expressed in Dell's Code of Conduct. The Board will not permit any waiver of any ethics policy for any director or executive officer." [28]

EXHIBIT 14.2

Dell Computers Governing Principles.

Of course, simply formulating a mission statement and code of ethics is just the beginning of effective venture governance. In addition to these statements, ventures need to establish lines of authority, accountability, and responsibility. This can be enhanced by developing an ultimate authority in a governing board of directors or board of advisors. We turn next to the role of governing boards in venture risk management.

Governing boards

Among the governance tactics that should be initiated in the early stages is an independent sounding board for strategic and financial decision making. Subchapter C- and S-corporations are required to establish a formal **board of directors**. Legally, a board of directors has **fiduciary responsibility** for the venture. That is, the board of directors is ultimately responsible for ensuring that the officers of the venture (the CEO, CFO, and others) are acting to maximize shareholder return on equity. The members of the board of directors have two principal duties that they must uphold: a **duty of care** and a **duty of loyalty**. Duty of care means the directors are appropriately focused on growing the value of the venture. Duty of loyalty means the directors will not disclose trade secrets or in any other way diminish the competitive advantage of the venture.

If the venture is organized as an LLC or partnership, a board of directors is not required by law. Instead, such a venture may elect to establish a **board of advisors** to assist in strategic and financial decision making. A board of advisors does not have fiduciary responsibility in the way that a board of directors formally does. In other words, there is no legal requirement for a member of a board of advisors to uphold the duties of loyalty and care. Nonetheless, a board of advisors can provide an independent "sounding board" for a venture's leadership team. It is up to the venture leaders to determine how much it will adhere to the advice of the board of advisors. This can be formalized by establishing board voting and decision-making protocols. With such protocols in place, the officers of the venture would be less likely to act contrary to board advice. Tech Tips 14.1 offers some advice for establishing an effective board of advisors for your growing venture.

Whether the venture is required to have a formal board of directors or it elects to operate with a board of advisors, there are techniques for effective board management that should be observed. These techniques include the following (on page 468):

TECH TIPS 14.1

Developing an Effective Board of Advisors

1. *Recruit advisors for short-term objectives*: Don't recruit advisors who will help you with future products or future markets. Focus on the short term and determine what skills, introductions, and knowledge you will need to accomplish your immediate business objectives.

2. *Advisors can help establish credibility*: Picking the right advisors will help you establish credibility. In fact, it is often easier to persuade industry luminaries and prominent experts to join your advisory board than it is to persuade operational executives who are not used to the idea of devoting personal time to serve on boards.

3. *Look for advisors in unusual places*: If your business has industry conferences or training workshops, this is one place to start looking. Ask your relatives and friends if anyone they know has started a comparable business. Talk to potential suppliers for introductions.

4. *A free lunch is often a better motivator than equity*: There is no standard compensation scheme for advisors because it depends on how many advisors you need, how much time they will devote, and what kind of company you have. For example, a rule of thumb for high-growth ventures is 1.0–2.5% of share capitalization for all advisors. If you are too early stage to put together an equity compensation plan, you should consider making a small cash payment to your advisors.

5. *Don't treat advisors like employees or suppliers*: Even if you are paying them, it is difficult to hold advisors accountable in practice. This is because most advisors have income from other sources and will treat your business as a part-time hobby or casual business interest.

6. *Set term limits*: Much like board members have term limits, advisory board roles should also have term limits. Some advisors will become very involved with your business, will take on the role of passionate advocates, and will want to renew their engagement.[14]

- Develop formal guidelines for board membership and turnover. Board members should be aware of their responsibilities and the length of time for which they have been appointed.
- Establish protocol for voting on matters brought to the board.
- Keep minutes of each board meeting and record the resolutions and actions of the board.
- Distribute the minutes of each meeting to board members for their independent review and approval.
- Retain copies of approved board meeting minutes in a safe place, such as with the venture's attorney.

Whether the venture develops a formal board of directors or a more informal board of advisors, it is always good practice to ensure that board member interests are aligned with those of the venture. One common technique for creating this alignment is through ownership incentives. Board members for start-up ventures normally will not require financial compensation, and for many qualified board members, that would have limited incentive anyway. More likely, board members will be motivated by equity in the venture. Technology entrepreneurs can use equity to entice key board members to join, and then use a vesting schedule to link them to the venture and its performance over an extended period. Often, technology entrepreneurs will provide a small equity grant to key individuals to entice them to join the board, and then put a vesting schedule in place for the board member to earn additional equity on the basis of individual and venture performance.

14.3.2 Investor Relations

Investor relations is commonplace in large, publicly traded firms. Go to the web site of nearly any large public corporation, and you will likely find a section dedicated entirely to investor relations. Typically, the site will include corporate information such as annual and quarterly reports and the latest price of the company's common stock. Often, such sites will also include biographical information about the company's officers and directors. It may also include special statements about the firm's governance philosophies and practices, its values and mission, and other things that would be of interest to current and potential investors.

Publicly traded companies undertake vigorous investor relations programs in part to comply with federal and other regulations regarding disclosure. Still, most firms go beyond compliance in their investor relations efforts because such efforts can help the firm reduce financial risk.

Although most start-up ventures do not have the same compliance burdens as do large, publicly traded companies, it is not a bad idea to establish some investor relations tactics as part of a comprehensive risk mitigation strategy. At the least, a start-up venture should dedicate a portion of its web site to investor relations if it is serious about managing its existing investors and concerned that it may need new investors in the future. Tech Tips 14.2 highlights some things a start-up venture might include in the "investor relations" section of its web site.

TECH TIPS 14.2

Items for Investor Relations

For start-up ventures, an investor relations section of its web site might include the following:

- A dated copy of the business plan or, as an alternative, a brief executive summary
- Information about who to contact about a possible investment
- Mission statement and governance philosophy
- Biographical information about the officers and advisors or directors
- An annotated timeline of the venture's history and milestones
- An overview of professional partners, such as legal and accounting firms

---

Of course, the investor relations section of the venture's web site must be in compliance with SEC regulations concerning stock offerings and fund raising. In very general terms, a venture must not publicly solicit investment. Still, it is within the boundaries of SEC regulations for a venture to list contact information in the case that someone has been investigating the venture and wants to inquire about potential investing opportunities.

Ventures can make their business plans available via their investor relations web site. However, consideration should be given to whether competitive advantage is compromised by doing so. For example, if the business plan contains proprietary information about technology or strategy, it should not be publicly disseminated. Business plans with that level of detail are usually only distributed as numbered copies to select, prescreened individuals. In lieu of the full business plan, ventures should at least have a short executive summary available for downloading from the web site.

14.3.3 Human Resources Management

Human resources management may very well be the most difficult challenge for leaders of technology ventures.[15] Growing ventures typically have employees who depend on it for career growth, including compensation, personal development, and, quite often, health and other benefits. Because business wields such great power over personal career paths, the federal government has enacted a series of laws to ensure that individuals receive fair and equal treatment in the workplace. It should be relatively easy to understand how a business with a bias against a certain type of employee could have a negative influence on the careers of people of that type. For example, a business that resisted working with people from a certain ethnic group could pass over such individuals for employment. Without legal recourse, such individuals would be powerless to change their situation.

In 1964, the U.S. federal government passed the **Civil Rights Act**. Title VII of that act expressly addresses employment issues and prohibits discrimination based on race, color, religion, sex, or national origin. Since that landmark act was adopted, the federal government in the United States has created additional laws that prohibit employment discrimination, including:

- Sexual harassment
- The Age Discrimination in Employment Act
- The Americans with Disabilities Act
- The Family and Medical Leave Act
- Hiring and firing employees

These various federal statutes have been developed explicitly to protect certain classes from discriminatory practices in the workplace. They are generally enforced by the **Equal Employment Opportunity Commission** (EEOC) and apply to most businesses with 15 or more employees.

Entrepreneurs should become familiar with each of these employment-related federal laws. For the most part, EEOC laws are based on common sense—treat everyone fairly. However, they can be complicated, and it is not uncommon for entrepreneurs to unwittingly run afoul of workplace antidiscrimination laws.

Another very sensitive area of employment law involves hiring and firing employees. Venture managers must take care that the employee selection process does not illegally discriminate against a protected class of individuals, as defined by the Civil Rights Act of 1964. For example, if a manager didn't want a certain type of person working in the organization, one criterion for employment might be whether a prospective employee could lift 100 pounds over his or her head. Clearly, some classes of individuals are less qualified to fulfill this requirement than others. Managers certainly can invoke such terms of employment; however, the terms must be related to actual requirements of the job. The EEOC refers to such criteria as **bona fide occupational qualifications** (BFOQ). Any employment criteria that are not BFOQs could lead to lawsuits based on employment discrimination. Thus, venture managers must develop job descriptions and hiring criteria that are clear and accurate reflections of actual job requirements.

The simplest way to summarize the intent of employment laws is fairness. Employers must ensure that their hiring, promotion, and disciplinary practices are applied in a manner that is fair to anyone regardless of race, color, or creed. As stated above, job descriptions and advertisements for positions must be highly correlated to actual job requirements.[16] With that in mind, it is a good idea for entrepreneurs to develop objective measures of an individual's conformance to those requirements. For example, if programming is involved in the job, a performance test to demonstrate competence would be objective and measurable. In contrast, merely observing an individual's ethnic background to determine whether they are capable of programming could lead to trouble.

In promoting individuals, managers must also use objective observations of past performance in assessing an individual's fitness for the new job. Promoting from within the company is similar to hiring new staff. Criteria used for making promotion decisions must be related to the job being filled, and the more objective criteria that can be applied, the better. Of course, all hiring and promotion decisions will involve some subjective element. Nonetheless, the threat of an EEOC lawsuit can be minimized by including objective criteria in the decision process.

Disciplining employees is another source of legal tangles for venture managers. As with hiring and promoting, the most important concept to keep in mind when disciplining employees is fairness. Some entrepreneurs get into difficulties when they discipline employees for behaviors that have not been proscribed. For example, disciplining an employee for making a personal call on company time could lead to problems if there is no clear organization-wide policy prohibiting such calls. Many firms have employee manuals and/or orientation sessions to inform employees of workplace norms and rules. Such communications can be supplemented with regular memos, policy statements, or other official means of keeping people informed of their professional obligations to the venture.

Terminating employees

Terminating employees is handled differently by different organizations. For example, one major software producer, in the middle of the 2008 economic recession, abruptly terminated a large number of staff. Their technique was to have security guards roam in pairs throughout the building, approach the cubicle of the person being terminated, and escort them to the parking lot. While humiliating and unnerving, there is nothing illegal about this approach. Of course, the company lost a lot of goodwill both among those terminated and those left behind. Smart managers realize that economic cycles may return in their favor and they will once again want to recruit good people. That is less likely to happen if the firm establishes a reputation for treating people in a humiliating manner.

The example above is one where termination was used to reduce costs during a financial crisis. Legal rules provide wide latitude for organizations to fire (or lay off) employees to avert a financial disaster. On the other hand, terminating employees for behavioral problems or workplace rules violations must be handled with care as wrongful termination lawsuits can be expensive. Some states follow "at will" employment laws, which heavily favor the employer. Others follow the **Model Employment Termination Act** that states: "An employer may not terminate the employment of an employee without just cause."[17] Just as managers must skillfully measure and collect data on an employee's performance prior to a promotion, they must use the same care in the case of employee termination. The more data a manager has when terminating an employee, the less the exposure to a damaging lawsuit. If proper progressive discipline procedures are followed, the data will have been collected and available in the employee's file.[18] Tech Tips 14.3 offers advice on what to do and what not to do when terminating employees.

14.3.4 **Legal Risk Management**

Every venture that operates anywhere in the world is exposed to legal risk. Simply engaging in commerce exposes the venture, its officers, and its representatives to

---

TECH TIPS 14.3

Dos and Don'ts for Terminating Employees

What to Do

1. Make the firing decision carefully, but once made, act quickly.
2. If firing due to performance-related reasons, ensure that multiple meetings have been held with the employee prior to making the firing decision to address and correct the performance issue.
3. Be respectful and discreet when having the termination meeting. The meeting should preserve the dignity of the employee.
4. Ensure that your severance or notice arrangements meet legal requirements.
5. Ensure that you have a written termination letter and release that documents the terms and conditions of the termination.

What Not to Do

1. Don't get personal. The reason for firing the employee should be business-related or due to performance issues.
2. Don't make a decision to terminate for cause without conducting a proper investigation. The decision should be based on objective facts, not one or two individual opinions or stories.
3. Don't hold the termination meeting in a public place.
4. Don't go into a termination meeting unprepared. Prepare the written letter, rehearse what you will say, how you will say it, and in what sequence.
5. Don't get defensive or debate the merits of the firing decision with the employee. Allow the employee to react and raise questions, be understanding and sympathetic, but reinforce that the firing decision is definite and final.[19]

---

legal risk. Lawsuits can be brought by nearly anyone against a venture for an increasingly wide range of transgressions. For example, according to an annual survey of legal activity in the United States, 83% of in-house counsel reported at least one fresh case commenced against their company in 2006–07, with 25% counting more than 20 new suits. Larger enterprises are targeted more often. Only 3% of billion-dollar companies managed to get through 2007 without being named a defendant; 50% were served with at least 20 new actions, including a third hit more than 50 times. Alternatively, 44% of companies under $100 million made it through 2007 without a single new suit and only 2% saw more than 20 new cases.[20]

A law is a standard or rule established by a society to govern the behavior of its members. Federal, state, and local governments, constitutions, and treaties all establish laws. So do court decisions. Laws have a direct and substantial impact on how business firms conduct various activities. In this section, we discuss some of the basic concepts of law, including the sources of law, the U.S. court system, and laws affecting business.

The U.S. Constitution specifies how the U.S. government must operate. It is also the foundation of U.S. law. Federal and state constitutions provide the framework for the various levels of government, which derive laws from three major sources: common law, statutory law, and administrative law.

Common law

The body of law created by judges through their court decisions is known as **common law**. On the basis of custom, usage, and court rulings of early England, common law came to America when the first colonies were established and has become a major body of law in the United States. Judicial decisions establish precedents, or standards, that later are used to help decide similar cases. Lawyers are trained to research and use precedents in their litigation and prosecution functions, and judges use precedents to help guide their decisions. The use of precedents provides a measure of certainty and stability to many areas of law. This evolutionary nature of common law gives business owners the confidence to innovate and take risks. They can do so because courts rarely upend precedents and take illogical actions. Business owners can be confident that their transactions will be interpreted according to established procedures and rulings. By way of contrast, imagine a legal system where laws were constantly changing. What do you think would happen to innovation in such a system?

Statutory law

A law created by a federal, state, or local legislature, constitution, or treaty is called a "statute." Together, the laws enacted by the various legislative bodies make up **statutory law**. A statute must be drawn up in a precise manner to be constitutional. However, courts must interpret a law's meaning and its actual consequences are played out in common law. Court decisions sometimes lead to statutes, being changed, clarified, or even dismissed entirely.

Many statutes pertain to the business environment or business practices. Congress, for example, passes laws establishing tax regulations for individuals and businesses. State legislatures and city councils also pass laws regulating general and specific business practices. For instance, many cities enact zoning laws to control the type of business activity that can legally take place in certain geographic regions.

An important part of statutory law is the **Uniform Commercial Code (UCC)**, a collection of statutory laws designed to simplify commerce by eliminating differences between state laws governing business. The UCC, consisting of 10 articles, covers the rights of buyers and sellers in transactions. Drafted in 1952, the UCC has been adopted in its entirety by every state except Louisiana (which has adopted about half of the code). Each state has unique adaptations of the code to fit its own common law findings. Generally, business owners can access a state's UCC through that state's Secretary of State office.

Administrative law

Regulations passed by state and federal administrative agencies are included in **administrative law**. Numerous administrative agencies enforce laws affecting business.

Federal agencies with extensive powers include the Federal Trade Commission, the Consumer Product Safety Commission, the Federal Communications Commission, and the Food and Drug Administration. Agencies on the state level include public utility commissions, licensing boards for various professions and trades, and other regulatory bodies. Some examples of local agencies are planning commissions, zoning boards, and boards of appeal.

14.4 LAWS AFFECTING START-UP VENTURES

Numerous and varied laws regulate the activities of all businesses and everyone involved in the business, from owner to manager to employee. In this section, we discuss the major business law categories, which involve torts, contracts, sales, property, bankruptcy, and negotiable instruments. Contract law was covered extensively in Chapter 8. The law of property, focusing primarily on intellectual property, was covered in Chapter 7. We will discuss the other major business law categories in this chapter, beginning with the law of torts.

14.4.1 Law of Torts

While criminal law deals with crime against society or the state, tort law is concerned with compensating the victims of noncriminal wrongs. Derived from the French word for "wrong," a tort is a noncriminal (civil) injury to other persons or their property or reputation. Torts can be intentional or they may result from negligence. Intentional torts are deliberate acts by a person or business firm. For example, slander (spoken defamation of character) and libel (written defamation of character) are intentional torts. A tort results from negligence when one party fails to exercise reasonable care and causes injury to another. Negligence torts arise from carelessness rather than intentional behavior. A hotel, for example, may be held responsible for injuries to a guest who trips on loose or worn carpeting on the premises. Product liability, mentioned in Section 14.2.1, is an important part of tort law.

14.4.2 Law of Sales

Sales law, which grew out of contract law, involves products sold for money or credit. Sales agreements are contracts subject to the requirements discussed above. Article 2 of the UCC provides that some sales contracts are binding even if all the requirements for a contract are not met. For instance, a sales agreement is legally binding even if the selling price is left out of the agreement; the buyer must pay the reasonable value of the goods.

Article 2 also establishes the law of warranty for sales transactions. An **express warranty** is any statement of fact or promise made by the seller to the buyer relating to the products sold, and which becomes an important part of the sales agreement. A warranty is in the nature of a guarantee, although no formal words such as "warranty" or "guarantee" need be used. If the products in fact are not as represented by the seller, the seller can be held responsible either to make the products as warranted or to give the buyer the money back. The law also imposes an **implied warranty**, not specifically expressed by the parties, ensuring that the business firm has clear title of the products it sells and that the products will serve the purpose for which they are sold.

14.4.3 Law of Bankruptcy

Bankruptcy is a legal procedure provided for individuals and firms that cannot pay their debts. By declaring bankruptcy, the individual or firm asks to be declared legally unable to satisfy creditors and to be released from financial obligations. Three types of bankruptcy are possible: Chapter 7, Chapter 11, and Chapter 13.

Under Chapter 7 bankruptcy, the business firm is dissolved and the assets are sold to pay off debt. Individuals filing Chapter 7 are allowed to keep a limited amount of assets, determined by federal or state law. Chapter 11 bankruptcy temporarily relieves a company from its debts while it reorganizes and works out a payment plan with its creditors. At the beginning of this chapter, we reviewed the collapse of Enron Corporation. To protect its assets and try to rebuild, the company filed for Chapter 11 bankruptcy protection. Chapter 13 bankruptcy allows an individual to establish a plan for repaying debts within 3–5 years. Note that only about 10% of all bankruptcy filings are by businesses; the rest are by private individuals.

14.4.4 Law of Negotiable Instruments

A negotiable instrument is a substitute for money. It is a written promise to pay a specified sum of money that can be transferred from one person or business firm to another. Examples of negotiable instruments include checks, drafts, and certificates of deposit. The UCC specifies that negotiable instruments must meet the following requirements:

- They must be in writing and signed by the maker or drawer (the party who orders the payment of the money).

- They must contain an unconditional promise to pay a certain sum of money.

- They must be payable on demand or at a specific date.

- They must be payable to a specific person or business firm or to the bearer (the party who possesses the instrument).

The payee (the one to whom the instrument is written) must endorse a negotiable instrument before it can be transferred. An endorsement is a person's signature on the back of a negotiable instrument. A blank endorsement is accomplished when the payee signs the back of the instrument. This type of endorsement can be unsafe because anyone can cash the instrument once it is endorsed. Restrictive and special endorsements protect the negotiable instrument should it be lost or stolen. Using the words "for deposit only" along with the signature constitutes a restrictive endorsement; it states what the instrument is for and is much safer than a blank endorsement. A special endorsement specifies to whom the instrument is payable by including the person's or firm's name on the back of the instrument along with the signature. Finally, a qualified endorsement—using the words "without recourse"—means the person who originally signed the instrument, not the endorser, is responsible for payment. The endorser does not guarantee payment if the instrument is not backed by sufficient funds.

14.5 OBTAINING AND WORKING WITH LEGAL COUNSEL

One of the hallmarks of a mature technology venture is an association with a reputable law firm. Rather than wait for necessity to dictate developing a relationship with such a firm, it is in the interests of most technology ventures to establish a working relationship from the beginning. The only reason that some ventures elect to avoid developing a working relationship with a law firm in the early stages is expense. For cash-strapped start-ups, every dollar must be carefully managed, and legal counsel is often expensive. Still, establishing a relationship with a reputable law firm is not a luxury expense, and most investors will feel more comfortable having a well-known firm associated with the venture.

As an entrepreneur, you will work with your attorney on a number of issues, including:

1. Drafting contracts
2. Negotiating human resource issues (including employment agreements)
3. Defending against lawsuits
4. Establishing policies
5. Establishing intellectual property and helping to protect it

Most attorneys charge using one of four billing schemes:

1. Hourly fees
2. Flat fees
3. Contingency fees
4. Retainer

Law firms will usually charge by the hour, and depending on the firm and seniority of the attorneys, they will charge fees that can range from $100 to $500 per hour. **Flat fees** can be arranged for discreet tasks such as drafting a specific contract. You should get an estimate of fees before you engage the attorney for the work. Contingency fees may be arranged when the attorney receives a fixed fee or percentage of the potential cash flow of some event. If you are pursuing a lawsuit for, say, collecting a past due account, the attorney may be willing to work for a percentage of the settlement. Finally, some law firms require an up-front payment or retainer to begin legal work. A modest retainer can ensure that both parties pay sufficient attention to the legal issue at hand. Consider that as the payer of the retainer, you will be motivated to dedicate time and attention to the matter. On the other side, your attorney will feel a greater sense of obligation since you have already paid for work to be done.

In addition to fees, attorneys charge for other non-time-related costs, such as filing fees, photocopying, courier services, and travel. To manage your costs, you should agree in advance with the firm what your budget limits are and whether they should contact you in advance about those non-time costs.

Another way entrepreneurs can minimize costs associated with legal work is to request that some work be handled by less expensive, but competent, junior attorneys. For example, not all contract drafting needs to be handled by a senior partner. Junior attorneys can be 50–75% less expensive and can do a credible job.

The entrepreneur's working relationship with his or her attorney can often be enhanced if the interaction isn't always about legal matters. It's a good idea to keep your attorney informed of important business issues that may or may not have legal implications. On the one hand, there may be legal issues involved that you don't recognize. More important, regular contact keeps the attorney engaged in your business and keeps it in the forefront of his or her attention. You can also stay informed on business law in your industry and provide your attorney with clippings and articles about developments.

Since we are advocating establishing a relationship with a reputable law firm at the venture's founding, it is helpful to know that there are techniques that can be used to minimize associated expenses. One of the surest techniques for limiting expenses associated with legal counsel is to choose a firm that has expertise in your industry. This saves costs in several ways. For starters, the firm with experience in your industry is likely to have templates for a variety of contracts, legal documents, and other items necessary to conduct business in the industry. This can save time and expenses that otherwise would be spent drafting new documents.

An additional benefit of working with an experienced firm is that legal documents evolve over time to meet changing conditions. For example, your venture likely will need an operating agreement that specifies decision-making protocols, shareholder rights and obligations, and many other things. The law firm that has been practicing in the industry has developed an understanding of events and issues that may arise to create problems. The legal documents they develop will reflect this knowledge gained and may prevent the start-up venture from having to learn about them all over again. Although the cost savings associated with problems avoided cannot be tallied, it is nonetheless arguable that an experienced firm can help a start-up avoid problems that a less experienced firm would not.

Another strategy for limiting the costs associated with legal counsel is to be aware of how to use a law firm. Many novice entrepreneurs make the mistake of thinking that legal counsel is also business counsel. Lawyers are trained to interpret and apply the law—they are not business advisors. The best way to use legal counsel is to be very clear about your legal needs. The more time spent in discussions or debates about what types of documents the venture will need and how specific they should be, the more expenses are recorded.

14.6 MANAGING FAILURE RISK

The odds are not in favor of the technology entrepreneur. The well-cited data that only about 1 in 20 ventures succeeds is, unfortunately, overly optimistic. In fact, most technology entrepreneurs do fail at that rate, and even those who succeed usually wind up making less money than they would be able to make by working for a large company.[21] With these depressing statistics, it is inadvisable for anyone to wade into entrepreneurial waters without an ability to cope with failure. Those who are equipped to deal with setbacks and failures will increase their odds of succeeding in the long run. Many successful technology entrepreneurs will attest that they had to suffer through multiple failed attempts before they finally succeeded.

In this section, we discuss several strategies the entrepreneur can use to manage failure risk, including developing a psychological trait called "resilience" and overcoming cognitive biases that are common among entrepreneurs. Each of these strategies should be followed simultaneously to develop and operate the technology venture. The entrepreneur who neglects to develop resilience and manage cognitive biases is at a greater risk of venture failure. Just as sales and marketing cannot be neglected in favor of operations and product development, the technology entrepreneur cannot neglect strategies of managing failure risk. We begin our discussion by examining the psychological trait of resilience and how it can be developed.

14.6.1 Developing Resilience

Although there are no guarantees for success in technology entrepreneurship, there are some well-known techniques for dealing with failure that can enhance its likelihood in the long run. For example, one of the hallmarks of successful entrepreneurs is a trait that is referred to as **resilience**.[22] Resilience is a term that has been used to refer to the ability to survive and even thrive under conditions of turbulence, change, or trauma. In general, it refers to an ability to absorb defeat and/or bad news without losing one's focus on goals and objectives.[23] This characteristic is especially useful for entrepreneurs since it has become common knowledge that entrepreneurship as a lifestyle will occasion failure.[24] The ability to rebound from entrepreneurial failure and continue the entrepreneurial lifestyle is a textbook example of what is referred to as "resilience."

The ability to continue in the wake of entrepreneurial failure includes confronting a range of obstacles. Among these are personal and internal obstacles, including emotional state, financial condition, family matters, and others. Resilience means being able to manage these various pressures in a manner that enables the continuance of the entrepreneurial lifestyle, whether or not a particular venture continues. Entrepreneurial failure also brings a number of external pressures to bear, including the entrepreneur's reputation among peers, investors, and others with potential influence. This reputation may influence the entrepreneur's future ability to launch a new venture, raise necessary funds, or acquire needed resources.

As a personality trait, resilience requires emotional intelligence as well as social awareness. Emotional intelligence includes the ability for one to recognize disappointment, frustration, and even depression as legitimate emotions associated with loss. When the entrepreneur loses his or her business, it should be expected that some negative emotional state will arise. The ability to accept a negative emotion, deal with it effectively, and move on to new challenges is a major component of resilience. So is the ability to move about in the social world during periods of challenge and difficulty. Withdrawing into some neutral corner or lashing out at forces beyond one's control as responses to entrepreneurial failure can damage the entrepreneur's social reputation. Resilience also includes the ability to maintain one's equilibrium in social settings in the event of a venture failure.

Resilience can be developed through a variety of techniques. One of these has been termed "learned optimism."[25] According to this perspective, optimism can be learned and practiced by entrepreneurs and others. Everyone has a tendency either toward optimism or pessimism. In fact, most ventures benefit by having both optimists and pessimists on hand. The optimists are the ones who focus on what's possible and the future. Pessimists tend to focus on the realities of the present and the constraint present reality places on visions of the future.

Resilience as a character trait, however, is aligned with optimism. Technology entrepreneurs can develop a tendency toward optimism by using some very straightforward techniques. For example, optimists tend to view the negative things that occur in business as within their power to influence. Pessimists tend to believe that forces beyond their ability to influence affect the outcome of business activities. This perspective can be changed, but the pessimist would need to recognize the thoughts behind the attitude and work to change them. Tech Tips 14.4 offers some final suggestions for developing and maintaining resilience as an entrepreneur.

14.6.2 Overcoming Cognitive Biases

Rather than focusing on inborn personality traits or trying to find some type of internal "spirit" or "essence" of entrepreneurs, scholars have begun to look at how successful entrepreneurs think. The logic is that successful entrepreneurs routinely exhibit a range of cognitive skills that, if followed and practiced by others, will lead to their success as well.

TECH TIPS 14.4

Developing Resilience

Make connections: Networking doesn't just make good business sense, it makes good psychological sense. Form a network of professional and personal resources who can step in with advice or even just a sympathetic ear when times get tough.

Avoid seeing crises as insurmountable problems: Whether your business has had to downsize or just changed markets to accommodate the economy, you should focus on the big picture and realize that these setbacks do not necessarily threaten the life of your business.

Accept that change is a part of living: While change can be painful, accept that your business will change to meet new circumstances—whether it is an updated business plan or a new niche of customers served.

Move toward your goals: Entrepreneurs tend to be very goal oriented, but sometimes the inactivity forced by a stagnant market can stall an entrepreneur. Develop some realistic goals and do something regularly—even if it seems like a small accomplishment—that enables you to move toward those goals.

Take decisive actions: Rather than detaching completely from problems and stresses and wishing they would just go away, take decisive actions. Problem solving is an active and ongoing process that can increase resilience considerably.

Look for opportunities for self-discovery: Entrepreneurs can focus so heavily on changes in the market that they forget that they also are evolving as entrepreneurs with each challenge they meet. People often learn something about themselves and may find that they have grown in some respect as a result of their struggles.

Nurture a positive view of yourself and your company: Entrepreneurs may have been hit hard by the economy, but they also have the ability to be nimble and flexible, something the larger companies often have trouble with.

Maintain a hopeful outlook: An optimistic outlook enables you to expect that good things will happen in your life. Try visualizing what you want, rather than worrying about what you fear.

Take care of yourself: Entrepreneurs spend so much time nurturing the company and their employees that they sometimes forget their own needs and feelings. Engage in activities that you enjoy and find relaxing. Exercise regularly. Taking care of you helps to keep your mind and body primed to deal with situations that require resilience.

Many books, audiotapes, CDs, videos, and other media have been dedicated to helping people develop success-oriented habits, behaviors, and mind-sets.[26] Sifting through this wealth of literature on the topic of success is difficult and time consuming. Only recently have scholars of entrepreneurship begun the process of analyzing how successful entrepreneurs think. A current fruitful line of research into what makes entrepreneurs successful examines them from a cognitive perspective. In other words, this research examines how entrepreneurs think and how they process information as a prelude to action.

Using the cognitive approach, entrepreneurship researchers have made great strides toward explaining why some people become entrepreneurs and others do not. As it turns out, differences in individuals' perceptions about the potential entrepreneurial venture play a major role in the decision to proceed. Entrepreneurs display distinct **cognitive biases** in decision environments characterized by uncertainty and complexity. Yet, it is often these biases that enable entrepreneurs to take action despite incomplete information. In general, entrepreneurial research has found a variety of cognitive tendencies among entrepreneurs at the opportunity evaluation phase of a new venture. These include:

- *Overconfidence*: This bias refers to the inability to recognize the limits of one's personal knowledge. This bias stems from entrepreneurs' relative ease of recalling their reasons for confidence in a new venture, despite the existence of new information. This bias also influences entrepreneurs to seek information supporting their original opportunity evaluation rather than to seek potentially disconfirming information.

- *Belief in the law of small numbers*: This refers to the entrepreneur's tendency to draw firm conclusions from a limited sample. They tend to believe that the small sample from which they have derived their conclusions is representative of the greater population. Entrepreneurs rarely have large random samples available and the information available to them is often derived from associates, friends, or family. A belief in the law of small numbers coupled with mostly positive information likely produces an overoptimistic view of the new venture's prospects.

- *Planning fallacy*: Entrepreneurs tend to have a future orientation. Thus, they tend to ignore past information and treat the current situation as isolated and unique. This is known as the "planning fallacy." Forecasts of future business outcomes are often based more on optimistic forecasts and scenarios rather than on hard evidence from the past. This orientation to the future and neglect of possibly countervailing past results also biases the entrepreneur to an overly optimistic view of the new venture.

- *The illusion of control*: An illusion of control occurs when an individual overemphasizes the extent to which personal skill can improve performance in situations where chance plays a major role. People with this bias have a higher expectation of personal success than warranted by reality. This bias may play a role in decisions such as making acquisitions or producing innovative new products or services. The active search process may produce this illusion because the entrepreneur generates confidence by virtue of the activity. A particular danger of this bias is underestimating competitive response. In turn, this belief might lead to ill-advised product launches or badly conceived new ventures. The illusion of control biases the entrepreneur to ignore the uncontrollable nature of competitive response.

- *Reasoning by analogy*: In an effort to understand a new situation, individuals often compare it to a more familiar one. This process is referred to as "reasoning by analogy." The use of personal information may lead entrepreneurs to this bias. Personal communications are usually more vivid and easily recalled than secondary resources and likely to be used more readily in entrepreneurial judgments. The danger is entrepreneurs may use this vivid information to the exclusion of potentially more salient information in decision making. This bias can be dangerous in that firms may falsely believe that the operational needs of their analogy apply to the new venture. Thus, this bias can lead to a misjudgment of the need for complementary assets to achieve business goals. For example, a firm that has the capacity to manufacture a new device may have inadequate distribution assets to reach primary markets.

The practical remedy to these biases is for entrepreneurs to conduct systematic research on their new venture, including searching for information that runs counter to their optimistic evaluation. Through such research, the entrepreneur is exposed to more information that can help them make better decisions and minimize their reliance on bias-influenced intuitions.[3] Below, we describe a few approaches that entrepreneurs can use to minimize the risk that their business decisions are based on biased cognitions:

- *Active versus passive searching*: "Active searching" is a term used to describe a process where an individual intrudes into the environment in an effort to understand it. In this type of search, the individual expends resources and cognitive effort in his or her search for answers. In contrast, individuals receiving information passively receive it as it flows by. This is an "automatic" information search mode that takes the environment for granted and allows the individual to direct attention elsewhere. In new ventures or new product launches, entrepreneurs need to take an active approach to their market environment.

- *Personal versus impersonal information*: A major difference in the quality of information is based on whether it is obtained from personal or impersonal sources. Personal sources include direct contact with other individuals or the environment; impersonal sources involve written documentation such as magazine articles or reports generated by others. Entrepreneurs with new ventures or new product launches tend to use the personal approach more than the impersonal one. On the other hand, impersonal information can be vital to developing a balanced view of the new venture opportunity.

- *External versus internal sources*: Entrepreneurs can also choose to use information generated internal to or external to the organization. Searching externally is usually a more active process since the external information is unlikely to be organized in the required manner. Thus, entrepreneurs using this technique are likely to spend more energy than those using internal sources. Ventures that are pioneering new products or services require greater attention to external information sources than do established firms.

The cognitive biases and techniques for overcoming them discussed above are often required to precipitate entrepreneurial action. Entrepreneurs develop a sense of self-efficacy by virtue of their market search, planning, and execution strategies. This sense of efficacy leads to actions that bring new products and services into the market. Still, the success rate of entrepreneurs is notoriously poor enough to warrant recognition of the various biases and to search for ways to improve their effects on the new venture success rate.

Sensitizing entrepreneurs to the need to search both external and internal sources of information may mitigate the effects of bias. For example, entrepreneurs may want to lessen their direct involvement in the personal search for information to reduce the illusion of control bias. Identifying specific information search processes and their potential biases is an important step in minimizing their potentially harmful effects.

Of course, given their action orientation, entrepreneurs may conclude that eliminating biases would inhibit them too much. However, gaining insight into potential biases and misperceptions may lead to more effective action. For example, entrepreneurs aware of the reasoning-by-analogy bias may look closely at a proposed new venture and develop appropriate alliances to avoid a misjudgment of the need for complementary assets.

SUMMARY

This chapter covered a number of topics associated with managing the risks that accompany the establishment and operation of a technology venture. While not every risk can be considered and discussed in a single chapter, the ones covered here are often encountered. In addition, not every risk mitigation strategy can be discussed and explored, but we have examined some key strategies for managing technology venture risks.

Perhaps, the overriding lesson that should be taken from this chapter is that venture risk management is a continuous process. The risks the venture will face as it grows and matures are not susceptible to being simply planned out of existence. Planning is essential and necessary, but it is not sufficient to ward off all the problems and difficulties that a venture inevitably will encounter. As such, the technology entrepreneur should establish sound venture governance policies and procedures, including an effective and active governing board. Whether the board is a formal board of directors or only serves in advisory capacity, it can play a vital role in risk management. If nothing else, the board provides an independent check on the decisions contemplated by the venture management team, and can validate or challenge the decisions. It is rarely a good idea to believe that the venture's management team sees the outcomes of its potential decisions with absolute clarity. Although the venture likely will need to move fast and it will want to avoid laborious decision-making processes, it should nonetheless allow for dissenting voices to be heard.

Several of the techniques for risk management highlighted in this chapter are essential to nearly every technology venture. Establishing a working relationship with a knowledgeable law firm and legal team is a good idea. A good legal team can not only help extract the venture from troubles that may arise, it can also help it avoid problems. It is the latter that is immeasurable and immensely valuable to the technology entrepreneur who wants to focus on the business and not on legal hassles.

Another important strategy for risk management is to be aware of the legal problems that often arise for start-up ventures. Contract law, sales law, tort law, and human resource laws do not have to be known in detail. Yet an understanding of their basic principles and how to comply with legal standards can help avoid problems. Simply treating all people fairly in the workplace can ensure that any legal issues that do arise will have minimal impact on the firm's reputation and on its ability to operate. Venture founders and leaders who establish compelling mission statements and codes of ethics must demonstrate their commitment to the mission and values through their actions. Further, they must communicate those values to the venture's employees regularly and in multiple ways for creating a culture that focuses on delivering value to customers.

Finally, managing failure risk means developing self-awareness and resilience. Every venture, no matter how well funded or well structured, will encounter unexpected challenges, setbacks, and potential failure. The resilience of the founder will be critical in surviving these challenges. This chapter highlighted several features of resilience as a character trait. Entrepreneurs should diligently and consistently work on becoming more steadfast and determined, even in the event of a failed venture. For the most part, failure is not the end of an entrepreneurial career even if it means the end of a business and money lost. A business may disappear, but the entrepreneur can survive to try again. In fact, research and experience are clear: Investors don't mind if an entrepreneur has a failed venture or two in his or her background. What matters is whether the entrepreneur has learned from failure and conducts business in a disciplined and determined manner.

STUDY QUESTIONS

1. Explain what is meant by "venture risk management." What are some things an entrepreneur can do to minimize venture risk?

2. Explain what is meant by the term "D&O liability." What should the entrepreneur do to minimize this risk?

3. What are the four operational risks discussed in this chapter?

4. What is the purpose of a mission statement? What should a mission statement say?

5. Explain the difference between a board of directors and a board of advisors. What two duties distinguish a board of directors?

6. Identify the elements of a thorough investor relations program for a technology venture.

7. Why should a technology entrepreneur be concerned about the laws pertaining to human resources management?

8. What is the Uniform Commercial Code? Explain how it is used to govern business transactions.

9. Explain the different types of bankruptcy.

10. What are the potential cognitive biases that entrepreneurs tend to display? How can a person actively overcome problems that may arise from these biases?

EXERCISES

1. *OUT-OF-CLASS EXERCISE*: This exercise is designed to get students out of the classroom and into the business world. Attorneys are important advisors to businesses and managers. Each student (or small group) is to select an attorney in the community who can spend about 1 hour for an information interview. The attorney should be able to address one of the following topics:
   - Contract law and how entrepreneurs and attorneys work together in creating contracts.
   - Human resource law and how entrepreneurs work with attorneys to manage liability.
   - Intellectual property law and how entrepreneurs work with attorneys to protect intellectual property.

   Each student (or group) should prepare a 2–3-page report on their meeting with the attorney. What are the key lessons for entrepreneurs? What was learned about the entrepreneur/attorney relationship that wasn't known before?

2. *OUT-OF-CLASS EXERCISE*: Part of what we have been learning about entrepreneurs is that, either by nature or nurture, they think differently about business opportunities and tend to be more resilient than nonentrepreneurs. For this exercise, you will identify and contact an entrepreneur in your community for a brief conversation.

   Entrepreneurs are not hard to find—nearly every store and shop in your neighborhood were founded by an entrepreneur (unless it is a "chain" store). Find out who owns the establishment by talking to employees. Most entrepreneurs are happy to meet with students and talk about their personal business journey.

   As part of your conversation, you should seek to find out more about the following:

   - How many ventures has the entrepreneur been involved with as an owner?
   - When did he or she first realize the dream of wanting to be an entrepreneur?
   - Why did he or she choose to be in the business he or she is in?
   - What support has the entrepreneur had along the way?
   - Has he or she used a business plan to guide the growth?
   - What mistakes has he or she made, and how were they overcome?

   Of course, you will want to go beyond these questions during your conversation. Complete the exercise by writing a brief (1–2 pages) essay about your conversation with an entrepreneur and what lessons you were able to learn.

KEY TERMS

Administrative law: Administrative laws are established by governmental agencies such as public utility boards.

Board of advisors: A board of advisors is an informal board that ventures use for advice and assistance in decision making. Such a board does not have legal responsibility for the venture's performance.

Board of directors: A board of directors is a formal entity that is required for all Subchapter S- and C-corporations. Such a board does have legal responsibility for the venture's performance.

Bona fide occupational qualifications: A BFOQ is a qualification for employment that is demonstrably related to actual job duties associated with that employment.

Civil Rights Act: The Civil Rights Act of 1964 ended discrimination in the workplace based on race, ethnic background, religion, and other protected categories.

Cognitive biases: These are habits of thinking that entrepreneurs tend to exhibit and that may lead to erroneous decision making if not recognized and managed.

Common law: Common law is law that is derived from the decisions made by judges and juries in trials held every day in the United States and encoded as precedents and standards in law books.

D&O liability: This is insurance that can be purchased to protect directors and officers of a venture from financial risks associated with a lawsuit.

Duty of care: A member of a board of directors has the duty to manage a venture in a responsible manner that is centered on enhancing shareholder value.

Duty of loyalty: A member of a board of directors has the duty to be loyal to the venture and not disclose strategies, intellectual property, or competitive advantages to other ventures.

Employer liability: Employer liability is also known as "worker's comp," and it is a mandatory form of liability insurance coverage that all businesses must carry.

Equal Employment Opportunity Commission: This is the federal government agency that acts to ensure that companies adhere to the Civil Rights Act of 1964.

Fiduciary responsibility: This is the responsibility that boards of directors and venture officers have to manage investor capital in a transparent manner and in a manner that enhances shareholder value over time.

General liability insurance: This form of business liability insurance is the main coverage to protect a venture from injury claims, property damages, and advertising claims.

Investor relations: This is a risk management strategy that engages investors and potential investors to keep them informed on key operating outcomes of the venture.

Law of bankruptcy: This is the set of laws that govern the manner in which firms can enter or be brought into bankruptcy, and how a bankruptcy proceeds over time.

Law of negotiable instruments: This term refers to the laws that govern non-monetary instruments, such as letters of credit, that govern transactions between companies.

Law of torts: A tort is a wrong committed against someone either through negligence or intent.

Law of sales: The law of sales governs transactions between businesses and customers.

Liability insurance: This type of insurance protects a venture against financial loss in the event of a lawsuit.

Model Employment Termination Act: The act states, "an employer may not terminate the employment of an employee without just cause."

Mission statement: This is a statement that governs the manner by which the venture engages with employees, customers, and others.

Product liability: This refers to the potential for a venture to be liable for any damages or injuries caused through the use of its products by customers.

Product liability insurance: This is a type of insurance that protects a venture from product liability lawsuits.

Professional liability insurance: This coverage protects a business against malpractice, errors, negligence, and omissions.

Resilience: This is a character trait that helps an entrepreneur deal with challenges and failures associated with starting and operating a venture.

Sarbanes-Oxley: This is an act of the U.S. federal government following several corporate scandals that requires extensive disclosure and reporting by companies that sell or intend to sell equity securities via public exchanges in the United States.

Serendipity: This term refers to the occurrence of an unexpected and auspicious event, such as encountering a venture capitalist on an elevator when you are seeking capital.

Social capital: The entrepreneur's social network and relationships is referred to as social capital.

Statutory law: Statutory law is established by state and federal elected legislators.

Strict product liability: In cases with strict product liability, everyone in a value chain who is responsible for delivering a product to a customer who is harmed or damaged by that product may be held liable.

Uniform Commercial Code: This is the legal code that governs business transactions. It provides what are known as "gap fillers" in the case of disputes and imprecise contracts.

WEB RESOURCES

The websites below are intended as destinations for your further exploration of the concepts and topics discussed in this chapter:

1. www.entrepreneur.com: This site has a great deal of information that is useful to the technology entrepreneur on the issue of venture risk management.

2. http://www.nvtc.org/tec/downloads/Insurance_Love_07.pdf: This site was developed by the Virginia Center for Innovative Technology. The particular link listed here is to a document that describes the types of insurance that a start-up venture should consider.

3. http://www.managementhelp.org/plan_dec/str_plan/stmnts.htm#anchor 519441: This site contains information about developing mission, vision, and values statements.

4. http://www.inc.com/resources/startup/articles/20050201/missionstatement .html: This is an article at Inc.com about developing effective vision and mission statements.

5. http://managementhelp.org/boards/boards.htm: This site has a wealth of information about developing and managing boards.

6. http://www.entrepreneur.com/humanresources/index.html: This site is hosted by Entrepreneur.com and has a great deal of information about managing human resources in entrepreneurial ventures.

7. http://www.nolo.com/: This site contains good information for entrepreneurs about legal issues of all sorts.

ENDNOTES

[1] Busenitz LW. Entrepreneurial risk and strategic decision making: it's a matter of perspective. *J Appl Behav Sci* 1999;**35**(3):325–40.

[2] Janney JJ, Dess GG. The risk concept for entrepreneurs reconsidered: new challenges to the conventional wisdom. *J Bus Venturing* 2006;**21**(3):385–400.

[3] Keh HT, Foo MD, Lim BC. Opportunity evaluation under risky conditions: the cognitive processes of entrepreneurs. *Entrepreneurship Theory Practice* 2002;**27**(2):125–48.

[4] Enterprise Risk Management Committee. Overview of enterprise risk management. *Casualty Actuarial Society* May 2003:8.

[5] Welch C. Victims plan to sue over Minnesota bridge collapse. *CNN.com* 2008;**November 13**.

[6] Himelstein L. The asbestos case of the 1990s. *Bus Week* 1995;**January 16**:82–3.

[7] Geyelin M. Product-liability groups take up arms. *Wall Street J* 1993;**January 29**:B1, B3.

[8] Efrati A, Lauricella T, Searcey D. Top broker accused of $50 billion fraud. *Wall Street J* 2008;**December 12**:A1.

[9] Salter M, George B. After Enron, little has changed. *Wall Street J* 2008;**December 15**:A16.

[10] Seth N, Range J, Arnand G. Corporate scandal rocks India: chairman of outsourcing giant resigns, saying he concocted financial results. *Wall Street J* 2009;**January 9**:A1, A9.

[11] Range J, Lublin JS. India's corporate governance will face stricter scrutiny. *Wall Street J* 2009;**January 9**:A9.

[12] Fletcher J. Super-efficient fridge among the innovations at builders show. *Wall Street J* 2009;**January 16**:B1.

[13] Sam Wyly: an entrepreneur looks at boards. *Corporate Board* November/December 2008;**29**(173):32.

[14] Advani S. Selecting an advisory board. *Entrepreneur.com* 2004;**July 6**.

[15] Ivancevich JM, Duening TN. *Managing Einsteins: leading high tech workers in the digital age.* New York: McGraw-Hill; 2001.

[16] Kohl JP, Miller AN. Recruitment practices of business firms: the ongoing battle of discrimination in help wanted ads. *Business Forum* 1990;**Spring**:8–11.

[17] Cihon PJ, Castagnera JO. *Employment and labor law.* Cincinnati: West Publishing Company; 1999; p. 20.

[18] Berger NL, Kleiner BH. Terminating employees legally. *Am Bus Rev* 1990;**January**:82–7.

[19] Carmichael E. How to fire an employee: the do's and don'ts of terminating employees to keep you out of hot water. Retrieved from http://www.evancarmichael.com/Human-Resources/773/How-To-Fire-An-Employee-The-Dos-and-Donts-of-Terminating-Employees-to-Keep-You-Out-of-Hot-Water.html.

[20] Fulbright & Jaworski, LLP. Fulbright's Fourth Annual Litigation Trends Survey Findings.

[21] Shane S. *The illusions of entrepreneurship.* New Haven: Yale University Press; 2008.

[22] Duening TN. Five minds for the entrepreneurial future. Presented at the 2007 United States Association for Small Business and Entrepreneurship conference, San Antonio, TX.

[23] Mangurian GE. Realizing what you're made of. *Harv Bus Rev* 2007;**85**(3):125–30.

[24] Timmons JA. Entrepreneurship and the creation of high-potential ventures. In: Sexton DL, Smilor RW, editors. *The art and science of entrepreneurship.* Cambridge, MA: Ballinger; 1986.

[25] Seligman M. *Learned optimism: how to change your mind and your life.* New York: Vintage Publishing; 2006.

[26] Kenneth H. Speaking of change. *Incentive* 2002;**September**:95–7.

[27] See this mission statement at: http://www.gene.com/gene/about/corporate/.

[28] See this statement of corporate governance principles at: http://www.dell.com/content/topics/global.aspx/about_dell/investors/corp_gov/corp_gov?~ck=ln&c=us&l=en&lnki=0&s=corp.

Your Entrepreneurial Career

After reading this chapter, students should be able to:

- Understand the entrepreneurial career
- Consider performing as an intrapreneur within a large company
- Realize that entrepreneurship is a lifetime pursuit
- Develop the characteristics of an entrepreneurial personality
- Recognize different entrepreneurial types
- Evaluate whether a franchise venture is right for you

TECH VENTURE INSIGHT

The Entrepreneurial Career of Amazon's Jeff Bezos

Entrepreneurs are said to be risk takers and Jeff Bezos is no exception. In 1994, he left his position as the youngest Vice President ever at D.E. Shaw and Company to found Amazon.com. Bezos had worked at D.E. Shaw and Company for a little over 4 years applying computer science models to the stock market when he began noticing a huge trend toward increased Internet use. With some additional research, in the latter half of 1994, he discovered that usage was increasing 2,300%! Seeing a great opportunity to use the Internet for more than simply charting stock market trends, he approached his colleagues with an idea: an online book catalog. Since there was not anything like it at the time, he assumed there was significant potential.

With little knowledge of the book market and to ensure being informed before approaching his colleagues, Bezos knew he needed to learn more about books and attended the American Booksellers' Convention. There he discovered that all he needed for his new idea was a location on the Internet. He also met several mentors who helped him learn more about what made the book business successful. With a new-found passion, Bezos approached D.E. Shaw and Co. for financial support. He received none.

At that moment, Bezos knew he was an entrepreneur at heart and wanted to take the risk of starting his own company. Following his passion and throwing caution to the wind, he and his wife, Mackenzie, drove cross-country to Seattle where they could have access to book wholesalers as well as the computer talent necessary to launch such a venture. It is said that during that cross-country drive, the business plan was completed. With steps in place, the company was named Amazon, after the South American river that is seemingly endless with its multiple branches. Bezos wanted his web site to be just like that, with any and all possibilities.

Bezos and his wife started the web site in their new two-bedroom house in Seattle, asking over 300 friends and family to test it. After a few months of testing across various computer platforms and making countless adjustments to ensure it was user friendly, Bezos launched the site to the public on July 16, 1995. Within the first 30 days, with no advertising, Amazon.com had sold books in all 50 states and over 40 foreign countries. Within months, and with the introduction of such additions as "one-click" shopping and e-mail order verification, the site had reached sales of $20,000 per week. Since its inception, it has been one of the few original dot-com companies to survive the bust and has gone on to make billions in profits.

Bezos founded his entrepreneurial career by being passionate about an idea, doing research, and not being afraid to take risks. Even when confronted with a challenging situation, he stuck to his instincts and managed to launch his entrepreneurial career. Along the way, he also prioritized developing good working relationships with mentors, friends, and colleagues, who helped him to be a successful entrepreneur.

Source: Adapted from Ramo JC. The fast moving Internet economy has a jungle of competitors: and here's the king. Time December 27, 1999:50–5; Quittner J. How Jeff Bezos rules the retail space. Fortune May 5, 2008:126–34.

INTRODUCTION

Entrepreneurship often involves a great deal of risk and personal sacrifice, especially in the first couple years of operating a new business. No one should even think about going into business for themselves unless they are prepared to work long hours, endure hardships, and persevere on the basis of a vision of future rewards. Running your own business requires an ability to deal with the unknown. It is well known that most business start-ups fail within their first 5 years. Entrepreneurs and small-business owners have nowhere to turn when failure occurs. They must rely on their own resolve and resourcefulness to pick themselves up and try again until they achieve success.

For example, Niklas Zennstrom and Janus Friis were blindsided by the recording industry when it attacked users of their music file sharing software KaZaA. KaZaA was one of the most downloaded software applications in the history of the Internet, with more than 370 million downloads to its credit. The recording industry sued people who downloaded MP3 music files in an effort to stem the practice of what it perceived to be a violation of its copyrights. Rather than fold up and go back to work for someone else, Zennstrom and Friis simply moved to London and launched another entrepreneurial volley at an old, established industry by founding Skype. This time, they took on the telecommunications industry and they were rewarded when Skype was purchased by eBay for several billion dollars. Their latest venture, Joost, is targeting the television and movie industries.

This chapter discusses the entrepreneurial spirit and entrepreneurial careers. First, we examine entrepreneurship and some of the habits and character traits that entrepreneurs will require. Then, we examine five independent entrepreneurial profiles, each revealing different personalities, talents, and motivations. Finally, we examine franchising, looking at both the advantages and disadvantages of being in the franchise business as a technology entrepreneur.

15.1 YOUR ENTREPRENEURIAL CAREER

The debate about what makes an entrepreneur continues. Years of scholarship to identify the essential ingredients of an entrepreneurial personality have yielded little consensus. Entrepreneurs, it turns out, come in all shapes and sizes: Some are young, others are old; some start out rich, others start out poor; and some are smart, others just work very hard.

Despite the lack of consensus on the essential traits of successful entrepreneurs, some characteristics of entrepreneurs and entrepreneurship are consistently portrayed and cited as important elements of entrepreneurial success.

15.1.1 The Individual Entrepreneur

The entrepreneur, according to French economist J.B. Say, "is a person who shifts economic resources out of an area of lower and into an area of higher productivity and yield."[1] But Say's definition does not tell us who this entrepreneur is. Some define the entrepreneur simply as one who starts his or her own new and small business. For our purposes, we define the entrepreneur as the person who organizes the resources, takes the risks, and receives the financial profits and other nonmonetary rewards of starting a new enterprise. (Note that we use the word "enterprise" to denote that an entrepreneur can start a business, a nonprofit organization, or some other organized activity.)

The person who opens a small pizza restaurant is in business, but is he or she an entrepreneur? The individual took a risk and did something, but did he or she organize the resources to start the business? If the answer is yes, then the individual is considered an entrepreneur. Steve Case is an example of an entrepreneur because he founded and established America Online (AOL). His Internet access business was not a new concept, but he applied new techniques, unique content, and took advantage of his customers' desire to connect to one another through chat and instant messaging. This is what entrepreneurs do; this is what entrepreneurship means.[2] Houston-based furniture magnate Jim McIngvale sells $180 million worth of furniture each year through Gallery Furniture, a single location store beside Highway 45 in Houston. How does he sell so much furniture? Simple: He promises people that he will deliver their furniture the same day they buy it. That simple change in the overall furniture-buying process has provided McIngvale a significant and sustainable competitive advantage in his market.[3]

Over the past decade, many of the sharp contrasts between the entrepreneur and the professional have faded away. Formerly, professionals such as doctors, lawyers, dentists, and accountants were not supposed to be entrepreneurial, profit driven, or market oriented. They were "above" the growth-oriented mind-set common to entrepreneurs. Entrepreneurs, on the other hand, were the mavericks of society. They were risk takers who aggressively sought to make substantial profits by responding to market opportunities. Long hours were about all the two worlds had in common. However, increased competition, saturated markets, and a more price-conscious public have changed the world of the professionals. Today, even the most buttoned-down professionals need to market their skills, talents, and competencies: Lawyers advertise their services; doctors specialize in one or another form of surgery; accounting firms join with other businesses (e.g., consulting and law) to serve clients. These professionals have had to become entrepreneurial in order to survive. Many have crafted unique services—such as catering to specific markets—or have come up with some other innovation that differentiates them from their competitors.

15.1.2 Conceptual Model of Entrepreneurial Careers

Do entrepreneurs decide to work for themselves? Do they wittingly assume the risk of organizing and managing businesses? Despite extensive research conducted on the careers of such self-starting individuals, a universal impetus cannot be pinpointed. Entrepreneurs embark on complicated, difficult, and precarious career paths for a myriad of reasons. According to the results of a 2007 Baylor University study, "Venture creation does not happen by accident; it requires directed effort exerted over time. Individuals are either pulled or pushed toward a career choice, such as becoming an entrepreneur. Negative push factors (e.g., being fired) and positive pull factors can direct an individual toward career choices."[4] Other sources suggest that the goal of self-employment is more a result of an individual's personality, self-confidence, education, environment, and exposure to role models,[2] while other researchers have concluded that "a triggering event and a business idea"[3] are the catalysts spurring the entrepreneurial career preference.

Culture also plays a role in the nurture of entrepreneurial creativity: A study comparing the career aspirations of Irish undergraduate and MBA students with those of their U.S. counterparts revealed a cross-cultural disparity in perception of the entre-preneurial career. Exhibit 15.1 illustrates the distinct attitudes of the two nations, and how affirmative societal opinions of entrepreneurialism are linked with the push toward an entrepreneurial career.

Irish people anticipating an entrepreneurial career must be able to withstand social pressure that discourages this choice. Selecting entrepreneurship as a career in Ireland may require rebelliousness, which is not needed in the United States where entrepre-neurship is a more conventional career choice, as indicated in Exhibit 15.2.

U.S. culture is distinguished by extremely high individualism. In the United States, success in individual pursuits is admired, as are choice, pursuit of seemingly impos-sible dreams, impatience with time, acceptance of mistakes, urge to improvise, and intrigue with what's new. Americans admire the maverick entrepreneur.[4]

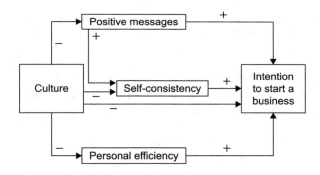

EXHIBIT 15.1

Ireland—Predictors of Entrepreneurial Intention.

Source: de Pillis E, Reardon KK. The influence of personality traits and persuasive messages on entrepreneurial intention. Career Dev Int 2007;12(4):393.

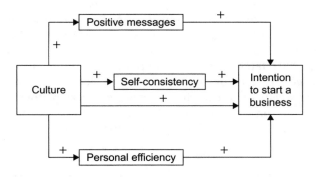

EXHIBIT 15.2

USA—Predictors of Entrepreneurial Intention.

Source: de Pillis E, Reardon KK. The influence of personality traits and persuasive messages on entrepreneurial intention. Career Dev Int 2007;12(4):393.

15.1.3 Finding a Mentor

Though entrepreneurs are often marked with confidence and self-determination, advice in the form of a mentor is often helpful in identifying and maintaining the right course. Entrepreneurship is still a relatively new and growing academic discipline, and often a mentor plays an integral role in the education of an entrepreneur. Personal mentoring can be immensely valuable to the career entrepreneur. In addition to providing direction, sharing experiences, and recommending resources, role models can help the nascent entrepreneur solidify confidence. "Role models can affect entrepreneurial intentions if they change attitudes and beliefs about a person's perceived ability to be successful in a new venture."[5]

Some forms of mentoring include shadowing, internships, and informational interviews. A 2006 study on entrepreneurial intentions revealed that working with mentors was the most influential factor in forming subjects' career objectives.[5] With a myriad of career options, mentors can help nascent and young entrepreneurs navigate to the option for which they are best suited. Entrepreneurs who work in companies prior to pursuing their own venture often find mentors in their employers. Dan Puchek, an Arizona medical device entrepreneur, who is the owner of Promedtek, claims he did not have or need a mentor when making his decision to go out on his own. However, prior to founding his own company, Puchek worked at such global medical corporations as Pfizer/Becton Dickinson and Kinetic Concepts and consulting firms such as Technomic Consultants. Though the choice to be his own boss came organically, Dan Puchek credits his first boss as helping him to develop himself further and move ahead in his career.[6]

A mentor usually is not an authoritarian influence on a nascent or young entrepreneur; mentors usually serve as sounding boards for ideas, providing much needed emotional support and guidance; mentors can provide a wealth of insightful anecdotes, which can provide perspective to the nascent and practicing entrepreneur.

Silicon Valley brand enhancer Sean Parker began his career as the cofounder of Napster in 1999, an online music-sharing program that allowed users to download music for free. The idea was popular, obtaining millions of dollars in venture capital funding, and garnering a large number of users. A few years later, Napster's founders were indicted on copyright infringement charges and the company shut down, filing for bankruptcy in 2002. Interestingly, Napster tweaked its business model in later years, and was purchased by Best Buy, a retail consumer electronics chain, in September 2008. Sean Parker weathered the storm and now serves as a partner in a venture capital firm, the Founders Fund, helping to guide young entrepreneurs. "I want to be something that no one was to me. A lot of people seek to be mentors who've never actually been entrepreneurs."[7]

15.2 THE ENTREPRENEURIAL PERSONALITY

Entrepreneurs exhibit many different personality types; searching for a specific personality pattern is very difficult. Some entrepreneurs are quiet, introverted, and analytical.

On the other hand, some are brash, extroverted, and very emotional.[8] As we have stated, there are probably as many personality varieties among entrepreneurs as there are entrepreneurs. However, all entrepreneurs exhibit one consistent trait: optimism. Across the different types of personalities, researchers have found that entrepreneurs have an optimistic way of thinking that leads them to deal with failure and change differently than do pessimists. Whereas pessimists view failure as personal and long-lasting, optimists view failure as due to causes beyond their control and short-lived. Optimists see failure as an opportunity to grow, whereas pessimists see failure as a reflection of their own shortcomings.[9] Can you see how this personality trait is important to someone starting a business of his or her own?

15.2.1 Characteristics of Growth-Oriented Entrepreneurs

Why do entrepreneurs continue to emerge when significant risks, time, and energy are needed to be successful? Researchers studying the entrepreneurial personality have made some interesting discoveries. Researcher David McClelland, a noted social psychologist, determined that founders of high-growth companies appear to share a distinct cluster of personal characteristics (no single trait but a pattern, or cluster).[10] As you review these characteristics, consider whether they can be applied to you. Do you have the characteristics of an entrepreneur?

High need for achievement

Growth-oriented entrepreneurs have a high need for achievement: They need to succeed, to achieve, and to accomplish challenging tasks. The strong desire for achievement leads to a desire for independence. Such entrepreneurs need to be free to set their own course, establish their own goals, and use their own style. The need for achievement may help explain why growth-oriented entrepreneurs are not satisfied with founding or working in one firm; they need to prove themselves again and again.

Low need to conform

Growth-oriented entrepreneurs listen, but they are able to ignore others' advice. Also, handling skeptics is easy for entrepreneurs. Taking the unpopular course of action, if they consider it best, is the way they do business.

Persistence

Growth-oriented entrepreneurs are persistent, doggedly doing what is best for the business to succeed. They work hard on the details and relentlessly attempt to find ways to become more profitable.

High energy level

The capacity for sustained effort requires a high energy level. The necessary work—planning, organizing, directing, creating strategy, and finding funds—can only be accomplished on a demanding schedule. The 60–80-hour workweek is common.

Risk-taking tendency

McClelland's findings suggest that people with a high need for achievement tend to take risks. Growth-oriented entrepreneurs believe so strongly in their ability to achieve that they do not see much possibility of failure. Thus, they accept risk and find it motivating.

In addition to these personality traits, much research has also been conducted into the emotions that entrepreneurs deal with and the manner in which they cope.

Emotional Intelligence

The emotional makeup of an entrepreneur is another area of great interest. Especially when entrepreneurs achieve fame and fortune, there is an inclination to search for psychological attributes that might have predisposed this individual to succeed. Many researchers maintain that emotional intelligence emerges from an entrepreneur's ability to mentally absorb, interpret, and categorize life events, and to subsequently express them. Also, the circumstances in which a blossoming entrepreneur experiences these occurrences will play a role in how such observations correlate to their personal narratives. "When individuals engage in intuiting particular characteristics of the immediate task environment, such as autonomy, nature of task, availability of feedback, etc., it affects their propensity to generate ideas."[11]

One researcher suggests that successful entrepreneurs have a keen ability to "connect the dots"[24] and gather knowledge from their own experiences. A February 2008 article in the *Harvard Business Review* about entrepreneurs who founded their own companies cited data that persistence, dogged determination, and persuasiveness—traits commonly associated with entrepreneurs—can impede an entrepreneur's ability to effectively manage employees and determine the appropriate course of action for a mature business. "Founders' overconfidence, ability to inspire people and passion may be necessary to get new ventures up and running, but these emotional strengths become liabilities."[12]

Noam Wasserman, in his 2008 *Harvard Business Review* article, suggested that entrepreneurs who launch their own ventures need to make a choice between getting rich or maintaining control of the company.[13] Some entrepreneurs are so loath to relinquish control of their companies they are resistant to hearing other points of view, accepting change, and attracting venture capital funds with which to grow the company.

A healthy appetite for risk and a strong stomach for adventure distinguish the entrepreneur from other members of society. According to Wasserman, "in founding a new venture, the entrepreneur risks financial failure and possible embarrassment. Compared to managers, entrepreneurs possess a significantly higher level of tolerance for ambiguity." Also, entrepreneurs' emotional intelligence is marked by an exceptional ability to communicate. "Entrepreneurs engage in explaining and defending the 'fuzzy' images of their insights." Being able to converse with employees, investors, stakeholders, and suppliers allows entrepreneurs to be effective leaders as well.

15.3 INTRAPRENEURSHIP

An enterprise need not be small or new to be entrepreneurial. For example, General Electric Company (GE), a large multinational business, has a reputation for being entrepreneurial. GE organized the GE Credit Corporation and numerous new manufacturing businesses. These firms made money for GE; they also provided ideas and concepts that raised the company's overall level of performance.

GE is one of a growing number of firms that employ and recognize **intrapreneurs**. An intrapreneur is a person with entrepreneurial characteristics who is employed within a large corporation. Companies such as 3M, Hewlett-Packard, Merck, and Rubbermaid encourage intrapreneurs. For example, Post-it™ notepads were developed at 3M by Arthur Fry, an employee looking for a better paper for marking pages in a book. At Hewlett-Packard, researchers are urged to spend 10% of their workweek on personal pet projects. Intrapreneurs are usually found in enterprises that encourage experimentation, tolerate failure, recognize success, and share the wealth. Tech Micro-case 15.1 highlights the intrapreneurship efforts at 3M.

Intrapreneurship requires many of the same skills as entrepreneurship. The major difference is that intrapreneurs generally must be astute at managing organizational politics, whereas entrepreneurs don't necessarily need political skills. When you think about the different contexts in which intrapreneurs and entrepreneurs

TECH MICRO-CASE 15.1

3M Programs Encourage Intrapreneurship

The "15% rule" at 3M allows inventor/employees to spend 15% of their time working on projects of their own choosing that they think might have merit for the company. Once the viability of the research is proven, the corporation may officially fund it. But "15% time isn't really a rule or a time allocation that requires approval," says Program Manager Colleen Horn Harris. "It's really part of the culture at 3M. We recognize that it's important for people to follow their own ideas."

Another 3M innovation and intrapreneurial program, the Genesis Grant, finances innovative projects that aren't readily funded through normal channels, offering researchers up to $85,000 to carry their projects past the idea stage. A formal panel of technical experts and scientists pass promising projects to yet another committee of senior technical people and management. They specifically look for:

1. Creative ideas that might lead to a competitive advantage
2. Preliminary experimental work that has already been completed
3. Resources identified both within and outside of 3M that are required

They award about 15 grants per year.[14]

operate, this skill set differential becomes clear. In large organizations, resources are normally allocated according to decisions that are made by executives and others. Like all human decision making, executive decisions about resource allocation often are influenced by personal relationships and other things. These elements of resource allocation within a large organization are usually not a part of the entrepreneurs' fund raising efforts.

Intrapreneurs must also constrain their imaginations, innovations, and inventions to products and services that fit with the organization's other products and services and with the overall organizational strategy. Entrepreneurs, by way of contrast, can dream up nearly any product or service and try it out in the marketplace. Intrapreneurs must consider how their new idea fits with the assets and production capabilities of the existing organization, and how it fits with the markets the organization currently serves. Thus, intrapreneurship is significantly different from entrepreneurship that entire books have been written about how to be an effective intrapreneur.[15] Tech Tips 15.1 suggests some techniques for fostering an intrapreneurial culture inside a large organization.

Despite the differences between intrapreneurship and entrepreneurship, the similarities are vast and anyone who is good at entrepreneurship likely will have many of the skills required to be successful as an intrapreneur. In fact, students of technical fields should be encouraged to spend the early parts of their careers working within an established organization in an industry they prefer. This serves the dual purpose

---

TECH TIPS 15.1

Fostering an "Intrapreneurial" Culture

- Staff members are allowed to make their own choice of which internal vendor they wish to use.
- Intrapreneurial employees receive "ownership rights" in the internal intraprises they create.
- Company-wide involvement is encouraged by insisting on truth and honesty in marketing and marketplace feedback.
- Intrapreneurial teams are treated as a profit center rather than a cost center.
- Team members are allowed a variety of options in jobs, innovation efforts, alliances, and exchanges.
- Employees are encouraged to develop through training programs.
- Internal enterprises have official standing in the organization.
- Contractual agreements between internal enterprises are defined and supported by the organization.
- A system for settling disputes between internal enterprises and between employees and enterprises is part of the intrapreneurship plan.[16]

of exposing the individual to the rigors of organizational life and, if the individual is observant, reveals many inefficiencies in the industry that may be potential entrepreneurial business opportunities at some future date. Remember, most entrepreneurs don't go into business to make a lot of money; most do it primarily because they are tired of working for someone else.[17] For many mid-career technical professionals, a life of working for a large organization can be replaced by one of working for oneself if the individual has been learning the industry and has been observing where the entrepreneurial opportunities lie.

15.4 FIVE ENTREPRENEURIAL CAREERS

So respected, in fact, is the entrepreneur, the U.S. cultural magazine *Vanity Fair* included in its July 2008 issue an oral history of the Internet's pioneering minds. To mark the 50th anniversary of the U.S. government's Advanced Research Projects Agency (ARPA), *Vanity Fair* journalists Keenan Mayo and Peter Newcomb interviewed some of the entrepreneurs who made it happen.

15.4.1 Paul Baran

I just did a little piece on packet switching and I get blamed for the whole Internet.[18]

An electrical engineer by trade, Paul Baran was employed by the Rand Corporation—a U.S. nonprofit think tank—when he first hatched his groundbreaking idea of packet switching in 1960. Packet switching divides data into separate packets and reroutes them down specific paths, thus decentralizing lines of communication. A diversified data-transfer system was vastly important during wartime, as potential missile attacks could snuff all means of communication[18]. Bringing an esoteric technological concept to the marketplace is no small feat. Paul Baran's brainchild was met with resistance from prevalent communications giants such as AT&T: "They fought it tooth and nail at the beginning. They tried all sorts of things to stop it."[18]

Baran believes the climate of the 1960s was especially conducive to technological pursuits, while currently, creativity is more stifled by corporate rivalry and global economic pressures. In its formative stages of development, Internet technology was less cutthroat and more collegial. "Now everyone is concerned about making money, or reputation. It was different then.... We all wanted to help one another. There was no competition, really...."[18]

In keeping with the spirit of this renaissance, Paul Baran is hesitant to even accept credit for packet switching. The invention of packet switching, he says, evolved organically: "Technology reaches a certain ripeness and the pieces are available and the need is there and the economics look good—it's going to get invented by somebody."[18]

15.4.2 Marc Andreessen

... They all think the Internet is the greatest thing. And I'm like, "Yeah, thanks to government funding."[19]

Marc Andreessen was an undergraduate student at the University of Illinois when he created the first web browser—Mosaic. Andreessen then teamed with an entrepreneur who designed silicon graphics to form the U.S. computer services corporation Netscape Communications,[9] which is known principally for its World Wide Web browser capabilities.

A computer aficionado, the young Andreessen could not have been in a more appropriate setting to stimulate his interest in technology. The University of Illinois at Urbana-Champaign is the cradle of the Internet and housed the first National Center for Supercomputing Applications (NCSA), funded by a grant from the U.S. government's National Science Foundation. Andreessen was a student employee of the center, and while grateful for the government's support of information technology, he remains humble about his own contribution to Internet history, suggesting the idea for the Mosaic browser "sounds obvious in retrospect."[19]

However, creation of the Mosaic was not a government-sanctioned assignment. The brainchild, says Andreessen, began as an antidote to boredom on the job. "Mosaic was a side project ... started in our spare time ... we didn't think the real project we were working on at the time was going to go anywhere."[19]

This "side project," unveiled in 1993, allowed laypersons easy access to the Internet for the first time; now school teachers and business owners could access the Internet for multimedia sources of information. The advent of Mosaic catapulted the World Wide Web into global popularity and out of its former sphere of influence dominated by universities and mammoth engineering hubs. Mosaic, with its uncomplicated installation and engaging graphics, extended the benefits of the Internet to nonscientists.

Still, Marc Andreessen claims his contribution is fleeting. "Any new technology tends to go through a 25-year adoption cycle."[19] With economic bubbles and with more sophisticated technological innovation ever present, Andreessen warns newcomers to keep their wits about them and to exercise foresight. Those Internet entrepreneurs focused on developing cost-cutting, streamlined solutions, Andreessen posits, will withstand turbulent market behavior. "In the late '90s, everyone got incredibly excited about the opportunities. A dynamic got created where there was more money than time, and that led to excesses: too many hires, too many servers, too much new software.... Then everyone had to clean up after the big party."[19]

15.4.3 Pierre Omidyar

I came on this notion that with the Web ... we could actually create ... a single market, where people from all over the world could ... do business with one another regardless of who they were.[19]

Much like Marc Andreessen and his World Wide Web browser, Pierre Omidyar's 1995 creation of eBay—the first Internet auction site—was initially "an experiment, a side hobby basically, while I had my day job."[19]

The only child of Iranian parents, Pierre Omidyar was 6 years old when he emigrated to the United States from France for his father's medical residency.[19] Omidyar found company in gadgetry and electronics and then computers. His first introduction to a computer occurred during his third grade school career; he would skip his gym class to sneak sessions with the school's computer. He then went on to Tufts University to major in Computer Science and started a career with Claris, an Apple Computer firm, where he began writing computer programs such as MacDraw. Omidyar then went on to launch a start-up called Ink Development, but the idea for eBay came about as a Labor Day lark in 1995. Also interested in economic market theory, he set out to write the code for an interactive web program wherein people could use the Internet for electronic commerce. "People thought it was impossible because how could people on the Internet—remember this is 1995—how could they trust each other? How could they get to know each other?"[19] Being an optimist, Omidyar believed people were essentially trustworthy, and he believed the Internet provided an ideal forum for people to communicate and conduct business. Within 6 months of its inception, his experiment was generating more income than his day job, and Omidyar's global flea market boasted 250 million users from 39 different countries.[19]

As for the tremendous popularity of his online auction, Omidyar offers no magical explanation. He is hesitant to give advice for young up-and-comers, believing the business sensation of his company was purely accidental. In his many interviews, Pierre Omidyar responsibly counsels entrepreneurs to honor their passions and intellectual curiosities, to work diligently, to contribute value, and to drown the criticism of naysayers:

> But I know that the job of a typical entrepreneur is a lot harder than that.... When you look at the accomplishments of accomplished people and you say, "Boy, that must have been really hard," ... that was probably easy. And conversely, when you look at something that looks easy, that was probably hard. And so you're never going to know which is which until you actually go and do it. So just go and do it, try it, learn from it. You'll fail at some things, that's a learning experience that you need.... And don't let people who you may respect ... tell you it can't be done ... because they don't have the courage to try.[19]

15.4.4 Craig Newmark

> I really did grow up as a nerd. In high school I really did have thick black glasses taped together.... I felt left out all the time.... Nowadays, I remember that feeling, and I want everyone to be included, and that's something we work on every day [at Craigslist].[20]

Craig Newmark was a 30-something software engineer at San Francisco's Charles Schwab office. In an effort to enhance his social life, he began to send group emails (ccs or carbon copies) to his friends about technology events. Over a period of 10 years, the site's reputation ballooned to now an international online community— 450 cities in more than 50 countries—for odd job postings, apartment listings, and personal ads.[20]

Very much like Pierre Omidyar, Craig Newmark built his business by looking to create a sense of community. "The real core here is that we've kind of lost our sense of neighborhood or community. In our culture, I think we crave that."[20] Also, like Omidyar, Newmark hatched his million-dollar idea by attempting to fulfill a community need, not by striving for a profit. He advises young capitalists to look to do the same when searching for entrepreneurial inspiration: "Listen to community; find out what they want. Don't waste time and resources on something your community doesn't want."[20]

On the subject of career paths, Newmark, like many great technology entrepreneurs, paid his dues working the corporate grind. While Craigslist was gaining momentum, he worked as a private software contractor to companies such as Bank of America, and in 1990 as a systems engineer for IBM. Rather than cursing his days as an employee, Newmark embraces his early work experience, saying it provided him an arsenal of ideas for creating his own business. "You see what's happening with your company, you see where the processes are broken, and you fix them. That's hopefully now hard coded into our company DNA."[20]

15.4.5 Robert Metcalfe

Flocks of MIT engineers come over here.... They say, "Wow! What a great house! I want to invent something like Ethernet...." I have to sit 'em down for an hour and say, "No, I don't have this house because I invented Ethernet. I have this house because I went to Cleveland and Schenectady and places like that. I sold Ethernet for a decade. That's why I have this house. It had nothing to do with that brainstorm in 1973."[21]

Robert Metcalfe's first foray into information technology innovation was, like that of Marc Andreessen, made possible by funding from the U.S. government. Having completed an undergraduate degree at MIT, Metcalfe started a PhD program at Harvard University, meanwhile finding student work at his alma mater on the other side of Boston's Charles River. The ARPANET (Advanced Research Projects Agency Network) was developed by the U.S. Department of Defense in the early 1960s. This project would connect terminals at various research locations across the country through Interface Message Processors (IMPs). The ARPANET system was designed to expedite technological innovation by allowing scientists to communicate with one another. After Harvard University rejected his doctoral dissertation on ARPANET, Robert Metcalfe took a position at the Xerox Corporation's Palo Alto Research Center (PARC) while simultaneously rewriting his thesis on the flaws of

University of Hawaii's radio-based computer data transmission system ALOHAnet. Metcalfe eventually earned his PhD from Harvard University in 1973.

At PARC, Bob Metcalfe wrote the protocol for Ethernet—a standard for connecting computers over short distances or local area networks (LANs)—on the basis of his analysis of the ALOHAnet model. Those LANs were in turn connected to the Internet to facilitate interinstitution communication. They were especially popular at universities where many workstations were connected using Ethernet. In this way, Ethernet was influential in the expansion of the Internet.[19]

In 1979, Metcalfe left PARC and started his own company, 3Com (computers, communication, compatibility), and continued to sell Ethernet to large computer companies such as Intel Corporation, Digital Equipment Corporation, and Xerox Corporation. Ethernet's computer networking technology now connects more than 100 million computers to the Internet.[22] Metcalfe emphasizes the integral role of salesmanship in bringing a technological innovation to market and attributes the resounding success of the Ethernet to his own efforts in propagating the technology.

> *Most engineers don't understand that selling matters.... They don't understand that nothing happens until something gets sold.... Not because of any flowery flash of genius in some academic hothouse.*[22]

15.5 THE FRANCHISING OPTION

Today, franchising is an international phenomenon. A franchise is a legal and commercial relationship between the owner of a trademark, service mark, trade name, or advertising symbol and an individual or group seeking the right to use that identification in a business. The franchise governs the method of conducting business between the two parties. Generally, a franchisee sells goods or services supplied by the franchisor or sells goods or services that meet the franchisor's quality standards. Franchising is based on mutual trust between the franchisor and franchisee. The franchisor provides the business expertise (i.e., marketing plans, management guidance, financing assistance, site location, training, etc.) that otherwise would not be available to the franchisee. The franchisee brings to the franchise operation the entrepreneurial spirit and drive necessary to make the franchise a success. Examples of franchise opportunities for technology entrepreneurs include such things as Jiffy Lube, Midas, Data Doctors, Wireless Toyz, and many others.

There are primarily two forms of franchising:

1. Product/trade name franchising
2. Business format franchising

In the simplest form, a franchisor owns the right to the name or trademark and sells that right to a franchisee. This is known as "product/trade name franchising." In the more complex form, "business format franchising," a broader and ongoing relationship exists between the two parties. Business format franchises often provide a full range of services, including site selection, training, product supply, marketing plans, and even assistance in obtaining financing. The International Franchise Association is a leading trade association in the franchise industry. Its web site lists "hot" franchises.

15.5.1 Franchising: A Brief History

An attractive way to become involved with business is to invest in a franchise. Legislation mandates that three components must be present for a franchise to exist:

1. The utilization of a uniform trade name or service mark (e.g., Kinko's, Jiffy Lube, Midas)
2. A uniform system of operations
3. A fee of at least $500 during the first 6 months of business

With these components, a franchise relationship is established between the franchisor (the company) and the franchisee (the small-business owner). A new franchised business opens every 15 minutes in the United States. Franchising is the fastest growing method of doing business, but it is not really new. Some believe that franchising dates back to the Middle Ages (A.D. 476–1453), when it was common practice for governments to offer people a license or franchise granting them the right to do business or to establish and collect tax revenues.

The first contractual commercial franchise in the United States was Singer Sewing Machine Company in 1851. To solve the "sewing machine war," in which four manufacturers were suing one another, Isaac Singer and his attorney-partner, Edmund Clark, organized the Sewing Machine Combination. It became the first patent pool in the United States, dividing the distribution rights among the manufacturers. The franchisees were granted worldwide rights, territories, and exclusive rights to sell and service their machines.

Two other early franchise operations in the United States were Coca-Cola, which first supplied its syrup to franchised bottlers around 1900, and Rexall Drug Stores, which started around 1902. Other early franchise pioneers included automobile dealerships and gasoline stations. Oil companies had been operating their own stations until about 1930. A few of them began to license dealers, and the practice spread from 1930 to 1935 until it became the sole method employed to distribute gasoline. In the 1950s, franchising spread across the United States. In fact, 90% of all franchise companies doing business today started in the early 1950s.

15.5.2 Uniform Franchise Offering Circular

Franchising is governed primarily by laws that require franchisors to inform prospective franchisees about the system. This information is contained in a document called the **Uniform Franchise Offering Circular**, or simply the UFOC. Under the federal and state rules, a franchisor cannot offer a franchise until the franchisor has prepared a UFOC.

In a nutshell, the UFOC provides information on the franchisor, the company's key staff, management's experience in franchise management, the franchisor's bankruptcy and litigation history, and the initial and ongoing fees involved in opening and running a franchise. Also included is information on the required investment, purchases you will be required to make from the franchisor or approved suppliers, and territory rights you will be granted. You also find information about the legal responsibilities of the franchisee and franchisor.

In addition, the UFOC presents information about the company-owned locations and the franchisees in the system, including the number of franchises opened, the number closed and transferred, and, most important, a list of existing and former franchisees with their contact information.

It's important that you fully understand the **franchise agreement** and any other agreements you need to sign. If you don't understand what you're signing, you may find yourself locked into a business relationship that doesn't wear well for you.

15.5.3 Ten-Day Rule

Included in the regulations governing the sale of franchises is a cooling-off period called **the ten-day rule**. Franchisors are required to wait a minimum of 10 business days after giving a prospect the UFOC before allowing the franchisee to sign the franchise agreement. As a potential franchisee, you are also legally entitled to have the final franchise agreement—with all the blanks filled in—for at least 5 business days before you are allowed to sign it. This gives you time to review and consider the terms of the agreement.

The franchise agreement provided in the UFOC may contain some blanks to be filled in, including who the franchisee is, where the franchise will be located, the size of the protected territory (if there is one), and other matters specific to the franchisee. On occasion, the franchise agreement must be changed as a result of changes made during negotiations. The franchisee must be given a copy of the final agreement—with the changes—at least 5 business days prior to signing the agreement.

15.5.4 Franchising Agreement

The franchise organization enters into a contractual agreement with each of its franchisees. These contracts may differ in a number of areas, such as capital needed, training provided, managerial assistance available, and size of the franchise territory. But most franchise contracts have points in common. The franchise buyer normally pays an initial fee to the company and agrees to pay the franchisor a monthly percentage of sales. In exchange, the franchisee has the right to sell a standard product or service.

In almost every franchise, the franchisee must invest some money, which is generally called the **franchise fee**. The amount can vary from a few thousand dollars to millions. Despite having to put money down, people still have to wait for certain franchises. Each year, more than 2,000 people apply for the approximately 150 franchises granted by McDonald's.

A McDonald's franchise can cost as much as $600,000. Although most of the fee can be borrowed from a bank, $66,000 would be the minimum cash down payment required. This payment covers landscaping and opening costs, license fee, site development fee, equipment down payment, signs, and security deposit. McDonald's Corporation does not lend money or guarantee loans. The company will provide a site, build the restaurant, and develop the parking lot. In return, the franchisee must pay 12% of gross sales to McDonald's and put at least 4% of gross sales annually into marketing and advertising.

Franchise purchasers enter the business to earn money; in fact, the owner and the company both want to earn money. Earnings are a measure of the franchise's success. To succeed, it must be well managed, provide good products or services, and obtain repeat customers. Exhibit 15.3 lists the franchises that were considered "hot" during 2009.

The franchisor begins to earn money when the cash down payment is made. In some cases, a franchise fee must be paid before certain rights of operation are granted. The franchisor also requires some type of royalty payment on gross sales (e.g., the 12% franchisees pay to McDonald's). The amount of royalty, or share of the proceeds, paid to the franchisor differs from company to company. Typically, a franchisee pays a royalty of between 3% and 15% on gross sales (total sales revenue).

No matter what the percentage of royalty, franchisees often dislike paying profits to someone else. It is only human nature for people to feel pain when they see part of the fruits of their labor go to someone else. Paradoxically, this pain may become more intense as the franchise operation becomes more successful. The franchisee may be making many thousands of dollars, but so is the franchisor.[23]

Being involved and working hard to make the franchise a success and then being required to share the earnings deflates the ego of some franchisees. If franchisees believe the sharing of profits is unfair, this perceived inequity is also a problem. The franchisee may believe that he or she does all the work and the franchisor takes too big a share of the earnings.

1. Subway
2. McDonald's
3. Liberty Tax Service
4. Sonic Drive In Restaurants
5. InterContinental Hotels Group
6. Ace Hardware Corp.
7. Pizza Hut
8. UPS Store, The/Mail Boxes Etc.
9. Circle K
10. Papa John's Int'l. Inc.

EXHIBIT 15.3

Top 10 Franchises of 2009.

Source: 2009 Franchise 500. Retrieved from Entrepreneur.com: http://www.entrepreneur.com/franchise500/index.html.

Many people who consider entering into a franchise contract assume they will be their own boss. This assumption is only partially accurate. The franchisor exercises a significant amount of control over the franchisee in such areas as:

- Real estate ownership
- Territorial restrictions
- Cancellation provisions
- Required exclusive handling

Some franchise companies own the real estate, select the site of the business location, and build the facility. These companies then lease the facility to a franchisee. Exxon, Shell, Gulf, and Texaco, for example, often select the location for a service station and build it according to their specifications.

The **cancellation provision** in a contract is a powerful control device. For instance, gasoline companies often issue operators of service stations a 1-year franchise. If the company does not consider the operator successful, the franchise agreement can be canceled after that time. This provision can force operators to run the business as directed by the franchisor.

Exclusive handling means that the franchisee will only sell items, products, or services that are acceptable to the franchisor. This means that the franchisor has control over what is sold directly to customers or clients.

15.5.5 Advantages of Owning a Franchise

Although owning a franchise business has its problems, there are reasons why it appeals to people. A person who has never owned or managed a business needs guidance to operate successfully. This guidance can be provided by a well-run franchise organization. Also, franchisors can provide a brand name, proven products or services, and financial assistance.

Guidance

A glaring weakness in small businesses is lack of managerial ability. A person with limited managerial skills may be able to get by in a large organization because he or she is just one of many managers. But no one can cover up for or "carry" a franchise manager. Many franchisors try to overcome managerial deficiencies or inexperience by providing some form of training. For example, Kwik Kopy operates its own training school for improving management skills. The initial training focuses on managing a business rather than on operating a printing press. Subjects include financial management, marketing, advertising and much more. Kwik Kopy also offers Personal Development Workshops to sharpen the people skills of trainees. Continuing education through seminars and periodic conferences covers all aspects of center operations.[24]

Brand name

The investor who signs a franchise agreement acquires the right to use a nationally or regionally promoted brand name. This identifies the local unit with a recognized product or service. Travelers recognize the Holiday Inn sign, the colors of a Pizza Hut building, and Century 21 real estate signs. National promotion brings these features and characteristics to the attention of potential consumers.

Proven product

The franchisor can offer the franchisee a proven product and method of operating the business. The product or service is known and accepted by the public: Customers will buy Baskin-Robbins ice cream, AAMCO transmissions, Athlete's Foot sneakers, and H&R Block income tax counseling.

Financial assistance

By joining a franchise company, the individual investor may be able to secure financial assistance. Start-up costs of any business are often high, and the prospective investor usually has limited funds. The sole owner generally has a limited credit rating, making it difficult to borrow needed funds. In some cases, association with a well-established franchisor—through its reputation and its financial controls—may enhance the investor's credit rating with local banks.

15.5.6 Disadvantages of Owning a Franchise

As does any business venture, franchising has some disadvantages. Many were mentioned briefly earlier in this chapter. Some of the more pressing negative features include costs, lack of control, and inadequate training programs offered by some unscrupulous promoters.

Costs

As already mentioned, franchisees must pay franchise fees. In return, the franchisor can provide training, guidance, and other forms of support that would otherwise cost money. Thus, the franchisee pays for the opportunity to share in these forms of support. If it were possible to earn the same income independent of the franchisor, the investor could save the amount of these fees.

External control

A person who signs a franchise agreement loses some independence. The franchisor, in order to operate all of the franchise outlets as a business, must exercise some control over promotional activities, financial records, hiring, service procedures, and managerial development. Although useful, these controls are unpleasant to the person who seeks independence. In the best of circumstances, the franchisee is semi-independent. In a sole proprietorship, by contrast, the owner is totally independent.

Overdependence

Often, new franchises become overdependent on the franchisor, and this can lead to an inability to make timely or even simple decisions without constantly checking and asking for advice. Franchisees need to use judgment and not rely on the franchisor to check every decision about the local market.

Poor local reputation can spread

In some communities where there are multiple franchise units with different franchisees, the performance and reputation of one can affect all the units. For example, Jack-in-the-Box restaurants were affected negatively nationwide because some franchises in the state of Washington improperly cooked food, allowing the spread of *E. coli* bacteria and resulting in the deaths of three young children.

15.5.7 Franchisor Disclosure

Because of the nature of franchising in the United States, evaluating an opportunity carefully before signing an agreement is important. The large number of franchise companies makes the task difficult. The Federal Trade Commission requires every franchisor to provide buyers with a full disclosure form. The form has 20 categories of information, including a financial statement, company history review, fees, investment requirements, and a litigation history. The best investment a prospective franchisee can make is to have an attorney and an accountant review the document and provide opinions.[25] Exhibit 15.4 presents a sample of the types of questions that need to be asked and answered before signing a franchise agreement.

Legal guidelines are crucial to the continued success of franchising. Currently, 35 states have laws on franchising issues such as disclosure, registration, and franchise-relationship definitions. The most stringent state law is the Iowa Franchise Act of 1992. The Iowa law requires that franchisors disclose financial information to prospective buyers. It also stipulates that franchisors agree to pay fair market value for any franchise location they decide to close. Moreover, franchisors are barred from

---

Because a franchise agreement can be a legally binding contract that defines the franchise relationship between the franchisor and franchisee, it should be carefully reviewed. Questions to address include:

1. What are the legal obligations of the franchisee? Of the franchisor?
2. Are there any legal suits currently pending or recently settled by the franchisor and other franchisees?
3. What is the financial history of the franchisor?
4. What amount of royalty must be paid to the franchisor? When?
5. Please provide me with a list of all current franchisees in my area.
6. What types of support services will the franchisor provide and how good are the services?
7. Does the franchise agreement conform to the requirements of the Federal Trade Commission?
8. Can the franchise agreement be terminated by either party? What are my (franchisee) obligations if the agreement is terminated by me? By the franchisor?

EXHIBIT 15.4

Questions to Ask Before Investing in a Franchise.

terminating, refusing to renew, or denying a transfer of a franchise to another owner except for "good cause."

15.6 LAUNCHING YOUR ENTREPRENEURIAL CAREER

Now that you have completed this textbook and, presumably, a course in technology entrepreneurship, you have taken the first step toward an entrepreneurial career. Congratulations. It must be pointed out again, here, at the end of the book as it was at the beginning, that entrepreneurship is a lifelong endeavor. We are under no illusions that this textbook—or any other book on entrepreneurship, for that matter—will somehow transform you into a successful entrepreneur. To the contrary, we have taken great pains to be as realistic as possible throughout this text about the difficulties of entrepreneurship and the need to develop a resilient and flexible mind-set.

Naturally, there are some individuals who will become entrepreneurs soon after completing a course in entrepreneurship, but the vast majority will not. You may have an idea in your mind already about a product or service you'd like to perfect and take to the market. Think carefully about whether you are ready for such a challenge and, indeed, whether your market is ready to buy what you are proposing to offer.

Your entrepreneurial career has begun, but you may not decide to launch your own venture until after you have gained some industry experience. Working for someone else for awhile does not mean you are not an entrepreneur—you are simply waiting for the right time and the right place to hatch your plans. Don't be discouraged if this is your path for as many as 10–15 years. It is pertinent here to point out that research has revealed the average age of the first-time technology entrepreneur to be 39.[18]

Building an entrepreneurial career is not just about building companies, it is about building a lifestyle. Entrepreneurs must shape themselves as much as they shape ventures to be able to handle uncertainty, ambiguity, and occasional failures. This text has provided innumerable ideas, tools, case studies, and suggestions about how to prepare you for a life of entrepreneurship. This preparatory work should have already begun. If not, then you should not hesitate to begin immediately. Some ways for you to begin developing your unique entrepreneurial personality and lifestyle are highlighted in Tech Tips 15.2.

These tips are just a beginning for the aspiring technology entrepreneur. There is no shortage of resources available to you from here on out. The average bookstore is brimming with stories about entrepreneurs and reference books about how to network, raise money, write a business plan, and myriad other things. You should resolve to continue your education on your own. Your education should include book learning and worldly learning. The latter refers to the many opportunities you will have in nearly every community in the world to interact with entrepreneurs, people who want to be entrepreneurs, investors, teachers, and mentors of all types.

TECH TIPS 15.2

Suggestions for Launching Your Entrepreneurial Career

1. *Don't work for less than you can afford to*, but do offer a discount to customers or clients who sign contracts with you.
2. *Find people who will refer jobs to you.* If they send you nightmare jobs, make sure they're balanced out with rewarding (profitable!) ones.
3. *Surround yourself with supportive people* and don't be discouraged by anyone. If your idea is good and you're determined to stick with it through the first few difficult years, your chances of success are great.
4. *Be flexible in your thinking.* Prepare to change the way you work, the products you use, and the services you offer, in order to meet the demands of your customers.
5. *Admit your mistakes, correct them, and carry on.* (For example, if you purchase a piece of equipment that does not meet your expectations, send it back, sell it, or exchange it!)
6. *Develop a good relationship with your bank manager and creditors.* Show a genuine interest in solving problems. Pay as much as you can afford to, to everyone to whom you owe money.
7. *Get trained!* You'll be spending a lot of time doing things that have nothing to do with your area of expertise, like bookkeeping, marketing, and IT support!
8. *Avoid isolation.* Even if you work closely with your clients, you won't be part of a gang anymore. Develop your own network of entrepreneurs that you see regularly and bounce ideas off. Ideally, they'll allow you to vent your anger and share your successes.
9. *Separate your work and personal life.* Set your working hours and stick to a strict timetable. When you're not available to clients, leave a message on your answer machine letting them know when they can expect a reply from you. Let them know how to reach you in an emergency.
10. *Plan some "thinking time" into every day.* If you pack your diary with back-to-back activities, your business will never grow.
11. *Plan time to do something you enjoy* at least a few times a week—recharge your batteries!
12. *Write a business plan* so you're clear about what you're doing, and update it every year.
13. *Develop an excellent telephone manner* and react quickly to any complaints or problems.
14. *Confirm orders* personally and immediately, especially those you receive on e-mail.
15. *Never lose sight of the big picture*—look for innovative, little-explored directions in which to take your *business*.
16. *When you find someone cleverer than you, employ him or her!*
17. *Solicit advice from people* who know, for example, other entrepreneurs and reputable small-business advisers—the *DTI* offers lots of information and support for new businesses.
18. *Don't enter a business or a venture that you know nothing about.* You'll be running to catch up for the rest of your business life.

19. *Have an existing, loyal customer base* and start locally.
20. *Be aware that you will get through any initial investment quickly*, so ensure that you are covered financially until at least the end of the second year.
21. *Focus on a specific goal* and work at it until it's achieved.
22. *Never worry about how to get things done when you are first developing your idea.* Money and resources will come together once you have set your goals and begun to work at them.
23. *Make quality in every aspect of your business your primary focus and aim.* If it isn't, you will eventually go out of business.
24. *Use the Internet.* Use e-mail. Build a web site (if you aren't familiar with web sites, try *HTML for Dummies*), send out e-mail newsletters, buy online banner advertisements, and register your site with all the major search engines.
25. *Delegate.* You might have to hire a good PA, lawyer, or marketing professional to ensure you'll be profitable in the future.[26]

It is up to you now. Don't wait a moment longer to begin your entrepreneurial career.

SUMMARY

This chapter has introduced you to the notion of entrepreneurship as a career. In this regard, you should be prepared to learn and grow as an entrepreneur throughout your lifetime. No doubt, many of the lessons you learn will be through your reading and studying, but the most important lessons likely are to be those you learn by trial and error. Entrepreneurs will fail more often than they succeed; there is no doubt about that. But failure can be viewed as an opportunity to learn. To the extent that you are able to separate business failure from personal failure, you will be able to try again.

We discussed the concept of the entrepreneurial personality. While it's true there are no specific personality traits that have always been linked to success as an entrepreneur, there are some traits that entrepreneurs tend to have or develop. These traits center on a strong need for achievement. This phenomenon was discovered by psychologist David McClelland in the 1960s, and remains a useful insight to this day. On the other hand, the need for achievement is a fairly general personality trait that can be linked to nearly any other traits. That is, one can be happy or sad, optimistic or pessimistic, extroverted or introverted, as long as there is a need for achievement. According to McClelland, the need for achievement trait actually is an important variable in business success and has been demonstrated to be critical to economic development around the world.[27]

Next, we looked at the careers of five well-known entrepreneurs. The point of examining their careers in some detail was to highlight the variability of starting points, ages, and speed to success each of them experienced. You can be sure that no two entrepreneurs have matching stories about their paths to success. As such, these stories are instructive to you. Your own entrepreneurial career will be unique. Some of the lessons conveyed in this text will be highly relevant and germane to your evolving career, and some will be extraneous. It's impossible to tell, now, which of the lessons will be the most valuable to you. We have tried to be honest about the challenge of writing a textbook for such a widely disparate career path. Nonetheless, it is certain that many of the things we discussed here will play a role in your success. In the end, it's up to you to find out which of the many tips, tools, and techniques we discussed are applicable to your unique path and industry.

Many entrepreneurs prefer to launch their careers by taking advantage of structures and systems that have been invented and perfected by others. For these individuals, the franchise option provides a unique opportunity. Many entrepreneurs have launched long and successful careers by taking the reins of one or more franchises, in one or more locations. The franchise option provides many of the attractive features of start-up entrepreneurship—including the freedom to run the venture without constant oversight, freedom to invest for future gains, and freedom to work according one's own schedule. At the same time, the franchise option can eliminate many of the difficulties of start-up entrepreneurship, including the uncertainty of markets, the need to develop scalable systems, and the need to develop brand awareness. This is an option that many aspiring entrepreneurs should consider.

Finally, we concluded with a reflection on the fundamentals of managing an entrepreneurial career. Essentially, the fundamentals can be summed up in a single phrase: Be prepared to learn at every turn. Entrepreneurship is a lifelong pursuit. Even those fortunate few entrepreneurs who manage to cash out early in their lives find themselves returning to entrepreneurship later, either as founders of new ventures or as investors. Your entrepreneurial career begins immediately, as you learn to spot potential opportunities, evaluate them, and consider whether you have the requisite resources to pursue them. In the end, entrepreneurs create value for other people. There simply is no other way to make profits in capitalist economies. Technology entrepreneurs are particularly adept at creating value—they must learn to be equally adept at protecting the value they create and capturing that value long term through equity ownership. We trust that this book has convinced you that all technology-oriented individuals are entrepreneurs in one way or another. Your job now is to find the path that is right for you.

STUDY QUESTIONS

1. Discuss the similarities and differences in the predictors of entrepreneurial intentions in Ireland and the United States.

2. Compare and contrast the five entrepreneurial career paths discussed in the text. Identify three similarities in most of the career paths and three differences.

3. What are the major elements of the "entrepreneurial personality"?

4. What are the advantages to starting your entrepreneurial career via a franchise?

5. What are the disadvantages of a franchise?

6. Discuss what is meant by the term "entrepreneurial lifestyle." What are some of the likely features of such a lifestyle?

7. What are some things that an individual can do immediately to begin an entrepreneurial lifestyle?

8. What role does learning play in an entrepreneurial lifestyle? How can an entrepreneur learn from a failed venture?

9. Explain what is meant by the term "intrapreneurship." What are some of the major differences between intrapreneurship and entrepreneurship?

10. If people are intrapreneurs at one point in their life, what lessons can they learn that would help them become entrepreneurs? Vice versa?

EXERCISES

1. *OUT-OF-CLASS EXERCISE*: Do you have what it takes to be an entrepreneur? The quiz on the web site listed below only takes a few minutes to complete. Try it out and be prepared to discuss what you learned about yourself during class. *Be sure to record your results and read the interpretation section of the web page.*

   http://www.liraz.com/webquiz.htm

   Use the following questions to guide the discussion in class about each student's entrepreneurial leadership capabilities:

   ❏ How do you interpret the results you achieved on the test?
   ❏ Do you think you have what it takes to be a successful entrepreneur?
   ❏ Do you think self-evaluation quizzes like this are useful for entrepreneurs? Explain.
   ❏ What did you learn about yourself that surprised you? Explain.

2. *OUT-OF-CLASS EXERCISE*: This activity is designed to enhance your understanding of the entrepreneurial personality and the motivations, challenges, and rewards of entrepreneurship.

   Complete the following assignment:

   1. Identify a successful entrepreneur in the community that you can interview. Concentrate on identifying a small business that is directed by its founder. The business section of recent issues of the community newspaper should be helpful in locating an entrepreneur.
   2. After identifying the entrepreneur, contact the person and arrange an interview.
   3. Conduct a 30–60-minute interview, including the following questions:

- What motivated you to start your own business?
- How would you describe yourself to a stranger? Are you self-confident, energetic, independent? Are you an optimist, a realist, a pessimist?
- Which personality characteristics and abilities are essential for success as an entrepreneur?
- How would you describe your leadership style?
- Describe a typical workday.
- What aspects of your work do you find the most satisfying? The most frustrating?
- Which aspects of your work are the easiest for you? The most difficult?
- In which of your business's activities (operations, finance, marketing, and personnel) are you most involved?
- How much emphasis do you place on motivating employees?
- What important lessons have you learned from your experience in creating and running your own business? What advice would you offer to a young, prospective entrepreneur?

4. After completing the interview, prepare a short (1–2-page synopsis).
5. Share and discuss your findings in class.

KEY TERMS

Cancellation provision: A clause in a franchise agreement that allows the franchisor to cancel its agreement with the franchisee if certain prespecified business milestones are not met.

Exclusive handling: A clause in a franchise agreement that states that the franchisee will only handle and sell franchisor products and services.

Franchise: A legal and commercial relationship between the owner of a trademark, service mark, trade name, or advertising symbol and an individual or group seeking the right to use that identification in a business.

Franchise agreement: The agreement that is signed indicating that a franchisor and franchisee have agreed to terms allowing the franchisee to operate a franchise.

Franchise fee: An amount of money paid up front by the franchisee for the right to operate a franchise.

Intrapreneur: A person with entrepreneurial characteristics who is employed within a large corporation.

Ten-day rule: A clause in all franchise agreements allowing the franchisee signatory to back out of a signed agreement within 10 days of signing.

Uniform Franchise Offering Circular: A document about the franchise that details the opportunity to potential franchisees.

WEB RESOURCES

The web sites below are intended as destinations for your further exploration of the concepts and topics discussed in this chapter:

1. http://www.score.org/entrepreneurship_career.html: This is a web site from the Service Core of Retired Executives (SCORE) that discusses entrepreneurship as an alternative career path.

2. http://www.princetonreview.com/careers.aspx?cid=60: *The Princeton Review* is one of the most respected career and higher education resources. This site features a discussion of entrepreneurship as a career.

3. http://www.career-intelligence.com/assessment/entrepreneurs-checklist.asp: This is a resource for women who are considering careers in entrepreneurship.

4. http://www.franchise.org/: This is the web site for the international franchise organization.

5. http://www.franchisegator.com/: This site provides search services for individuals seeking to purchase a franchise.

6. http://www.redhotfranchises.com/: This site is one of many that aggregate information about the wide range of franchise opportunities that are available.

7. http://www.smallbusinessnotes.com/choosing/intrapreneurship.html: This is a small-business web site that discusses various aspects of intrapreneurship.

ENDNOTES

[1] Hisrich RD, Peters MP. *Entrepreneurship*. Burr Ridge, IL: Richard D. Irwin Publishing; 1989.

[2] Bygrave WD. The entrepreneurial process. In: Bygrave W, editor. *The portable MBA in entrepreneurship*. New York: John Wiley & Sons; 1994. pp. 1–23.

[3] McIngvale J, Duening T, Ivancevich J. *Always think big*. Chicago: Kaplan Publishing; April 2002.

[4] Schjoedt L, Shaver KG. Deciding on an entrepreneurial career: a test of the pull and push hypotheses using the panel study of entrepreneurial dynamics data (B. University, Ed.). *Entrepreneurship Theory Pract* 2007;**31**(5):733-52.

[5] Van Auken H, Fry FL, Stephens P. The influence of role models on entrepreneurial intentions. *J Dev Entrepreneurship* 2006;**11**(2):157-67.

[6] Personal conversation with one of the authors (Hisrich).

[7] Lacy S. Meet the anti-VC VC. *Fortune* 2007;**156**(11):108.

[8] Christy JH. The Americanization of Matthias Zahn. *Forbes* 1995;**March 13**:123-4.

[9] Seligman M. *Learned optimism*. New York: Alfred A. Knopf; 1991.

[10] McClelland D. Achievement motivation can be developed. *Harv Bus Rev* 1965;**November/December 43**(6):6-17.

[11] Dimov D. Beyond the single-person, single-insight attribution in understanding entrepreneurial opportunities. *Entrepreneurship Theory Pract* 2007;**31**(5):713-31.

[12] Krueger NF. What lies beneath? The experiential essence of entrepreneurial thinking. *Entrepreneurship Theory Pract* **January** 2007;**31**(1):123-38.

[13] Wasserman N. The founder's dilemma. *Harv Bus Rev* May 2008;**86**(2):102-9.

[14] King C. Intrapreneurship: Heady Business. Retrieved from http://developers.sun.com/toolkits/articles/intrapreneur.html

[15] Pinchot .G, Pellman R. *Intrapreneuring in action: a handbook for business innovation.* San Francisco: Berrett-Koehler Publishers; **January** 2000.

[16] Switzer C. Creativity and motivation: the corporate intrapreneur. Retrieved from http://ezinearticles.com/?Creativity-and-Motivation—The-Corporate-Intrapreneur&id=423288.

[17] Shane S. *The illusions of entrepreneurship*. New Haven: Yale University Press; 2008.

[18] Karan B. How the Web was won. (Mayo K, Newcomb P. Interviewers, July 2008.)

[19] Mayo K, Newcomb P. How the Web was won. Retrieved from www.vanityfair.com/culture/features/2008/07/internet200807?currentPage=all.

[20] Craig Newmark. Craig list isn't a media menace. Retrieved from Media Interviews In Their Own Words: www.iwantmedia.com/people/people66.html.

[21] Kirsner S. The legend of Bob Metcalfe. *Wired Digital*; **November** 1998. Retrieved from: www.wired.com/wired/archive/6.11/metcalfe.html.

[22] Metcalfe B. Invention is a flower, innovation is a weed. *Technology Review,* **November** 1999. Retrieved from www.technologyreview.com/web/11994/pg2/.

[23] Steinberg C. Turnaround. *Success* 1995;**January/February**:71-7.

[24] Hadfield B. *Wealth within reach*. Nashville: Word Books Publishers; August 1992.

[25] Howard PK. *The death of common sense*. New York: Random House; 1995.

[26] Cynthia Schmae. 25 tips by entrepreneurs for entrepreneurs. Retrieved from http://www.ivillage.co.uk/workcareer/ownbiz/bizprep/articles/0,196_156910,00.html

[27] McClelland DC. Business drive and national achievement. *Harv Bus Rev* **July/August** 1962;**40**(4):99-112.

Model Business Plan

USuggest.com

Control Copy Number: _____   Issued To: _____

DISCLAIMER

This Confidential Business Plan (the "Plan") is being furnished through USuggest, LLC, as exclusive agent, solely for the use by prospective investors in connection with their consideration of a possible investment in the Company.

None of USuggest, LLC or any of its respective affiliates, employees, or representatives make any representation or warranty, expressed or implied, as to the accuracy or completeness of any of the information contained in the Plan or any of the information (whether communicated in written or oral form) transmitted or made available to prospective buyers. Each of the aforementioned parties expressly disclaims any and all liability relating to or resulting from use of the Plan or such other information by a prospective investor or any of its employees, affiliates or representatives. Only those particular representations and warranties, if any, which may be made in a definitive written agreement regarding an investment in, or acquisition of USuggest, LLC, when, as and if executed, and subject to such limitations and restrictions as may be specified therein, will have any legal effect.

By receipt, each recipient of this Plan agrees that the information contained herein is not to be used for any purpose other than in connection with its consideration of an investment in the Company, that the information contained herein is of a confidential nature and that the recipient will treat it in a confidential manner, and that the recipient (i) will not, directly or indirectly, disclose or permit its affiliates or representatives to disclose any information contained herein to any other person or reproduce this Plan, in whole or in part, without the express prior written consent of the Company and USuggest, LLC and (ii) will promptly return any and all copies of this Plan to USuggest, LLC upon request.

The market analysis and financial projections presented in this Plan represent the subjective views of the Company and the Company's current estimates of future performance based on the various assumptions that may or may not prove to be correct. There can be no assurance that the Company's views are accurate or that the Company's projections will be realized. Industry experts may disagree with these assumptions and with the Company's view of the market and the prospects for the Company. Prospective investors should make their own independent inquiries and investigations of the assumptions, calculations and accounting policies upon which the projections are based.

THIS PLAN DOES NOT CONSTITUTE AN OFFER TO SELL OR A SOLICITATION OF AN OFFER TO BUY ANY SECURITIES. The sole purpose of this Plan is to assist prospective investors in deciding whether to proceed with further investigation of the Company in connection with their consideration of an investment in or acquisition of the Company. This Plan does not purport to contain all information, which may be material to an acquirer of the Company, and recipients of this Plan should conduct their own independent evaluation of the Company.

The Company reserves the right, in its sole discretion, to reject any and all proposals made by or on behalf of any prospective investor with regard to an investment in the Company and to negotiate with any prospective investor at any time.

The Company does not intend to update or otherwise revise this Plan following its distribution.

Questions relating to the information contained in this Plan and requests for additional information should be directed to one of the USuggest, LLC representatives listed below. **Under no circumstances should the Company or any of its respective employees be contacted directly.**

For more information please contact:

Prospective investors who decide not to pursue this matter are requested to return this Plan (together with any other material relating to the Company which may have been received from the Company) as soon as practicable.

EXECUTIVE SUMMARY

USuggest, LLC is a pioneer software company commercializing its patent-pending technology that harnesses the power of free markets, self-interest, and self-organizing systems to spread product and services **deal suggestions** with unprecedented speed and effect. Marketers know that referral or "word of mouth" advertising is a powerful marketing tool. People trust the opinions of advisors or friends when seeking to purchase goods or services. USuggest.com is the first Internet shopping service that **rewards** people for **finding** and **sharing** the best deals on the net within a **private**, **secure** online **deal-suggestion portal.**

The USuggest technology produces a **cascading effect** by creating incentives for shoppers to enter suggestions into its portal. Using a patent-pending software tool called **InstantValue**, shoppers merely highlight the deal they find, tag it with keywords from a drop-down menu, and hit the "Enter" key. The USuggest portal **instantly lists** their suggestion and begins to track its attractiveness to other shoppers. Each suggested deal will be **ranked** according to click-through and conversion (sale). The best deals will rise to the top of the list, providing the suggester with **commission rewards** for each sale.

Research indicates that U.S. online retail sales are expected to reach $172 billion in 2005. Excluding travel, Internet sales are expected to reach **$109.6 billion in sales in 2005**, that's 5.5 percent of all retail sales and up from $89 billion in 2004. The study, conducted by Forrester Research, indicates that online sales will grow fastest in categories such as cosmetics, jewelry, and flowers. The U.S. Department of Commerce reported that Q1 2005 Internet sales totalled $19.8 billion, up 24 percent from 2004. The survey found that **online retailers stepped up their use of**

search-related advertising, which they now consider their best source of new customers.

Over the past four years, the USuggest management team has mastered the affiliate marketing business and has developed **unique insights** into mobilizing a wide network of **major online retailers**. It is anticipated that USuggest will become the world's premiere **deal-suggestion portal** within three years. Operating under a series of non-descript names, the team is currently drawing **more than 5,000 visitors per day** to its affiliate shopping sites. If these numbers hold through start-up and into the first three years, USuggest will grow to more than 50,000 loyal customers with over five million monthly visitors, generating more than $12 million in annual revenue by the end of year three.

InstantValue and USuggest.com are flagship offerings of USuggest, LLC, for establishing its position as **the premier online deal-suggestion portal**. The unique combination of a browser embedded toolbar for uploading deal suggestions, and a **privacy-preserving** portal that **organizes, ranks, and rewards** suggestions, will generate unparalleled traffic and sales. InstantValue also boosts the ease of finding products for consumers by prompting deal-suggesters to annotate products with search-hot keyword tags. For example, a "white sneaker" can also be tagged as a "running shoe". InstantValue will be **freely distributed** to online shoppers from high-traffic affiliate shopping portals. Currently, affiliates pay 10 to 45 percent commissions for each sale.

USuggest, LLC is seeking seed start-up funding of $250K to develop and deploy its products, and to secure the related intellectual property. By Q3 2005, the company will release the alpha versions of InstantValue and the USuggest.com portal. By Q4 2005 a market-ready and tested product will be released with **projected monthly revenue in excess of $20K.** During Q1 and Q2 2006, USuggest plans to raise an additional $3 million to boost marketing and consumer adoptions. The company will establish a Board of Advisors consisting of decorated technologists, investors, and management experts.

Upon widespread acceptance in the market, the company will explore the option of a merger or acquisition by a large Internet company that derives revenue from traffic on its shopping network. Logical candidates for such a deal are Amazon Associates Network, Google Ad Sense Publishers, EBay Stores, Yahoo Shops, or MSN Shopping Networks.

PRODUCT DESCRIPTION

The InstantValue toolbar and the USuggest.com deal-suggestion portal are the flagship products of USuggest, LLC. InstantValue and USuggest.com provide the methods and incentives for enabling millions of Internet shoppers to shop, share their deals with each other, and earn significant rewards.

Online shoppers spend inordinate time and effort to find items that match their needs. Most commonly, an online shopper will attempt a range of keyword searches and follow various links—often to destinations that offer no opportunities to compare deals with others on the net. In addition, the shopper is often required to complete online forms at different shopping web sites. USuggest.com is the only deal-suggestion portal that combines ease of deal listing with deal ranking, keyword tagging, and suggester rewards. The USuggest.com process is illustrated in the graphic below:

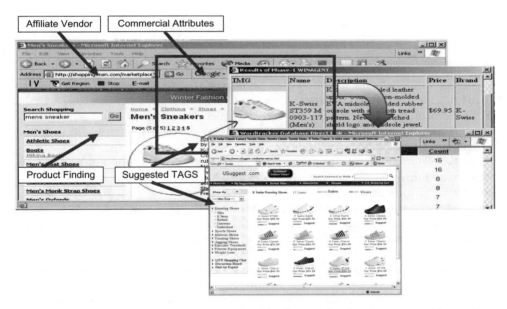

InstantValue featuring the InstantValue Toolbar is the first Internet-wide application that allows shoppers to annotate their findings with all relevant search-hot tags, thereby significantly boosting the ease of matching for their deal suggestions. The figure above demonstrates a "men's sneaker" suggestion tagged with "men's athletic shoes" and "running shoes" etc.

Shoppers can also tag their suggestions with existing or new "spin off" group names such as "runners club". Such groups evolve into "hubs of influence" providing relevant contexts for rapid diffusion and sales conversion. USuggest, LLC has filed provisional patents on buying-through-tags and cash reward mechanisms for suggestions.

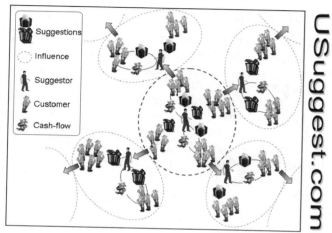

InstantValue Commerce Model: Rapid Propagation of Boosted Suggestions to Relevant Web Shoppers.

InstantValue will be distributed to shoppers at no cost from USuggest.com's high-traffic affiliate shopping sites. Revenue is derived from a performance based 8 to 25 percent commission for each sale. In order to encourage shoppers to adopt InstantValue, USuggest.com shares its commission on each sale with the shoppers providing an immediate rewards incentive.

InstantValue can rapidly populate large and effective social shopping networks to capture and turn numerous stray shoppers into qualified leads and sales for its affiliate network, thus generating large sales revenue for its vendors and commissions for suggesters, shoppers, and USuggest, LLC. Over the past four years, the USuggest management team has mastered the affiliate marketing business and has extensive experience in mobilizing a wide network of major vendors in the on-line consumer market.

Following on-line consent, InstantValue is deployed seamlessly from USuggest.com affiliate marketing portals into a shopper's browser. It provides cash rewards of up to 25% on sales at all affiliated vendors whenever a suggestion triggers a sale. Currently USuggest.com's vendor list includes most of the major online retailers through affiliate networking sites such as Commission Junction™ and LinkShare™.

InstantValue provides unique value-added shopping and online collaboration experience through the following features:

- **Interest-Based Group Messaging:** InstantValue facilitates the creation of instantaneous interest based shopping groups through profile-based matching and messaging. Through real-time messaging, shoppers can exchange critiques and recommendations.

- **Co-Browsing Connections:** InstantValue helps to connect friends and acquaintances with similar needs and enables them to shop together using its co-browsing technology.

- **Product Reviews:** InstantValue facilitates up-to-date product annotations by shoppers, so that they can share their reviews about any online product with all other InstantValue shoppers.

- **Product Suggestions:** InstantValue offers shoppers the convenience of connecting with product suggestions from other shoppers to help locate hard-to-find bargain products and to obtain expert advice on not-so-simple product decisions, such as configuring and buying a laptop or buying supplies for a home improvement project.

- **Rewarding the Suggestion:** InstantValue allows any shopper the option to register as a suggestor and collect cash commission from sales of their suggested products. Suggestors can login and view detailed performance and cash commission reports for their suggestions. Both suggestors and their suggestions are ranked by the number of sales that they facilitate hence; a shopper can easily validate the expertise of any suggestor and suggestion instantly.

Essentially, the InstantValue suggester network acts like a collection of loosely federated salesmen for the InstantValue affiliated vendors, thus contributing to "hubs of influence" online to increase the traffic and conversion at InstantValue affiliated vendors.

COMPANY BACKGROUND

USuggest, LLC will operate in its start-up phase out of existing offices located in Stony Brook, New York and Phoenix, Arizona. The company's official headquarters will be in the Phoenix office located at:

INSERT CONTACT INFO HERE

USuggest has been established as a limited liability company registered in the State of Arizona. There are currently five members of the company, as noted below:

INSERT LLC MEMBERS HERE

Each member has extensive technology company and new venture start up experience. The team is solely responsible for the core technologies that lie behind both the USuggest.com deal-suggestion portal and the InstantValue deal-uploading software. They jointly and working collaboratively invented and developed the core technologies, and they are the chief signatories on each of the patents filed for the core technologies with the U.S. Patent and Trademark Office.

MARKET ANALYSIS

People love shopping in their underwear

This statement, although a bit crude, sums up the potential of online shopping. Retailing has changed with the rapid development of new online sales and distribution tools that can literally be used from anywhere, anytime—from work, school, a hotel, car, or airplane. These developments have impacted retailing as much as the advent of strip malls, catalog retailing, and TV-based home shopping.

With more households shifting to faster broadband connections and online shopping increasingly friendly, Americans are purchasing ever more products over the Internet. Analysts at Forrester Research, Inc., say U.S. online sales will grow from $47.8 billion in 2002 to over $130 billion in 2006. Annual spending per buyer is expected to increase from an average of $457 in 2001 to $784 in 2006.

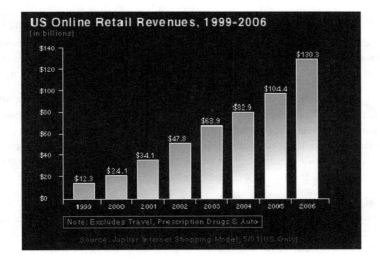

VeriSign, Inc., the leading provider of intelligent infrastructure services for the Internet and telecommunications networks, released e-commerce data showing that the first week of the 2004 prime holiday shopping season accounted for $2.2 billion of online purchases. The two full weeks preceding the holidays are traditionally the top shopping days during the season. In tracking online spending and transactions during this period in 2004, VeriSign research indicates that significantly more consumers turned to online shopping. The total volume of sales transactions recorded during the week of December 6, 2004 reached 28.9 million, a 21 percent increase as compared to the same period in 2003.

Although it is difficult to predict with precision the extent of the market opportunity in online shopping, most analysts agree that the market is "substantial" and growing rapidly. Below are some indicators of the growth of online shopping, and the centrality of the "aggregator" model that USuggest.com will use:

- 59% of online shoppers start at aggregator sites rather than going directly to a merchant's site and taking the first price offered. That represents a 13-percentage-point shift towards aggregator sites in less than three years, when only 46% of online shoppers started at aggregator sites and 54% went straight to the merchant's site.

- 87% of online shoppers are now comparing the offerings of online retailers against catalog merchants and retail stores to find the best deals and items that are in stock. 71% of online shoppers report they were able to find better sales and discount offers online than offline via a retail or catalog merchant.

- New research shows that 70% of US adults use the Internet as an information source when shopping locally for products and services—up from 60% in October 2003. These figures put the Internet on par with newspapers as a local shopping information resource, and suggest that the Internet is on track to surpass newspapers as a consumer influencer in the very near future.

- E-commerce transactions in 2004, 2005, 2006, 2007, 2008, 2009 and 2010 to reach $145 billion, $168 billion, $198 billion, $228 billion, $258 billion, $288 billion and $316 billion, respectively. The estimates include travel, typically the biggest e-commerce sales driver. By 2010, e-commerce will have a 14 percent compounded annual growth rate over six years. Excluding travel, e-commerce is expected to grow 14 percent in 2005 from last year.

- Nearly 100 million adults made purchases after doing online research last year, coming close to the number of adults who purchased through catalogs, direct-mail ads and telemarketing calls combined. 114.1 million adults searched for product information on the web last year, and that 98.9 million of this group went on to make purchases either online or offline.

- 74% of the Internet population over the age of 13 will shop online this year, up a bit from 73% last year and 7 percentage points from 2001. That equals 115.1 million consumers over 13, up from 82.3 million in 2001. In 2007, 77% of the online population over age 13—131.3 million people—will shop online.

- In total, members of Generation Y—people born since 1977—have annual income of $211 billion, a Harris Interactive poll determined. Each year, this group spends about $172 billion of that income. And the amount spent online is growing: This year, people aged 8 to 21 spent 14.5 percent of their income online, versus 11.8 percent last year.

- U.S. e-commerce will grow at a 19 percent compound annual growth rate over the next five years. Most significant, online retail will reach nearly $230 billion and account for 10 percent of total U.S. retail sales by 2008.

- The average male spends more money shopping online per month than the average female — $204 to $186, respectively. The numbers are bigger, as is the margin, when looking at holiday gift shopping. The average male plans to spend more of his holiday shopping money online than the average female, by a 15% margin — $326 to $284, respectively.

The USuggest.com deal-suggestion portal will quickly become the destination of choice for price-conscious online shoppers. Not only will they be able to conduct quick and efficient keyword searches to find the product that suits their needs, they will be able quickly to evaluate which deals are the most popular among their shopping peers. USuggest.com's intuitive interface and unique deal ranking algorithms ensure that online shoppers will be able to find what they need, and trust that the listed deals are real.

The incentives created by the InstantValue software ensures that hot deals will always be listed on USuggest.com and that they will be cross-listed and tagged to maximize their ease of matching. In essence, the deals listed on USuggest.com are there for a simple reason—they are selling. USuggest.com is driven by fundamental market mechanisms and the self-interest of shoppers and deal suggesters. The history of global economics amply bears witness to the power of these forces. USuggest.com merely seeks to harness them to improve the online shopping experience for consumers—no matter what deals they are seeking.

MARKET ENTRY STRATEGY

The USuggest.com portal will be launched in tandem with distribution of the InstantValue software and associated toolbar. The latter will be distributed for free through USuggest.com affiliated vendors and at the USuggest.com website. Currently, there are more than 5,000 vendors and over 2 million products affiliated with USuggest.com. The USuggest.com deal-suggestion portal will initially be marketed through search-related advertisements on leading search engines, such as Google and Yahoo. When a critical mass of members has registered with the USuggest.com portal (more than 5,000 members) the company will undertake a more aggressive marketing campaign.

InstantValue will be distributed over the Web in a similar manner as the Google Toolbar or Microsoft Messenger. It will be promoted as a better, faster, more secure, and more rewarding way to shop online. After initial development, the software has no recurring costs to USuggest. All sales initiated through USuggest.com will earn a commission for USuggest, and a cash-back reward for the customer and the deal suggester.

Based on market forces and the self-interest of shoppers and suggesters, respectively, USuggest.com will rapidly become populated with a myriad of products and vendor offerings. Customers searching for deals on the Internet will be attracted to the USuggest.com portal where they will be offered the opportunity to download InstantValue and participate in the shop, share, earn online community. Revenues will be derived from commissions generated by product sales in the affiliate network. As USuggest.com becomes known worldwide, it will be a 24/7 revenue generator.

As determined by the USuggest executive team, the keys to success in the deal-suggestion portal market include:

Adoption by Shoppers: The Company owns and runs a cluster of shopping portals which have high traffic volume and good search engine placements. InstantValue will be distributed to the shoppers who are visiting these portals.

Partnership with Vendors: InstantValue is designed to satisfy the proper "code of conduct" requirements of most affiliate networks and vendors. In particular, InstantValue (i) does not interfere with or seek to influence improperly the referral of a potential customer or visitor (ii) does not generate non-end user initiated clicks or transactions, (iii) does not alter any publisher's site in any way, and, (iv) does not utilize any stealth software install or uninstall procedures.

Opportunity to Find Items Using Keyword Searches: Most online shopping search engines enable individuals to search for items using keywords. Significantly, however, these keywords are determined by the online shopping service, not by consumers. Thus, in many cases, the keywords associated with goods are not intuitive to the online shopper. For example, shoppers wishing to find products suitable for a romantic dinner will not be able to find products using those keywords at online retail websites. There are no categories for "romantic dinner" gear. However, a deal suggester to USuggest.com can tag his or her found deals with a "romantic dinner" tag, making the item locatable using those exact keywords.

Opportunities to form Hubs of Influence: Many shoppers and people with avid interests in hobbies, recreation, or sports enjoy shopping and sharing ideas with others who share their interests. Most competitor online shopping sites do not allow for the natural formation and dissolution of hubs of influence. USuggest.com not only enables but encourages formation of such groups. Using the romantic picnic example, USuggest consumers may choose to form and participate within a hub of influence that is labelled as "romantics". Individuals within the hubs of influence will be motivated to find deals and share them with each other. The shop, share, earn philosophy of USuggest.com gives everyone a chance to benefit, and to benefit others.

COMPETITIVE ANALYSIS

The amount of money spent online increased by more than 33 percent in 2004 over the previous year—and more and more websites are using cash-back incentives to attract users. Butterflymall.com has joined Ebates.com, Fatwallet.com, and Rebateshare.com to reward shoppers for surfing to their sites to find products they need. Typically merchants pay portals a fee for sending buyers to their sites; the rebate sites pass part of that fee on to the customer. The typical rebate is about 4 percent of purchases. But since individuals can combine rebates with any other coupons or discount codes, they can enjoy real savings. Most rebate sites won't pay-out to customers until an individual earns a minimum ($5 to $25). Butterflymall.com partners with more than 500 retailers like Gap and Barnes & Noble. Fatwallet.com lets members of PayPal cash out at any amount. Ebates.com, with more than 700 partners, is the easiest to navigate. Rebateshare.com also offers coupons. Each of these competitors offers only a few of the features of USuggest.com, without the benefit of its social networks and only from within their own Web sites.

Shopping.com which filed an IPO recently and others such as Pricegrabber.com, Kelkoo.com, Mysimon.com, and Amazon.com can also be regarded as competitors since they provide comparative shopping and user reviews. A recent related acquisition by Buy.com is Yub.com. Yub.com enables a "buy-thru-friends-profile" suggestion mechanism. In the USuggest.com model there are no profiles in order to preserve the privacy of suggesters. Also, the USuggest.com model organizes product and services suggestions by their search-hot tags. Through this approach all similar products are easily findable-thru-tags, boosting the visibility of the products. In Yub.com there is no organization of similar products since various products are grouped only under personal profiles.

It has also been rumored that Google is building its own browser and incorporating search and instant messaging features into it. Friendster, a social networking company from Google, has lost the attention of its earliest adopters since they failed to provide any added value or an incentive mechanism. Another interesting model is letsbuyit.com from Europe, which adopts a co-buying model where the shoppers interested in certain products, are aggregated to obtain lower negotiated prices.

Below is a comparison of the major competitors to USuggest.com on the major features inherent to online shopping and shopping search engines. As the graphic clearly shows, USuggest.com is the ONLY competitor to combine consumer tagging (self organization) with suggester rewards (self interests) and hubs of influence (free markets).

	Web Wide Reach	Buyer Rewards	User Tagging	Ranking System	Privacy Preserving	Suggester Rewards	Group Formation
FatWallet	X	X	0	X	X	0	0
Yub	0	X	0	0	0	X	0
Overstock	0	X	0	0	X	0	0
Amazon	X	0	0	X	X	0	0
Google	X	0	0	0	X	0	0
Ebates	0	X	0	0	X	0	0
Butterflymall	0	0	0	0	X	0	0
Shopping.com	0	X	0	X	X	0	0
Yahoo	X	0	0	0	X	0	0
MyPoints	0	X	0	0	X	0	0
RebateShare.com	0	X	0	0	X	0	0
USuggest.com	X	X	X	X	X	X	X

MANAGEMENT TEAM

The company is to be a wholly owned enterprise managed by seasoned technologists and entrepreneurs with an established partnership record and no financial liabilities.

Management Team and Advisory Committee

INSERT MANAGEMENT TEAM BIOS HERE

STRATEGY

USuggest.com's experience with online affiliate marketing indicates that the revenue of any online retailer is determined by three important parameters: *traffic, conversion and loyalty*. Traffic refers to the number of people visiting a site. Conversion is calculated as the percentage of shoppers who actually buy a product. Loyalty is the percentage of repeat visitors. USuggest.com's experiential evidence indicates that the average conversion rate on the Web is approximately 1% (1 sale for every 100 clicks) and that traffic can be boosted and retained by a series of measures such as online advertising and search engine optimization. InstantValue improves both traffic and conversion and retains customers through its rewards program.

InstantValue will be disseminated for free from USuggest.com's high traffic affiliate shopping portals. Currently, USuggest.com's portals enjoy consistently high rankings at major Web search engines, thousands of online shopper visits per day, and a very rich product catalog of more than 5 million products provided by its affiliate vendors. The revenue model for USuggest.com portals is based on pay-per-performance (or CPO) driven commission. Commission rates vary from 5% to 45% per sale. In order to entice shoppers to download and install InstantValue™, USuggest.com plans to share the commission on each sale with the shopper. An immediate savings benefit, in the form of a cash-back, is the most powerful incentive in online shopping and USuggest.com believes this strategy will lead to widespread adoption of the product.

Once installed, InstantValue begins to work for shoppers, affiliated vendors and USuggest.com:

- A shopper, while shopping outside USuggest.com's network, may find a great deal. With a few clicks the shopper can annotate and post that deal at USuggest.com's shopping portal so it can be seen by all relevant InstantValue users and USuggest.comcustomers.

- Another shopper, while shopping outside USuggest.com's network, may want to check the reviews of an item and suggestions of similar products from the InstantValue network. If the shopper buys a suggested item, both the buyer and the owner of the suggestion collect cash-back commission from that sale.

- Another stray shopper may want to connect with other shoppers who are searching for a product. InstantValue™ will instantly send suggestions from its network and invitations to online shoppers in its network and facilitating them to band and shop together while they exchange their findings and opinions. If they make a decision to co-browse together, they would be likely to pull the stray shopper into the USuggest.com network.

- A shopper may need online expert suggestions during shopping for a complex or hard-to-find product. Upon connecting with an online InstantValue™

suggestor who has great relevant suggestions, by the click of a button, a suggestor can band together with the shopper and guide her to the right buying decision within the network. In such a guided scenario, where the shopper uses the assistance of an expert to buy a product, the suggestor keeps part of the commission as the reward.

OPERATIONS

USuggest intends to operate using a hybrid organizational structure during its start-up phase. That is, the company will establish its headquarters in Phoenix, AZ, and utilize offshore labor to develop its software.

USuggest is operating its existing shopping portals from its operations office in Stony Brook, New York running the shopping portals as a sole proprietorship. This unit will continue to operate the shopping portals during and through the start-up phase. Once the InstantValue software has been downloaded by a critical mass of users, these ancillary shopping portals will be discontinued to allow the entire executive team to focus on building the value of USuggest, LLC.

In addition to its highly experienced and qualified executive team, USuggest will establish a Board of Advisors comprised of business management and technology experts. It will hire a legal counsel to assist with patent preparation and submission as well as other intellectual property and trademark issues. By Q3 2005, the company will release the alpha version and perform pilot runs. By Q4 2005 it plans to release a market ready and tested product with projected monthly revenue in excess of $20K. By Q1 2006 USuggest.com plans to raise $3 M VC funding in order to aggressively deliver its product to the market.

INTELLECTUAL PROPERTY

The company has filed two provisional patents to protect its intellectual property. Consequently it plans to file two patents on the technologies, followed by a series of trademarks and intellectual property security measures. The provisional patents detail and secure the following technological innovations:

- A Reward-driven Suggestion-Portal™ Creation and Management Method for Online Products and Services

- A Privacy-Preserving Method for Dynamic Grouping of Web Users into Shared Interest Communities based on Matching Web Location and Web Page Content.

FINANCIAL PROJECTIONS

USuggest is anticipating rapid growth in the dynamic, global, and fast-moving online shopping marketplace. While business models from the days of the dot-com bubble often were based on phantom metrics such as "clicks per thousand" or "unique users", USuggest already is generating significant revenues through its range of search-engine linked online shopping portals.

The five-year earnings projections detailed in the following pages are based on empirically derived numbers based on more than two years of experience operating prototype online shopping websites. The daily visitors, click-through rate to affiliate web sites, and conversion rates (i.e., conversion to an actual sale and commission) are empirically derived and linearly projected.

Detailed five year financial projections are presented in the attached Excel document. Below is a summary of the data, based on the most conservative assumptions.

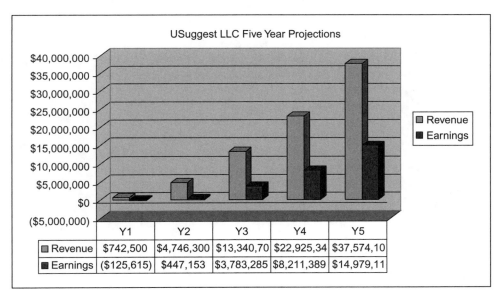

USuggest LLC Five Year Projections

	Y1	Y2	Y3	Y4	Y5
Revenue	$742,500	$4,746,300	$13,340,70	$22,925,34	$37,574,10
Earnings	($125,615)	$447,153	$3,783,285	$8,211,389	$14,979,11

Index

A

AAA. *See* American Accounting Association (AAA)

Absolute advantage, 46

Abstraction-filtration-comparison process, 225

Acceptance, 252
  revocation of, 253

Accounting, 389
  cycle, 390, 391f
  defined, 389–393
  equation, 392

Accounts receivable, 404, 410
  aging of, 413
  average net computation, 408
  credit and collection, 411

Accounts receivable turnover, 408

Accredited investors, 148

Acid-test ratio. *See* Quick ratio

Acquisition, 81, 173–178
  bootstrap purchase, 178
  deal, 177–178
  direct purchase, 178
  due diligence process, 174–177
  financial analysis, 175
  management and key personnel, 177
  market analysis, 176
  operations, 177
  product/service line, 175
  research and development, 176–177
  synergy, 176
  warning signs in, 176

Activity ratio, 408

Administrative law, 473–474

Adventiq, 64–66

Advertising, 10
  print, 370–371
  in radio, 370
  in television, 369–370
  through Internet, 371–372
  of video game, Massive, Inc., 383–385

Affordable Loss Principle, 438

Aftermarket, 187

Agreements, 249
  confidentiality. *See* Confidentiality agreements
  consulting, 268–269
  development, 268–269
  employment, 266

  franchise, 272
  license. *See* License agreements
  maintenance and support, 269
  manufacturing, 269–270
  noncompete, 266, 267f
  nonuse, 209
  OEM, 273
  purchase, 273
  technical services, 272
  VAR, 273

Agriculture, 63

AIM. *See* Alternative Investment Market (AIM)

AIPA. *See* American Inventors Protection Act of 1999 (AIPA)

AITC. *See* Alabama International Trade Center (AITC)

A.L. Wilson Company, 71

Alabama International Trade Center (AITC), 30

Allen, Paul, 70–71

AllPoints Research, Inc., 422

Alternative Investment Market (AIM), 80–81

Amazon.com, 491

Amended contract, 255

American Accounting Association (AAA), 389

American Inventors Protection Act of 1999 (AIPA), 215

America Online (AOL), 493–494

Analytical skills, 428–429
  affecting venture performance, 428
  in multiple complex variables, 429
  venture's key technology, 429

Andreessen, Marc, 502

Angel Capital Association (ACA), 143

Angel financing, 143

Angel investors, 143–144, 156

AOL. *See* America Online (AOL)

Apple Computer, 7–8, 317, 425
  logo, 234f
  networking opens up opportunities, 317–318

Arbitrary determination method, 375

Arizona Technology Investor Forum (ATIF), 140, 143

ARPA. *See* U.S. government's Advanced Research Projects Agency (ARPA)

ARPANET system, 504–505

Articles of Incorporation, 108

545

Articles of organization, 116
Assets, 392
Assignment agreements, in contracts, 270
Authorized shares, 120–121
Aycock, Sherrie, 422

B

Balance of payments, 47
   United States, 48f
Balance of trade, 47
Balance sheet, 402–405
   review equity position of company, 331
   templates of, 402–405
Bank. See Institutional lender
Baran, Paul, 501
Bargain, in good faith, 297
Bargaining gambits, 300–301
Barriers to entry, 64
   capital intensity as, 64
   in competitive analysis, 324
Bartering, 84
Basche, Todd, 425
Basic 7(a) loan guaranty, 155
BATNA. See Best alternative to a negotiated
         agreement (BATNA)
B2B. See Business-to-business (B2B)
B2C. See Business-to-consumer (B2C)
83(b) election, 125
Belle Baby Carriers, quality standards for, 341
Bellwether customers, 298
Best alternative to a negotiated agreement
         (BATNA), 291
Best-cost provider strategy, 447
Bezos, Jeff, 491
BFOQ. See Bonafide occupational qualifications
         (BFOQ)
Bilinkis, Santiago, 174
Billboards, 372
Bird-in-Hand Principle, 438
Biznik, 334
Blank check company, 148
Blogs, 5
Board of directors, 108
Bombay Stock Exchange, 80–81
Bonafide occupational qualifications (BFOQ), 470
Bonneville Power Administration, 291
Bootstrap financing, 161–162. See also Debt
         financing, Equity financing
Bootstrap purchase, 178
BPO. See Business process outsourcing (BPO)
Brands, establishing, 20–21
Breach, 248, 254–258

Breakeven analysis, 405–406
Break the bank, 158
Broad differentiation strategy, 447
Buhr, Michael, 171–172
Business. See also specific entries
   as an organized activity, 7
   as a continuous affair, 8
   defined, 6–7
   fundamentals, learning, 6
   as a purposeful activity, 7–8
   resources, 9–10
Business model, 8, 78
Business plan
   basic outline for, 321f
   company information, 322–323
   competitive analysis, 324
   FedEX, 323–324
   financial projections, 330–331
   industry analysis, 326–328
   for launching technology venture, 319
   management team, 328–329
   market analysis, 324–326
   marketing plan, 329–330
   product/service description, 323
   tips for writing executive summary, 322
   writing, 320–322
Business processes
   repeatable, 15–16
   scalable, 16–17
Business process outsourcing (BPO), 21–22, 61
Business-to-business (B2B), 78
Business-to-consumer (B2C), 78

C

Call loan, 153–154
Camp, Garrett, 171–172
Cancellation provision, 509
Cannon, negotiation fundamentals, case study,
         310–312
Capital and deal structuring, 139
   angel investors, 143–144
   bootstrap financing, 161–162
   case study, 168–169
   debt financing. See Debt financing
   defined, 141
   elevator pitch, 159
   equity financing. See Equity financing
   fund-raising, tools and techniques, 156–159
   loan rates, payment methods, and lender
         types, 152–156
   private placement memorandum, 158
   role, in technology ventures, 141

SBIR program, 159–160
small business administration loans, 155–156
sources, 142–146
STTR program, 160–161
subscription agreement, 158–159
venture capital, 144–146
web resources, 167–168
Capital assets, purchase of, 414
Capital budgeting, 415
Capital intensity, 63–66
as barrier to entry, 64
Capital-intensive, 63
Capital investment, United States, 64f
Capitalism, 12, 42. *See also* Mixed economy
Capital requirements, across businesses, 64, 65f
Capital resources, 335
defined, 31
Cash breakeven point
cash flow statement, 405
Cash flow statement, 402
cash breakeven point, 405
templates for, 403f
C-corporation, 110–111
characteristics of, 114f
profits and losses, 119–120
shareholders and, 119
taxation, 119
CEO. *See* Chief executive officer (CEO)
Certificate of limited partnership, 105
Certified public accountant (CPA), 392
CFO. *See* Chief financial officer (CFO)
Check-the-Box Regulations, 116
Chief executive officer (CEO), 442
survey of, 68, 69f
Chief financial officer (CFO), 393–394
CIP. *See* Continuation-in-part (CIP) application
Cisco, 78
CISG. *See* Contracts for the International Sale of Goods (CISG)
Civil Rights Act, 469
of 1964, 460
Clauses, restrictiva, 127–128
Clemmons, Philip, 29
Closing gambits, 302
CNBC, 5
Coauthors, 227–228
Coca-Cola, Inc., 83, 213
employees try to sell secrets to PepsiCo, 213
Cognitive biases, 479–483
belief in law of small numbers, 481
illusion of control, 481

overconfidence, 481
planning fallacy, 481
reasoning by analogy, 481–482
remedy for entrepreneurs, 482
COGS. *See* Cost of goods sold (COGS)
Cohler, Matt, 144–145
Collateral, 152
Collection policy, for collecting from customers, 411
Commercial impracticability, 256
Commodity, 363
Common law, defined, 473
Common law of state, in contracts, 249
Common stock, 121–122
voting rights related to, 124
Communication skills, 431
Communism, 12, 41–42
Company information, 322–323
Compaq computer, 178–179
Comparative advantage, 46
Compensatory damages, 256–257
Competition, 18
economy systems and, 43–44
Competitive analysis
barriers to entry, 324
for firm's top competitors, 324
Competitive benchmarking, for obtaining information, 342
Competitive parity method, 375
Competitive strategy model, 445, 445f
Complete performance, 254
Complex project contracts, for negotiation, 305
Conception, 209
Conceptual skills, 432
Confidential information, 213
Confidentiality agreements
in contracts, 266
example of, 214f
in record keeping, 209
in trade secrets, 213
Consideration, 253
Consortia, 205
Consulting agreements, in contracts, 268–269
Consumer Product Safety Commission, 473–474
Continental Lite, 446
Continuation application, in patent, 220
Continuation-in-part (CIP) application
in patent, 221

Contracts
 boilerplate provisions, 264–265
 breach, 254–258
 CISG and UCC relationship, 250f
 consideration, 261–262
 defined, 249
 enforcement against defences, 248–250
 formation of, 250–253
 indemnity, 263
 for intellectual property, 261
 law, 249–250
 license agreements, 270–271
 operating agreements, 265–270
 overview, 248–250
 ownership, 261
 performance matrix, 254–258, 260–261
 preamble, 259
 recitals, 249
 remedies for unexcused breach of, 255f
 representations of, 262
 term of, 264
 warranties, 262
 writing an effective written, 258
Contracts for the International Sale of Goods
  (CISG), 249
Convertible preferred stock, 123
Cooperative negotiation, 294–295
Copyrights, 224–228
 abstraction-filtration-comparison process, 225
 of authorship, 224
 example of, 224
 notice of, 226
 ownership of, 227–228
 registration for, 226–227
 with respect to software, 225–226
 term of, 226
Corporate finance, 394
Corporate governance
 mission statement, 465
 Sarbanes-Oxley regulations, 464
Corporate veil, 99–100
Corporation
 defined, 108
 equity interests in, 108
 maintaining status, 111–115
 structure of, 108
 tax laws, 110–111
 types of, 110–111
Corrective action, defined, 338–339
Corroboration, 210
Cost leadership strategy, 446
Cost of goods sold (COGS), 38

Cost-plus-pricing, 34, 362
Cost/price leadership, 376
Counteroffer, 251–252
 battle of forms, 251
 defined, 251
 gap fillers in, 252
 mirror image rule, 251
Countertrading, to enter international market, 84
CPA. *See* Certified public accountant (CPA)
Crazy Quilt Principle, 438
Creativity in business, 18–21
Credit and collections policies, for maximize
  revenue, 411
Credit regime, suggestions for, 411–412
CRM systems. *See* Customer relationship
  management (CRM) systems
3Cs
 competition, 363
 consumer, 363
 cost, 361–362
 for price, 361
Currency swaps, 52
Current ratio, 407
Customer facing jobs, 369
Customer relationship management (CRM)
  systems, 74

D

Damages, 256–257
 compensatory, 256–257
 incidental, 256
 reliance, 257
Davis, Mitch, 383
Days receivable, 412–413
Debt financing, 120, 151–152
 alternatives to, 159–162
 defined, 146–152
 institutional lender requirements, 152
Debt ratios, 408
Debt-to-equity ratio, 409
Decision-making skills, 430
 crafting an effective elevator pitch, 432
 pitfalls to avoid, 430–431
 in start-up ventures, 430
Deficiency letter, 189–191
Deflation, defined, 38–39
Dell, Michael, 436–437
Dell Computers, 12, 436–437
 governing principles, 466f
Demand
 defined, 33
 law of, 35f

Demographics, 325
  in target market, 356-357
Descriptive mark, in trademark, 233
Design patent, 215
Development agreements
  in contracts, 268-269
  in record keeping, 208
Differentiation strategy, for higher-priced product,
    445-446
Diffusion rate, of different technologies, in United
    States, 67, 68f
Dilution, 65-66, 150
Directed discovery, 443-444
Direct mail, effective in target market, 372
Direct ownership, 84-85
Direct purchase, 178
Direct salespeople, 368-369
Discounted cash flow, 180-181
Discount rate, 181-183
Disk operating system (DOS), 70-71
Disney, Walt, 83
Disney World, 341
Dispute Settlement Board (DSB), 50
Distinctive competence, 445
Distribution agreement, 272
Distributive bargaining, 295-297
  vs. integrative, 294-297
D&O liability, 463
DOS. *See* Disk operating system (DOS)
Double taxation, 111, 119-120
Drug companies, 18
DSB. *See* Dispute Settlement Board (DSB)
Due diligence process, 143, 174-177. *See also*
    Exit strategies, for technology ventures
  acquisition deal, 177-178
  financial analysis, 175
  management and key personnel, 177
  market analysis, 176
  operations, 177
  product/service line, 175
  research and development, 176-177
  synergy, 176
  warning signs in, 176
Dumping stratergy, in compromised market, 39
Duty of care, 466
Dynamic capabilities, 77

E

Early adopters, 71
  W-Systems Corporation and, 71
Earnings before interest, taxes, depreciation, and
    amortization (EBITDA), 330

Earn-out strategy, 179
EBay, 171-172
Eberhardt, Todd, 460
EBITDA. *See* Earnings before interest, taxes,
    depreciation, and amortization
    (EBITDA)
Econnect.entrepreneur, 334
Economic freedom, 13-13
Economic resources
  allocation for, 31-33
  capital resources, 31
  goods and services, 31
  knowledge resources, 31
  labor resources, 31
  natural resources, 30-31
  product distribution, 33
Economics
  COGS, 38
  cost-plus pricing, 34
  defined, 30
  deflation, 38-39
  demand, 33-40
  dumping stratergy, 39
  elastic market, 35
  Federal Reserve Board, 40
  GDP, 38
  gross margin, 37-38
  inelastic market, 35
  inflation, 38
  interest rate, 40
  price gouging, 39
  price sensitive, 36-37
  price wars, 36
  pricing, 34-36
  pricing power, 38
  resources for, 30-33
  supply, 33-40
Economic system, 41-46
  communism, 41-42
  comparing, 43f
  competition and, 43-44
  mixed economy, 42-44
  socialism, 41
  United States, 44-46
  world top 10 largest companies in 2008, 44f
Economies of scale
  in production capabilities, worldwide market,
    355
Economy, global. *See also* Globalization
  acquisitions in, 81
  value capture in, 78-81
  value creation in, 72-78

EEOC. *See* Equal Employment Opportunity Commission (EEOC)
Effectuation, 437
  logic, 438, 438f
Efficiency, economic, 73–74
Egonomics, 441
  warning signs, 441
Elastic market, 35
E-learning company, 139
Elevator pitch, 159, 431
Elk River Safety Belt Inc., 29
Emotional intelligence, 498
Employee stock option (ESO), 127–128
  restrictive clauses, 127–128
  tax issues, 128
  vesting, 127
Employer liability, 463
Employment agreements
  in contracts, 266
Enforceable contract, 253, 254
Engineering creative solutions, 18–19, 21
Engineers, challenges faced by, 18–19
Enterprise risk management (ERM), 461–462
Enterprise value, 428
  capturing, 80–81
  creation of, 73–75
  protecting, 77–78
Entrepreneurial career, 493–496
  Andreessen, Marc, 502
  Baran, Paul, 501
  conceptual model of, 494–495
  launching your, 512–514
  mentor, 496
  Metcalfe, Robert, 504–505
  Newmark, Craig, 503–504
  Omidyar, Pierre, 502–503
Entrepreneurial ethics, 439–441
Entrepreneurial expertise, 437–439
Entrepreneurial leadership, 434–437
  defined, 435
  vs. management, 436–437
Entrepreneurial personality
  growth-oriented entrepreneurs, 497–498
Entrepreneurial strategy, 442–445
  competitive strategy model, 445, 445f
  cost leadership strategy, 446
  differentiation strategy, 445–446
  niche strategy, 446
  real options approach, 443
  resourcebased theory, 444
Entrepreneur managers, 427–428

Equal Employment Opportunity Commission (EEOC), 470
Equity, 120–125
  distribution, 125–128
  interests in corporation, 108
  and stocks, 120–121
Equity financing, 120, 146–152
  alternatives to, 159–162
  costs of, 150–151
  growth strategy for, 151
  institutional lender requirements, 152
  SEC rules, 146–152
Equity interest, 97
ERM. *See* Enterprise risk management (ERM)
ESO. *See* Employee stock option (ESO)
Exchange rates, currency, 52
Exclusive handling, 509
Exclusive right, patent, 217–218
Execution, in startup ventures, 427
Executive summary, tips for writing, 322
Exercise period, 127
Exit event, 145
Exit strategies, for technology ventures, 171
  acquisition, 173–178
    bootstrap purchase, 178
    deal, 177–178
    direct purchase, 178
    due diligence process, 174–177
    value drivers. *See* Value drivers
  case study, 196–198
  going public, 183–191
    advantages of, 183–184
    disadvantages of, 184–185
    investment bank, selecting, 187
    prospectus, 188–189
    red herring, 189–191
    registration statement and timetable, 188
    reporting requirements, 191
    timing, 186
    underwriter selection, 186–188
  mergers, 178–179
  purpose of, 172–173
  venture valuation, 180–183
    discounted cash flow, 180–181
    multiples technique, 180
  web resources, 195–196
Expenses, 392
Export
  to enter international market, 82
  in international trade, 47
Export Assistance Centers, 82
Export Trading Company Act 1982, 84
Express warranties, 474–475

defined, 263
indemnity, 263

F

Facebook.com, 144–145, 180, 247
  tech venture of, 3–4
Failure risk, management, 478–483
Fanciful and arbitrary marks, 233–234
  Apple Computer logo, 234f
FDA. *See* Food and Drug Administration (FDA)
Federal Communications Commission, 473–474
Federal Reserve Board, 40
Federal Trade Commission, 473–474
FedEX
  business plan, C grade in school, 323–324
"FF & F," friends, family, and fools, 142–143
FICA. *See* Social Security and Medicare Tax (FICA)
Finance, defined, 393
Financial accounting, 389
Financial analysis, 175
Financial management
  accounts receivable, 410
  credit and collections policies, 411
  inventory management, 414
  payment of debt, 414–415
  payment of dividends, 415
  purchase of capital assets, 414
  start-up costs, 409
  working capital, 409–410
Financial manager, 393–394
Financial plan, 394–395
Financial projections, 330–331
  for income statements, 330
  for sales forecast, 330
Financial statements, 395–405
  analysis for, 405–409
  breakeven analysis, 405
  company's financial position, 391
  ratio analysis, 406
  sales forecast, 396–399
Fixed costs
  defined, 400
Flat fees, 476–477
Focused low-cost strategy, 447–448
Focused strategy based on differentiation, 448
Food and Drug Administration (FDA), 80,
  473–474
Force majeure clause, 256
Ford Motor Company, 7, 83–84
Formalities, procedural, failure to observe, 100
Form D, 149
Form 8-K report, 191
Form 10-Q, 191

Form S-1, 189, 191
Form S-18, 189
Founders, individual entities, 97
Founder's stock, 124–125
Four P's, for marketing plan, 353
Franchise agreements, 272
Franchise fee, 507
Franchising
  agreement for, 507–509
  brand name, 509–510
  costs of, 510
  defined, 505
  disclosure of, 511–512
  external control, 510
  financial assistance, 510
  guidance, 509
  history of, 506
  overdependence, 510
  poor reputation, 511
  proven product, 510
  ten-day rule, 507
Frankfurt Stock Exchange, 80–81
Fraud, 100
Free Trade Agreement (FTA), 51
Free trade areas, 51–52
Freire, Andrés, 174
Friedman, Thomas, 61
FTA. *See* Free Trade Agreement (FTA)
Fund-raising, tools and techniques,
  156–159
  elevator pitch, 159
  private placement memorandum, 158
  subscription agreement, 158–159
Future value (FV), defined, 182

G

G&A. *See* General and Administrative (G & A)
  expenses
GAAP. *See* Generally accepted accounting
  principles (GAAP)
Gambits, negotiating
  bargaining gambits, 300–301
  closing gambits, 302
  opening gambits, 299–300
Gap fillers, 252
Gates, Bill, 70–71
GATT. *See* General Agreement on Tariffs and Trade
  (GATT)
GDP. *See* Gross domestic product (GDP)
GE. *See* General Electric (GE)
Genentech mission statement, 465f
General Agreement on Tariffs and Trade
  (GATT), 50

General and Administrative (G & A) expenses, 400
General Electric (GE), 61, 499
General ledger, 390–391
General liability insurance, 463
Generally accepted accounting principles (GAAP), 389
General Motors (GM), 84
General partnerships, 98–99, 103–105
    advantages and disadvantages of, 106f
Generic mark
    Kleenex logo, United states, 233f
    in trademark, 232–233
Giesellschaft mit beschrankter Haftung (GmbH), 116
Gladwell, Malcolm, 71–72
Globalization, technology ventures and, 61–72
Global trade. *See* International trade
GM. *See* General Motors (GM)
GmbH. *See* Giesellschaft mit beschrankter Haftung (GmbH)
GoBigNetwork, 334
GO Corporation, 72–73
Going public, 183–191
    advantages of, 183–184
    defined, 172
    disadvantages of, 184–185
    investment bank, selecting, 187
    prospectus, 188–189
    red herring, 189–191
    registration statement and timetable, 188
    reporting requirements, 191
    timing, 186
    underwriter selection, 186–188
Goldman Sachs, 186–187
Goods and services, 31
    circular flow of, 32f
    distribution of product, 33
Google, 75, 78, 179
    acquisition volume of, 81, 82f
    acquisition with Omnisio, 196–198
Gordon, Zorik, 387
The Grameen Bank Project, 20
Graphical user interface (GUI), 7–8
Gross domestic product (GDP), 38
Gross margin, defined, 37–38
Growth-oriented entrepreneurs, 497–498
    high energy level, 497
    high need for achievement, 497
    low need to conform, 497
    persistence, 497
    risk-taking tendency, 498
Growth strategy, for equity financing, 151

H

Hazard risks, 462–464
Hewlett-Packard, 178–179, 499
Higher-priced goods, 9
Hold-harmless, 263
Homebrew Computer Club, 317
Human capital. *See* Human resources
Human resources, 335–336
    management, 469
    system, 75
Humility, 440–441
    illustration of, 442f

I

IBM, 21, 68
IBRD. *See* International Bank for Reconstruction and Development (IBRD)
ICSID. *See* International Center for Settlement of Investment Disputes (ICSID)
IDA. *See* International Development Association (IDA)
IFC. *See* International Finance Corporation (IFC)
IMF. *See* International monetary fund (IMF)
IMP. *See* Interface Message Processors (IMP)
Implied warranty, 474–475
Import, in international trade, 47
Inadequate performance, 255
Incentive stock option (ISO), 128
Incidental damages, 256
Income statement, 399–400
    profit breakeven point, 405, 406f
    templates for, 401f
Indemnity, contracts, 263
2008 Index of Economic Freedom, 13f *See also* Economic freedom
Individual entrepreneur, 493–494
    Ireland-predictors of, 495f
    USA-predictors of, 495f
Industry analysis, 326–328
    life cycle for, 327f
    in terms of market size, 317
Inelastic market, 35
Inflation
    defined, 38
    GDP, 38
Influence
    of leadership behavior, 435
    seven strategies for entrepreneurial leaders, 436
Information disclosure statement, patent, 220
Information technology systems, 74

Initial public offering (IPO), 80, 111. *See also* Going public
Institutional lender, 151. *See also* Loan requirements, 152
Integrative bargaining, 294–295
  vs. distributive, 294–297
Intel, 21
Intellectual property (IP), 11, 18, 75
  Microsoft attempts, in China, 242–243
  overview, 204
  protection for, 205–211, 206f
  recognition of, 207
  record keeping, 207–210
  technology ventures, 205
  tips for protecting, 76
Interactive Alchemy, 139–140
Interest rate, defined, 40
Interface Message Processors (IMP), 504–505
Interference proceeding, 209
Internal Revenue Code, 128
Internal Revenue Service (IRS), 102
International Bank for Reconstruction and Development (IBRD), 49
International Center for Settlement of Investment Disputes (ICSID), 49
International Development Association (IDA), 49
International Finance Corporation (IFC), 49
International market, strategies to enter, 81–85
International monetary fund (IMF), 48–49
International monetary system, 47–49
  IMF, 48–49
  World Bank, 49
International Organization for Standards, 74–75
International trade, 46–47
  balance of payments, 47
  balance of trade, 47
  elements of, 49–52
  export and import, 47
Internet World Stats, 18
Intrapreneurs, 499–501
Invention
  disclosure forms, 210
  making of, 209
  reducing practice, 209
Invention assignment agreement for employees, 222
Inventory management, 414
Investment horizon, 172–173
Investor relations, 468
Investors, 97
IP. *See* Intellectual property (IP)
IPO. *See* Initial public offering (IPO)

IRS. *See* Internal Revenue Service (IRS)
ISO. *See* Incentive stock option (ISO)
Issued shares, 120–121

J

JetBlue airlines, 433
Joint ventures, to enter international market, 83

K

KaZaA, 493
Key ratios, 406–407
Kleenex logo, United states, 233f
Know-how licenses, 271
Knowledge-intensive, 66–67
Knowledge resources
  defined, 31
Knowledge workers
  loyalty of, 66–67
10-K reports, 191
  defined, 148

L

Labor-intensive, 63
Labor resources, 31
LaFrance, Justin, 171
LAN. *See* Local area networks (LAN)
Land-intensive, 63
Law
  of bankruptcy, 475
  of negotiable instruments, 475–476
  of sales, 474
  of torts, 474
Laws affecting start-up ventures, 474–476
  law of bankruptcy, 475
  law of negotiable instruments, 475–476
  law of sales, 474
  law of torts, 474
L.C.L. *See* Less than carload (L.C.L.)
Leadership. *See* Entrepreneurial leadership
Legal risk management, 471–474
Legal structure, choice of, 101–117
Lemonade Principle, 439
Lenders, 97
Less than carload (L.C.L.), 366
Less than truckload (L.T.L.), 366
Liabilities, 392
Liability
  limited *versus* unlimited, 98–99
  ownership and, 97–101
Liability insurance, 463

License agreements, 270–271
    distribution agreement, 272
    franchise agreements, 272
    know-how, 271
    OEM, 273
    for patent, 271
    purchase agreements, 273
    in record keeping, 208
    technical services agreements, 272
    for trademark, 271–272
    VAR, 273
Licensing, to enter international market,
        82–83
Licensure rights
    INTEL and NVIDIA battle over, 279–282
Limited liability
    extent of, 99–101
    versus unlimited liability, 98–99
Limited liability companies (LLC), 98, 116–117,
        118–120
    advantages and disadvantages of, 117f
    characteristics of, 116–117
    LP and, 120
    single-member, 99–100
Limited liability entities, 98, 118–120
    defective creation of, 100
    expenses, 119
Limited partnerships (LP), 98, 105–108
    advantages and disadvantages of, 109f
    certificate of, 105
    LLC and, 120
Line of credit, 154–155
LinkedIn, 334
Liquidity event, 172–173
Liquidity ratio, 407
    current ratio, 407
    quick ratio, 407
LLC. *See* Limited liability companies
        (LLC)
Loan
    call, 153–154
    from family member, tips for, 152
    lender types, 152–156
    line of credit, 154–155
    payment methods, 152–156
    rates, 152–156
    restrictive covenants and, 153–154
    revolving, 154–155
    SBA program for, 155–156
        basic 7(a) loan guaranty, 155
        loan prequalification, 155
        microloan 7(m) loan program, 155
        for war veteran, 156

short-term, 153
Loan prequalification, 155
Local area networks (LAN), 505
Logistics systems, 74
London Stock Exchange, 80–81
Lose-lose negotiating, 297–298
Lose-win negotiating, 297
Low-cost provider strategy, 446–447
LP. *See* Limited partnerships (LP)
L.T.L. *See* Less than truckload (L.T.L.)

M

3M, encourage intrapreneurship, 499
Macroeconomics, 30
Madoff, Bernard, 464
Mahindra Holidays and Resorts, red herring for,
        189–191
Mailbox rule, 251
Maintenance and support agreements, 269
Management and key personnel, due diligence
        process and, 177
Management team, for new technology venture,
        328–329
Managerial accounting, 389
Manufacturing agreements
    in contracts, 269–270
Market analysis, 324–326
    acquisition and, 176
    demographics, 325
    psychographics, 326
    sociographics, 326
Market cap, 73–74
Market capitalization, 73–74
Market equilibrium, 34–36
Marketing *See also* specific entries
    concept of, 355
    defined, 353
    four P's, 353
    overview, 352–353
    psychological profile of, 357
    segmentation of, 355–357
Marketing budget, 374–376
    arbitrary determination method, 375
    characteristics of, 374
    competitive parity method, 375
    objective and task method, 375
    percent-of-sales method, 375
    quantitative method, 376
Marketing collateral, 372
Marketing message, in target market, 376
Marketing mix, 353
    billboards, 372
    business location in, 372–373

customer facing jobs, 369
defined, 368
direct mail, 372
direct salespeople, 368-369
elements of, 354f
Internet, 371-372
marketing collateral, 372
print advertising, 370-371
radio advertising, 370
signage, 373
television advertising, 369-370
trade shows, 373
Marketing plan, 329-330, 373-374
in advertising, 330
for introduce its products, 329
Market share, 76
Market value
capturing, 78-80
creation of, 72-73
protecting, 76-77
Mask works, 228
Massive, Inc., 383
with in-game advertising network,
383-385
Material breach, 255
McDonald's, 79
McKinsey global survey, of executives, 67
Mentor, 496
Merchant account, 410-411
Merck, 362, 499
Mergers, 178-179. *See also* Exit strategies, for
technology ventures
Metcalfe, Robert, 68-70, 504-505
Metcalfe's law, 68-70, 70f
Microeconomics, defined, 30
Microloan 7(m) loan program, 155
Microsoft, 70-71, 79-80, 180
acquisitions and, 174
attempts Intellectual property protection,
China, 242-243
MIGA. *See* Multilateral Investment Guarantee
Agency (MIGA)
Min/max concept, 158
Mirror image rule, 251
Mixed economy, 41, 42-44
Model Employment Termination Act, 471
Money
flow of, in United States economy, 45f
MoneyTree™, report, 64f
Moore, Geoffrey, 71
Moore, Gordon, 68
Moore's Law, 68, 69f
Moscow City Council's Food Service, 83

Multilateral Investment Guarantee Agency (MIGA),
49
Mundie, Craig, 79-80
Murray, Seth, 341
Mutual assent, 250-251
Mutual rescission, 255

N

NAFTA. *See* North American Free Trade Agreement
(NAFTA)
NASDAQ. *See* National Association of Securities
Dealers Automated Quotation
(NASDAQ)
National Association of Securities Dealers
Automated Quotation (NASDAQ),
80, 80
National Association of Securities Dealers (NASD),
185, 188
National Center for Supercomputing Applications
(NCSA), 502
National Transportation Safety Board, 462
Natural resources, 30-31
NCSA. *See* National Center for Supercomputing
Applications (NCSA)
NDA. *See* Nondisclosure agreement (NDA)
Neeleman, David, 433
Negative covenants, 153-154
Negotiation fundamentals
approaches for, 293-294, 293f
BATNA, 292
behavior, 288
case study, Cannon, 310-312
contracts for, 303-305
defined, 287
emotions, 290
fact, 303
gambit, 299
integrative *vs.* distributive bargaining,
294-297
lose -lose, 297-298
lose-win, 297
mindset, 288
overview, 286-287
position *vs.* interest, 290-291
preparation for, 288, 289f
process, 288
substance, 288
tips for effective, 301
vendor tactics, 302-303
win-win, 298
Nelken, Yori, 168
Netscape Communications, 502
Network effect, 68-72

Networking
  defined, 331
  effective techniques, 332
  primary objective of, 334
  tips for, 333f
  using internet to, 333–334
Newmark, Craig, 503–504
New York Stock Exchange (NSE), 45, 80
Niche strategy, 446
  best-cost provider strategy, 447
  broad differentiation strategy, 447
  for ethnic group, 376
  focused low-cost strategy, 447–448
  focused strategy based on differentiation,
    448
  low-cost provider strategy, 446–447
NIKKEI, 80–81
Nonaccredited investors, 148–149
Noncompete agreements
  in contracts, 266
  sample of, 267f
Nondisclosure agreement (NDA), 266
  turnoff to investors, 268
Nonobvious subject matter, 216
  requirements for, 216–217
Nonqualifying stock options, 128
Nonuse agreements, in record keeping, 209
North American Free Trade Agreement
    (NAFTA), 49–50, 62
Notebooks, record keeping, 210–211
Novation, 256
Novel subject matter, 216
  defined, 216–217
  requirements for, 216–217

O

Obama, Barack, 41
Objective and task method, 375
Objective metrics, 338
OEM. *See* Original equipment manufacturer
    (OEM) agreements
Offer, 250–251
  proposal to constitute an, 251
  revocation of, 253
Officenet, 174
Olson, Tara, 422
Omidyar, Pierre, 502–503
Omnisio, 196–198
Opening gambits, 299–300
Operational risks, 464–474
Opportunity register, 443
Ordinary loss, 111

Organic growth, 161–162
Organizational resources, 336–337
Oriental Land, 83
Original equipment manufacturer (OEM)
    agreements, 273
Outsourcing, 21
Overcapitalization, 141
Owners' equity, 392

P

Pace of change, accelerating, 67–68
Palo Alto Research Center (PARC), 504–505
PARC. *See* Palo Alto Research Center (PARC)
Participating preferred stock, 123–124
Partnership agreement, 103
Patent, 77, 215–222
  application process, 218–221
  claims, 220
  for digital bookmark, 219f
  examination process, 221–222
  exclusive right for, 217–218
  information disclosure statement, 220
  international law for, 222–223
  licenses, 271
  ownership of, 222
  pending notice, 222
  written description, 218–220
Patent Cooperation Treaty (PCT), 77, 222–223
Payment
  of debt, 414–415
  of dividends, 415
PCT. *See* Patent Cooperation Treaty (PCT)
People's Republic of China, The, 42
PepsiCo, 82–83, 213
Percent-of-sales method, 375
Performance matrix, contracts, 260–261
  example of, 260f
Performance standards
  corrective action, 338–339
  for technology venture, 338
Personal guarantee, 152
Phantom income, 119
Physical distribution, product, 363
  company inventory sheet on various aspects
    of, 365f
  functions of, 364
  managerial practices in, 366–367
  related to customer service, 364
Piercing the corporate veil, 100
Pierson, Don, 139–140
Pilot-in-the-Plane Principle, 439
Planned economy, 31

Plant patent, 215–216
Point-of-sale (POS), 373
Policy systems, 74–75
Ponzi-scheme investment, 464
Porter, Michael, 445
POS. *See* Point -of-sale (POS)
Positive covenants, 153–154
Positive-sum game, 298
Post-it™, 499
Post-money valuation, 151
Preferred stock, 122
    convertible, 123
    distributions, 122–123
    participating, 123–124
    voting rights related to, 124
Pre-money valuation, 150–151
Present value (PV), 181
Price, 9
    competition, 363
    consumer, 363
    cost, 361–362
    3Cs of, 361
    economic dimension of, 361
    gouging, 39
    sensitive, 36–37
Price wars, 36
Pricing, in economics, 34–36
    dumping stratergy, 39
    elastic market, 35
    gouging stratergy, 39
    gross margin, 37–38
    inelastic market, 35
    interest rate for, 40
    power of, 38
    wars for, 36
Pricing power, 38
Prime rate, 152–153
Prime time, highest advertising charges, 369–370
Principal register, in trademark, 231
Print advertising, 370–371
Prior art, 216–217
    ordinary skill in art, 217
Private Placement Memorandum (PPM), 146–147, 148, 158
Procedural formalities, failure to observe, 100
Procter & Gamble (P & G), 11
Product
    cost considerations of, 361–362
    curve, life cycle, 360f
    defined, 358
    differentiating of, 376
    distribution of, 33

life cycle for, 358f, 360
physical distribution of, 363
planning and development process, 358, 358f
promotion of, 367–373
test marketing for, 360
tips for generating new ideas, 359
Production systems, 74
Productivity, applies to service businesses, 428
Product liability, 463
    insurance, 463
Product life-cycle S-curve, 175
Product markets, 45
Product/service description, 323
Product/service line, 175
Professional liability insurance, 463
Profitability ratio, 408
Profit breakeven point, income statement, 405, 406f
Profit making, 43
Profit motivation, 21
Promissory estoppel, 253
Promissory note, 146–147
Promotion, product, 367–373
Promotion mix, 367
Provisional application, in patent, 221
Psychographics, 326
Public stock offering. *See* Initial public offering (IPO)
Purchase agreements, 273

Q
Qualified stock option, 128
Qualitative metrics, 338
Quality leadership, in brands, 376
Quality standards, 339–340
    for Belle Baby Carriers, 341
    dimensions of, 340f
    engineering and design, 342
    field support, 343
    information, 342
    managing for, 341–343
    materials, 342
    for people, 342
    policy, 341–342
    for television set, 339
Quantitative method, of marketing budget, 376
Quasi-contract, 258
Quick assets, 407
Quick ratio, 407

R
Radio advertising, 370
Raising capital, 141–142. *See also* Capital and deal structuring

Rand Corporation, 501
Rate sheet, for radio advertising, 370
Ratio analysis, 406
   activity ratio, 408
   debt ratios, 408
   liquidity ratio, 407
   profitability ratio, 408
RDBMS. *See* Relational database management
   system (RDBMS)
ReachLocal, Inc., 387
Real options approach
   to entrepreneurial strategy, 443
Record keeping, 207–210
   confidentiality agreements, 209
   development agreements, 208
   development defence, 209–210
   license agreements, 208
   nonuse agreements, 209
   procedure for, 210–211
Red herring, 189–191
Reformation, 258
Registration statement and timetable,
   IPO and, 188
   prospectus portion of, 188–189
Regulation D
   Form D, 149
   rule 504 of, 148
   rule 505 of, 148–149
   rule 506 of, 149–150
Relational database management system (RDBMS),
   351–352
Relational Software Inc. (RSI), 352
Reliance damages, 257
Repeatable business processes, 15–16
Reporting requirements, for going public, 191
Rescission, 257
Resilience, 478
   developing of, 479–480
Resource aggregation, 334–337
   capital resources, 335
   human resources, 335–336
   organizational resources, 336–337
   technology resources, 337
Resourcebased theory, 444
Resource markets, 45
Restitution, 257
Restricted securities, 148
Restrictive covenants, 153–154
   negative, 153–154
   positive, 153–154
Retained earnings, 111
Return on equity (ROE), 408

Return on investment (ROI), 144
Return on sales, 408
Revenues, 392
Revolving loan, 154–155
RFID Industry Patent Consortium, 205
Risk-adjusted discount rate method
   (RADR), 182
Risk-free rate of return, defined, 182–183
ROE. *See* Return on equity (ROE)
RSI. *See* Relational Software Inc. (RSI)

S

Sales and marketing systems, 74
Sales contracts, for negotiation, 303–304
Sales cycle, 399
Sales forecast, 396–399
   components of, 397
   defined, 397
   templates of, 397, 398f
   tips for development, 399
SAM. *See* Served available market (SAM)
Sarbanes-Oxley regulations
   corporate governance, 464
SBA. *See* U.S. Small Business Administration (SBA)
Scalable business processes, 16, 17
Scarcity, 46
   defined, 287
S-corporation, 110
   characteristics of, 112f
   profits and losses, 119–120
   taxation, 119
SDL. *See* Software Development Laboratories
   (SDL)
Securities Act of 1933, 183
Securities and Exchange Commission (SEC),
   146–147, 148, 188
   rules, 148–150
      Regulation D. *See* Regulation D
Securities Exchange Act of 1934, 148
Security interest, 152
Seed-stage ventures, 143
Segmentation techniques
   for market, 356f
Self-employment tax, 119
Semiconductor Chip Protection Act, 228
Semiconductor Test Consortium (STC), 205
Serendipity, 333
Served available market (SAM), 356
Service mark. *See* Trademark
Settlement range, of other party, 296
Shanghai Stock Exchange, 80–81
Shareholders, 119

Shaw, D.E., 491
Shop rights, 222
Short-term loan, 153
Signage, 373
Silicon Storage Technology (SST), 185
Silicon Valley Bank, 153
Small Business Administration (SBA) loan,
     155–156
  basic 7(a) loan guaranty, 155
  loan prequalification, 155
  microloan 7(m) loan program, 155
  for war veteran, 156
Small Business Corporation, 111
Small Business Innovation Development
     Act, 159
Small Business Innovation Research (SBIR)
     program, 141, 159–160
Small Business Technology Transfer Program
     (STTR), 160–161
Smith, Fred, 323
Smith, Geoff, 171
Social capital, 332
Socialism, 12–13, 41
Socialist countries, 12–13
Socialist economies. *See* Socialism
Social network
  defined, 331
  positioning for, 332
Social obligation, 14–15
Social reaction, 15
Social responsiveness, 15
Social Security and Medicare Tax (FICA),
     116
Sociographics, 326
Software, copyrights protection, 225–226
Software Arts, Inc., 203
Software Development Laboratories (SDL),
     351–352
Sole proprietorships, 98–99, 102–103
  advantages and disadvantages of, 104f
Southern California Edison Co., 291
Specific performance, 257
Spinks, Richard, 60–61
Spot exchange, 52
SQL. *See* Structured query language (SQL)
Standard operating procedures, 74–75
Start-up costs
  creating start up budget, 409
  templates of, 410f
Statute of frauds, 254
Statutory bars, 216–217
Statutory entities, 98, 102

Statutory law, defined, 473
STC. *See* Semiconductor Test Consortium (STC)
Stocks, equity and, 120–121
Strategic alliances, to enter international market,
     83–84
Strict product liability, 463
Strike price, 127
Structured query language (SQL), 352
StumbleUpon, 171–172
Subprime crisis, 395–396
  factors for, 396
Subscription agreement, 158–159. *See also*
     Fund-raising, tools and techniques
Substantial performance, 255
Substituted contract, 255
Suggestive mark, 233
Sullivan, Pat, 140
Supplemental register, in trademark,
     231–232
Supply
  defined, 33
  law of, 35f
Surplus, defined, 33
Sustainable competitive advantage, 76
  in market, 445
Sweat equity, 124
Symbian Foundation, 205
Synergy, 176
Systems establishment, 15

T

TAM. *See* Total available market (TAM)
Target market, 355
  demographics in, 356–357
  geographic location of, 357
  marketing message for, 376
Taxation
  C-corporation and, 119
  double, 111
Tax laws, for corporation, 110–111
Team building skills, 433–434
Technical services agreements, 272
Technological change, 17–18
  and globalization, 18, 18
Technology entrepreneurs, 11, 51, 53, 61, 141, 142,
     145, 155, 159, 173, 179, 184, 186, 296, 335,
     395, 416, 428, 434, 478
Technology resources, 337
Technology ventures
  benefits matrix, 324, 325f
  business plan, 319
  capital, role of, 141

Technology ventures (*Continued*)
  capital resources, 335
  company information, 322–323
  competitive analysis, 324
  corrective action, 338–339
  CRM systems and, 74
  drivers, 62–63
  exit strategies for, 171
    acquisition, 173–178
    going public, 183–191
    mergers, 178–179
    venture valuation, 180–183
  financial projections, 330–331
  globalization and, 61–72
  human resources, 335–336
  human resource system and, 75
  industry analysis, 326–328
  information technology systems and, 74
  for intellectual property, 205
  logistics systems and, 74
  management team, 328–329
  market analysis, 324–326
  marketing plan, 329–330
  networking, 331–334
  networking opens up opportunities, 317–318
  objective metrics, 338
  organizational resources, 336–337
  performance standards, 338
  policy systems and, 74–75
  production systems and, 74
  product/service description, 323
  qualitative metrics, 338
  quality standards, 339–340
  sales and marketing systems and, 74
  serendipity, 333
  strategies to enter international market, 81–85
  technology resources, 337
  value protection in, 75–78
Television advertising, 369–370
Ten-day rule, 507
Terminating employees, 471
  dos and don'ts for, 472
Term of loan, 153–154
Test marketing, for products, 360
TimeBridge , Inc., 168–169
Times-interest-earned ratio, 409
Tipping point, 71–72
Top 10 Franchises of 2009, 508f
Total available market (TAM), 355
Trade blocs, 51–52
Trademark, 228–236
  acquiring rights, 229
  descriptive mark, 233

fanciful and arbitrary marks, 233–234
generic mark, 232–233
guidelines to choosing, 234–235
intent to use, 232
licenses, 271–272
maintaining rights, 235–236
in principal register, 231
registering, 229–230
restrictions for registration, 231
strength of mark, 232–234
suggestive mark, 233
supplemental register, 231–232
term of registration, 235
use™ symbol in lieu of, 230
Trade publications, 371
Trade secrets, 211–213
  maintaining, 212–213
  protection of, 212
Trade shows, for new products, 373
Trading companies, to enter international
    market, 84
Transportation, modes of, 364–366
  types of, 366

U

UCC. *See* Uniform Commercial Code (UCC)
Underwriter, 186–187
Underwriting syndicate, 186–187
Uniform Commercial Code (UCC), 249, 473
Uniform Franchise Offering Circular, 506
Unilateral mistake, 253
United Parcel Service (UPS), 74
United States, 44–46
  adoption of marketing concept, 355
  balance of payments, 48f
  capital investment 2001–2007, 64f
  comparison of intellectual property protection
    mechanisms, 206f
  diffusion rate, of different technologies, 67, 68f
  export, 47
  flow of products and money in economy, 45f
  free market system, 361
  import, 47
  trade secrets, 211
Units of investments, 158
Unlimited liability
  companies, 98
  limited *versus*, 98–99
UPS. *See* United Parcel Service (UPS)
U.S. government's Advanced Research Projects
    Agency (ARPA), 501
U.S. National Academy of Engineering (NAE),
    18–19

U.S. Small Business Administration (SBA), 30
Utility patent, 215

V

Value, 8–9
    calculation, 8–9
    capture, 11
    creation, 9–10
    defined, 8–9
    proposition, 11–12
    protection, 11
Value-added reseller (VAR) agreements, 273
Value capture, in global economy, 78–81
Value creation, in global economy
    enterprise. *See* Enterprise value
    market. *See* Market value
Value drivers, 173–174
    generic categories of strategic, 173–174
        adding markets and distribution channels,
            173
        broader product lines, 173
        customer base, increasing, 173–174
        economies of scale, creating, 174
        internal skills, extending, 174
        technology base, expanding, 173
Value proposition, 71–72, 74
    for customer intimacy, 377
    for operational excellence, 377
    for product leadership, 377
VAR. *See* Value-added reseller (VAR) agreements
Variable costs, 400
Venture capital (vc) firms, 144–146
Vendor tactics, 302–303
Venture Capital (VC), 144–145
    right target, 146
    VC firms, 144–146
Venture control, basic elements of, 340f
Venture governance, 464–466
Venture management. *See also* Venture risk
        management (VRM)
    analytical skills, 428–429
    communication skills, 431
    competitive strategy model, 445, 445f
    conceptual skills, 432
    decision-making skills, 430
    effectuation, 437
    entrepreneurial ethics, 439–441
    entrepreneurial leadership, 434–437
    entrepreneurial strategy, 442–445
    entrepreneur managers, 427–428
    skill, management, 428
    team building skills, 433–434
    vs. leadership, 436

Venture risk management (VRM)
    administrative law, 473–474
    board of advisors, 467
    board of directors, 466
    cognitive biases, 479–483
    common law, 473
    failure risk, 478–483
    fiduciary responsibility, 466
    hazard risks, 462–464
    human resources management, 469
    investor relations, 468
    law of negotiable instruments, 475–476
    law of sales, 474
    law of torts, 474
    legal risk management, 471–474
    liability insurance, 463
    operational risks, 464–474
    overview, 460–461
    product liability, 463
    puzzle interviews, 459–460
    resilience, 478
    statutory law, 473
    terminating employees, 471
    working with legal counsel, 476–477
Venture's burn rate, 395
Venture's managers, users of accounting
        information, 390
Venture valuation, 180–183. *See also* Exit
        strategies, for technology ventures
    discounted cash flow, 180–181
    multiples technique, 180
Vesting, 124
    ESO and, 127
Vicariously liable, 98
VisiCalc, 203–204
Voice over IP (VOIP), 67
Voidable contract, 253
VOIP. *See* Voice over IP (VOIP)
Voting rights, for stock, 124
VRM. *See* Venture risk management (VRM)

W

Walk-away alternative, 291
Wall Street Journal, The, 370
Wal-Mart Stores, 44
Warranties, contracts, 262
War veteran, SBA loan for, 156
Welch, Jack, 61
Wharton, Peter, 64–66
Win-win negotiating, 298
    rules for, 298
Work for Hire agreement
    in contracts, 268

Working capital, of firm's operations, 409–410
World Bank, 49
World Intellectual Property Organization, 77
World is Flat, The, 61
World Trade Organization (WTO), 50–51
Wozniak, Stephen, 317
W-Systems Corporation
    case study, 71
WTO. *See* World Trade Organization (WTO)

X

Xerox Corporation, 504–505

Y

You, Inc., 429
YouTube, 10, 11, 78

Z

Zero sum game, 297
Zuckerberg, Mark, 247